ESO ASTROPHYSICS SYMPOSIA
European Southern Observatory

Series Editor: Jacqueline Bergeron

Springer-Verlag Berlin Heidelberg GmbH

Lex Kaper   Alexander W. Fullerton (Eds.)

Cyclical Variability in Stellar Winds

Proceedings of the ESO Workshop
Held at Garching, Germany,
14 – 17 October 1997

 Springer

Volume Editors

Lex Kaper
Astronomical Institute
"Anton Pannekoek"
University of Amsterdam
Kruislaan 403
NL-1098 SJ Amsterdam
The Netherlands

Alexander W. Fullerton
Department of Physics
and Astronomy
University of Victoria
P.O. Box 3055
Victoria, BC
Canada V8W 3P6

Series Editor

Jacqueline Bergeron
European Southern Observatory
Karl-Schwarzschild-Strasse 2
D-85748 Garching, Germany

Cataloging-in-Publication Data applied for

Die Deutsche Bibliothek - CIP-Einheitsaufnahme

Cyclical variability in stellar winds : proceedings of the ESO workshop, held in Garching, Germany, 14 - 17 October 1997 / Lex Kaper ; Alexander W. Fullerton (ed.). - Berlin ; Heidelberg ; New York ; Barcelona ; Budapest ; Hong Kong ; London ; Milan ; Paris ; Singapore ; Tokyo : Springer, 1998
   (ESO astrophysics symposia)
   ISBN 978-3-662-11392-9   ISBN 978-3-540-68597-5 (eBook)
   DOI 10.1007/978-3-540-68597-5

Typesetting: Camera-ready by authors/editors
Cover design: *design & production* GmbH, Heidelberg
SPIN: 10651976   55/3144 - 5 4 3 2 1 0 - Printed on acid-free paper

Preface

Although it has long been known that stellar winds are variable, in recent years it has become increasingly clear that the fluctuations are not completely chaotic, but are often cyclical in nature. This is a property that seems to be shared by the winds of cool and hot stars, even though their outflows are driven by fundamentally different physical mechanisms. A key ingredient in both cases appears to be the presence of deep-seated structures due to phenomena like surface magnetic fields or pulsations, which alter the emergence of the wind from the stellar surface and impart a measure of cyclical regularity to the outflow.

Since very similar models have been proposed to explain the cyclical wind variations observed in a wide variety of stars, the time was ripe for astrophysicists from many different sub-disciplines to come together and compare their various approaches. This was a primary motivation for the ESO Workshop on *"Cyclical Variability in Stellar Winds"*, which was held at ESO Headquarters in Garching bei München from October 14 to 17, 1997. The timeliness of the meeting was demonstrated by the attendance of more than 100 astronomers from all over the world. A full program was developed by the Scientific Organizing Committee: Keith MacGregor (Chair), Torsten Böhm, Andrew Cameron, Claude Catala, Lee Hartmann, Huib Henrichs, Lex Kaper, Henny Lamers, Stan Owocki, Joachim Puls, and Otmar Stahl. A special feature of this programme was the presentation of the latest results from the MUlti-SIte COntinuous Spectroscopy (MUSICOS) campaign of November 1996.

We hope that these proceedings will provide a useful and up-to-date overview of the observations, interpretation, and modelling of the time-dependent mass outflows from all sorts of stars, ranging from young pre-main sequence objects to cool, solar-type stars to red supergiants and hot, massive stars. What the proceedings do not record is the conference dinner in the (rotating) Olympia Turm at the Olympic Park in Munich, or the well-attended visit to the Augustiner pub.

We would very much like to thank Christina Stoffer, Britt Sjöberg, and Saskia de Haas for their help in organizing and running the workshop, and Pamela Bristow for the preparation of the proceedings for publication. Ed Janssen designed the beautiful workshop poster. We are also very grateful for the generous financial support received from the European Southern Observatory. One of us (AWF) is similarly grateful to the DFG (through grant Pu 117/3-1) for financial support during the organization of the workshop.

Finally, we would like to thank all the workshop participants for the stimulating atmosphere they fostered during the workshop and their contributions to the proceedings.

Garching, May 1998   *Lex Kaper*
                                                        Alex Fullerton

Conference poster by Ed Jansen

Contents

Part 1. SETTING THE STAGE: WIND ACCELERATION MECHANISMS

X

Part 2. OBSERVATIONS OF CYCLICAL WIND VARIABILITY

Part 3. PROCESSES AFFECTING THE EMERGENCE OF THE STELLAR WIND

Part 4. MODELLING CYCLICAL VARIABILITY IN STELLAR OUTFLOWS

Part 5. THE MUSICOS 1996 CAMPAIGN

Part 6. FUTURE PERSPECTIVE

List of Participants

Name	Institution
Nahum Arav	Theoretical Physics, Caltech `arav@tapir.caltech.edu`
Bernhard Aringer	Institut für Astronomie, Universität Wien `aringer@astro.ast.univie.ac.at`
Thorsten U. Arndt	Inst. f. Astronomie und Astrophysik, Technische Universität Berlin `Arndt@physik.tu-berlin.de`
Dietrich Baade	European Southern Observatory `dbaade@eso.org`
Nicole Berruyer	Lab. G.-D. Cassini, Observatoire de la Cote d'Azur `nicole@obs-nice.fr`
Nina Beskrovnaya	Pulkovo Observatory `beskr@pulkovo.spb.su`
Ronny Blomme	Astrophysics, Royal Observatory Belgium `Ronny.Blomme@oma.be`
Jean-Claude Bouret	Lab. d'Astrophysique de Toulouse, Observatoire Midi-Pyrenees `bouret@obs-mip.fr`
Huilai Cao	Beijing Astronomical Observatory `caohl@bao01.bao.ac.cn`
Joseph P. Cassinelli	Astronomy Dept., Univ. of Wisconsin `cassinelli@astro.wisc.edu`
Drahomir Chochol	Astron. Inst., Slovak Academy of Sciences `chochol@auriga.ta3.sk`
Francesco Damiani	Osservatorio Astronomico di Palermo `damiani@oapa.astropa.unipa.it`
Gennaro D'Angelo	Dept. Physics, Univ. of Naples "Federico II" `dangelo@astrna.na.astro.it`

Jeroen A. de Jong   Astron. Inst., Univ. of Amsterdam
jdj@astro.uva.nl

Jean-François Donati   Lab. d'Astrophysique,
Observatoire Midi-Pyrenees
donati@obs-mip.fr

Andrea K. Dupree   Center for Astrophysics, Cambridge, Mass.
adupree@cfa.harvard.edu

Joel A. Eaton   CEIS, Tennessee State University
eaton@coe.tnstate.edu

James P. Emerson   Physics Dept.,
Queen Mary & Westfield College
j.p.emerson@qmw.ac.uk

Matilde Fernández   Max-Planck-Institut für Extraterr. Physik
matilde@mpe.mpg.de

Bernard H. Foing   Space Science Dept., European Space Agency
bfoing@estec.esa.nl

Yves Frémat   Groupe d'Astrophysique et de Spectroscopie,
Université de Mons-Hainaut
fremat@umh.ac.be

Alex Fullerton   Universitäts-Sternwarte München
alex@usm.uni-muenchen.de

Jorge Gameiro   Centro Astrofisica da Universidade do Porto
jgameiro@astro.up.pt

Andreas Gauger   Max-Planck-Institut für Radioastronomie
gauger@speckle.mpifr-bonn.mpg.de

Krzysztof Gesicki   Physics and Astronomy,
Centrum Astronomii UMK, Torun
gesicki@astri.uni.torun.pl

Jack T. Gosling   Space and Atmospheric Sciences,
Los Alamos National Laboratory
jgosling@lanl.gov

Thomas Granzer   University of Vienna, Institute for Astronomy
granzer@astro.ast.univie.ac.at

Vladimir Grinin   Astron. Inst., St. Petersburg University
grinin@VG1723.spb.edu
grinin@crao.crimea.ua

Martin Groenewegen   Max-Planck-Institut für Astrophysik
groen@mpa-garching.mpg.de

Detlef Groote   Universität Hamburg, Hamburger Sternwarte
dgroote@hs.uni-hamburg.de

Erik L. Gullbring — Center for Astrophysics, Cambridge, Mass.
egullbring@cfa.harvard.edu

Jinxin Hao — Beijing Astronomical Observatory
hjx@bao01.bao.ac.cn

Lee W. Hartmann — Center for Astrophysics, Cambridge, Mass.
hartmann@cfa.harvard.edu

Huib F. Henrichs — Astron. Inst., Univ. of Amsterdam
huib@astro.uva.nl

Susanne Höfner — Niels Bohr Inst., Univ. of Copenhagen
hoefner@stella.nbi.dk

Anne Marie Hubert — DASGAL, Obs. de Paris, Section de Meudon
anne-marie.hubert@obspm.fr

Wolfgang Hummel — Universitäts-Sternwarte München
hummel@usm.uni-muenchen.de

Richard Ignace — Dept. of Physics and Astronomy, University of Glasgow
rico@astro.gla.ac.uk

Nazar Ikhsanov — Institut für Astronomie und Astrophysik, Universität München
ikhsanov@usm.uni-muenchen.de

Garik Israelian — Instituto de Astrofisica de Canarias
gil@iac.es

István Jankovics — Gothard Astrophysical Observatory, Eötvös Loránd University
ijankovi@gothard.hu

Vera Jatenco-Pereira — Instituto Astronomico e Geofisico - USP, São Paulo
jatenco@orion.iagusp.usp.br

Christopher M. Johns-Krull — Space Sci. Lab., Univ. California, Berkeley
cmj@sunburst.ssl.berkeley.edu

Alexander Kakouris — Physics Department, University of Athens
akakour@atlas.uoa.gr

Lex Kaper — European Southern Observatory
lkaper@eso.org

Maria Katsova — Sternberg Astronomical Institute, Moscow
maria@sai.msu.su

Andreas Kaufer — Landessternwarte Heidelberg-Königstuhl
A.Kaufer@lsw.uni-heidelberg.de

Indrek Kolka — Tartu Observatory
indrek@aai.ee

Jiri Kubát

Astronomical Institute, Ondřejov
kubat@sunstel.asu.cas.cz

Wolfgang Kundt

Institut für Astrophysik, Universität Bonn
wkundt@astro.uni-bonn.de

Henny Lamers

Astronomical Instititute, Utrecht University
hennyl@sron.ruu.nl

Egil Leer

Inst. of Theoretical Astrophys., Univ. of Oslo
egil.leer@astro.uio.no

Joao J.G. Lima

Centro de Astrofisica, Universidade do Porto
jlima@astro.up.pt

Jeffrey L. Linsky

JILA, University of Colorado, Boulder
jlinsky@jila.colorado.edu

Irene R. Little-Marenin

Center for Astrophysics & Space Astronomy,
University of Colorado, Boulder
ilittle@casa.colorado.edu

Leon B. Lucy

ST-ECF
llucy@eso.org

Keith B. MacGregor

National Center for Atmospheric Research,
High Altitude Obs., Boulder, Colorado
mac@hao.ucar.edu

Derck Massa

NASA/GSFC
massa@xfiles.gsfc.nasa.gov

David McDavid

Limber Observatory
dmd@astro.as.utexas.edu

Thierry Morel

Dépt. de Physique, Univ. de Montréal
morel@astro.umontreal.ca

Dermott J. Mullan

Bartol Research Institute,
University of Delaware, Newark
mullan@bartol.udel.edu

Ignacio Negueruela

Astrophysics Group, EEP,
Liverpool John Moores University
ind@staru1.livjm.ac.uk

Andrzej Niedzielski

Centre for Astronomy,
N. Copernicus University, Torun
aniedzi@astri.uni.torun.pl

Tiit Nugis

Tartu Observatory
nugis@aai.ee

Atsuo Okazaki

Hokkai-Gakuen University, Sapporo
okazaki@elsa.hokkai-s-u.ac.jp

Joana Oliveira　　　　　　　Solar System Division, ESTEC/ESA
　　　　　　　　　　　　　　joana@so.estec.esa.nl

Rachid Ouyed　　　　　　　Dept. of Physics & Astronomy,
　　　　　　　　　　　　　　St. Marys University, Halifax, Nova Scotia
　　　　　　　　　　　　　　rouyed@ap.stmarys.ca

Stanley P. Owocki　　　　　Bartol Research Institute,
　　　　　　　　　　　　　　University of Delaware, Newark
　　　　　　　　　　　　　　owocki@bartol.udel.edu

Luca Pasquini　　　　　　　European Southern Observatory
　　　　　　　　　　　　　　lpasquin@eso.org

Kresimir Pavlovski　　　　　Faculty of Geodesy, Hvar Observatory,
　　　　　　　　　　　　　　Zagreb Univeristy
　　　　　　　　　　　　　　kresimir@geof.hr

Geraldine J. Peters　　　　　Space Sciences Center,
　　　　　　　　　　　　　　Univ. of Southern California, Los Angeles
　　　　　　　　　　　　　　gjpeters@mucen.usc.edu

Peter Petrenz　　　　　　　Institut für Astronomie und Astrophysik,
　　　　　　　　　　　　　　Universität München
　　　　　　　　　　　　　　uh101bv@usm.uni-muenchen.de

Mikhail Pogodin　　　　　　Astrophys. Dept., Pulkovo Observatory
　　　　　　　　　　　　　　pogodin@pulkovo.spb.su

Andy Pollock　　　　　　　ISO Observatory, Villafranca del Castillo
　　　　　　　　　　　　　　apollock@iso.vilspa.esa.es

Nina Polosukhina-Chuvaeva　Crimean Observatory, Nauchny
　　　　　　　　　　　　　　polo@crao.crimea.ua

Raman K. Prinja　　　　　　Department of Physics and Astronomy,
　　　　　　　　　　　　　　University College London
　　　　　　　　　　　　　　rkp@star.ucl.ac.uk

Joachim Puls　　　　　　　Universitätssternwarte München
　　　　　　　　　　　　　　uh101aw@usm.uni-muenchen.de

Thomas Rivinius　　　　　　Landessternwarte Heidelberg-Königstuhl
　　　　　　　　　　　　　　T.Rivinius@lsw.uni-heidelberg.de

Mark Runacres　　　　　　Astrophysics, Royal Observatory Belgium
　　　　　　　　　　　　　　mrunacre@oma.be

Lili Sapar　　　　　　　　Tartu Observatory
　　　　　　　　　　　　　　lilli@aai.ee

Gerrit Jan Savonije　　　　Astron. Inst., Univ. of Amsterdam
　　　　　　　　　　　　　　gertjan@astro.uva.nl

Yvonne Simis　　　　　　　Leiden Observatory
　　　　　　　　　　　　　　simis@strw.leidenuniv.nl

Otmar Stahl — Landessternwarte Heidelberg-Königstuhl
O.Stahl@lsw.uni-heidelberg.de

Stanislav Štefl — Astronomical Institute, Academy of Sciences of the Czech Republic, Ondřejov
sstefl@sunstel.asu.cas.cz

Raphael Steinitz — Dept. of Physics, Ben Gurion University
raphael@bgumail.bgu.ac.il

Larisa Tambovtseva — Laboratory of the Physics of Stars, Central Astronomical Observatory, Pulkovo
tamb@pulkovo.spb.su

John H. Telting — Isaac Newton Group of Telescopes, ASTRON/NFRA, Santa Cruz de La Palma
jht@ing.iac.es

Yvonne C. Unruh — Institute of Astronomy, University of Vienna
ycu@venus.ast.univie.ac.at

Jeroen van Gent — Astronomical Instititute, Utrecht University
gent@fys.ruu.nl

Eva Verdugo — ISO Project, VILSPA - ESA
ev@vilspa.esa.es

Albert Washüttl — Institute for Astronomy, University of Vienna
wasi@astro.ast.univie.ac.at

Michael Weber — Institute for Astronomy, University of Vienna
weber@astro.ast.univie.ac.at

Kerstin Weis — Inst. für Theor. Astrophysik, Heidelberg
kweis@ita.uni-heidelberg.de

Ulf Wessolowski — Max-Planck-Institut für Extraterr. Physik
uuw@mpe-garching.mpg.de

Lee Anne Willson — Department of Physics and Astronomy, Iowa State University
lwillson@iastate.edu

Walter Windsteig — Institut für Astronomie, Universität Wien
windsteig@auro.ast.univie.ac.at

Peter Woitke — Institut für Astronomie und Astrophysik, Technische Universität Berlin
woitke@physik.tu-berlin.de

Bernhard Wolf — Landessternwarte Heidelberg-Königstuhl
bwolf@lsw.uni-heidelberg.de

Kutluay Yüce — Astronomy and Space Dept., Ankara Univ.
kyuce@astro1.science.ankara.edu.tr

Part 1

SETTING THE STAGE: WIND ACCELERATION MECHANISMS

Acceleration Mechanisms for Cool Star Winds

K.B. MacGregor

High Altitude Observatory, National Center for Atmospheric Research[**],
P.O. Box 3000, Boulder, C0 80307, USA

Abstract. We examine the acceleration of winds from late-type stars, using simple, one-dimensional, stationary flow models to investigate different mechanisms for driving atmospheric expansion. We focus on the winds from low-mass, main sequence stars like the Sun, first determining the basic physical properties of thermally driven mass loss, and subsequently considering some of the ways in which rotation and magnetism can influence flow dynamics.

1   Introduction

Any discussion of variability (cyclical or otherwise) in stellar winds must necessarily begin with consideration of the one such outflow whose physical characteristics can be directly measured: namely, the solar wind. The flow associated with the Sun's expanding corona is known to be variable on a variety of time and length scales. There are indications that during the Sun's main sequence lifetime, solar wind properties have changed in response to the changes in internal and atmospheric conditions caused by stellar evolution. Flow attributes also vary over the course of the solar cycle, reflecting the influence of changes in the magnetic structure of the corona and in the overall level of solar activity. On the time scale of the Sun's rotation period, the longitudinal structure of the coronal source regions of the wind leads to the presence of high-speed streams and attendant co-rotating interaction regions in the flow. On yet shorter time scales with durations of a few days or so, wind properties are perturbed as a result of interactions between the flow and coronal mass ejections. Finally, on time scales of hours or less, *in situ* measurements of solar wind plasma and magnetic parameters reveal significant variability, produced by propagating hydromagnetic waves and disturbances.

The Sun remains the single known cool dwarf star with a directly observable wind-type outflow. This state of affairs is a consequence of the difficulties involved with the detection and measurement of radiative signatures of a tenuous, high-temperature flow like the solar wind. There is, however, compelling circumstantial evidence suggesting that winds are common (perhaps even intrinsic) among main sequence stars of solar spectral type (see, e.g., MacGregor 1996, and references therein). Among other things, such stars are known to be emitters of radiation at X-ray wavelengths, an indication of the presence of atmospheric regions containing plasma with temperatures and densities like those

---

[**] The National Center for Atmospheric Research is sponsored by the National Science Foundation

found in the solar corona ($T \sim 10^6$ K, $N \sim 10^8$ cm^{-3}). For stars with surface gravitational accelerations comparable to that of the Sun, these thermodynamic conditions imply the existence of a thermal pressure gradient sufficient to initiate coronal expansion. In addition, numerous observations have established that the rotational velocities of solar-type stars decrease with advancing age. The loss of angular momentum implied by these observations is plausibly interpreted as being a result of the torque applied to a rotating star by a magnetized wind emanating from the stellar corona. Finally, studies of the local interstellar medium along lines of sight toward several nearby late-type dwarf stars have yielded evidence for the interaction between the winds of these stars and the surrounding interstellar gas (Wood & Linsky 1998).

This brief survey of results pertaining to the existence of winds from cool dwarf stars suggests that the occurrence of flow variability of the kind described above may be widespread among Sun-like stars. The present paper will not directly address the origins, nature, and consequences of this variability, but will instead attempt to depict the dynamical state of the wind from a solar-type star in the absence of variations. In particular, the acceleration of steady winds will be examined by using simple, one-dimensional, stationary flow models to investigate how thermal, rotational, and magnetic forces can drive atmospheric expansion. In this way, a picture of the time- and space-averaged flow dynamics can be established, against which variability can be recognized and interpreted.

2   Thermally Driven Winds

Prior to its 'discovery' by means of spacecraft observations in the early 1960's, the existence of the solar wind had been hinted at by a variety of geophysical and astronomical phenomena. Auroral displays, geomagnetic storms, the modulation of cosmic rays, and the orientation and acceleration of comet tails were variously ascribed to charged, 'corpuscular radiation' emitted by the Sun. In 1958, Parker (1958, 1963; see also Hundhausen 1972) demonstrated that the dynamical expansion of the outer layers of the solar atmosphere was an inescapable consequence of the high temperature of the Sun's corona and the low pressure of the local interstellar medium. He noted that although the gas pressure distribution $p(r)$ within a spherical, hydrostatic, thermally conducting corona initially decreases with increasing radial distance r from the Sun, it tends toward a finite value p_∞ as $r \rightarrow \infty$. This asymptotic coronal gas pressure is significantly larger than the estimated confining pressure that can be supplied by the interstellar gas, the galactic magnetic field, and galactic cosmic rays, a fact that led Parker to suggest that the Sun's atmosphere cannot be maintained in hydrostatic equilibrium throughout. Instead, he reasoned that it must be steadily expanding into interplanetary space, driven by the force arising from the thermal pressure gradient in the hot coronal material.

In the most basic representation of a steady, thermally driven wind from a spherical, non-rotating star of mass M_*, the radial flow of a fully ionized, ideal gas is described by equations expressing the conservation of mass and momentum (see, e.g., Hundhausen 1972); adopting spherical polar coordinates, these are

$$\frac{\mathrm{d}}{\mathrm{d}r}\left(r^2\rho u\right) = 0, \tag{1}$$

and

$$u\frac{\mathrm{d}u}{\mathrm{d}r} = -\frac{1}{\rho}\frac{\mathrm{d}p}{\mathrm{d}r} - \frac{GM_*}{r^2}, \tag{2}$$

respectively. In equations (1) and (2), $\mathbf{u} = u(r)\mathbf{e}_r$ is the wind velocity, $\rho(r)$ is the mass density, and

$$p = \rho kT/\mu \tag{3}$$

is the pressure of a gas having temperature $T(r)$ and average mass per particle μ (assumed constant). The subsequent analysis is simplified considerably by replacing the equation of energy conservation (needed to determine T) with a polytrope relation of the form $p \propto \rho^\alpha$, where the index α has a constant value in the range $1 \leq \alpha < 5/3$. Equations (1) and (2) can then be combined to obtain the wind equation of motion

$$\frac{\mathrm{d}u}{\mathrm{d}r} = \frac{2a^2 - (GM_*/r)}{u^2 - a^2}, \tag{4}$$

in which $a = (\alpha p/\rho)^{1/2}$ is the sound speed.

If values are assigned to the flow and sound speeds at a reference level $r = r_0$ in the stellar corona, then the equation of motion (4) can be integrated to obtain $u(r)$ throughout the domain $r_0 \leq r < \infty$. For a given value of a_0 ($\equiv a(r_0)$), the totality of solutions to equation (4) is found to consist of several different types of flows, the properties of which have been studied extensively (see, e.g., Parker 1963; Holzer & Axford 1970; Hundhausen 1972). The most frequently considered among these are the two, unique, transonic solutions for which u is everywhere either a monotonically increasing or decreasing function of r. Both attain the speed $u = a$ at the so-called sonic critical point, $r_c = (GM_*/2a^2)$, at which location the numerator and denominator of equation (4) simultaneously vanish. These solutions are distinguished from one another on the basis of their behavior at the boundaries of the domain. The 'wind' solution describes an outflow that is subsonic at r_0 and accelerates to a constant, supersonic velocity u_∞ with $(p, \rho) \to 0$ as $r \to \infty$. The 'accretion' solution describes an inflowing gas with asymptotic properties $u_\infty = 0$, $(p_\infty, \rho_\infty) \neq 0$ that arrives at r_0 with a supersonic velocity. In addition to the transonic wind and accretion solutions, four families of solutions exist for which u as a function of r is either double-valued or non-monotonic. Two families of transonic solutions have the former property, with one composed of flows that are confined to the region $r_0 \leq r < r_c$, and the other of flows contained entirely within $r_c < r < \infty$. The latter property is shared by the two remaining solution families, one of which consists of flows that are subsonic for all $r > r_0$, and the other of flows that are everywhere supersonic.

Since all of the references cited in the preceding paragraph contain lengthy discussions of the properties and applications of solutions to equation (4), in the remainder of this section we focus on a particular problem, intended to illustrate both the connection between winds and accretion *and* the possibility of variability in flows for which the thermal pressure gradient and gravity are the principal forces. Specifically, we examine the accretion of a polytropic gas by a spherical, gravitating star of mass M_* and radius R_*. The inflow originates at infinity where the pressure and mass density of the ambient gas have the values p_∞ and ρ_∞, respectively, and where the flow velocity is assumed to be $u_\infty = 0$. The initial state of the system is a supersonic accretion flow, with $u(r)$ given by the transonic accretion solution described above. Because the central star represents a physical barrier to the inflowing material at $r = R_*$, a standing, spherical shock wave exists in the flow, located at a position r_{sh} that is determined by the requirement that the post-shock flow satisfy a pressure boundary condition at the stellar surface (see, e.g., Holzer & Axford 1970). If r_0 is chosen such that $r_0 = R_*$, then we require that p_0, the pressure in the flow at r_0, have the particular value p_* of the pressure in the surface layers of the star. In the problem under consideration, we inquire about the evolution of the flow properties in response to gradual changes in the pressure p_* imposed at this inner boundary. A similar problem has recently been treated by Velli (1994).

Figure 1 depicts results pertaining to an accretion flow with polytrope index $\alpha = 4/3$, and for which the sonic critical point has been assumed to have the location $r_c/r_0 = 10$. Shown in the Figure is the scaled pressure p_0/p_∞ that prevails in the flow at r_0 as a function of the position r_{sh}/r_0 of the accretion shock. The two curves present in the Figure correspond to results obtained for two different assumptions regarding the physical nature of the shock. The results labelled (*a*) apply to the case of an 'adiabatic' shock, for which the jump conditions connecting the upstream and downstream values of ρ, p, and u are the standard Rankine-Hugoniot relations for a perfect gas with constant ratio of specific heats γ chosen to be equal to the polytrope index α (see, e.g., Landau & Lifshitz 1975). Under these conditions, the energy per unit mass of the gas

$$E = \frac{1}{2}u^2 + \frac{a^2}{\alpha - 1} - \frac{GM_*}{r}, \tag{5}$$

(obtained by direct integration of equation [2]) is conserved throughout the flow, having the constant value $a_\infty^2/(\alpha - 1)$ everywhere in $r_0 \leq r < \infty$. The results labelled (*b*) apply in the case of an 'isothermal' shock, for which $a_2/a_1 = 1$, $u_1/u_2 = p_2/p_1 = M_1^2$, and $M_2^2 = 1/M_1^2$, where $M \equiv u/a$ is the flow Mach number, and the subscripts 1 and 2 denote values on the upstream and downstream sides of the shock, respectively. These relations can be formally derived by taking the $\gamma \rightarrow 1$ limit of the Rankine-Hugoniot relations, and describe a gas-dynamical discontinuity in which the pre- and post-shock temperatures are identical. This property implies that energy is lost from the flow as it crosses the shock, presumably the result of efficient radiative cooling of the gas immediately behind the shock. Therefore, in this case the magnitude of E as given by equation (5) changes across the shock, from the value $a_\infty^2/(\alpha - 1)$ appropriate to the upstream flow to a smaller (but still constant) value in the downstream flow.

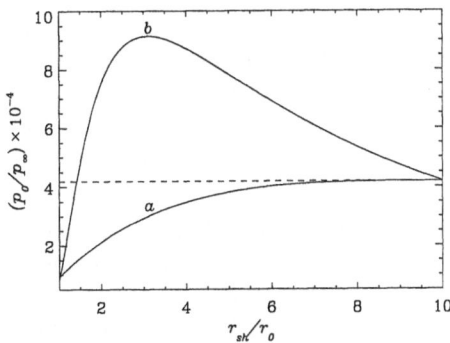

Fig. 1. The reference level pressure as a function of shock position for a supersonic accretion flow with sonic critical point location $r_c/r_0 = 10$. The curves labelled (a) and (b) correspond to flows containing adiabatic and isothermal shocks, respectively, as described in the text.

As is apparent in the Figure, there are significant differences between the r_{sh} dependences of the base pressure ratios corresponding to adiabatic and isothermal shocks. For an adiabatic shock, p_0/p_∞ increases monotonically as the shock is moved outward from the reference level to the critical point location. In the case of an isothermal shock, however, p_0/p_∞ is a non-monotonic function of shock position, having a maximum (for the present choice of parameters) at $r/r_0 \approx 3.1$. Moreover, the magnitude of the base pressure ratio for an isothermal shock is greater than that of an adiabatic shock for almost any shock location (except for $r_{sh} \approx r_0, r_c$). This behavior can be understood in part by noting that for both kinds of shocks, the ram pressure of the nearly freely falling pre-shock inflow is balanced by the thermal pressure of the gas just behind the shock. In the case of an adiabatic shock, the enhanced pressure of the post-shock gas results from the compression and heating of the flow as it crosses the shock, while for the isothermal shock, any increase in the post-shock gas pressure is a consequence of increased density only. Since the flow for radii between r_0 and r_{sh} is subsonic, the density distribution therein can be approximated by that of a hydrostatic, polytropic atmosphere. Because the temperature of the gas behind the isothermal shock is lower than that behind the adiabatic shock, the scale height within this atmosphere is smaller for a discontinuity of the former type than for the latter. Hence, for the same shock position, the reference level density ρ_0 and pressure p_0 ($\propto \rho_0^\alpha$) are both larger in the case of an isothermal accretion shock.

That the quantity p_0/p_∞ is a non-monotonic function of r_{sh}/r_0 for an isothermal shock can be understood from similar considerations. Any increase in p_0 necessarily arises from an increase in ρ_0. As we have just seen, one way to produce such an increase is to increase r_{sh}, thereby increasing the number of scale heights in the (roughly) static atmosphere between r_0 and r_{sh}. An alternative way is to increase the magnitude of the density jump $\rho_2/\rho_1 = M_1^2$ occurring in the gas as it crosses the shock. For $r_{sh} < r_c$, the velocity of the pre-shock flow is supersonic and nearly equal to the local gravitational escape speed ($\propto r_{sh}^{-1/2}$). This would suggest that ρ_2/ρ_1 (and therefore ρ_0) can be increased by decreasing r_{sh}. On the basis of this analysis, we conclude that the maximum in p_0/p_∞ occurs at the shock location for which the density increases associated with each of these two competing tendencies are optimized.

Now consider a supersonic accretion flow containing a standing shock wave with position $r_{sh} < r_c$ such that the pressure p_0 of the infalling gas at r_0 matches the pressure p_* of the material already present at the surface of the star. We seek to determine how the properties of the flow change as (say) the stipulated boundary pressure p_* is made larger. We assume that this variation is accomplished by increasing p_* incrementally, with the time interval between consecutive pressure enhancements taken to be of sufficient duration that the evolution of the flow can be traced using a succession of stationary solutions of the type described above. The results for $r_c/r_0 = 10$ shown in Figure 1 indicate that if the accretion shock is adiabatic, any increase in the magnitude of p_* will be accompanied by the relocation of the shock to a position farther from the reference level. In this way, the inflow can continue to accommodate an increasing pressure exerted at its inner boundary, until such a time as the shock position coincides with the critical point location. When p_* attains a value sufficient to make $r_{sh} = r_c$ (denoted by the horizontal dashed line in the Figure), the flow for $r_0 \leq r \leq r_c$ is nearly the same as the subsonic portion of the wind solution to equation (4) (except in the $-\mathbf{e}_r$-direction), while for $r > r_c$, it consists of the subsonic portion of the accretion solution to (4). An additional positive increment to p_* will cause a complete reversal of the accretion inflow, and the formation a supersonic wind outflow containing a shock located just beyond r_c. This latter feature is required in order to match the asymptotic pressure of the now outwardly flowing gas to the stipulated pressure p_∞ of the originally inflowing gas at infinity. Further increases in p_* cause the shock position to move further from the star (see Velli [1994] for discussion concerning the evolution of isothermal flows).

If the accretion shock is isothermal, a pattern of gradual increases in p_* produces a rather different evolution of the flow. Assume, for specificity, that the shock is located interior to the position corresponding to the maximum in p_0 as a function of r_{sh} (see Fig. 1). In this case, any increase in the external pressure supplied at r_0 leads to a new flow configuration in which the shock is located a greater distance from the star, as before. Note, however, that because the shock is isothermal, larger imposed base pressures can be accommodated by a steady inflow than were possible for an adiabatic accretion shock. In particular, new equilibrium accretion flow configurations can continue to be found for base pressures exceeding the value required to push an adiabatic shock beyond the critical point and establish a supersonic outflow. Yet inspection of Figure 1 reveals that such adjustments to the flow can only take place so long as p_* is less than the maximum value of p_0/p_∞. Indeed, Figure 1 indicates that for base pressures in excess of this maximum value, steady, supersonic accretion is not possible for an inflow containing an isothermal shock. How does the system evolve in response to the application of an external pressure at r_0 that is larger than the greatest pressure that can be produced by the flow at the same location? Under such circumstances, we conjecture that the unfavorable pressure gradient that develops within the subsonic gas near r_0 must ultimately reverse the direction of the flow in this region (i.e., from inflow to outflow), and cause the shock to be expelled from that portion of the flow interior to r_c. We suggest that this outward movement of the shock through material located upstream of its starting position probably leaves the original pre-shock inflow in an outflowing state. If it is possible for the system to attain a stationary configuration after these changes have been effected, it is likely to consist of a supersonically expanding wind together with a shock transition somewhere in $r > r_c$ in order that the outflow match the pressure p_∞ that prevails in the ambient gas at large distances. The validation or rejection of these suppositions necessarily awaits the results of a more detailed, quantitative analysis than has been conducted herein.

3 Thermally Driven Winds: Effects of Rotation

The basic picture of a thermally driven wind presented in the preceding section is modified somewhat when provision is made for the effects of stellar rotation on the dynamics of the radial component of the flow. Some insight into the nature of these changes can be gained by considering the expanding corona of a star that rotates uniformly with angular velocity Ω, but which is otherwise subject to similar assumptions and governing equations as given at the outset of §2 (see, e.g., Weidelt 1973; Mufson & Liszt 1975; Hartmann & MacGregor 1982). In this case, the flow velocity for $r_0 \leq r < \infty$ in the equatorial plane of the star is assumed to have the form $\mathbf{u} = u_r(r)\mathbf{e}_r + u_\phi(r)\mathbf{e}_\phi$, where, as a consequence of the constancy of the specific angular momentum of the wind, $u_\phi(r) = \Omega r_0^2/r$. As before, the mass and momentum conservation equations can be combined to obtain a single equation for the radial velocity of the wind,

$$\frac{r}{u_r}\frac{du_r}{dr} = \frac{2a^2 + u_\phi^2 - (GM_*/r)}{u_r^2 - a^2}. \tag{6}$$

The only apparent difference between equation (6) and the analogous equation (4) in the absence of rotation is the presence of u_ϕ^2 in the numerator, a manifestation of the inclusion of the centrifugal force in the description of the radial flow dynamics.

For specified values of M_*, r_0, and Ω, an acceptable wind solution to equation (6) has subsonic expansion speed at r_0 and asymptotically vanishing gas pressure. The single solution having these properties is the one that passes smoothly through the sonic critical point where the numerator and denominator of equation (6) are simultaneously equal to zero. In Figure 2, we show the radial and azimuthal velocity profiles for several such critical solutions, obtained assuming isothermal flow (i.e., $\alpha = 1$) with $T = 1.5 \times 10^6$ K from a star having $M_* = M_\odot$ and $r_0 = 1.25$ R$_\odot$. Each curve is characterized by a different value of the parameter $\epsilon \equiv \Omega/\Omega_{crit}$, where $\Omega_{crit} = (GM_*/r_0^3)^{1/2}$ is the angular velocity for which the centrifugal force balances gravity at r_0; for the curves labelled (a)-(d) in Figure 2, $\epsilon = 0.00$, 0.70, 0.82, and 0.86, respectively.

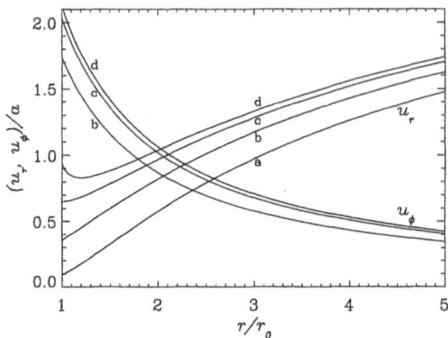

Fig. 2. Radial and azimuthal velocity profiles for thermally driven winds in the presence of rotation. Curve (a) applies to the case of an outflow from a non-rotating star, while curves $(b) - (d)$ depict the effects of increasingly rapid rotation, as discussed in the text.

As is evident from the Figure, increasingly rapid rotation of the wind-emitting star leads to a significant enhancement of the radial flow speed for $r \approx r_0$. It can be readily shown that

$$\frac{u_{r0}}{a} \approx \left(\frac{r_c}{r_0}\right)^2 \exp\left\{\frac{1}{2} - 2\left(\frac{r_{c0}}{r_0}\right)\left[1 - \frac{\epsilon^2}{2}\left(1 - \frac{r_0^2}{r_c^2}\right) - \frac{r_0}{r_c}\right]\right\}, \qquad (7)$$

where r_{c0} [$= (GM_*/2a^2) \approx 3.08\ r_0$, for parameter values appropriate to the results of Figure 2] and r_c are, respectively, the locations of the sonic critical point corresponding to $\epsilon = 0$ and $\epsilon \neq 0$. For ϵ in the range $0 \leq \epsilon < [1 - (r_0/r_{c0})]^{1/2} \approx 0.82$, a single sonic critical point occurs in the flow at a distance

$$r_{c1} = (r_{c0}/2)\ \{1 + [1 - 4\epsilon^2(r_0/r_{c0})]^{1/2}\} \qquad (8)$$

that decreases from $r_{c1} = r_{c0}$ to $r_{c1} \approx 2.08\ r_0$ as ϵ is increased. For rates of rotation such that $[1 - (r_0/r_{c0})]^{1/2} \leq \epsilon < (r_{c0}/4r_0)^{1/2} \approx 0.88$, a second, non-critical zero exists in the numerator of the equation of motion (6), with a location

$$r_{c2} = (r_{c0}/2)\ \{1 - [1 - 4\epsilon^2(r_0/r_{c0})]^{1/2}\} \qquad (9)$$

that increases from $r_{c2} = r_0$ for increasing ϵ within the indicated range. Finally, when $\epsilon = (r_{c0}/4r_0)^{1/2}$, $r_{c1} = r_{c2} = (r_{c0}/2)$.

Some insight into the dynamical reasons behind the wind properties discussed above can be gained by examination of Figure 3, which shows the accelerations produced by the gravitational, thermal pressure gradient, and centrifugal forces as functions of r for solutions a, b, and d of Figure 2. These results indicate that as the rate of rotation is increased, the radial dynamical structure of the flow for r near r_0 is modified. In the absence of rotation ($\epsilon = 0.0$), the thermal pressure gradient is solely responsible for driving coronal expansion, while for $\epsilon = 0.70$, the thermal and centrifugal forces provide comparable contributions to the outward acceleration of material at the base of the wind. For rates of rotation such that $\epsilon \geq [1 - (r_0/r_{c0})]^{1/2}$ (≈ 0.82), the magnitude of the centrifugal force exceeds that of the thermal pressure gradient for $r \approx r_0$. As a result, the numerator of the equation of motion (6) changes from negative to positive at r_0, and the wind is decelerated throughout a small range of heights just above the coronal base. Inspection of the wind solutions shown in Figures 2 and 3 reveals that rotational modifications to the basic thermally driven flow are largely confined to the vicinity of the reference level, a consequence of the fact that the centrifugal acceleration varies as r^{-3} and is therefore unimportant at larger distances.

4   Magnetic/Centrifugal Acceleration: Evolution of the Solar Wind

The general magnetic field of the Sun is transported into interplanetary space by the ionized, electrically conducting solar wind. This advected magnetic field can influence wind dynamics through the Lorentz force it exerts on the outflowing material in which it is embedded. Although the magnitude of any magnetic contribution to the radial acceleration of the present-day solar wind is known to be negligible (see, e.g., Barnes 1974), a different dynamical state of affairs might have existed at an earlier stage in the Sun's main sequence lifetime. In particular, the higher rates of rotation and (possibly) stronger magnetic fields that prevail among some young (i.e., ages $\sim 10^7 - 10^8$ years) solar-type stars suggest that the winds from these objects might be driven predominantly by magnetic and rotational (rather than thermal) forces. In the present section, we briefly explore this conjecture by using the magnetohydrodynamic (MHD) wind model of Weber & Davis (1967; see also Belcher & MacGregor 1976) to trace the evolution of solar wind properties during the course of the Sun's residence on the main sequence. The interested reader is referred to previously cited work (MacGregor 1996) for some additional details of these calculations.

The steady-state wind model describes an axisymmetric, perfectly conducting, polytropic outflow in the equatorial plane of a star that rotates uniformly with angular velocity Ω. It was used to construct a representation of the present-day, quiet solar wind, with properties similar to those of the model given by Withbroe (1988). Adopting $\alpha = 1.13$ and $\Omega = 3 \times 10^{-6}$ s^{-1}, the gas temperature, proton number density, and radial magnetic field strength were assigned the values $T_0 = 1.5 \times 10^6$ K, $N_{p0} = 5 \times 10^7$ cm^{-3}, and $B_{r0} = 2$ G, respectively, at a reference level having location $r_0 = 1.25$ R$_\odot$. The resulting solar wind model had radial flow speed $u_r = 264$ km s^{-1}, electron number density $N_e = 15$ cm^{-3}, gas temperature $T = 2.1 \times 10^5$ K, and proton flux $N_p u_r = 3.9 \times 10^8$ cm^{-2} s^{-1} at a distance 1 AU (≈ 215 R$_\odot$) from the Sun, in good agreement with the aforementioned model of Withbroe (1988). Examination of the radial dynamical

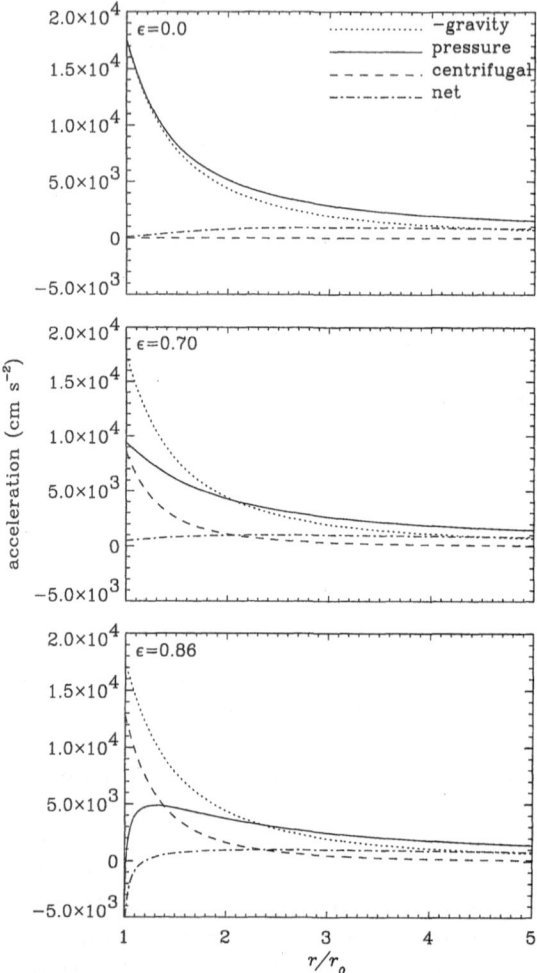

Fig. 3. Profiles of the radial accelerations produced by the gravitational, thermal pressure gradient, and centrifugal forces, for solutions (*b*), (*c*), and (*d*) of Figure 1.

structure of the model indicates that virtually all of the acceleration of the flow is provided by the thermal pressure gradient, with perhaps a few percent of the expansion near r_0 driven by the Lorentz and centrifugal forces. In this case, the asymptotic speed of the wind is $u_\infty \approx 350$ km s^{-1}, and is well approximated by

$$u_\infty \approx \left(\frac{2a_0^2}{\alpha - 1} - \frac{2GM_\odot}{r_0} \right)^{1/2}, \tag{10}$$

a result that follows from the constancy of the energy per unit mass E (see eq. [5]) in the absence of rotational and magnetic effects.

Models intended to represent to solar wind at earlier times were constructed by holding α, T_0, and N_{p0} fixed at the values given above, and prescribing the time variations of Ω, B_{r0}, and the solar structure. Values of Ω at particular times were estimated from averages of measured rotational velocities of G stars in the α Persei (age $\approx 5 \times 10^7$ years), Pleiades (age $\approx 7 \times 10^7$ years), and Hyades (age $\approx 6 \times 10^8$ years) clusters. For ages between 10^9 years and the age of the Sun, the Skumanich (1972) relation was used to establish the value of Ω at a given time. The value of the solar radius at any age was obtained from an unpublished evolutionary track for a non-rotating star with $M_* = M_\odot$ computed by D. VandenBerg. The coronal magnetic field strength for the model was evaluated by scaling the value adopted for the present-day Sun using a phenomenological dynamo relation of the form $B_{r0}(t) \propto \Omega(t)\, R_\odot^{-2}(t)$.

The resulting models for the wind from the young Sun differ significantly from the present-day model described above. For example, the solar radius at age 5×10^7 years is found to be 0.885 of the current value; for an average rotation rate $\Omega = 6 \times 10^{-5}$ s^{-1}, the dynamo relation yields a reference level radial field strength of about 50 G. For these parameter values, the computed wind properties at 1 AU are $u_r = 3917$ km s^{-1}, $N_e = 1.24$ cm^{-3}, $T = 1.54 \times 10^5$ K, and $N_p u_r = 4.84 \times 10^8$ cm^{-2} s^{-1}. The greatly enhanced radial flow speed is a consequence of the acceleration produced by magnetic and centrifugal forces, both of which dominate the wind dynamics at large distances from the Sun. In contrast to the basically thermally driven model considered previously, the wind in the present magnetically/centrifugally driven model attains an asymptotic speed $u_\infty \approx 5300$ km s^{-1}. As before, it is possible to derive an approximate expression for the wind terminal velocity

$$u_\infty \approx \left(\frac{\Omega^2 r_0^2 B_{r0}^2}{4\pi \rho_0\, u_{r0}} \right)^{1/3}, \qquad (11)$$

a result which makes explicit the dependence of wind properties on the magnitudes of B_{r0} and Ω.

One aspect of the wind evolution is explored in Figure 4, wherein the mass-loss rate $\dot{M}$ is depicted as a function of age for solutions obtained using two different assumptions regarding the time-dependence of the temperature T_0 at the coronal reference level. For the results labelled (a), T_0 had the constant value $T_0 = 1.5 \times 10^6$ K, while for the results labelled (b), T_0 was permitted to vary linearly with Ω between $T_0 = 2.0 \times 10^6$ K and $T_0 = 1.5 \times 10^6$ K. This latter prescription is intended to crudely simulate the thermal history of a magnetically heated corona. The quantity $\dot{M}$, estimated by extrapolating the equatorial plane solution to other latitudes, is given by

$$\dot{M} = 4\pi r_0^2 \rho_0 u_{r0}, \qquad (12)$$

where the initial radial velocity is approximately

$$u_{r0} \approx a \left(\frac{r_c}{r_0} \right)^2 \exp\left\{ -\frac{1}{2} - \frac{GM_\odot}{r_0 a_0^2} \left(1 - \frac{r_0}{r_c} \right) - \frac{1}{2} \left(\frac{\Omega r_0}{a_0} \right)^2 \left[\left(\frac{r_c}{r_0} \right)^2 - 1 \right] \right\}.$$

$$(13)$$

14   K.B. MacGregor

In equation (13), r_c is the location of the slow magnetosonic critical point (see, e.g., Weber & Davis 1967); when $\Omega = 0$, r_c becomes equal to the usual sonic critical point location $(GM_\odot/2a_0^2)$.

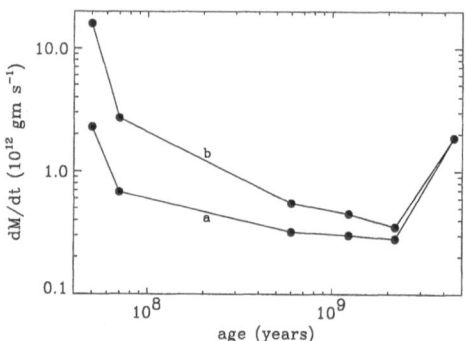

Fig. 4. The mass loss rate of the evolving solar wind, between about the time of the Sun's arrival on the zero-age main sequence and the present-day. The results labelled (a) were obtained for a coronal temperature that was constant in time. Those labelled (b) were obtained for a corona that was somewhat hotter at earlier times, before cooling to the same temperature as (a) at the current solar age.

Throughout the evolution, the coronal temperatures of the case (b) solutions remain higher than the temperature of the corresponding case (a) solutions. Because of this, the $\dot{M}$ values of the case (b) models are greater than those of the case (a) models. Significant spin-down occurs during the first few $\times 10^7$ years shown in the Figure, with the rate of rotation slowing from 20 times the present solar rate at age 5×10^7 years to about 8 times the present rate at age 7×10^7 years. As a result, acceleration of the flow close to r_0 by magnetic and centrifugal forces is markedly reduced at later times, and $\dot{M}$ is concomitantly decreased. Note, however, that for both solution sequences, $\dot{M}$ increases with time during the last several $\times 10^9$ years of evolution. This behavior is largely a consequence of the fact that the solar radius increases with age on the main sequence. Although the magnitude of this increase is modest (e.g., from about 0.9 $r_\odot$ to 1.0 $R_\odot$), the exponential dependence of u_{r0} on r_0 (see eq. [13]) can lead to a substantial increase in $\dot{M}$ over time.

I am grateful to T.E. Holzer for discussions concerning the subject matter of this paper, as well as for comments on the manuscript.

References

Barnes, A. (1974): ApJ, **188**, 645
Belcher, J.W., MacGregor, K.B. (1976): ApJ, **210**, 498
Hartmann, L.W., MacGregor, K.B. (1982): ApJ, **259**, 180

Holzer, T.E., Axford, W.I. (1970): ARA&A, **8**, 31
Hundhausen, A.J. (1972): *Coronal Expansion and Solar Wind* (Springer, Berlin, Heidelberg)
Landau, L.D., Lifshitz, E.M. (1975): *Fluid Mechanics* (Pergamon, New York), 331
MacGregor, K.B. (1996): in *Solar and Astrophysical Magnetohydrodynamical Flows*, ed. K.C. Tsinganos (Kluwer, Dordrecht), 301
Mufson, S.L., Liszt, H.S. (1975): ApJ, **202**, 183
Parker, E.N. (1958): ApJ, **128**, 664
Parker, E.N. (1963): *Interplanetary Dynamical Processes* (Interscience, New York)
Skumanich, A. (1972): ApJ, **171**, 565
Velli, M. (1994): ApJ, **432**, L55
Weber, E.J., Davis, L. (1967): ApJ, **148**, 217
Weidelt, R.D. (1973): A&A, **27**, 389
Withbroe, G.L. (1988): ApJ, **325**, 442
Wood, B.E., Linsky, J.L. (1998): ApJ, **492**, 778

Thirty Years of Radiation-Driven Hot Star Winds

Leon B. Lucy

Space Telescope – European Coordinating Facility,
Karl-Schwarzschild-Str. 2, D–85748 Garching, Germany

Abstract. Thirty years ago rocket-borne UV spectrographs discovered the high-speed winds of the OB supergiants in Orion and thereby revived interest in selective radiation pressure as a stellar mass-loss mechanism. Some highlights of this and subsequent observational and theoretical developments are recalled. In addition, recent progress in solving the momentum problem of Wolf-Rayet winds with the same mechanism is reviewed.

1   Introduction

In the last decade or so, mass loss by hot stars has evolved from being an intriguing dynamical phenomenon of interest to a few specialists into a topic of major importance of interest to a broad community of astronomers. This has happened not because of the phenomenon's intrinsic interest but because of its direct relevance to major themes of current astrophysical research. Thus, the evolution of massive stars, the chemical evolution of galaxies, the dynamical state of the interstellar medium, and the age-dependent integrated spectra of extragalactic star bursts are all active and inter-related research areas where quantitative investigations are now inconceivable without input defining the mass loss characteristics of luminous stars. Moreover, to this uncontroversial list, we can tentatively add the measurement of cosmological parameters since the precision with which the circumstellar spectra of luminous stars can now be predicted allows their luminosities and therefore distances to be determined, yielding a theory-based route to H_0 that is completely independent of classical methods and of their calibrators (McCarthy et al. 1997).

In view of these ramifications, preparing a complete review of radiatively-driven winds is now a formidable undertaking. Fortunately, the Organizing Committee invited me to select topics biased to my own research interests over the years, and these have been motivated entirely by the challenge of understanding the phenomenon itself, irrespective of its wider relevance. Accordingly, this mini-review will recall the origins of the modern resurgence of interest in this topic, look back critically at some of the early theoretical efforts, and relive the exciting era when X-ray emission from hot-star winds was predicted, discovered, and then explained. Finally, current attempts to construct physically and dynamically consistent models of W-R winds will be discussed.

2   Early Work

Ever since light pressure was recognized as a physical effect, astronomers have been fascinated by the rôle it might play in the dynamics of dust and gas in various astrophysical environments. Much of the early theoretical work treated the levitation and expulsion of individual atoms by selective radiation pressure, especially of Ca II ions to form the solar chromosphere. The most memorable papers from this era are those of Milne (1924, 1926, 1927), who also predicted the high-speed ejection of atoms from stars and remarked on the instability of line-driven outflows and on their susceptibility to line-locking.

The most memorable observational papers from this time are those of Beals (1929, 1941, 1944) reporting his spectroscopic investigations of W-R stars. Not only did he correctly conclude that these stars are losing mass in high speed outflows but he also discovered their winds' severe stratification of ionization and, in particular, that the recombination He II → He I occurs within the outflow.

3   The Modern Era

Following years of quiescence, mass loss by hot stars re-emerged as an active research field with Morton's discovery of P Cygni profiles for the resonance lines of C IV, Si IV and N V in the spectra of Orion supergiants (Morton 1967a,b; Morton et al. 1968). This discovery was made in flights of Aerobee sounding rockets on 13 Oct 1965 and 20 Sept 1966, flights which mark the beginning of far-UV stellar astronomy. Additional importance was attached to this discovery by the presence of these same UV resonance lines in the spectra of the then newly-discovered quasars, leading to the speculation that similar physical processes might be at work (Morton 1967a), a thought still being pursued for the BAL quasars.

Although the Aerobee spectra were the direct stimulus of modern theoretical work on line-driven winds, there was in fact earlier evidence that hot stars other than W-R stars were losing mass. From co-added photographic spectra, Wilson (1958) discovered that the C III $\lambda5696$ Å emission line in the spectrum of α Cam (09.5 Ia) had wings extending up to 1500 km/sec, and he interpreted these as due to high-speed outflow. The series of papers from the Royal Observatory Edinburgh to which Wilson's belongs should be of interest to anyone curious about early attempts to secure high S/N stellar spectra.

4   Implications of the Aerobee Spectra

At the time of the Aerobee flights, mass loss by hot stars in the form of the W-R winds was already a long-known phenomenon. It might therefore seem odd that the Aerobee UV spectra should have initiated theoretical developments rather than optical spectra of W-R stars. The crucial point here is that access to the far-UV resonance lines allowed us to observe mass loss by hot stars in the weak-wind limit. In this limit, simple zero-order approximations such as modelling the circumstellar continuum radiation field as the geometrically diluted continuum emitted by a static photosphere are adequate and are indeed directly suggested by the observed spectra. This partial decoupling of radiative transfer from the equations of statistical equilibrium and from the equations of gas dynamics allowed progress to be made; and the resulting successes provide the basis for current attacks on the strong-wind limit, to which the W-R winds surely belong.

Specific implications of the Aerobee spectra, which directly guided theoretical developments, were the following:

1) OB supergiants have spherically-symmetric winds with terminal velocities well in excess of escape velocities and are thus losing mass.

The strong emission components of the P Cygni profiles imply that the outflows producing the absorption troughs cannot be confined to the observer's direction but must indeed be essentially isotropic. The terminal velocities are of course given directly by the Doppler displacements of the violet edges of the absorption troughs.

2) A theoretical description of these winds can be based on the single-fluid approximations.

The terminal velocities for the different ions agreed to within errors – i.e., there was no evidence of differential streaming. This argued against theoretical developments based on the expulsion of individual ions or on a multi-component plasma and in favour of a single-fluid description in which the gradient of radiation pressure acts as a body force.

3) The winds of OB supergiants are not analogues of the solar wind.

The widths of the P Cygni absorption troughs implied that the ions Si IV, C IV and N V exist in the flow as it accelerates from $\sim$ 1000 km/sec to the terminal velocity at $\sim$ 1500 km/sec. For a scaled solar wind, the enthalpy term would imply $T \sim 10^7$ K at $v \sim 1000$ km/sec and the observed ions would be quickly ionized away. The winds are therefore dynamically cold, and this strongly suggested selective radiation pressure as the driving mechanism.

4) Non-radiative heating occurs in the winds of OB supergiants.

If the outflow were laminar, the gas temperature would be everywhere close to that corresponding to thermal equilibrium with the diluted photospheric radiation. In this circumstance, N V ions would not exist in the winds of the Orion supergiants. Accordingly, the very presence of the N V P Cygni line indicated that some dissipative, non-radiative process was operative in their winds.

5 Theory of Line-driven Winds

Stimulated by Morton's discovery and guided by the above implications of his spectra, Lucy & Solomon (1970; LS) initiated modern work on line-driven winds. Rather than review this work in detail, it is of interest, with benefit of hindsight, to comment on the strengths and weaknesses of their paper.

The strong points are as follows:

1) The onset of winds for upper main-sequence stars was attributed to the non-existence of static reversing layers, a conclusion reached by finding negative effective gravities $g_{\mathrm{eff}} = g - g_R$ in the outer parts of simple static models. These sign reversals were due to the strong contributions of such lines as C IV $\lambda 1548$ Å to g_R.

2) The mass flux through non-static reversing layers appeared in the theory as an eigenvalue to be determined by demanding regularity at the sonic point. Thus, the equation of motion for steady, plane-parallel isothermal flow was written as

$$\left(v^2 - a^2\right) \frac{1}{v} \frac{dv}{dx} = g_R - g$$

and solutions found such that $g_R - g \to 0$ as $v \to a$. At the time, this prospect of deriving mass-loss rates entirely from first principles was rather exciting. For the solar wind, we had not – and still have not – reached the point of being able to predict its mass-loss rate theoretically.

3) The single-fluid description suggested by the P Cygni line profiles was justified *a posteriori* by showing that the driven ions – e.g., C IV – share their momentum gains with the non-driven components – e.g., the protons.

4) A qualitative discussion of the instability of line-driven winds was given, and the growth of the instability was conjectured to lead to shocks and thus to the dissipation of some of the work done by radiative forces.

The weak points of the LS paper are the following:

1) The line list was restricted to the Balmer continuum ($\lambda > 911$Å), consistent with the expected negligible fluxes below the Lyman limit for the Orion supergiants. As a result, the high mass-loss rates for early O stars due to the high density of lines in the Lyman continuum were not predicted. The importance of line-driving at $\lambda < 911$Å, which can increase mass loss rates by $\sim 10^2$, was first appreciated by Castor et al. (1976) and by Lamers & Morton (1976).

2) The wind solutions were not continued beyond the sonic point and hence terminal velocities were not predicted – apart from brief details for one case. The reason for this lies in point 4) above. These winds were clearly seen to be unstable, so there seemed little point in computing steady-state supersonic solutions which, additionally, would certainly fail to predict the observed N v line.

3) In computing the radiative acceleration, the equation of radiative transfer was solved for a grid of rest frequencies spanning each line and then g_R was obtained by numerical integration. This was needlessly accurate for an initial investigation and computationally too expensive for the huge line lists later recognized to be necessary.

Subsequent to the work of LS, the paper of Castor et al. (1975; CAK) is widely recognized as a landmark contribution, especially in devising procedures for the efficient computation of physically realistic wind solutions that could then be used in the quantitative interpretation of spectroscopic evidence of mass loss. The basic point of departure from LS was to replace their accurate but cumbersome calculations of g_R by an approximate but rapid calculation based on an analytical expression for each line's contribution given by Sobolev theory (Lucy 1971; Castor 1974).

With g_R thus approximated and sphericity allowed for, the equation of motion for steady, isothermal flow has the mathematical form

$$\left(v^2 - a^2\right) \frac{1}{v} \frac{dv}{dr} = g_R \left(\rho, v, \frac{dv}{dr}\right) - g + \frac{2a^2}{r}$$

As written, this equation reflects the fact that, when evaluated in the Sobolev approximation, g_R is a function only of local variables including, in particular, the velocity gradient dv/dr, which therefore, remarkably, appears twice in the equation of motion. Of course, in reality, g_R is a functional of the entire solution, though in practice a velocity interval of a few Doppler widths of a typical metal line is what counts. Accordingly, the $\frac{dv}{dr}$ in the argument of g_R, which we may call the Sobolev derivative, should be regarded as a mean gradient defined as that value of $\frac{dv}{dr}$ for which the Sobolev approximation for g_R reproduces the exact value. In contrast, the derivative on the left-hand side, Newton's derivative, is of course not a mean gradient.

With the derivatives thus identified, we can describe the CAK solution tactic as setting the Sobolev mean gradient equal to Newton's gradient and thereby reducing an integro-differential equation to an ordinary differential equation. An immediate consequence is that $v = a$, the sonic point, is no longer a critical point since the value of $\frac{dv}{dr}$ for which the rhs vanishes at $v = a$ is determinable algebraically. Instead, a quite separate critical point appears in the supersonic flow, being the point at which solving for $\frac{dv}{dr}$ does not in general yield a finite value.

The strong points of the CAK paper are as follows:

1) Computational efficiency. By simplifying the calculations of g_R, extensive line lists with $\gtrsim 10^5$ transitions could effectively be used and thus the implications investigated of the forest of lines at $\lambda \lesssim 1100\,\text{Å}$.

2) The mass-loss rate is predicted from first principles, appearing in the theory as an eigenvalue determined by demanding regularity at the CAK critical point.

3) The entire solution is determined including the terminal velocity, an important observable quantity.

4) Scaling laws are predicted, giving the dependence of mass loss rate and terminal velocity on stellar parameters.

The weak points of the CAK paper are the following:

1) In applying Sobolev theory to calculate g_R, an essential assumption is that throughout the flow each line interacts with an unattenuated continuum. Unfortunately, because of the strong clumping of lines in frequency space – due particularly to multiplets – this assumption is strictly speaking always violated

and certainly its adoption raises doubts about any CAK solution with mass-loss rate comparable to the single-scattering limit, L/cv_∞.

2) The CAK critical point is an artefact of setting the Sobolev mean gradient equal to the Newtonian derivative. If, for example, the Sobolev mean is approximated by a backward difference formula with step length $\simeq$ the Doppler width of a typical metal line, then the CAK critical point vanishes and the sonic point regains its status as the critical point that determines the mass-loss rate.

This suggestion that the CAK critical point is unphysical was made immediately following the appearance of their paper (Lucy 1975). Subsequently, several investigators have privately remarked that they remain convinced of its physical reality because of Abbott's (1980) demonstration that the CAK critical point is the furthest point downstream that is able to communicate information to all points of the flow. But as Abbott himself remarks this correspondence between critical points and speed of propagation of information is characteristic of transonic flows. Accordingly, any approximation affecting wave propagation would preserve the above-mentioned correspondence; and, consequently, that correspondence does not itself prove the physical significance of a critical point.

This concern over the physical reality of the CAK critical point is voiced again not as an objection to the use of this formalism but to remind users of the potential for error. If physical conditions change little between the sonic point and the CAK critical point, then the CAK solution may well provide an excellent approximation to the exact steady-state solution. But one should positively expect this not to be so if this condition is not met.

The most notable advances subsequent to CAK are described in a major series of papers by the group at the München Sternwarte (Pauldrach et al. 1986; et seq). By incorporating NLTE treatments of ionization and excitation into dynamical wind models, they have created a powerful diagnostic tool for the quantitative analysis of the circumstellar spectra of early-type stars.

6   X-ray Emission

As noted earlier, the Aerobee detection of N v in the circumstellar spectra of Orion supergiants implied a non-radiative heating mechanism. The later discovery of O vi in the winds of hotter O stars was further evidence, and the general phenomenon became known as one of superionization. Since this effect was not restricted to a few stars and was not predicted by models of steady, line-driven winds, its investigation promised to deepen our understanding of the mass loss mechanism. To this end, a workshop was held at JILA (Cassinelli et al. 1978) at which four models were discussed in depth.

Of the models considered at that time, the one of interest here is the hot corona – cool wind model of Cassinelli & Olson (1978) and Olson (1978). Motivated by earlier work of Hearn (1975), these authors proposed that a thin coronal zone with $T \sim 5 \times 10^6$ K existed at the base of the cool wind. X-ray emission from this hot gas could then explain superionization via the two-step Auger mechanism in which an X-ray photon ejects a K shell electron and then a further electron is ejected in consequence of the energy released when an L shell electron drops into the hole in the K shell. In this way, the dominant ions N III and O IV yield trace amounts of N V and O VI, thus potentially explaining superionization.

Because Cassinelli & Olson's hot corona – cool wind model was not (and could not be) developed from first principles, the implication that upper main-sequence stars should be X-ray sources cannot strictly be regarded as a prediction. Rather two distinct aspects of their work should be recognized. First, the inference of ambient X-rays within the winds from the observed high ions; and, second, the astrophysically-reasonable conjecture that these X-rays are emitted by a coronal zone.

The discovery of OB stars as a new class of X-ray emitters was made serendipitously with the *Einstein* Observatory a month or so after launch during observations of the known strong source Cygnus X–3. In reporting this discovery, Harnden et al. (1979) noted the rough agreement of the observed X-ray luminosities with those required by the hot corona – cool wind model. Nevertheless, they stopped short of announcing a confirmation of this model because it required a higher column density of absorbing gas than implied by the data. Subsequently, with more stars observed, Long & White (1980) stated categorically that the X-ray data in fact contradicted the hot corona – cool wind model. They noted that most of the observed X-ray emission was in photons with energies less than 1 keV whereas the model predicts severe attenuation of soft X-rays by the cool wind that lies between the observer and the hot corona.

This apparently definitive rejection of the Cassinelli-Olson model proved to be premature. A revised version due to Waldron (1984) – see also Wolfire et al. (1985) – in which the transition from hot corona to cool wind was treated in detail was no longer ruled out by the Einstein IPC data. This astonishing resurrection of the hot corona idea led Baade & Lucy (1987) to search for the [Fe XIV] λ5303 Å coronal line in the spectrum of ζ Puppis (O4If), repeating at higher S/N an earlier attempt by Nordsieck et al. (1981). No emission was detected and the resulting model-dependent upper limits on an emission measure – coronal temperature diagram excluded the model of Wolfire et al. (1985) by a wide margin.

An alternative interpretation of the X-ray emission was as a clue to the finite amplitude state reached when dissipative effects halt the growth of the instabilities of line-driven winds. Following up on the LS suggestion of shocks and shock-heated gas, Lucy & White (1980) and Lucy (1982a) developed phenomenological models of the small-scale chaotic hypersonic motions within these winds. These models provided crude quantitative support to the idea that the X-ray emission originated in the cooling zones of numerous shock fronts distributed throughout these winds and that the irradiation of cool, inter-shock gas by these X-rays explained superionization, thus incorporating an important feature of the hot corona – cool wind model.

Subsequent work on the instability of line-driven winds and the resulting X-ray emission has been dominated by the direct numerical approach pioneered by Owocki et al. (1988) – see Feldmeier (1995) for a recent example. Inspection of these numerical simulations shows poor correspondence with the small-scale structure envisaged in the phenomenological modelling. Although the physical effects incorporated in those models are real enough, the richness and complexity of these unstable outflows, even when investigated with the constraint of spherical symmetry, defy simple description. Clearly, there is no alternative to gas dynamic code simulations if we wish to understand these unstable winds in detail.

The current state of the shock model of X-ray emission both as a tool for fitting observed spectra and as a theory capable of predicted X-ray luminosities of OB stars is well summarized by Feldmeier et al. (1997a,b).

7 Black Troughs

The theory of steady, line-driven winds predicts velocity laws $v(r)$ that are a monotonically increasing function of r, and such velocity laws have conventionally been assumed when computing P Cygni line profiles (e.g., Castor & Lamers 1979). In contrast, outflows containing shocks share the generic characteristic that the instantaneous velocity field is non-monotonic. This implies that a beam of radiation can come into resonance with a given transition at many points along its path through the wind and not just at one point as is the case for monotonic flow. Thus, when the discovery of X-ray emission provoked ideas about shocks distributed throughout the winds, the theory of P Cygni line formation needed to be reworked to treat non-monotonic flows.

In the limit of multiply non-monotonic flow, strong resonance lines such as those of C IV and N V completely backscatter the photospheric radiation and thus form P Cygni profiles with black absorption troughs (Lucy 1982b; Puls et al. 1993). Such troughs are in fact commonly observed, and the failure of conventional theory to reproduce extended black troughs had already been emphasized by Castor & Lamers (1979). Moreover, this failure is not eliminated when these winds' small-scale motions are modelled classically as microturbulence – see plots in Hamann (1981). This diagnostic power of the UV line profiles has not been fully exploited in the testing of the numerical simulations of unstable wind, especially with respect to the information resulting from the doublet structure of the C IV and N V resonance lines (Lucy 1983; Prinja & Howarth 1986).

8 Wolf-Rayet Winds – the Ultimate Challenge

In addition to interpreting W-R spectra as implying high-speed mass ejection, Beals (1929) conjectured that the W-R phenomenon was due to the action of selective radiation pressure. Accordingly, all those who participated in the 1970's and 80's in developing line-driven wind models for OB stars, with their ever lengthening line lists and ever more realistic treatments of excitation, ionization and radiative transfer, have been aware that the ultimate challenge was to produce dynamically consistent models for the winds of the W-R stars.

However, despite this widely-shared ambition and despite the successes achieved for the OB stars, satisfactory line-driven models for W-R winds were not produced, and this led to a growing suspicion that such models might in fact be infeasible. In particular, attention focussed on the "momentum problem" for W-R winds, which emerges when the momentum fluxes of their winds are expressed as a ratio ϕ of the momentum fluxes of their emitted radiation (Barlow et al. 1981). Observational estimates (Willis 1991) of ϕ for W-R winds extend up to 72, with median $= 7.1$, whereas the highest values for the successfully-modelled winds of O stars are $\simeq 0.6$. An entirely reasonable reaction to these numbers is to adopt the working hypothesis that the O-star winds represent the extreme of what line-driving can sustain and that some other mechanism dominates for the W-R winds (Maeder 1985; Poe et al. 1989).

In the light of this situation, Lucy & Abbott (1993; LA) emphasized that, although line-driving had fallen out of favour for W-R winds, it had not in fact been excluded. They pointed out that a momentum flux ratio of ~ 10 corresponds to a modest 5% of the star's nuclear power being converted into wind power, so that a line-driven solution would surely not violate fundamental limits. Secondly, they noted that no alternative mechanism had achieved any compelling successes in predicting observed properties.

Having thus justified further work on the line-driving mechanism, LA went on to describe and use an improved version of their earlier Monte Carlo treatment of multiline transfer in winds. Now, for line-driven winds, the asymptotic flux of mechanical energy derives from the decreases on average of photons' rest energies as they undergo transitions in the differentially expanding flow. Accordingly, for a fixed terminal velocity, a high mass-loss rate can be sustained only if photons undergo a sufficiently large number of interactions before escaping from the wind – i.e., photon trapping must be effective.

In their earlier investigation, Abbott & Lucy (1985) found that gaps in the line list limit photon trapping and therefore also mass-loss rates. In their second paper, simply by imposing a semi-empirical ionization formula that crudely reproduces the severe stratification of ionization observed in W-R winds, they dramatically increased the effectiveness of photon trapping. This occurs because the list of effective lines now changes significantly with radial distance, so that a gap in the line list at one radius is often filled by lines at another radius, thus preventing escape and allowing additional energy to be transferred to the bulk flow. With photon trapping thus enhanced, LA were able to compute a model W-R wind with momentum flux ratio $\phi = 9.2$ and for which the asymptotic flux of mechanical energy was fully accounted for by photon energy losses within the wind – i.e., global dynamical consistency was achieved.

This strong-wind effect of enhanced photon trapping due to ionization stratification appeared to be the key to understanding the high mass-loss rates of W-R stars. However, a major difficulty has been raised by Schmutz (1994), who has also suggested how it might be overcome. The difficulty is that the strong photon trapping found by LA is not confirmed when their ionization stratification is replaced by that predicted by a non-LTE ionization code. Interestingly,

Schmutz does not then reject the LA solution but instead conjectures that current ionization codes are missing a major physical effect whose inclusion would give an ionization stratification close to that of LA and thus would retain its dynamical success.

For the conjectured missing effect, Schmutz postulates a line coincidence between the $L\alpha$ line of He^+ at $\lambda 304\,\text{Å}$ and a strong line of some other species. In the absence of any such coincidence, the trapping of the $L\alpha$ photons maintains a significant population of He^+ ions in the $n = 2$ level, to the point that most photoionizations $He^+ \rightarrow He^{++}$ occur from this level and not from the ground state. Clearly, then, some other species with a non-resonance line at $\simeq 304\,\text{Å}$ could destroy this line trapping by absorbing the $L\alpha$ photons and re-radiating via other transitions. The result would be a collapse of the population of the $n = 2$ level, with consequent reduction in the photoionization rate. As He is the dominant element in W-R winds, the implied shift inwards of the $He^{++} \rightarrow He^+$ recombination zone would have major impact on the ambient radiation field and thus on the ionization stratification of all other elements.

In a subsequent paper, Schmutz (1997) has illustrated his idea by parameterizing the conjectured loss of He^+ $L\alpha$ photons. He finds the expected change of ionization stratification and obtains a dynamically consistent model of HD 50896 that also reproduces its He emission-line spectrum. In addition, he identifies transitions that are closely coincident with He^+ $L\alpha$ and which may all be contributors to the required photon loss. Evidently, the next step is the removal of the parameterization to see if the effect survives a calculation from first principles.

Although further work is thus required, the papers of LA and Schmutz have re-established radiative driving as the most likely acceleration mechanism for W-R winds. Moreover, this assessment is reinforced by the severe observational difficulties faced by the Luminous Magnetic Rotator model of Poe et al. (1989). In W-R spectra, the flat-topped profiles of the emission lines from ions with low ionization potentials (Kuhi 1973) are inconsistent with a latitude-dependent terminal velocity. Also a search for the polarization signature expected for this model found no evidence of departures from sphericity for 32 out of 34 W-R stars observed (Schulte-Ladbeck 1994).

9   Conclusion

In the thirty years since the Aerobee rockets allowed us to observe the relatively weak winds of OB stars, the theory of radiatively-driven winds has been developed into a sophisticated diagnostic tool capable of providing essential input data for studies at the forefront of modern astrophysical research. Nevertheless, fascinating technical challenges remain, especially in the strong-wind limit relevant for W-R stars. Moreover, in view of the rôle that these stars have played in determining the mix of chemical elements in the present-day Universe, unlocking the secrets of their mass-loss mechanism is undoubtedly of fundamental importance. However, many other wind phenomena, though intrinsically interesting, do not similarly relate to bigger questions and are thus of far lesser importance. Nevertheless, skills developed in computing and analysing the spectra of stellar winds are undoubtedly applicable to other dynamical phenomena whose investigation will be facilitated by the VLT and other 8-metre-class telescopes.

References

Abbott, D.C. (1980): ApJ, 242, 1183
Abbott, D.C., Lucy, L.B. (1985): ApJ, 288, 679
Baade, D., Lucy, L.B. (1987): A&A, 178, 213
Barlow, M.J., Smith, L.J., Willis, A.J. (1981): MNRAS, 196, 101
Beals, C.S. (1929): MNRAS, 90, 202
Beals, C.S. (1941): in Observation des Novae, ed. A.J. Shaler (Paris: Hermann), 113
Beals, C.S. (1944): MNRAS, 104, 205
Cassinelli, J.P., Castor, J.I., Lamers, H.J.G.L.M. (1978): PASP, 90, 496
Cassinelli, J.P., Olson, G.L. (1979): ApJ, 229, 304
Castor, J.I. (1974): MNRAS, 169, 279
Castor, J.I., Abbott, D.C., Klein, R.I. (1975): ApJ, 195, 157 (CAK)
Castor, J.I., Abbott, D.C., Klein, R.I. (1976): in *Physique des mouvements dans les atmosphères stellaires*, eds. R. Cayrel & M. Steinberg (Paris: CNRS), p. 363
Castor, J.I., Lamers, H.J.G.L.M. (1979): ApJS, 39, 481
Feldmeier, A. (1995): A&A, 299, 523
Feldmeier, A., Kudritzki, R.-P., Palsa, R., et al. (1997a): A&A, 320, 899
Feldmeier, A., Puls, J., Pauldrach, A.W.A. (1997b): A&A, 322, 878
Hamann, W.-R. (1981): A&A, 93, 353
Harnden, F.R. Jr., Branduardi, G., Elvis, M., et al. (1979): ApJ (Letters), 234, L51
Kuhi, L.V. (1973): in IAU Symp. 49, eds. M.K.V. Bappu & J. Sahade (Dordrecht: Reidel), p. 205
Lamers, H.J.G.L.M., Morton, D.C. (1976): ApJ Suppl., 32, 715
Long, K.S., White, R.L. (1980): ApJ (Letters), 239, L65
Lucy, L.B. (1971): ApJ, 163, 95
Lucy, L.B. (1975): Mem. Soc. Roy. Sci. Liège, 8, 359
Lucy, L.B. (1982a): ApJ, 255, 286
Lucy, L.B. (1982b): ApJ, 255, 278
Lucy, L.B. (1983): ApJ, 274, 372
Lucy, L.B., Abbott, D.C. (1993): ApJ, 405, 738
Lucy, L.B., Solomon, P.M. (1970): ApJ, 159, 879
Lucy, L.B., White, R.L. (1980): ApJ, 241, 300
Maeder, A. (1985): A&A, 147, 300
McCarthy, J.K., Kudritzki, R.P., Lennon, D.J., et al. (1997): ApJ, 482, 757
Milne, E.A. (1924): MNRAS, 85, 111
Milne, E.A. (1926): MNRAS, 86, 459
Milne, E.A. (1927): MNRAS, 87, 697
Morton, D.C. (1967a): ApJ, 147, 1017
Morton, D.C. (1967b): ApJ, 150, 535
Morton, D.C., Jenkins, E.B., Bohlin, R.C. (1968): ApJ, 154, 661
Nordsieck, K.H., Cassinelli, J.P., Anderson, C.M. (1981): ApJ, 248, 678
Olson, G.L. (1978): ApJ, 226, 124

Owocki, S.P., Castor, J.I., Rybicki, G.B. (1988): ApJ, 335, 914

Pauldrach, A.W.A., Puls, J., Kudritzki, R.-P. (1986): A&A, 164, 86

Poe, C.H., Friend, D.B., Cassinelli, J.P. (1989): ApJ, 337, 888

Prinja, R.K., Howarth, I.D. (1986): ApJS, 61, 357

Puls, J., Owocki, S., Fullerton, A.W. (1993): A&A, 279, 457

Schmutz, W. (1994): in IAU Symp. 163, eds. K.A. van der Hucht & P.M. Williams (Dordrecht: Kluwer), p. 127

Schmutz, W. (1997): A&A, 321, 268

Schulte-Ladbeck, R.E. (1994): IAU Symp. 163, eds. K.A. van der Hucht & P.M. Williams (Dordrecht: Kluwer), p. 176

Waldron, W.L. (1984): ApJ, 282, 256

Willis, A.J. (1991): in IAU Symp. 143, eds. K.A. van der Hucht & B. Hidayat (Dordrecht: Kluwer), p. 265

Wilson, R. (1958): Publ. Roy. Obs. Edinburgh, II, No. 3

Wolfire, M.G., Waldron, W.L., Cassinelli, J.P. (1985), A&A, 142, L25

Owocki: Regarding your criticism of the CAK model with respect to the "Sobolev" vs. "Newtonian" velocity gradient, and the implied artificial nature of the CAK critical point, I would suggest that all discussion of critical points is artificial in that those are artifacts of analytically solving a reduced set of *ordinary* differential equations for an idealized *steady state* flow. If you compute time-dependent wind simulations using the CAK/Sobolev line-force, you don't need to worry about such subtleties of critical points, yet these simulations inevitably relax to very nearly the CAK solution. Further, such relaxation to the CAK solution is also found in *non*-Sobolev models computed with fully integrated forms of both the direct and diffuse line-forces. In effect, I think this shows clearly that the basic scalings predicted by CAK theory (including modern extensions, to calculate the line-list and include the finite stellar disk) are really very robust, and not at all sensitive to the subtleties of Sobolev theory implied by your comments.

Lucy: To me, it reflects badly on stellar wind theorists that the questionable physical significance of the CAK critical point is not stated whenever CAK solutions are computed and used.

Kundt: Your cartoon on Wolf-Rayet-wind radiative acceleration is inconsistent with radial-momentum balance: photons would have to scatter $\gtrsim \phi$ times from opposite hemispheres.

Lucy: The unphysical idealization of scattering to and from opposite hemispheres achieves a given energy transfer with the minimum number of scatterings. With realistic path lengths between scatterings, a vastly greater number of scatterings is required. I found Monte Carlo simulations with synthetic line lists instructive in understanding the relation between photon trapping, energy degradation and momentum transfer. You could carry out the same exercise or consult the paper of U. Springmann (1994, A&A, 289, 505).

A Thermo-Radiatively Driven, Analytical 2-D Model for Stellar Outflows

Alexander Kakouris[1,2]

[1] Section of AA&M, University of Athens, Athens, Greece
[2] Hellenic Air Force Academy, Dekelia, Attiki, Greece

Analytical, steady-state, 2-D solutions for thermally driven stellar winds were found by Kakouris & Moussas (1996). These solutions were extended to include the differential rotation of the fluid and the influence of the radiative force in an optically thin stellar atmosphere (Kakouris & Moussas 1997). The solutions use a new thermo-radiative mechanism in order to describe self-consistently outflows from early-type supergiants, for which the radiative force is important. The solutions can either describe a monotonically accelerated outflow or the existence of a deceleration region in the stellar envelope. The deceleration can lead to a density increase (blob or shell formation), depending upon the radiative force parameters (Kakouris & Moussas 1998). Fig. 1 illustrates the logarithmic density contours for our model of the wind of the hypergiant P Cygni, which was computed by using the optically thin radiative force given by Chen & Marlborough (1994) and the acceleration in the envelope determined by Lamers (1986). The results resemble the three-zone envelope of Nugis et al. (1979).

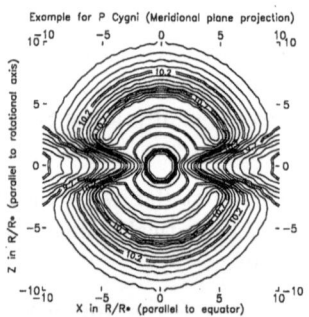

Fig. 1. Logarithmic density contours for our model of the wind of P Cygni.

References

Chen, H., Marlborough, J.M. (1994): ApJ 427, 1005
Kakouris, A., Moussas, X. (1996): A&A 306, 537
Kakouris, A., Moussas, X. (1997): A&A 324, 1071
Kakouris, A., Moussas, X. (1998): A&A 333, 178
Lamers, H.J.G.L.M. (1986): A&A 159, 90
Nugis, T., Kolka, I., Luud, L. (1979): in P.S. Conti & C.W.H. de Loore (eds.), Proc. IAU Symp. 83, Mass-Loss and Evolution of O-Type Stars, Reidel, Dordrecht, p. 39

Using ISO to Probe the Acceleration of Wolf-Rayet Winds

Richard Ignace[1], Joseph P. Cassinelli[2], Patrick Morris[3], and John C. Brown[1]

[1] Dept. Physics and Astronomy, U. Glasgow, Kelvin Bldg, Glasgow G12 8QQ, UK
[2] Dept. Astronomy, U. Wisconsin, 5534 Sterling Hall, Madison, WI, USA
[3] ESA ISO-Science Operations Centre, Madrid, Spain

Brown et al. (1997) and Ignace et al. (1997) have developed an inversion method to derive the wind velocity distribution from optically thin line profiles formed in spherically symmetric outflows. The method has been applied to three He II lines (9–7 at 2.83 μm, 7–6 at 3.09 μm, and 8–7 at 4.76 μm) observed with the ISO SWS for the stars WR 134 (WN6) and WR 136 (WN6). Of these, the 2.83 μm line probes deepest into the wind owing to the dominant free-free opacity. The winds of both WR 134 and WR 136 have attained only 75% of their respective terminal speeds at the radii where the 2.83 μm line is formed.

The optically thin [Ca IV] 3.207 μm line has also been detected for both WR 134 and WR 136 (see Fig. 1); it yields terminal speeds of 1965±123 and 1750±147 km s^{-1}, respectively. Note that the profile for WR 136 is reasonably flat-topped, which indicates a spherically symmetric wind, whereas the profile for WR 134 is double-peaked, which suggests an equatorial density enhancement that is consistent with the intrinsic polarization of this star.

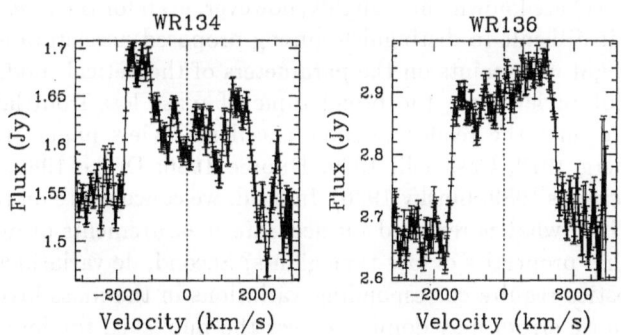

Fig. 1. The [Ca IV] 3.207 μm line observed in WR 134 and WR 136 by ISO. Note the double-peaked profile of WR 134.

References

Brown, J.C., Richardson, L.L., Cassinelli, J.P., Ignace, R. (1997): A&A 325, 677
Ignace, R., Brown, J.C., Richardson, L.L., Cassinelli, J.P. (1998): A&A 330, 253

A Critical Evaluation of Mass-Loss Rates and Wind Properties of Evolved Late-Type Stars

Jeffrey L. Linsky[1], Graham M. Harper[2], Philip D. Bennett[2], Alex Brown[2], and Jeffrey Valenti[1]

[1] JILA, University of Colorado and NIST, Boulder, CO, 80309–0440, USA
[2] CASA, University of Colorado, Boulder, CO, 80309–0389, USA

Abstract. We evaluate the accuracy with which the properties of late-type giant star winds can be estimated by fitting ultraviolet line profiles and radio continuum emission. Our test cases include the stars α TrA (K4 II), λ Vel (K4 Ib–II), and ζ Aur (K4 Ib + B5 V). Of particular interest is the sensitivity of the line profiles to the wind parameters $\dot{M}$, v_∞, v_{turb}, β, and T_{wind}. We are finding that while accurate and reliable values of $\dot{M}$ are difficult to obtain, the most reliable values come from the application of two or more independent techniques.

1 Introduction

Blue-shifted absorption features in optically thick resonance lines demonstrate that mass loss is ubiquitous among giant stars later than spectral type K3, and centimeter wave continuum emission plays an important role in confirming the wind properties. Accurate physical parameters describing the mass loss (e.g., the mass-loss rate $\dot{M}$, terminal velocity v_∞, velocity law $v(R/R_\star)$, geometry, and time variability) are known only roughly, however, even for the best studied stars. This makes it difficult to distinguish among proposed acceleration mechanisms or to place tight constraints on the parameters of theoretical models.

We cannot review here the broad topic of mass loss from late-type giant stars, and we refer the reader to some excellent review papers (e.g., Reimers 1975; Goldberg 1979; Cassinelli 1979; Dupree 1986; Drake 1986; Holzer 1987; Dupree & Reimers 1989; Linsky 1996). Instead, we concentrate on three essential questions. First, what is required for accurate measurements of mass-loss rates and other wind properties of late-type giants? Second, do variations in empirical wind diagnostics require corresponding variations in the mass-loss rate? Third, what systematic errors may dominate over the published random errors? Only after these questions are answered satisfactorily, can one evaluate the accuracy of simple $\dot{M}$ scaling laws like that proposed by Reimers (1987),

$$\dot{M} = 4 \times 10^{-13} \frac{\eta(L/L_\odot)}{(g/g_\odot)(R/R_\odot)} \quad M_\odot/\mathrm{yr} \tag{1}$$

with $1/3 < \eta < 3$, and the alternative laws proposed by Goldberg (1979) and Drake (1986). Although most of the work to date on these topics has involved the analysis of ultraviolet and optical spectra, radio continuum observations, which provide a direct measurement of the ionized plasma in a stellar atmosphere, can be critical in measuring accurate mass-loss rates.

2   Physical Processes and Phenomena to be Considered

Why is it so difficult to determine accurate values for the wind properties of late-type giants given the very high quality spectra that have been obtained with the GHRS on HST? The reasons almost certainly lie in the simple approximations that one must make when modeling the complex physical conditions in these winds. Consider the following challenges:

Geometry While winds likely have many structures with different properties controlled by the geometry of magnetic fields or different boundary conditions in the convective photosphere, wind models generally assume spherical symmetry with properties that depend only upon radial position.

Velocity law For simplicity one generally approximates the wind speed as a smooth function of radial position of the form

$$v(R/R_\star) = v_\infty [1 - (R_\star/R)^\gamma]^\beta, \tag{2}$$

where γ is usually set to unity but can be a useful additional parameter (cf. Bennett and Harper 1997). This approximation ignores shocks or other discontinuities, nonmonotonic flows, circulation patterns, and corotating interacting regions. Inversion techniques discussed in §6 allow one to derive the velocity law semi-empirically without assuming a functional form.

Magnetic fields When $B^2/8\pi > P_{\text{gas}}$, the magnetic field determines the dynamics and energetics of the wind and the wind geometry. Since there are no magnetic field measurements for these stars, we can only estimate upper limits to the field fluctuations from the widths of optically thin lines.

Acceleration mechanism The wind acceleration mechanism(s) in these stars are not yet known, although various authors have proposed that acoustic waves driven by convection or stellar oscillations (radial or nonradial), Alfvén waves, radiation pressure on dust (for the coolest stars), or Parker-type thermal pressure gradients for hot winds may be responsible. The acceleration mechanism may not be steady but rather stochastic (e.g., Cuntz 1990).

Ionization The ionization equilibria for hydrogen and the metals are critically important for inferring the total mass loss rate $\dot{M}_{\text{total}}$ from either the ionized metal lines or the radio continuum emission, which measures only the ionized fraction of $\dot{M}_{\text{total}}$. LTE ionization and excitation are invalid for the low density winds of these stars and will incorrectly predict the radiative cooling rate (cf. Willson 1998). Coronal ionization equilibrium may also not be valid as photoionization can be important and the ionization, and recombination timescales may be long compared to the transport timescales. Then the ionization structure of the inner wind is "frozen in" and advected outward.

Radiative transfer One typically tries to fit the P Cygni-shaped profiles of optically thick lines such as the Mg II h and k lines. Photons in the wind are scattered out of the blue side of the line profile and reemitted typically near line center. Since the terminal velocity of the wind is usually not more than a few times the thermal and turbulent broadening, one should solve the transfer equation using a co-moving frame formulation consistent with the solution of the transfer equation in the chromosphere at the base of the wind. Since a code for this type of solution does not yet exist, the models that we will present below are based on observer's frame complete redistribution (CRD) with the chromospheric emission lines assumed as a lower boundary condition. Sobolev-type (SEI) solutions, which are valid for the fast winds of hot stars, may also provide useful approximate solutions for late-type stars with slow winds (cf. Lamers et al. 1987).

Turbulence For simplicity one assumes constant isotropic microturbulence or perhaps allows v_{turb} to depend on radial position, but as yet there is no evidence that the turbulence is isotropic or depends only on radial position. Turbulence plays a role in line formation and in the dynamic support of the atmosphere (cf. Hartmann et al. 1985). Large turbulent pressure gradients can also drive the wind.

Binary systems In some close binary systems, the winds of each star may not be isolated but rather interact along a shock front.

3 Lessons Learned from Studying the Solar Wind

The Ulysses space probe, which is studying the solar wind *in situ* over a wide range of solar latitudes, has provided the first detailed three-dimensional picture of a late-type star's wind. Phillips et al. (1995) showed that the Sun has not one but at least three different winds as summarized in Table 1, where the ion densities are extrapolated to 1 AU.

Table 1. Typical parameters for the three solar winds

Region	v_∞ (km s^{-1})	n_{ion} (cm^{-3})	$\dot{M}_{total}$ ($10^{-14} M_\odot$/yr)	Coverage at solar (max, min)
Open magnetic fields	700–800	2.7	≈ 1.5	20%, few %
Closed magnetic fields	≈ 400	7	2	$\approx 80\%$, $\approx 95\%$
Coronal mass ejections	≤ 1000	0.2	?	small

This example of one star with three winds provides some lessons that may be useful when studying late-type giant winds. First, the presence of three quite different winds associated with different magnetic field structures highlights the inhomogeneity of late-type star winds and one source of the inhomogeneity. Corotating interaction regions (CIRs), which occur when a fast wind overtakes a slow wind, can produce characteristic signatures in the observed line profiles formed in inhomogeneous winds. Second, the properties of a stellar wind may differ depending on the viewing angle: for the Sun one would see a high speed wind when viewing from the poles, a lower speed wind when viewing from near the equator, and a cyclic wind when viewing from intermediate latitudes as high and low speed streams rotate into and out of the field of view. The most important lesson is that there is **no significant difference** in $\dot{M}$ between the different solar winds and apparently no more than a factor of 1.5 difference in $\dot{M}_{\text{total}}$ over the solar cycle, despite large changes in the magnetic field.

4 Analysis of Radio Emission from Late-type Stars

The brightness temperature T_B is the equivalent blackbody temperature for an optically thick surface with radius R. The radio luminosity (erg s^{-1} Hz^{-1}) is

$$L_\lambda = 1.3 \times 10^6 \, (6\text{cm}/\lambda)^2 \, (R_\star/R_\odot)^2 \left[(R/R_\star)^2 T_B \right]. \qquad (3)$$

Thermal free-free emission (also called bremsstrahlung) can be emitted by partially ionized plasma in a chromosphere, corona, and wind; and nonthermal gyrosynchrotron emission can be important in a corona with relativistic electrons and magnetic fields. Nonthermal emission is detected only from those rapidly-rotating late-type giants in close binary systems, and thermal coronal emission is only detected from warmer giants like Capella. For late-type giants with optically thick winds, the right-hand term in Eq. (3) is given by $(R/R_\star)^2 T_B = (R_{\text{wind}}/R_\star)^2 T_{\text{wind}}$, where R_{wind} is the effective radius of the wind. On the other hand, when the wind is optically thin we have $(R/R_\star)^2 T_B = (R_{\text{chr}}/R_\star)^2 T_{\text{chr}} (1 - e^{-\tau_{\text{wind}}})$, where R_{chr} is the effective radius of the chromosphere and τ_{wind} is the wind optical depth.

For optically thick thermal emission from a constant velocity, spherically symmetric, partially ionized wind with ionized mass loss rate $\dot{M}_{\text{ion}}$, the flux density (cf. Wright & Barlow 1975) is

$$S_\nu \sim \dot{M}_{\text{ion}}^{4/3} v_{\text{wind}}^{-4/3} D^{-2} \nu_5^{0.6} T_4^{0.1}, \qquad (4)$$

where $\nu_5 = \nu/5$ GHz, $T_4 = T/10^4$, and D is the distance in parsecs. M giants and supergiants typically have optically thick cool winds. For example, Drake and Linsky (1986) detected the M3 III star μ Gem as a 0.18 mJy source. This flux corresponds to $\log L_6 = 14.7$ and $(R/R_\star)^2 T_B = 6 \times 10^4$ K. Since $T_{\text{chr}} \leq 10^4$ K, the emission from the wind dominates with $R_{\text{wind}}/R_\star = 5 - 7$ for an assumed wind temperature in the range 1000–2000 K or smaller sizes for higher temperatures. By comparison, $\log L_6 = 11.0$ for the quiet Sun and $(R/R_\star)^2 T_B = 8 \times 10^4$ K. Since $T_{\text{chr}}(\tau_6 = 1) = 20,000$ K from models and the solar wind is optically thin, free-free emission from the transition region and corona dominate.

Eq. (4) predicts that the radio spectral index α given by $S_\nu \sim \nu^\alpha$ will have the value $\alpha = 0.6$ for an optically thick wind with constant velocity and ionization. Drake & Linsky (1986) found that for late-type giants $0.80 \leq \alpha \leq 1.26$, which can be explained by decreasing ionization or increasing wind velocity with radial distance. Both changes from the simple Wright-Barlow approximations are likely to be valid, as are departures from steady spherical flow.

5   Results from Modeling GHRS Spectra of α TrA (K4 II)

Alpha TrA is a member of the class of hybrid-chromosphere stars that shows both high temperature emission lines (e.g., Si IV and C IV) and high velocity winds identified by the P Cygni shape of the Mg II resonance lines. GHRS echelle spectra of the Mg II lines obtained in 1993 and 1994 (see Fig. 1) provided the high-S/N line profiles which Harper et al. (1995) fit with a simple wind model assuming a spherical flow, a specified chromospheric emission line at the base of the wind, pure scattering for a two-level Mg II atom, constant v_{turb} and $\dot{M}$, and the simple functional form for the velocity law (Eq. 2) with β and v_∞ as parameters and γ set to unity. The best fit parameters for this model are: $v_\infty = 104$ km s^{-1}, $v_{\text{turb}} = 19$ km s^{-1}, and $\dot{M} \geq 1.6 \times 10^{-10}$ $M_\odot$/yr (assuming $\beta = 1$ and $R_\star = 97 R_\odot$, but β is uncertain).

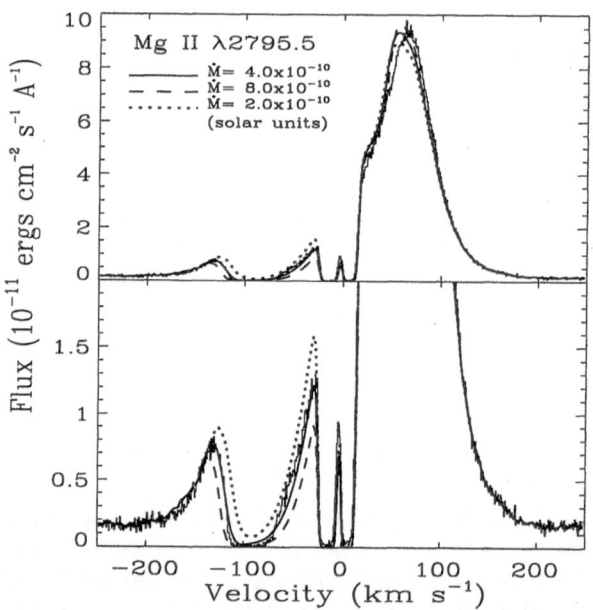

Fig. 1. GHRS echelle spectrum of the Mg II λ2795.5 line of α TrA (expanded version of the line in the lower portion of the figure). The solid line is the best fit model with a mass loss rate $\dot{M} = 4.0 \times 10^{-10}$ $M_\odot$/yr, and the dashed lines are models with $\dot{M}$ changed by a factor of 2.

Harper et al. (1998) have improved on their initial analysis by allowing the non-LTE hydrogen ionization to be advective and by using a new value for the radius ($R = 131 R_\odot$) inferred from the Hipparcos parallax. The best fit values of v_∞ and v_{turb} do not change, but now $\beta = 1.5$ and $\dot{M} \geq 4 \times 10^{-10}$ $M_\odot$/yr. The mass-loss rate is a lower limit because a significant fraction of magnesium may be in ionization states other than Mg II and β is uncertain. They also ran models with parameters different from their best fit values to test the sensitivity of different portions of the line profile to these parameters:

- $v_\infty = 104 \pm 5$ km s^{-1} is measured by the velocity at the blue edge of the wind absorption feature;
- $v_{\text{turb}} = 19 \pm 3$ km s^{-1} is measured by the shape of the blue edge of the wind absorption feature and by the shape of the reemission near 0 km s^{-1};
- $\beta = 1.5 \pm 0.5$ is also measured by shape of the wind absorption feature and the reemission near 0 km s^{-1}, but the value of β could lie outside of this range if the assumed chromospheric intensity profile is inaccurate or if the velocity law has a more complex shape (for example, if $\gamma \neq 1.0$ in Eq. 2);
- $\dot{M}_{\text{total}} \approx 4 \times 10^{-10}(N_{Mg}/N_{MgII})$ $M_\odot$/yr is measured by the entire shape of absorption feature and is uncertain by about a factor of 1.5. Figure 1 shows the changes in the computed line profile when $\dot{M}$ is changed by a factor of 2 about the best fit value.

Note that the uncertainty in the magnesium ionization equilibrium is now explicitly included in the estimate of $\dot{M}_{\text{total}}$. Harper et al. (1998) estimate that $T_{\text{wind}} = 16,000 - 20,000$ K by fitting the Si III 1206 Å line. For these high wind temperatures magnesium is mostly Mg III and $\dot{M}_{\text{total}}$ must be significantly larger than its lower limit. A further constraint is provided by the observed radio flux, $S_{3.5cm} = 0.26 \pm 0.05$ mJy. Table 2 shows how the total mass-loss rate $\dot{M}_{\text{total}}$, radio fluxes at 3.6 and 6.0 cm, and the radio spectral index α depend on the assumed wind temperature. The radio emission thus supports a model with $T_{\text{wind}} = 20,000$ K. Table 3 shows the sensitivity of the radio emission to the wind parameters for $T_{\text{wind}} = 20,000$ K. Radio observations at 3 and 6 cm with the Australian Telescope Compact Array will further refine the model. However, at this point it appears that $\dot{M}_{\text{total}} \approx 2.0 \times 10^{-9}$ $M_\odot$/yr. This can be compared with the value $\dot{M}_{\text{total}} \approx (4-5) \times 10^{-9}$ $M_\odot$/yr that Hartmann et al. (1981) computed with their Alfvén-wave-driven wind model, and the value $\dot{M}_{\text{total}} = 8 \times 10^{-10}$ $M_\odot$/yr that Brosius & Mullan (1986) find best fits the observed line strengths and widths and the electron density.

Table 2. Computed radio emission parameters for α TrA

T_{wind} (K)	16,000 K	20,000 K
$\dot{M}_{\text{total}}$	5.4×10^{-10}	2.0×10^{-9}
$S_{3.6cm}$ (mJy)	0.10	0.28
$S_{6.0cm}$ (mJy)	0.05	0.15
α	+1.5	+1.2

Table 3. Sensitivity of the radio emission parameters for the 20,000 K model

Parameter	$S_{3.6cm}$ (mJy)	$S_{6.0cm}$ (mJy)	α
for $T = 20,000$ K model	0.276	0.150	1.192
if $v_\infty = 114$	0.258	0.139	1.215
if $v_\infty = 94$	0.298	0.164	1.168
if $\dot{M} = 2 \times 10^{-10}(N_{Mg}/N_{MgII})$	0.175	0.087	1.371
if $\dot{M} = 8 \times 10^{-10}(N_{Mg}/N_{MgII})$	0.490	0.289	1.036
if $\beta = 1.0$	0.221	0.124	1.129
if $\beta = 2.0$	0.332	0.176	1.238

6   Analysis of GHRS Spectrum of λ Vel (K4 Ib-II)

The reverse approach of fitting observed line profiles with theoretical profiles computed from semi-empirical models, called the "wind inversion technique," is very useful for estimating wind properties and provides an important reality check on the accuracy of the inferred wind parameters. Carpenter et al. (1998) have obtained an estimate of $v(R/R_\star)$ in the wind of λ Vel (K4 Ib-II) by analyzing GHRS spectra (cf. Carpenter et al. 1995) of Fe II lines with very different opacities. Bennett & Harper (1997) outlined the exact solution of the computationally demanding Fe II radiative transfer problem for the wind of λ Vel.

For unsaturated lines Carpenter et al. (1998) measured the velocity at minimum flux, $v(\tau)$, where τ is the optical depth where the line minimum is formed. For saturated lines they measured the velocity of the blue edge of the absorption feature and corrected for turbulence and opacity broadening. The values of $v(\tau)$ for each line were then placed on a common $v(R/R_\star)$ scale for an assumed model and non-LTE excitation of Fe II. For λ Vel the data fit Eq. (2) with the parameters $v_\infty = 29.70$ km s^{-1}, $\beta = 2.06$, and $\gamma = 0.80$. This velocity law is roughly similar to the two-parameter relation with $\beta \approx 3$, and $\gamma = 1.0$. Both velocity laws show a very gradual rise in v with $v = 0.5v_\infty$ at $R/R_\star = 4$. Lambda Vel is a close spectral proxy for the primary of ζ Aur (see below). Schematic models for λ Vel with varying T_{wind} have been constructed based on the detailed structural model of ζ Aur. This model predicts the observed value of the radio flux, $S_{3.5cm} = 0.14 \pm 0.03$ mJy for $T_{\text{wind}} = 8000 \pm 500$ K.

7   Results from the Analysis of Ultraviolet Spectra of ζ Aur (K4 Ib + B5 V)

Reimers (1975), Eaton (1996), Baade et al. (1996) and their collaborators have called attention to the very useful role that a main-sequence B companion can play in studying the winds of K-giant and supergiant stars. A good example is the 972-day eclipsing binary system ζ Aur (K4 Ib + B5 V), for which the B star is a small but very bright UV background source against which the K star chromosphere and wind can be readily observed in absorption at different phases. We have now obtained and analyzed HST/GHRS spectra of ζ Aur at 11 orbital phases covering two orbits. Bennett et al. (1996) have derived very accurate parameters for this system: age $(80 \pm 15) \times 10^6$ yr and K-star radius $R_\star = 148 \pm 3 R_\odot$. The K4 Ib star is thus near the tip of the red-giant branch.

This very rich data set will take time to digest and publish. However, several important conclusions can already be identified in the analysis so far:

Ionization stratification At most epochs the velocity distributions of the singly and doubly ionized ions are quite different, indicating that the degree of ionization changes along the line of sight. However, the absence of ionizing radiation from a hot companion may simplify the ionization structure of single star winds, and the ionization state far from the star may be "frozen in" and thus homogeneous.

Time variability The distribution of ion densities with velocity is very different at epochs 3 and 7, which have the same orbital phase (0.95) but are one full orbit apart. Thus the properties of the wind are not just dependent on orbital phase but also on time, and therefore suggestive of intrinsic variability in the K supergiant wind.

Asymmetry The distributions of ion densities with velocity are very different at epochs 7 (phase 0.95) and 9 (phase (0.05), which are symmetric about eclipse. The data at epoch 9 suggest an ejection event. Thus either the wind properties are asymmetric relative to eclipse, or the data could be explained by time variability.

Ionization time variability In February 1995 (orbital phase 0.71) the radio fluxes were $S_{3.6cm} = 0.35 \pm 0.02$ mJy, and $S_{6.0cm} = 0.25 \pm 0.02$ mJy. These data are consistent with the power law $S_\nu \sim \nu^{0.60 \pm 0.18}$, where the spectral index is the same as predicted by the Wright-Barlow constant wind-velocity model even though the wind does NOT have a constant speed or ionization. However, in March 1997 (phase 0.48 near apastron) we observed a significant decrease in $S_{3.6cm} = 0.25 \pm 0.02$ mJy, and thus the mass-loss rate of ionized gas. Whether this is due primarily to a decrease in $\dot{M}_{total}$ or to a decrease in the ionization fraction is not yet known.

Mass-loss rate Despite the considerable differences in the distribution of ion densities with velocity at each phase, we find that $\dot{M}_{total} \sim 5 \times 10^{-9}$ $M_\odot$/yr is a good estimate of the mass-loss rate at all phases within a factor of 2. Thus as is found for the Sun, $\dot{M}$ appears to be relatively independent of time and orbital phase.

8   Conclusions

While accurate and reliable values of $\dot{M}_{total}$ and other wind properties of late-type giants are difficult to estimate even for well-studied and characterized late-type stars, we are finding that the most reliable values are obtained when two or more techniques give similar results. Drake (1986) previously came to the same conclusion, and argued for the use of completely independent techniques.

Examples of winds as different as those of the Sun and ζ Aur show that the observed variability, whether cyclic or not, in a wind parameter like v_∞ or in the shape of the absorption feature in a line profile, does **not** require large changes in $\dot{M}_{\text{total}}$. On the contrary, the examples presented in this review indicate that mass-loss rates may vary by only a factor of 2 with time. Future analyses of excellent quality UV spectra and radio-flux data for a more complete sample of late-type giant and supergiant stars are needed to determine whether or not the mass-loss rates vary with time by only a factor of 2 and to obtain accurate values for other the wind parameters needed to infer the wind acceleration mechanism(s) operating in these stars.

This work is supported by NASA grant numbers S-56460-D, S-56500-D, NAG5-3033, NAG5-4804, GO-3626, GO-6069, and AR-06383 to the University of Colorado. We thank the HST, VLA, and AT for observing time.

References

Baade, R., Kirsch, T., Reimers, D., et al. (1996): Ap.J., **466**, 979

Bennett, P.D., Harper, G.M. (1997): in *12th Kingston Meeting: Computational Astrophysics.* ed. D.A. Clarke & M.J. West, ASP Conf. Series, **123**, 87

Bennett, P.D., Harper, G.M., Brown, A., Hummel, C.A. (1996): Ap.J., **471**, 454

Brosius, J.W., & Mullan, D.J. (1986): Ap.J., **301**, 650

Carpenter, K.G., Robinson, R.D., Harper, G.M., et al. (1998): in preparation

Carpenter, K.G., Robinson, R.D., Judge, P.G. (1995): Ap.J., **444**, 424

Cassinelli, J.P. (1979): Ann. Rev. Astro. Astrophys., **17**, 275

Cuntz, M. (1990): Ap.J., **353**, 255

Drake, S.A. (1986): in *Cool Stars, Stellar Systems, and the Sun*, ed. M. Zeilik & D.M. Gibson (Springer, Berlin), 369

Drake, S.A., Linsky, J.L. (1986): A.J., **91**, 602

Dupree, A.K. (1986): Ann. Rev. Astro. Astrophys., **34**, 377

Dupree, A.K., Reimers, D. (1989): in *Exploring the Universe with the IUE Satellite*, ed. Y. Kondo et al. (Kluwer, Dordrecht), 321

Eaton, J. (1996): in *Cool Stars, Stellar Systems, and the Sun*, 3., ed. R. Pallavicini & A.K. Dupree (San Francisco, Astr. Soc. Pacific), 503

Goldberg, L. (1979): Q.J.R.A.S., **20**, 361

Harper, G.M., Wood, B.E., Linsky, J.L., et al. (1995): Ap.J., **452**, 407

Harper, G.M., Skinner, S.L., Ayres, T.R., et al. (1998): in preparation

Hartmann, L., Dupree, A.K., Raymond, J.C. (1981): Ap.J., **246**, 193

Hartmann, L., Jordan, C., Brown, A., Dupree, A.K. (1985): Ap.J., **296**, 576

Holzer, T.E. (1987): in *Circumstellar Matter*, ed. I. Appenzeller & C. Jordan (Kluwer, Dordrecht), 289

Lamers, H.J.G.L.M., Cerruti-Sola, M., Perinotto, M. (1987): Ap.J., **314**, 726

Linsky, J.L. (1996): in *Radio Emission from the Stars and the Sun*, ed. A.R. Taylor and J.M. Parades (Astron. Soc. Pacific, San Francisco), **93**, 439

Phillips, J.L. (1995): Science, **268**, 1030

Reimers, D. (1987): in *Circumstellar Matter*, ed. I. Appenzeller & C. Jordan (Reidel, Dordrecht), 307

Willson, L.A. (1998): this volume

Wright, A.E., Barlow, M.J. (1975): MNRAS, **170**, 41

Willson: I don't think you were provocative enough in dealing with the scaling laws for mass loss. I will argue in my talk that you can't even apply them to the stars from which they were derived.

Eaton: (1) What is the nature of the high turbulence you measure in the wind of ζ Aur (they are much higher than I found for 31 Cyg)? (2) Are you arguing that these high turbulent velocities represent actual local variations in wind velocity?

Linsky: (1) I would speculate that the physical explanation is that the flow includes perturbations (slower and faster streams) which we are approximating as a steady mean flow and turbulence about this mean flow. (2) I doubt that the inferred "turbulence" represents local variations of the wind velocity, because the "turbulence" would then be highly superior. Instead the "turbulence" may be variations in the flow speed along the line of sight.

Mullan: In 1995 (Florence Cool Star workshop) Harper claimed that winds in isolated cool giants accelerate rapidly as a function of radius where winds from cool giants in ζ Aur systems accelerate slowly. Is this claim still believed now that more extensive modelling has been performed? A comment: Brosius and Mullan (1986) found $\dot{M}$ (α Tra) $= 10^{-9}$ $M_\odot$/yr on the basis of a fully energetic model of the Alfvén wave driving (cf. Harper et al. 1995).

Linsky: We are now determining velocity laws for a number of late-type giants and supergiants using line profile fitting (e.g. α Tra) and inversion techniques (e.g. λ Vel and γ Cru). There is no clear picture yet, but we do find that the wind velocity reaches close to its terminal value at 4 stellar radii for ζ Aur (K4Ib+B5V) but near 100 stellar radii for λ Vel (K4Ib-II). Note that the K stars have very similar spectral types. These results are preliminary.

Variable Winds in Yellow Supergiants

Krzysztof Gesicki[1] and Miroslaw Schmidt[2]

[1] Centrum Astronomii UMK, ul.Gagarina 11, PL-87-100 Torun, Poland
[2] NCAC, ul.Rabianska 8, PL-87-100 Torun, Poland

In Fig. 1 we present spectra of ϱ Cas, a classical example of a yellow supergiant. The observations from year 1970 (thin lines) were discussed by Gesicki (1992). They are repeated here to compare with the unpublished spectrum from 1978 (thick lines). Although not cyclical, the variability in the wind is apparent. Non-LTE analysis (see Gesicki 1992) results in a mass-loss rate of about 5×10^{-3} $M_{\odot}$ yr^{-1} in 1978, which is higher than in 1970. These rates were calculated by assuming the solar abundance for barium. Our mass-loss rates are sensitive to this assumption, because the abundance of s-processed elements like barium can be significantly increased after the third dredge-up phase.

In 1996 we performed a search of other yellow supergiants with circumstellar shells visible in the Ba II lines. Of 20 objects observed with the ESO CAT/CES, we found 7 with a similar phenomenon. In Fig. 1 we illustrate one example, the F2 Ia-0 supergiant HD 172481. The atmosphere of this star is poorly known, but our preliminary fits of the Ba II lines imply a mass-loss rate of 5×10^{-4} $M_{\odot}$ yr^{-1}. Our analysis is continuing; new observations will help to answer questions about wind variability.

This work was supported by KBN grant No.2.P.03D.026.09.

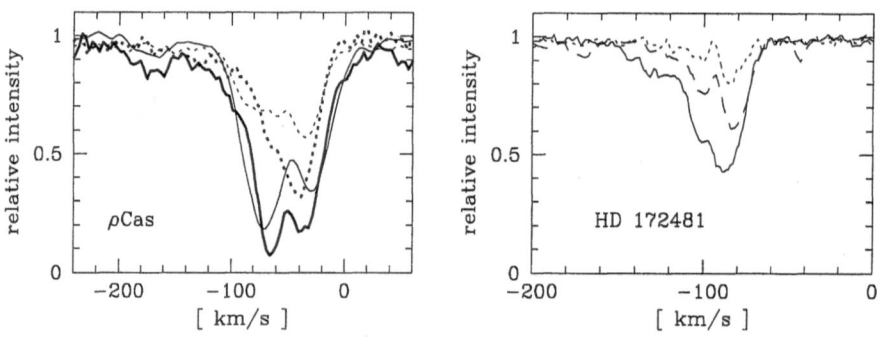

Fig. 1. The Ba II lines of two yellow supergiants: solid – 4934 Å; dots – 5853 Å; dashes – 6496 Å. See the text for details.

References

Gesicki, K. (1992): A&A 254, 280

Numerical Simulations of Dust-Driven AGB Winds: A Two-Fluid Model

Yvonne J.W. Simis

Leiden Observatory, P.O. Box 9513, 2300 RA Leiden, The Netherlands

Radiation pressure on dust grains is a main driving mechanism for the strong winds of AGB stars (e.g., Habing 1994). Numerical simulations can be used to explore the parameter space associated with these winds. Since the behaviour of both the dust and gas have to be followed in order to achieve the desired self-consistent description of the outflow in two dimensions, I have written a two-fluid hydrocode. This code uses spherical symmetry. Centered differencing and a two-step predictor-corrector method that apply the FCT (Boris 1976) and LCD (Icke 1991) algorithms are also used. These numerical techniques were chosen because of their suitability for a two-dimensional extension of the code. The code solves the equations of continuity, momentum, and energy for an (ideal) gas and a single dust component. Both sets of equations are coupled by the condensation of gas onto dust grains and the transfer of momentum between dust and gas. Condensation is described by the source term in the continuity equations, S_{cond}, in which C_1 is the condensation rate and $C_2 > 0$:

$$S_{\text{cond}} = C_1 \exp(C_2 |T - T_{\text{cond}}|) \ . \tag{1}$$

Momentum coupling between the gas and dust is described by the frictional drag force, where a is the grain radius and m_{d} is the grain mass:

$$F_{\text{drag}} = \frac{\pi a^2}{m_{\text{d}}} \rho_{\text{g}} \rho_{\text{d}} (v_{\text{d}} - v_{\text{g}}) |v_{\text{d}} - v_{\text{g}}| \ . \tag{2}$$

By using the Henyey method, I have calculated a stationary gas flow profile as the initial condition for the time-dependent calculation. Dust starts to form according to (1) when the dynamical calculation is started. First results show that in the early stages of dust formation momentum transfer according to (2) has a negligible influence on the gas and only a small influence on the dust flow.

References

Boris, J.P. (1976): NRL Mem. Rep. 3237
Icke, V. (1991): A&A 251, 369
Habing, H.J., Tignon, J., Tielens, A.G.G.M. (1994): A&A 286, 523

YSO Winds and Jets:
Mechanisms and Variability

Lee Hartmann

Center for Astrophysics, 60 Garden St., Cambridge, MA 02138

Abstract. I review the current knowledge concerning the powerful outflows of young stellar objects. These outflows are bipolar and, at least initially, appear highly collimated. The energy source for these strong winds or jets is accretion; the mass-loss rate is $\sim 10^{-1}$ of the mass-accretion rate over several orders of magnitude in mass fluxes. The wind has been shown to arise directly from the circumstellar disk in one case, FU Ori, and the presumption is that jets arise from the surfaces of accretion disks in all cases. The most likely mechanism for accelerating these winds is rotating magnetic fields rooted in the accretion disk, and the most probable effect accounting for the flow collimation is toroidal magnetic fields, although the details of the magnetic field geometries are controversial. The time-variability of the outflows is poorly understood; observations of the interactions of jets with the circumstellar medium suggest that the outflows are variable on almost all time scales, but are punctuated by substantial bursts which may be related to FU Ori-like events of high disk accretion.

1 Introduction

Young stellar objects (YSOs) are known to have spectacular, powerful, bipolar, and often highly-collimated outflows. A vast literature now exists on this subject; a recent review is given by Bachiller (1996), while the recent IAU Symposium 182 was entirely given over to understanding these outflows (Reipurth & Bertout 1997). Much of the published discussion concerns the increasingly well-observed interactions of outflows/jets with the ambient circumstellar medium. Space prevents a thorough discussion of many of these issues. Here I will touch briefly on what seem to me the most important points for understanding the causes of these outflows.

2 Winds = Jets

One of the earliest and most contentious issues that arose concerning YSO outflows was the relationship between the outflows seen in molecular species and jets seen in ionized material. The molecular flows, particularly seen in low-excitation species such as CO, probably trace the cold gas of the molecular cloud that has been swept up by the (higher-velocity) original outflow. Frequently this swept-up material, though usually showing bipolar lobes, is not highly collimated. In contrast, the optical jets seen primarily in lines from ionized gas ($H\alpha$, [O I], [S II]) generally exhibit very narrow structure (some spectacular examples of this can be seen in the reviews by Reipurth & Heathcote (1997) and Bally & Devine

(1997); see Figure 1). The distinction between these two types of geometry (often in the same object; for example, L1551 IRS 5, Snell, Loren, & Plambeck 1980), coupled with initial estimates that jets were not powerful enough to account for the molecular outflows, led to the notion of a *cold, less-collimated* "second wind", which was not visible in optical emission but was responsible for accelerating the molecular gas.

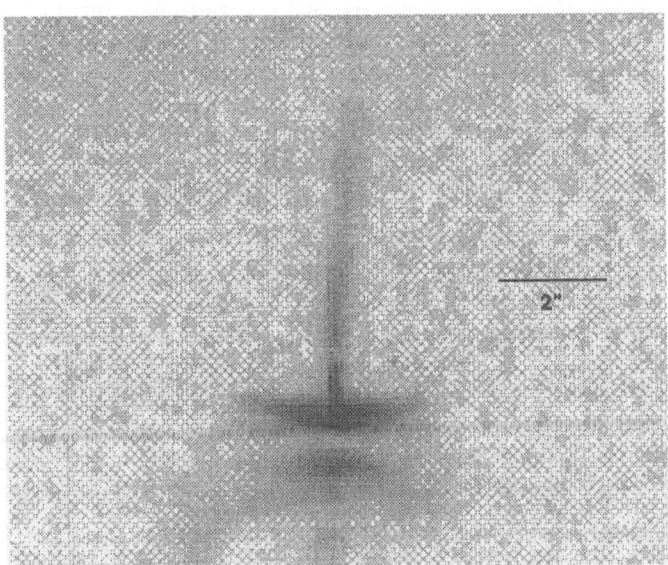

Fig. 1. *HST* optical image of the YSO HH 30, shown as a negative image (dark structures are intrinsically bright). The double-concave reflection nebula is produced by light from a hidden central star scattering off the upper surfaces of a dusty disk seen nearly edge-on (see text). The dust absorption in the disk plane (central lane between the upper and lower reflection nebulae) completely obscures the central star. The roughly vertical linear structure is a bipolar high-velocity jet, which is observed in shock-excited emission lines. At the distance of Taurus (140 pc), 2 arcsec corresponds to 280 AU. From Burrows et al. (1996).

Although there is still some support for the idea of wide-angle winds not seen in "jet" emission (see Shu & Shang (1997) and §5 below), the case for a cold, dispersed wind has weakened substantially in the last few years. Although it is impossible to eliminate the possibility of some wide-angle unseen wind, in my opinion the preponderance of the evidence supports highly-collimated jets as the true outflows, especially if one is not burdened by theoretical preconceptions. The evidence may be summarized as follows:

Jets can power outflows. Assuming that the jets or winds, moving at hundreds of $km\,s^{-1}$, dissipate most of their kinetic energy when they collide with the stationary external medium, the jets or winds still must account for the momentum in the molecular outflows. The earliest estimates indicated that jet momentum fluxes were roughly an order of magnitude too small to account for the molecular outflows. These estimates assumed that the gas in the jet was completely ionized, and thus equated the estimated electron densities from line fluxes and ratios to the hydrogen densities. However, most jets show emission from relatively low-excitation species, such as [O I] and [S II], which can easily result from regions of partial ionization. More detailed calculations (Hartigan et al. 1994) found that the typical emitting regions in several jets had ionization fractions more like 10^{-1}, which implied that the total mass flux (and momentum flux) is larger than that estimated assuming complete ionization by a similar factor. While the initial studies of this problem used detailed shock-wave models, more recent analyses which do not use shock models give basically the same result (Bacciotti 1997). Thus, it now seems that the optically-observed jets *can* power molecular outflows.

Outflows are extended. One of the peculiarities of molecular outflows which was realized early on is their apparent short lifetimes. A typical molecular outflow might have an extent of 0.1 pc and be moving at $\sim 10\,\mathrm{km\,s^{-1}}$, implying a dynamical lifetime of 10^4 yr. However, increasingly sensitive surveys began to indicate that nearly all (or maybe all) heavily-extincted YSOs (i.e., all YSOs surrounded by dense natal material) exhibited molecular outflows. Since the lifetime of the embedded YSO is thought to be closer to 10^5 yr, this finding is obviously inconsistent with the dynamical time (see Padman et al. 1997 and references therein for further discussion).

Recently, large-scale CCD imaging has provided evidence for spectacular optical jets and Herbig-Haro objects collimated over distances of parsecs (see Bally & Devine 1997 and references therein). The clear implication is that observed molecular outflows often do not trace the full extent of the moving material; presumably, if the jet/wind breaks free of the molecular cloud, there is little swept-up material to be detected. Since outflows can be most easily detected near the YSO, the net result is an observational bias toward estimating less collimation than is really present.

Friends and family. It is becoming increasingly clear that stars generally do not form in isolation but in groups and clusters. Jets from more than one YSO may therefore contribute to pushing around molecular gas in the local region. Unless the jets are all precisely aligned, which is demonstrably untrue in some cases, the net result will be a molecular outflow which is less collimated than the individual jets powering it.

Highly-collimated CO flows from very young YSOs. Improvements in interferometry have led to the detection of molecular outflows that are extremely highly collimated. A good example of this is shown on the cover of the IAU Symposium 182 volume (Reipurth & Bertout 1997), showing Plateau de Bure Interferometer observations of the HH 211 jet by Gueth & Guilloteau. Other examples are discussed in the review by Bachiller (1996). It is difficult to account for this kind of structure with a wide angle wind. Excitation/shock conditions seem irrelevant to this situation, since CO can emit strongly at low temperatures. These observations strongly suggest that, when YSOs are very young and heavily enshrouded in circumstellar gas, their winds are clearly seen to be jets; as time proceeds, the cavity produced by the jet expands, making the outflow appear less collimated on small scales than it really is.

3   Accretion-powered Outflows

It has been known for some time that the energy and momentum of molecular outflows correlate with the luminosities of their sources (Lada 1985). What has become clearer now is that there is a good correlation between the mass-loss rates of low-mass YSO winds and their *disk-accretion* rates (Hartigan et al. 1995). This analysis requires determining the mass-loss rates near the YSO, not from piled-up molecular material, but from the rapidly-expanding jet/wind gas; in addition, the disk-accretion rates must be derived, separating accretion-powered emission from the stellar luminosity.

The most recent analysis is that of Calvet (1998), which is reproduced in Figure 2. It should be recognized that there is considerable uncertainty in estimating both mass loss and mass-accretion rates for T Tauri stars, and that the scatter among the low-accretion rate objects is consistent with expected errors. Accretion and ejection rates for the three rapidly-accreting FU Ori objects are probably a bit more certain, within probably a factor of three. Within the errors, it is clear that there is a strong correlation between mass accretion and mass ejection.

The FU Ori objects are YSOs in which the disk-accretion rate suddenly increases by two or more orders of magnitude (Hartmann & Kenyon 1996). It is thought that the central stars of FU Ori objects are low-luminosity T Tauri stars. The energy fluxes in the winds of FU Ori objects far exceed the luminosities of the pre-outburst stars; thus, it is quite clear that accretion energy is powering the outflow, consistent with the strong correlation between mass loss and accretion rates. These results, coupled with the observation that the non-accreting young T Tauri stars, the so-called "weak" or WTTS, do not show any evidence for detectable winds, makes it clear that disk accretion is necessary to produce powerful outflows - at least in low-mass YSOs, where radiation pressure from the central star is insufficient to drive mass ejection (see §5).

4   Disk Winds

As discussed in the previous section, disk accretion powers the observed outflows in low-mass YSOs. In one special case, that of the rapidly-accreting object FU Ori, it is possible to demonstrate observationally that the outflow is arising from the surface of the disk. This can be done because at the rapid accretion rates in FU Ori objects, the disk is orders of magnitude brighter than the central star; the wind motion can be detected directly from the doppler shifts of lines seen in absorption against the bright disk continuum emission.

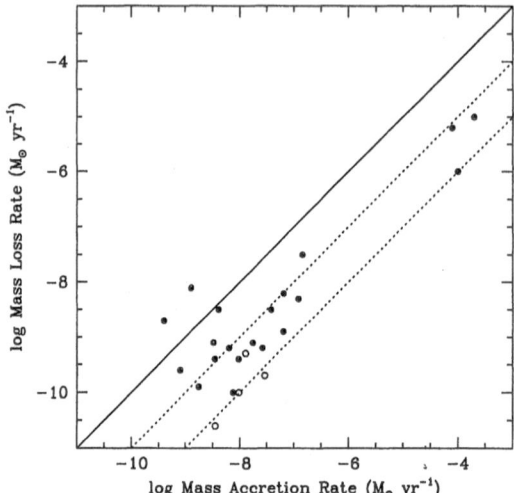

Fig. 2. Mass accretion rates vs. wind mass loss rates for T Tauri stars and FU Ori objects. Taken from Calvet (1998).

Absorption lines formed in a disk wind should show increasing blueshifts with increasing line strength. Weak absorption lines, formed near the disk photosphere, should display the "doubled" line profiles produced by disk rotation, as shown by the uppermost theoretical profile (dotted line) shown in the upper panel of Figure 3. Stronger lines will be formed further up in the disk atmosphere, farther from the disk midplane. The atmosphere is cooler in these outer layers, and therefore the absorption profiles of these lines will be deeper, as in a normal stellar atmosphere. Since the gas density is lower at higher levels in the atmosphere, conservation of mass for a (roughly) steady wind implies a larger expansion velocity, resulting in a larger blueshift of the absorption.

The sequence of profiles shown in the upper panel of Figure 3 was calculated by Calvet et al. (1993) for typical Fe I lines in the atmosphere of FU Ori, with each line having a different absorption strength. The sequence of calculations quantitatively demonstrates the profile evolution described above. An interesting feature of the calculation is that the absorption component on the red side of the line becomes increasingly blueshifted, as would be expected intuitively, but the blue absorption component does not appear to shift as much in velocity. This is because the blue component is not an actual feature in the wind-velocity profile, but instead is due to the convolution of the rotational broadening profile with the wind-expansion profile (Calvet et al. 1993).

A schematic way of characterizing this evolution of line profiles with increasing line strength is to consider the positions of the two absorption components

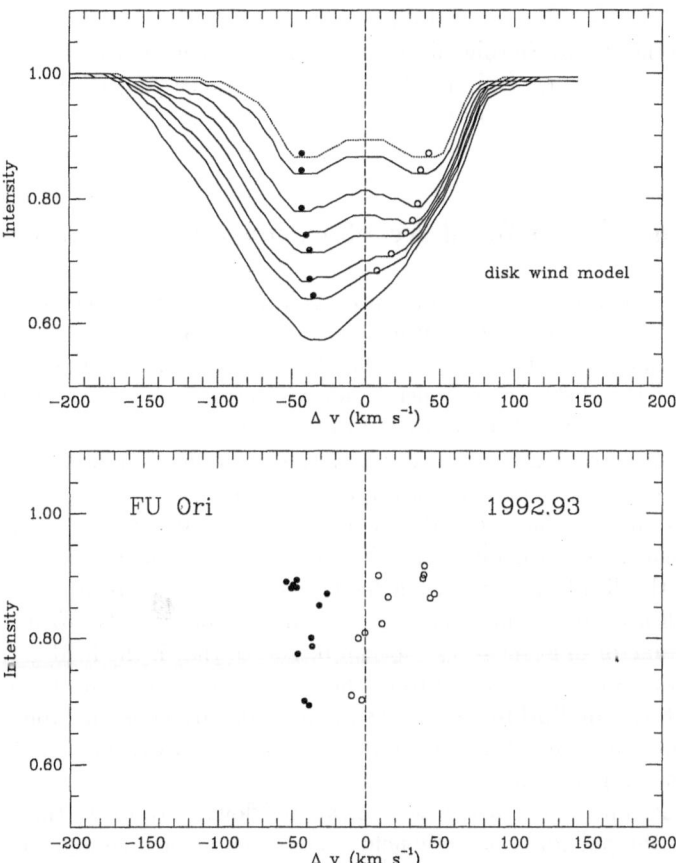

Fig. 3. Upper panel: Predicted disk wind profiles for a series of lines with different strengths, taken from Calvet et al. (1993). Filled and open circles mark the positions of the absorption dips for the blue and redshifted components, respectively. Lower panel: velocity shifts of absorption components in FU Ori, from Hartmann & Calvet (1995). See text.

in the line profiles, marked in the upper panel of Figure 3 by the open circles for the redshifted component and filled circles for the blueshifted component. As the line strength increases, the line becomes deeper and so the two absorption components appear at lower residual intensities; the line becomes more blueshifted, and the two absorption features move together. Eventually, for lines strong enough to be formed at a sufficiently high atmospheric level where the expansion velocity dominates the rotation, the two absorption components merge into one blueshifted feature.

This qualitative behavior of line profiles has been observed in FU Ori, ini-

ent measurements of the absorption dip positions by Hartmann & Calvet (1995), confirming the Petrov-Herbig effect. The correspondence between the model predictions and the observed dip positions clearly demonstrates the evolution of the flow from pure Keplerian rotation to outflow as material is ejected from the disk surface.

5   Magnetocentrifugal Acceleration and Collimation

In both FU Ori objects and T Tauri stars radiation pressure is insufficient to drive the mass loss; moreover, the outflows are demonstrably cold, so that thermal pressure cannot be important. This has led to the notion that magnetic fields anchored in the disk are responsible for accelerating YSO winds (Pudritz & Norman 1983, 1986; cf. Blandford & Payne 1982).

The basic idea is straightforward. Material in the Keplerian disk, though rotating at break-up velocity, is still gravitationally bound. If the magnetic field lines threading the disk are sufficiently strong, and if they tilt outward, away from the central star, then gas following outward along the field line will acquire energy. In the limit of very strong fields, where strict co-rotation is enforced, the gas will acquire escape energy by the time it moves outward in cylindrical radius by a factor of at most $2^{1/2}$. It can be shown that if the field is rigid at the disk surface, and is tilted away from the rotation axis by more than 30°, then the hydrostatic equilibrium of the gas at the disk surface is unstable; any small perturbation will result in the gas flowing freely outward. This is the essence of magnetocentrifugal acceleration models.

This mechanism also produces a *collimated* flow. Exterior to the Alfvén surface, where the poloidal magnetic field is no longer strong enough to completely dominate the flow, the angular momentum of the material causes the build-up of toroidal field (Figure 4). This toroidal field exerts a hoop stress which tends to channel the ejected gas toward the rotational axis, in principle explaining highly-collimated outflows (§2).

All current models for outflows from low-mass YSOs rely upon this qualitative picture (Pudritz & Norman 1983, 1986; Königl 1989; Pelletier & Pudritz 1993; Shu et al. 1994; Shu & Shang 1997; although Lovelace et al. 1993 present a slightly different view, in which magnetic pressure due to diverging field structure is more important). Where the models mainly differ is in the *assumed* magnetic field structure of the disk - i.e., the boundary conditions. For example, Pudritz & Norman (1983, 1986) and Königl (1989) assumed that the disk magnetic field is sufficiently strong and its geometry sufficiently favorable that the wind emanates from the entire disk surface; in addition, it is assumed that the angular-momentum loss represented by the wind is responsible for allowing the disk material to accrete. This view appears to be too extreme. There is essentially no observational evidence for a slow, dense outflow from the outer disk regions (Hartmann 1995). Moreover, it is clear that winds cannot take away all of the angular momentum for accretion, because then they take away all of the energy of accretion as well; this means that there would be no accretion energy

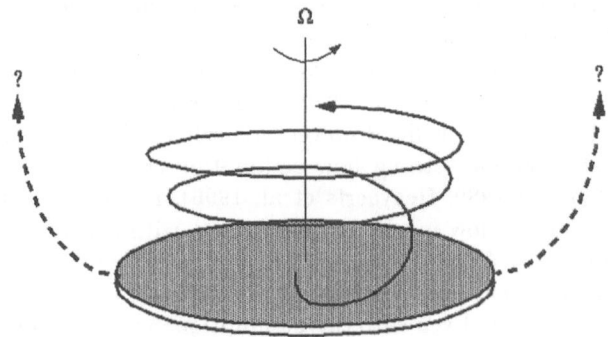

Fig. 4. Schematic diagram of magnetocentrifugal acceleration and collimation of a disk wind. Material is initially flung outward, away from rotational axis as well as out above the disk. As it does so, the gas acquires sufficient kinetic energy to escape from the potential well. Eventually, as the flow proceeds outwards, the magnetic field tends to be wound up as a result of the angular momentum of the wind; this toroidal field produces a hoop stress which tends to push the outflow toward the rotation axis. The amount of this collimation depends in detail upon the magnetic-field structure and also upon the assumed boundary conditions for the outflow (e.g., external magnetic fields (dashed lines)). See text.

energy of accretion as well; this means that there would be no accretion energy left for viscous dissipation, i.e. that the disks could not be self-luminous, and this is clearly not the case in the FU Ori objects.

At the other extreme of models, Shu and collaborators (Shu et al. 1994; Shu & Shang 1995) assume that the wind emanates only from the innermost annulus of the accretion disk. This model assumes the presence of a stellar magneto-sphere, which truncates the disk. Part of the magnetosphere "bulges out" over the disk, so that any vertical field lines in the disk are bent over away from the rotation axis, as required for wind acceleration. Moreover, the model also assumes that there are no magnetic fields elsewhere in the disk which could accelerate outflows. The problem with this extreme variant is that the FU Ori objects have powerful disk winds, yet show no evidence for an inner magneto-sphere (Hartmann 1995). Indeed, it would not be surprising if the accretion rate in FU Ori objects is so high that the disk pressure is sufficient to crush the stellar magnetosphere (Shu et al. 1994). In addition, many line profiles of FU Ori objects exhibit very strong absorption at low velocities, and this is difficult to accomplish without assuming that the wind emanates essentially everywhere from the (finite) optically-emitting disk surface, not just from a narrow disk annulus (Calvet et al. 1993). Theoretical models also suggest that the disks of FU Ori objects may not be very geometrically thin in the innermost regions,

make it difficult to produce any narrow *radial* structure, since any structure will tend to be smoothed radially over the same scale as the vertical thickness.

The assumed disk magnetic-field structures have important implications for the collimation of the large-scale flow. The full-disk models, which were similarity solutions (e.g., Blandford & Payne 1982), beg the question of outer boundary conditions on the magnetic fields. However, non-self-similar, extended disk winds probably can still produce strong flow collimation (Pelletier & Pudritz 1993; Heyvaerts & Norman 1989; Heyvaerts et al. 1996). In contrast, the models of Shu and collaborators show much less collimation, with substantial amounts of material moving away from the rotation axis. This effect is produced by the assumption that the field lines accelerating the flow are all pinched together at the inner edge of the disk. Even with this assumption, however, which is probably one of the least favorable geometries for collimation, the outflow still becomes remarkably organized along the rotation axis. Shu & Shang (1997) argue that "the jetlike appearance of YSO outflows is somewhat of an optical illusion", because in their model the isodensity contours are more cylindrical than the flow streamlines. Nevertheless, even in this extreme limit of a disk wind, 70% of the mass flux is concentrated within a half-angle of $\sim 24°$ from the rotational axis after proceeding only $\sim 10^2$ AU from the central source, and the flow will become even more collimated at larger distances.

It is clear that, whatever the disk field geometry, YSO outflows produced by magnetocentrifugal acceleration will be collimated to a substantial degree. In my view, the likeliest situation is one in which the flow is launched over a finite area of the disk, and most of the mass flux is contained within a flow that is well-collimated within ~ 10 AU from the central star. In this picture, the solution of Shu & Shang (1997) could be appropriate for the "outermost" disk field lines, but these streamlines may not contain the bulk of the wind.

6   Variability

There is evidence for variability in YSO winds on almost every time scale. It has been recognized for some time that there must be large-scale variability, probably in mass ejection rates as well as in flow velocity, as indicated by widely-separated, well-developed bow shocks of jets interacting with the external medium; and that this large-scale variability is probably related to variations in disk accretion, possibly in many cases FU Ori outbursts (Reipurth 1990). FU Ori outbursts of rapid disk accretion at $10^{-4}\,M_\odot\,\mathrm{yr}^{-1}$ may last for hundreds of years and recur at thousands of year time scales (Hartmann & Kenyon 1996), with corresponding mass-ejection rates of $\sim 10\%$ of the accretion rates. However, these outbursts probably simply represent the extreme (and rare) variability; much of the "emission knots" seen in optical jets may well represent less-dramatic wind variability on timescales of years to decades (Reipurth & Heathcote 1997). Much remains to be done in understanding variability, particularly as YSO timescales are often quite long compared with normal observing periods. It may be that the record of mass and velocity fluctuations can best be inferred, at least over long timescales, from reconstructions based on the spatial shock structure along jets (Bally & Devine 1997).

References

Bachiller, R. (1996): Ann. Rev. Astr. Ap., **34**, 111

Bacciotti, F. (1997): in *Herbig-Haro Flows and the Birth of Low-Mass Stars*, IAU Symp. 182, eds. B. Reipurth & C. Bertout (Dordrecht: Kluwer), p. 73

Bally, J., Devine, D. (1997): in *Herbig-Haro Flows and the Birth of Low-Mass Stars*, IAU Symp. 182, eds. B. Reipurth & C. Bertout (Dordrecht: Kluwer), p. 29

Burrows, C.J., Stapelfeldt, K.R., Watson, A.M. et al. (1996): ApJ, **473**, 437

Calvet, N. (1998): in *Proc. of the 8th Annual Astr. Conf. on Accretion Processes in Astrophysical Systems*, American Institute of Physics, ed. S. S. Holt (in press)

Calvet, N., Hartmann, L., Kenyon, S.J. (1993): ApJ, **402**, 623

Hartigan, P., Edwards, S., Ghandour, L. (1995): ApJ, **452**, 736

Hartigan, P., Morse, J.A., Raymond, J. (1994): ApJ, **346**, 125

Hartmann, L. (1995): in Circumstellar Disks, Outflows, and Star Formation, Rev. Mex. Astr. Ap. (Serie de Conferencias), 1, 285.

Hartmann, L., Calvet, N. (1995): AJ, **109**, 1846

Hartmann, L., Kenyon, S.J. (1996): ARAA, **34**, 207

Heyvaerts, J., Norman, C.A. (1989): ApJ, **347**, 1055

Heyvaerts, J., Priest, E.R., Bardou, A. (1996): ApJ, **473**, 403

Königl, A. (1989): ApJ, **342**, 208

Lada, C.J. (1985): Ann. Rev. Astr. Ap., **23**, 267

Lovelace, R.V.E., Romanova, M.M., Contoupoulos, J. (1993): ApJ, **403**, 158

Padman, R., Bence, S.J., Richer, J.S. (1997): in *Herbig-Haro Flows and the Birth of Low-Mass Stars*, IAU Symp. 182, eds. B. Reipurth & C. Bertout (Dordrecht: Kluwer), p. 123

Pelletier, G., Pudritz, R.E. (1992): ApJ, **394**, 117

Pudritz, R.E., Norman, C.A. (1983): ApJ, **274**, 677

Pudritz, R.E., Norman, C.A. (1986): ApJ, **301**, 571

Reipurth, B. (1990): in *Flare Stars in Star Clusters*, IAU Symp. 137, eds. L.V. Mirzoyan, B.R. Petterson, M.K. Tsvetkov (Dordrecht: Kluwer), p. 229

Reipurth, B., Heathcote, S. (1997): in *Herbig-Haro Flows and the Birth of Low-Mass Stars*, IAU Symp. 182, eds. B. Reipurth & C. Bertout (Dordrecht: Kluwer), p. 3

Shu, F., Najita, J., Ostriker, E., et al. (1994): ApJ, **429**, 781

Shu, F.H., Shang, H. (1997): in *Herbig-Haro Flows and the Birth of Low-Mass Stars*, IAU Symp. 182, eds. B. Reipurth & C. Bertout (Dordrecht: Kluwer), p. 225

Kundt: Whether or not the jets from YSOs are focussed stellar winds is controversial (cf. Lecture Notes in Physics 471 (1996) or Dec. 1996 Torino workshop).

Hartmann: It is difficult to maintain that YSO winds are uncollimated, based on (1) the observation of very long collimated jets seen in optical shock-excited emission; and (2) the evidence for highly collimated CO outflows in very young stars. The latter does not depend on shock emission to be detected and one would expect to see the swept-up interstellar CO from the unfocussed wind – which is not detected. And they are not *stellar* winds, for reasons outlined in §3.

Chochol: You mentioned that the variability of FUORs winds is of the order of days. We found hourly variations in Z CMa's wind (see our poster).

Foing: Is there any evidence for a temporal correlation between sporadic accretion events occuring on T Tauri stars and the variability of wind features such as the blue-shifted absorption components? What do you expect to be observed?

Hartmann: There is no evidence for this currently on short timescales for lack of data. One might expect to see enhanced forbidden line emission at or shortly after accretion events indicated by an increase in the hot continuum emission produced as the infalling gas shocks at the star.

Owocki: My comment follows on that by Cassinelli, namely that the issue of asymptotic "collimation" of field and jet outflow depends crucially on the assumed boundary conditions, with collimation most easily achieved by alignment with an asymptotic interstellar field. But the beautiful HST picture you showed of HH30 clearly shows that the jet is not along an arbitrary direction defined by an ISM field, but instead is clearly perpendicular to the equatorial disk emission. Doesn't this mean that you need to find a way to collimate a field generated intrinsically (vs. externally) by the system?

Hartmann: I think that the HH30 picture shows clearly that the jet is ejected along the rotation axis. There has been some suggestion that magnetic braking of the protostellar cloud might lead to alignment of the angular momentum with the magnetic field. However, it is not at all clear that the disk field is interstellar; it might be due to a disk dynamo, in which case the rotation axis is the only preferred direction in the system.

Savonije: What is known about the relative masses of the disks in YSOs compared to those of central stars?

Hartmann: In typical T Tau stars disk masses are thought to be $\sim 0.01\,M_\odot$, or a few percent of the central star's mass. Disks around younger T Tau stars or FU Ori objects might be more like 10% of the central mass.

On the Origin of Relativistic Winds from Accreting Stars

Nazar R. Ikhsanov and Guido T. Birk

Institut für Astronomie und Astrophysik der Universität München,
Scheinerstr. 1, 81679 München, Germany

We propose that the energy of the non-thermal component in the radiation of pre-main sequence stars is supplied by free magnetic field energy stored in the sheared component of the magnetospheric field of the accreting star. This energy is released by the reconnection of magnetic field lines, which is governed by the dissipative tearing mode of instability. During this process, charged particles are accelerated in the magnetopause by magnetic-field aligned electric fields. In the case of the T Tauri stars, electrons are accelerated to energies over 10 MeV.

The interaction between the accreting star's poloidal field and infalling material leads to the generation of the sheared component of the magnetic field in the magnetopause, B_{sh}. The value of B_{sh} can be estimated from (Ikhsanov & Pustil'nik 1996) $B_{sh} \lesssim V_{rel}(R_m) \sqrt{4\pi\, \rho(R_m)}$, where $V_{rel}(R_m)$ and $\rho(R_m)$ are the relative velocity of the magnetosphere surface with respect to the surrounding plasma and the mass density of the plasma at the boundary, respectively.

According to Otto (1991), such a magnetic field configuration is unstable to the dissipative tearing mode. By assuming that the collisional electrical resistivity is dominated by ion-dust collisions, we have found that a magnetic field-aligned electric potential structure with

$$U \simeq 2 \cdot 10^4 \text{ statvolt cm}^{-1} \left[\frac{B_{sh}}{10\,\text{G}}\right] \left[\frac{\delta_m}{10^9\,\text{cm}}\right]^{-1} \left[\frac{\lambda_{acc}}{2 \cdot 10^{10}\,\text{cm}}\right]$$

evolves (see also Schindler et al. 1991). The length of the acceleration region, λ_{acc}, is taken to be on the order of the wavelength of the most unstable tearing mode (Birk 1998). In such potential structures electrons can be accelerated up to 10 MeV.

This model allows the non-thermal GHz-radio emission detected from several pre-main sequence stars (White et al. 1992; André 1996) to be interpreted.

References

André, P. (1996): PASPC 93, 273
Birk, G.T. (1998): A&A 330, 1070
Ikhsanov, N.R., Pustil'nik, L.A. (1996): A&A 312, 338
Otto, A. (1991): Phys. Fluids B 3, 1739
Schindler, K., Hesse, M., Birn, J. (1991): ApJ 380, 293
White, S.M., Pallavicini, R., Kundu, M.R. (1992): A&A 259, 149

Part 2

OBSERVATIONS
OF CYCLICAL
WIND VARIABILITY

The Solar Wind in Three Dimensions

J.T. Gosling

Los Alamos National Laboratory, Los Alamos, New Mexico, USA

Abstract. Ulysses, which is in a polar orbit about the Sun, is providing the first direct observations of the solar wind at high heliographic latitudes. Among the new results from Ulysses summarized here are the following: 1. On the declining phase of the solar activity cycle and near solar minimum solar wind variability is largely confined to a relatively narrow latitude band centered on the heliographic equator. 2. Solar wind speed far from the Sun is anti-correlated with ionization temperature in the low corona. 3. A polytrope of the form $T = (2.0 \times 10^5) N^{0.57}$, where T and N are the proton density and temperature, describes the free expansion of the high-speed wind at high latitudes. 4. Equipartition of plasma and magnetic field energy does not occur in the high-speed wind, the plasma beta there typically being ~3.0. 5. Corotating interaction regions have opposed north-south tilts in the northern and southern hemispheres. 6. A new class of forward-reverse shock pairs associated with over-expanding coronal mass ejections has been identified at high heliographic latitudes.

1 Introduction

The solar wind expansion is strongly modulated by the coronal magnetic field. The complex interplay between expansion and the magnetic field is what produces a highly structured solar corona and a variable solar wind, even in the absence of transient solar activity. The left panel of Figure 1 illustrates the connection between coronal structure and the quiescent solar wind expansion. Where regions of opposite magnetic polarity abut in the low corona, the field is usually sufficiently strong to constrain the plasma from expanding outward. This produces closed field arcades of relatively dense coronal plasma bound to the Sun. The field weakens with increasing height above the solar surface so that at higher altitudes the arcades are opened up by the pressure of the coronal plasma, and the plasma is free to expand outward. The resulting expansion produces characteristic helmet-like streamers in the corona and a relatively dense, low-speed solar wind flow. Within the interiors of large-scale unipolar regions on the Sun the expansion is relatively unconstrained by the coronal magnetic field. The resulting expansion produces regions of low plasma density in the solar atmosphere known as coronal holes and high-speed solar wind flows. The heliospheric current sheet (HCS), which separates flows of opposite magnetic polarity in interplanetary space, is embedded within the low-speed wind.

Well above the photosphere the Sun's magnetic field is often approximately that of a dipole. The tilt of the dipole relative to the rotation axis varies as the Sun's magnetic field evolves through the solar activity cycle. As illustrated in the left panel of Figure 1, the dipole tends to be aligned nearly with the solar rotation axis near solar activity minimum, but is inclined substantially to

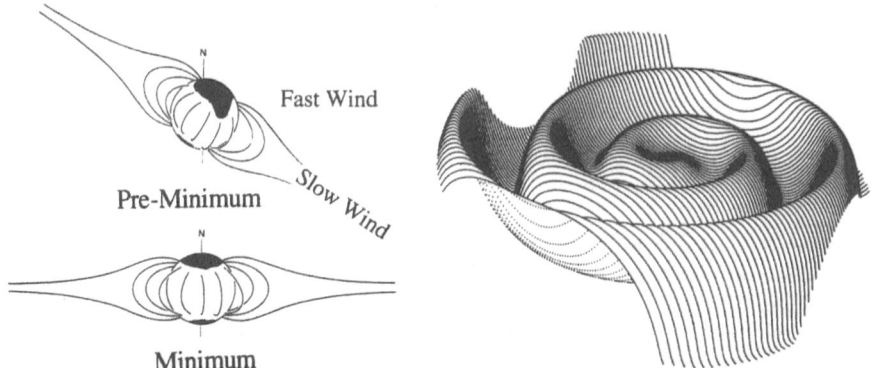

Fig. 1. (left) Schematic illustrating the changing tilt of the solar magnetic dipole and coronal structure relative to the rotation axis of the Sun, as well as the origin of high and low-speed solar wind flows. (right) The configuration of the heliospheric current sheet in interplanetary space when the tilt of the solar magnetic dipole is substantial. Adapted originally from Hundhausen (1977) and Jokipii & Thomas (1981).

the rotation axis in the years leading up to solar minimum. Thus, near solar minimum the HCS tends to coincide roughly with the solar equatorial plane. On the other hand, during the approach to solar minimum solar rotation and the solar wind flow cause the HCS to become warped into an overall structure that resembles a ballerina's twirling skirt, as illustrated in the right panel of Figure 1. In this simple picture the maximum solar latitude attained by the current sheet is equal to the tilt of the magnetic dipole axis relative to the rotation axis of the Sun. In practice, the solar field is never actually a simple dipole, particularly near solar activity maximum, and the shape of the HCS in interplanetary space is usually more complex than illustrated.

2   Latitudinal Variation of the Solar Wind Flow

Both low and high-speed flows are commonly observed at low heliographic latitudes as the Sun rotates (with a period of about 26 days as observed from a fixed point in space). Near solar minimum solar wind variability is almost entirely confined to a relatively narrow latitude band centered on the heliographic equator (Gosling et al. 1995c), while at high latitudes a nearly constant wind with speed of about 750 km s^{-1} and density (scaled to Earth's orbit) of about 2.5 cm^{-3} prevails (Phillips et al. 1994). Figure 2 shows solar wind speed as a function of solar latitude as measured by the Ulysses space probe on the declining phase and near the minimum of the most recent solar activity cycle. (Ulysses is in a unique polar orbit about the Sun and reaches heliographic latitudes of ± 80 degrees during its ~6-year solar orbital period.) The latitude effect obvious in Figure 2 is a consequence of two factors: (1) solar wind properties change rapidly with distance from the heliospheric current sheet, with flow speed being lowest and

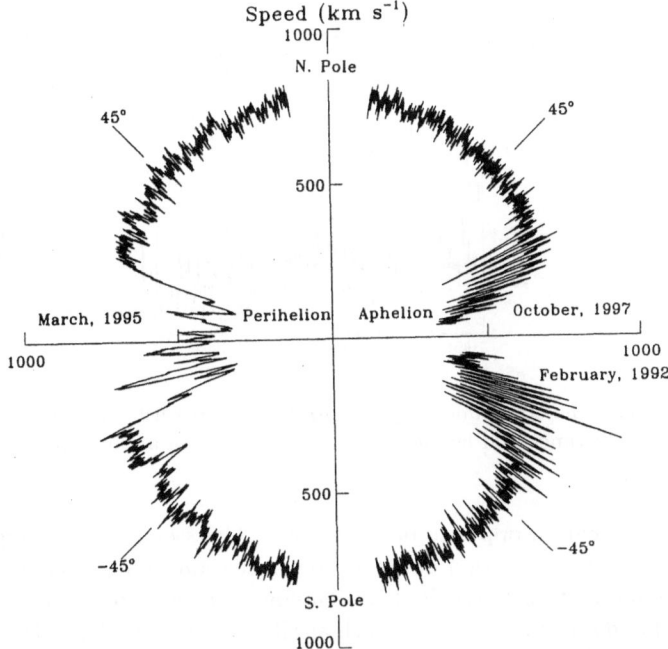

Fig. 2. Solar wind speed as a function of heliographic latitude measured by Ulysses. Data shown in the left portion of the figure are centered on orbit perihelion at 1.4 AU, while data on the right are centered on orbit aphelion at 5.4 AU. The apparent difference in the latitude scale of structure in the flow speed at perihelion and aphelion is an artifact associated with the fact that the spacecraft changed latitude very rapidly near perihelion but very slowly near aphelion.

number density highest in the vicinity of the current sheet; and (2) the tilt of the dipole relative to the solar rotation axis on the declining phase of the solar cycle and near solar minimum is generally less than about 30 degrees. The width of the band of solar wind variability changes as the tilt of the dipole changes. In the years leading up to and near solar minimum the band of variability appears to range from about ± 20 to ± 35 degrees wide. Near solar activity maximum the band of variability probably extends to significantly higher heliographic latitudes, perhaps almost to the poles of the heliosphere, and high-speed flows are less prevalent at low latitudes at that time.

3   The Relationship Between Flow Speed and Coronal Temperature

In Parker's original theory of the solar wind, flow speed far from the Sun was directly related to temperature in the corona (e.g., Parker 1963). That is, the highest coronal temperatures produced the fastest flows. Ulysses observations,

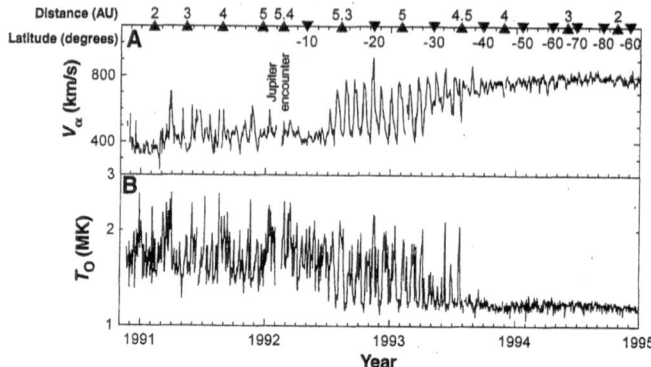

Fig. 3. Solar wind speed and oxygen ionization temperature versus time. The changing heliocentric distance and latitude of Ulysses are shown at the top. From Geiss (1995).

however, reveal an opposite behavior (e.g., Geiss et al. 1995). Figure 3 demonstrates an anti-correlation between flow speed and oxygen ionization temperature; similar relationships exist between flow speed and ionization temperature for other abundant solar wind ions such as silicon and iron. Transitions between different ionization temperatures are relatively steep and indicate relatively sharp boundaries between low and high-temperature plasma in the corona where the wind originates. The Ulysses measurements have also demonstrated that, when compared to photospheric values, the low-speed wind is over-abundant in elements having low first ionization potentials, or short ionization times. Given the magnetic control of the coronal expansion outlined above, these effects must be related to the large-scale magnetic structure of the solar atmosphere.

4   Energetics of the Expansion of the High-Speed Wind

Ulysses' long immersion in the nearly uniform high-speed wind at high latitudes has provided a unique opportunity, free of the variability present at low latitudes, to test how the solar wind kinetic temperature evolves with distance from the Sun. The left panel of Figure 4 shows that both proton kinetic temperature and ion enthalpy are well behaved functions of density, suggesting a polytropic relation between proton density, N, and temperature, T, in the high-speed wind of the form $T = aN^{(\gamma-1)}$ where a is the entropy. The best fit to the data gives $\gamma = 1.51$. For all of the high-speed (>600 km s^{-1}) wind data obtained at high and intermediate northern and southern latitudes the best fit is obtained for $\gamma = 1.57$. The right panel of Figure 4 shows that the resulting histogram of values of entropy for this value of γ is a simple, narrow-peaked distribution. A choice of the adiabatic value ($\gamma = 1.67$) provides a noticeably broader entropy histogram. It is thus apparent that a single polytrope law of the form $T = (2.0 \times 10^5)N^{0.57}$ adequately describes the free expansion of the high-speed wind (Feldman et al. 1997).

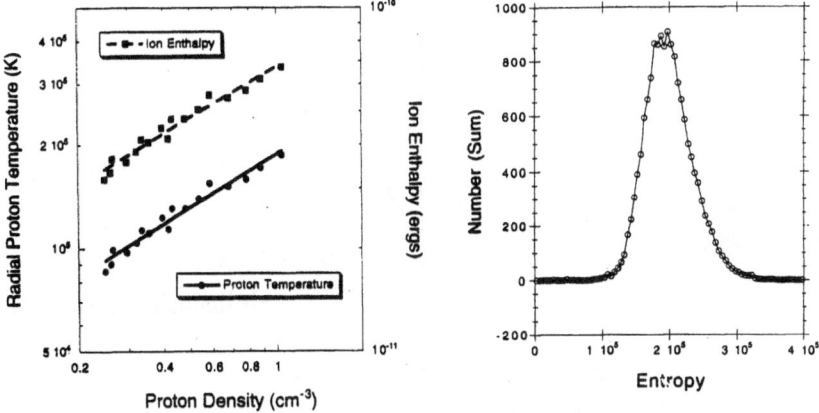

Fig. 4. (left) Proton temperature and ion enthalpy in the northern high-speed wind averaged in 0.1 AU bins over the heliocentric distance from 1.5 to 3.2 AU plotted versus proton density. (right) Histogram of proton entropy in the high-speed wind assuming ($\gamma = 1.57$. Adapted from Feldman et al. (1997).

Near the ecliptic plane at 1 AU the most probable value of the total plasma beta (ratio of plasma to magnetic field pressure) is of the order of 1.0. However, beta varies from less than 0.1 to greater than 10.0 near 1 AU. Moreover, as Figure 5 demonstrates, the mean and median values of beta in the high-speed wind at high latitudes are 3.3 and 2.7, respectively, reasonably independent of latitude and heliocentric distance (McComas et al. 1995). It is thus clear that, in general, plasma and magnetic field pressures are not equal in the solar wind, nor is there any fundamental reason why they should be approximately equal in other stellar winds.

5   Corotating Interaction Regions

Within the low-latitude band of variability flows of different speed become radially aligned as the Sun rotates. Faster wind overtakes slower wind ahead while simultaneously running away from slower wind behind. Since these radially aligned parcels of plasma originate from different longitudes on the Sun at different times, they are prevented from interpenetrating by the inclined magnetic field that is frozen into the flow. As a result, a compression commonly forms on the leading edge of a high-speed "stream" at low latitudes and a rarefaction forms on the trailing edge. The compression, commonly referred to as an interaction region, is a region of high pressure. Its leading edge, which propagates into the slower wind ahead, is called a forward wave and its trailing edge, which propagates backward into the high-speed flow, is called a reverse wave. The propagation of these waves produces an acceleration of the slow wind ahead of the stream and a deceleration of the fast wind within the stream with increasing heliocentric distance. The net effect of the interaction is to transfer momentum

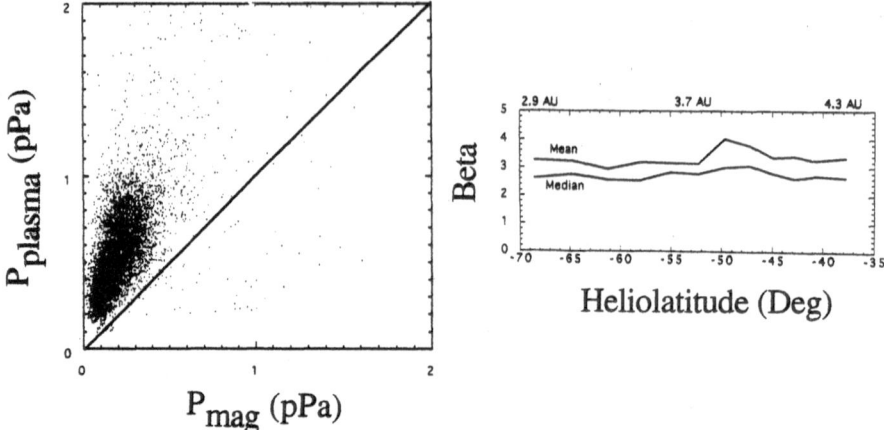

Fig. 5. (left) Scatter plot of plasma versus magnetic field pressures in the high-latitude wind. (right) Mean and median beta averaged over complete solar rotations in the high-latitude wind plotted versus heliolatitude. Heliocentric distance is indicated at the top of the panel. Adapted from McComas et al. (1995).

from the fast wind to the slow wind. When the stream amplitude is sufficiently large (about twice the local fast mode speed), the forward and reverse waves bounding the interaction region eventually (typically by 2-3 AU) steepen into shocks.

When the coronal expansion is quasi-stationary but spatially variable, the above process proceeds in an identical fashion at all longitudes; however, because of solar rotation the state of evolution is a function of longitude. As a result, interaction regions become aligned with Archimedean spirals that have pitches relative to the radial direction that are intermediate between those of the fast and slow flows that produce them. The entire pattern of interaction rotates with the Sun, and the region of compression is known as a corotating interaction region, or CIR. Because CIRs are inclined relative to the radial direction, the forward and reverse waves have both radial and azimuthal components of propagation. In particular, the forward waves propagate antisunward and in the direction of planetary motion about the Sun (westward), while the reverse waves propagate sunward (in the plasma rest frame) and eastward. Thus the slow wind is accelerated outward and deflected westward and the fast wind is decelerated and deflected eastward as a result of the interaction.

Ulysses has provided the first measurements of CIRs well out of the ecliptic plane; this in turn has led to new insights about the 3-dimensional structure of CIRs. Figure 6 provides a synthesis of CIR shock observations during Ulysses' first polar orbit. Forward shocks are plotted above the horizontal line in each panel and reverse shocks are plotted below the line. Note that the shocks often occur as forward-reverse shock pairs at intermediate latitudes as would be expected from our previous discussion. However, during the southern transit (left panel) Ulysses encountered only two forward shocks poleward of S26°, while it

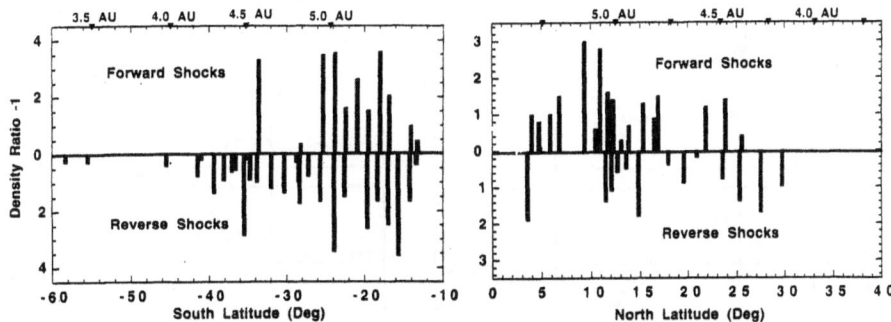

Fig. 6. Bar diagrams of shock "strength" versus latitude for corotating shocks observed (left) during Ulysses' initial transit to high southern latitudes and (right) during Ulysses' recent approach to the solar equatorial region from high northern latitudes. Heliocentric distance is indicated at the top of the panels. Adapted from Gosling (1996) and McComas et al. (1997).

continued to encounter reverse shocks regularly up to a latitude of S42°, and sporadically thereafter. In general, these high-latitude shocks were weaker than their low-latitude counterparts. A somewhat similar, but less dramatic, pattern was observed during the northern transit near solar minimum (right panel) when the solar dipole was less steeply inclined. As Ulysses moved equatorward (right-to-left) it observed reverse shocks before it began observing shock pairs. Below about N10° all but one of the CIR shocks observed were forward shocks.

Flow deflections observed downstream from the corotating shocks are crucial for understanding the physical origin of the above effects (Gosling et al. 1993). Because the forward and reverse shocks are both convected away from the Sun by the high bulk flow of the solar wind, Ulysses samples the region downstream of the forward shocks after shock passage and the region downstream of the reverse shocks prior to shock passage. As illustrated in Figure 7, Ulysses commonly observed positive (westward) azimuthal (ϕ) flow deflections downstream from the forward shocks and negative (eastward) azimuthal flow deflections downstream from the reverse shocks in both hemispheres, as expected. More interesting, the meridional (θ) deflections downstream from the shocks were of the opposite sense in the northern and southern hemispheres. In the northern hemisphere the deflection pattern was south/north (+/-) while in the southern hemisphere it was north/south (+/-). Thus in both hemispheres the forward shocks were propagating equatorward while the reverse shocks were propagating poleward. This explains why the reverse shocks were observed at higher latitudes than the forward shocks in both hemispheres.

The Ulysses observations thus indicate that CIRs have substantial north-south tilts that are opposed in the northern and southern hemispheres. Figure 8 illustrates how these opposed tilts arise (e.g., Pizzo 1991). Slow wind emanates from the region around the solar magnetic equator, which commonly is tilted relative to the heliographic equator on the declining phase of the solar cycle, as

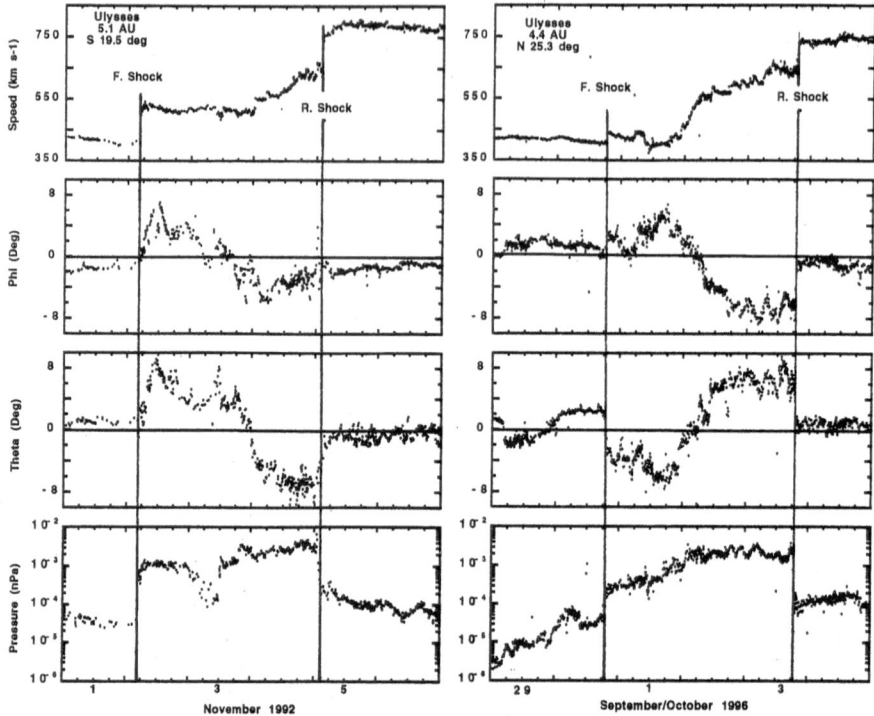

Fig. 7. Typical corotating interaction regions observed in the southern (left) and northern (right) hemispheres by Ulysses. Parameters plotted from top to bottom are solar wind speed, azimuthal and meridional flow angles (positive for flow in the direction of planetary motion about the Sun and northward, respectively), and proton thermal pressure. Adapted from Gosling et. al. (1995b) and Gosling et al. (1997).

noted earlier. As the Sun rotates, the fast wind overtakes the slow wind in interplanetary space along interfaces that are inclined relative to the solar equator in the same sense as is the band of slow wind. (At the other interfaces, where fast wind runs away from the slower wind, rarefactions are produced.) The interfaces, and the CIRs in which they are embedded (shaded in Figure 8), have opposite north-south tilts in the northern and southern hemispheres. The forward and reverse waves bounding the CIRs always propagate roughly perpendicular to the interfaces; thus, the forward waves in both hemispheres propagate antisunward, westward, and equatorward while the reverse waves propagate sunward, eastward and poleward. The geometry illustrated in Figure 8 has been used to specify the inner boundary conditions for a simulation using a 3-D MHD code developed to model corotating flows in the solar wind (Pizzo and Gosling 1994). Space limitations prevent showing the results of this simulation here; however, the simulation generates CIRs having opposed north-south tilts in the northern and southern hemispheres and reproduces the basic observational effects shown in Figures 6 and 7.

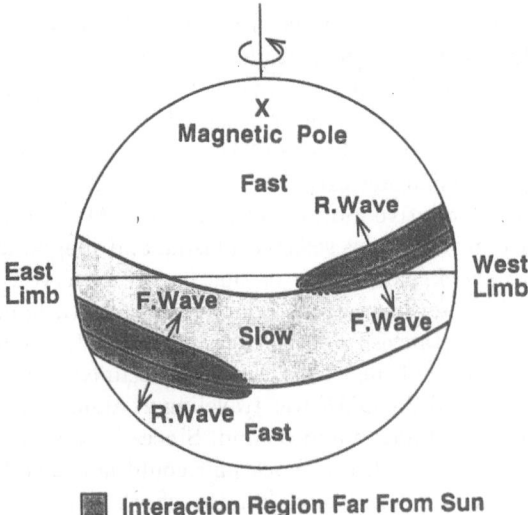

Fig. 8. Sketch illustrating the origin of tilted corotating interaction regions in interplanetary space. The interaction regions form well away from the Sun. From Gosling (1996).

6   Coronal Mass Ejections and Solar Wind Disturbances

The most dramatic changes in the solar corona occur during events known as coronal mass ejections, CMEs, (Crooker et al. 1997) in which 10^{+15} to 10^{+16} g of solar material are ejected into the heliosphere. CMEs originate in closed magnetic field regions in the corona where the field normally is sufficiently strong to constrain the plasma from expanding outward. Usually these closed field regions are found in the coronal streamer belt surrounding the solar magnetic equator. Outward speeds in CMEs as observed within about 5 solar radii of the Sun range from less than 50 km s^{-1} to greater than 2000 km s^{-1}. Regardless of their outward speeds, all CMEs depart from the Sun and become a part of the solar wind, where they usually can be identified by one or more anomalous signatures such as counterstreaming suprathermal electrons (which indicate a closed field topology) or low plasma beta (e.g., Gosling 1990).

At low heliographic latitudes the faster CMEs commonly drive shock disturbances in the solar wind. (About 1/3 of all CMEs observed at low latitudes have sufficiently high speeds to produce such shock disturbances.) These shock disturbances are a consequence of relative motion between the CMEs and slower wind ahead. A region of high pressure (a compressive interaction region) develops on the leading edge of a CME as it overtakes slower wind ahead. This compression is bounded on its leading edge by a forward shock that propagates into the slow wind. Only rarely are reverse shocks observed in CME-driven disturbances at low latitudes at any distance from the Sun.

Since Ulysses' first polar orbit about the Sun occurred on the declining phase and near the minimum of the solar activity cycle, CMEs were observed relatively infrequently at high heliographic latitudes. Nevertheless, a sufficient number of events were observed to identify a class of CME-driven disturbances at high latitudes that has not previously been observed at low latitudes (Gosling et al. 1994). In this newly identified class of CME events shock pairs are produced by over-expansion (i.e., expansion driven by a high initial internal pressure) of the CMEs rather than by relative motion between the CMEs and the ambient wind. The left panel of Figure 9 shows selected plasma and magnetic field parameters for one of several shock pair events of this nature observed by Ulysses. Note that the forward and reverse shocks bounding the disturbance were offset by approximately equal distances from the edges of the CME, the pressure reached a minimum near the center of the CME, the speed declined from the forward shock to the reverse shock, and the CME was traveling at about the same speed as the unshocked ambient wind ahead and behind. Since the CME was moving slower than the ambient wind ahead, the shock pair could not have been produced by relative motion. Instead, it is likely that the shock pair was produced as the CME over-expanded into the surrounding ambient wind.

The right panel of Figure 9 presents the result of a simple 1-dimensional gas-dynamic simulation that illustrates how these high-latitude shock pairs arise. The simulation was constructed to mimic the ejection from the Sun of a dense CME whose internal pressure initially is higher than that of the surrounding solar

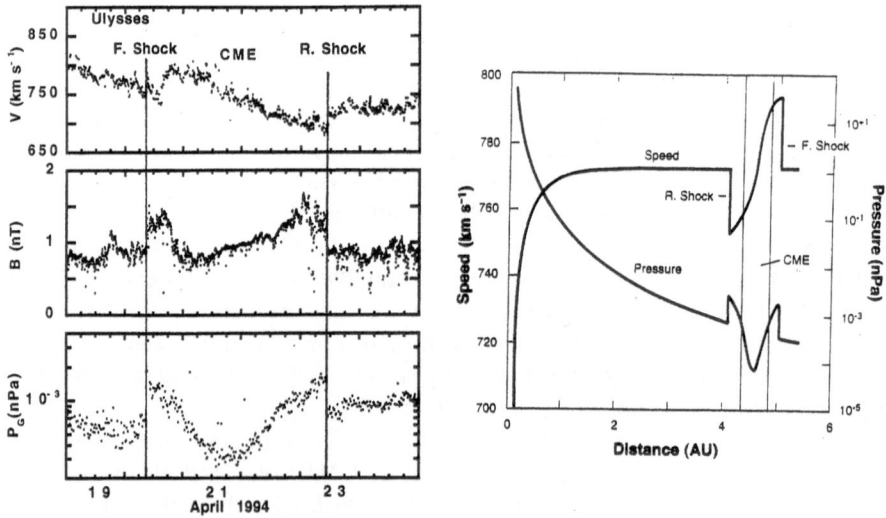

Fig. 9. (left) Selected plasma and magnetic field parameters for a CME-driven disturbance observed by Ulysses at 3.2 AU and S61. (right) Solar wind speed and pressure as functions of heliocentric distance for a simulated disturbance that has just arrived at 5 AU. The simulation utilizes a 1-D gas-dynamic code and was initiated at 0.14 AU by increasing the density by a factor of four in a bell-shaped pulse 10-hours wide. From Gosling et al. (1994).

wind but whose speed is the same. Owing to its initial high internal pressure, the CME expands as it travels out from the Sun. This expansion drives forward and reverse waves that propagate into the ambient plasma ahead and behind while simultaneously producing a region of low pressure at the center of the CME. At large heliocentric distances the forward and reverse waves steepen into shocks. The temporal signature produced at a fixed point in the outer heliosphere by this simulation bears a remarkable resemblance to the overall appearance of the disturbance shown in the left panel of Figure 9.

CMEs are often sufficiently broad that they extend into both the low and the high-latitude wind. Ulysses observed one such event in February 1994 when it was situated at 3.53 AU and S53° and nearly at the same longitude as Earth (Gosling et al. 1995a). Whereas the disturbance observed at Ulysses was nearly identical to that shown in Figure 9 and thus was driven by over-expansion of the CME, the disturbance observed near Earth by the IMP 8 satellite included a strong forward shock associated with a high relative motion between the CME and a slower ambient wind ahead. Thus the same CME can produce graphically different disturbances at high and low latitudes. Such differences have been simulated remarkably well using both 1 and 2-dimensional gas-dynamic codes (Riley et al. 1997) and highlight the importance of initial conditions within both the CME and the ambient wind in determining the eventual evolution of the resulting disturbances in the solar wind.

Acknowledgement

This work was supported by a Los Alamos Research and Development grant.

References

Crooker, N., Joselyn, J.A., and Feynman, J. (1997): *Coronal Mass Ejections, Geophysical Monograph 99*, (American Geophysical Union, Washington DC)

Feldman, W.C., Barraclough, B.L., Gosling, et al. (1997): J. Geophys. Res., submitted

Geiss, J., et al. (1995): Science **268**, 1033

Gosling, J.T. (1990): *Physics of Magnetic Flux Ropes, Geophysical Monograph 58* (American Geophysical Union, Washington DC), 343

Gosling, J.T. (1996): Annu. Rev. Astron. Astrophys. **34**, 35

Gosling, J.T., Bame, S.J., McComas, D.J., et al. (1993): Geophys. Res. Lett. **20**, 2789

Gosling, J.T., McComas, D.J., Phillips, J.L., et al. (1994): Geophys. Res. Lett. **21**, 2271

Gosling, J.T., McComas, D.J., Phillips, J.L., et al. (1995a): Geophys. Res. Lett. **22**, 1753

Gosling, J.T., Bame, S.J., McComas, D.J., et al. (1995b): Space Sci. Rev. **72**, 99

Gosling, J.T., et al. (1995c): Geophys. Res. Lett. **22**, 3329

Gosling, J.T., Bame, S.J., Feldman, W.C., et al. (1997): Geophys. Res. Lett. **24**, 309

Hundhausen, A.J. (1977): in *Coronal Holes and High Speed Wind Streams* (Colorado Associated University Press, Boulder), 225

Jokipii, R.J., Thomas, B. (1981): Astrophys. J. **243**, 1115

McComas, D.J., Barraclough, B.L., Gosling, J.T., et al. (1995): J. Geophys. Res. **100**, 19,893

68   J.T. Gosling

McComas, D.J., et al. (1997): Geophys. Res. Lett., in press
Parker, E.N. (1963): *Interplanetary Dynamical Processes* (J. Wiley and Sons, New York)
Phillips, J.L., et al. (1994): Geophys. Res. Lett., **21**, 1751
Pizzo, V.J. (1991): J. Geophys. Res. **96**, 5405
Pizzo, V.J., Gosling, J.T. (1994): Geophys. Res. Lett. **21**, 2063
Riley, P., Gosling, J.T., Pizzo, V.J. (1997): J. Geophys. Res. **102**, 14,677

Owocki: That was a wonderful discussion of how the solar wind variations can be understood by simple models. I'm struck particularly that in discussing the variations in the *wind* (as opposed to coronal origin), you never even mentioned magnetic fields. As someone working on stellar winds that is often being pushed to consider the dynamical role of magnetic fields, I find that such in situ measurements can be generally understood without fields.

Gosling: Most aspects of solar wind dynamics can be understood using simple gas dynamic models. The magnetic field plays an essential role in making the solar wind plasma behave like a fluid since it is the field that prevents different plasma elements from interpenetrating one another. However, magnetic forces are generally comparable to or less than gas pressure forces and those forces can be mimicked without including them specifically in the models. In general, gas dynamic models give results that are qualitatively correct although some of the quantitative details change when one includes the field explicitly.

Kakouris: The Ulysses solar wind data have shown a mean (almost constant) wind velocity at high latitudes, but also variability on short timescales. Does this variability suggest turbulence in the high-latitude solar wind?

Gosling: Yes, but the variability that is observed is not due solely to turbulence. Some of the variability is Alfvénic in character (correlated variations in vector velocity and vector magnetic field) while some of the variability is associated with microstreams that appear to be related to small changes in the expansion speed of the corona.

Foing: Based on SOHO and Ulysses measurements, what is the contribution of coronal mass ejections to the solar mass loss as a function of latitude, and its variation over the solar cycle?

Gosling: The second part of your question is easier to answer than the first. In the ecliptic plane CMEs account for about 15% of the solar wind near solar activity maximum and a little less than 1% of the solar wind near solar activity minimum. Averaged over the solar cycle they account for about 7% of the solar wind observed in the ecliptic plane. Near solar minimum CMEs are largely confined to latitudes below about 30 degrees and are quite rare at higher latitudes. For example, not a single CME was detected by Ulysses during its 1.5 year traverse of high northern latitudes from April 1995 until October 1996. On the other hand, coronagraph observations suggest that CMEs are relatively common at high latitudes near solar activity maximum.

Observations of Wind Variability in Cool Stars

A.K. Dupree[1]

Harvard-Smithsonian Center for Astrophysics,
60 Garden Street, Cambridge, MA 02138, USA

Abstract. Spectroscopic signatures of winds are briefly reviewed for the Sun and more luminous cool giants and supergiants using specific examples of a hybrid star (α Aqr) and *Betelgeuse*. The relation of winds and mass flow to (cyclic) magnetic activity and atmospheric pulsation is explored using recent ultraviolet, optical, and infrared spectroscopic observations.

1   Introduction

Winds of cool stars (spectral type F or later; $T_{eff} \lesssim 7700K$) exhibit a diversity of characteristics: terminal velocities spanning 5 to 800 km s^{-1}; velocity profiles showing acceleration and deceleration; temperatures spanning three orders of magnitude; and mass-loss rates spread over nine orders of magnitude (10^{-14} to 10^{-5} M$_\odot$ yr^{-1}). While evidence of mass-loss episodes from cool stars in the historic past have been inferred from circumstellar material (planetary nebulae, circumstellar shells, etc.), this review focuses on new current observations of cool stellar winds. Three increasingly difficult levels of understanding of cool star winds must be achieved to address the topic of this Workshop, Cyclical Variability of Stellar Winds, namely:

Existence How can winds of cool stars be detected? Which stars have winds? What are the characteristics of these winds? And what are the processes driving the winds?

Variability Are the winds variable in velocity, acceleration, temperature, mass-loss rate? How are these parameters related?

Periodicity Is the wind variability periodic, or only "periodic" at times, or not at all?

Winds of cool stars are more subtle in their appearance and harder to detect than the winds of hot stars. Recall that the solar wind was not discovered by the sort of spectroscopic observations available for stars. The existence of a "particle flux" from the Sun was inferred from the motions of the tails of comets circling the Sun and from the magnetic storms affecting the Earth's atmosphere. Space craft making *in situ* measures in interplanetary space verified that indeed there was a wind from the Sun; particle sampling preceded any spectroscopic detection of a solar wind.

And even when stellar spectra are available, their interpretation is not always straightforward. To conclude that actual mass loss, or a wind, is present (as distinct from a mass *flow*), velocities comparable or greater than the escape velocity at the appropriate atmospheric level must be detected. At the photosphere, the escape velocity is 600 km s^{-1} for a dwarf star of $1R_\odot(M_\odot)$, 300 km s^{-1} from a giant of $10R_\odot(4M_\odot)$, and 90 km s^{-1} from a supergiant of $1000R_\odot(20M_\odot)$. These values decrease at $2R_*$, not untypical for a chromospheric dimension in a low-gravity atmosphere, to 425 (V), 214 (III), and 64 (I) km s^{-1} for a dwarf, giant, and supergiant respectively. In the solar chromosphere and transition region, outside of explosive flaring events, mass motions and flows occur frequently, and it is not obvious that these flows develop into full fledged mass *loss* except in areas where the magnetic field is radial and coronal holes are found. We may anticipate similar mass motions to be present in cool stars where the surface is not yet resolved. Velocities detected in cool-star spectra are typically less than the photospheric escape velocity, and only in a few cases (particularly the hybrid stars, see below) do the velocities become commensurate with the escape values at several stellar radii (cf. Dupree & Reimers 1989).

The spectral diagnostics themselves must be evaluated carefully. Most of the strongest measured transitions are resonance lines and subject to optical depth effects that can compromise the interpretation of a wavelength shift. Emission in a resonance transition is susceptible to self-absorption on the short wavelength side of a line profile creating apparent redshifts in an outwardly accelerating atmosphere. Forbidden transitions and transitions from metastable levels such as He I ($\lambda 10830$) do not suffer such problems.

In these observations we search for clues to the driving mechanisms of the winds. A variety of physical conditions and augmented, or entirely new, mechanisms as compared to the solar example must occur in cool stars. If we scale the solar mass-loss rate $(2 \times 10^{-14} \ M_\odot \ \mathrm{yr}^{-1})$ to the surface area of a giant star, a value of $\dot{M}$ of 2×10^{-12} is predicted whereas values of $10^{-8} - 10^{-9} \ M_\odot \ \mathrm{yr}^{-1}$ are observed; similarly a prediction of $2 \times 10^{-8} \ M_\odot \ \mathrm{yr}^{-1}$ for a supergiant contrasts with observed values of $10^{-6} \ M_\odot \ \mathrm{yr}^{-1}$. There must be several ways to drive winds, and apparently more vigorous means of energy and momentum deposition present in luminous stars than in the Sun, causing the discrepancies with simple solar scaling considerations.

In addition to detecting a wind, and establishing if periodicities are present, a final goal is determining a reliable mass-loss rate. This is not easy for cool stars, since detailed atmospheric models must be constructed to interpret the observed profiles that carry the signature of mass flow. And, as the quality of observations improves, it appears that time-variable phenomena are present and ever more sophisticated models are necessary. Even for an optically thin line, where a Doppler shift might be measured directly, the density in the atmosphere and the height of formation are needed to complete the arithmetic: $n \times v \times r^2$ and derive the mass-loss rate. Careful construction and evaluation of atmospheric models is a major effort which really has not yet been addressed in a systematic sophisticated way across the cool half of the color-magnitude diagram.

In the following sections, three different cool stars are discussed as examples in the detection of cool star winds and mass loss: the Sun, a hybrid supergiant star (α Aqr), and the cool supergiant Betelgeuse (α Ori). Each in its way illustrates the challenges and recent advances in studying cool stars for evidence of winds and periodicity in both winds and mass-loss rate.

2   The Sun

The Sun presents a challenging spectroscopic problem in order to detect the signature of winds. Here of course, we *know* that the Sun has a wind, albeit a weak one with an average total mass-loss rate of 2×10^{-14} $M_{\odot}$ yr^{-1}. Extreme ultraviolet observations from a rocket (Rottman et al. 1982), with effective resolution of ≈ 1 arcmin across a coronal hole on the disk, detected a wavelength shift in a coronal- and transition-region line corresponding to an outflow velocity of 12 km s^{-1} for Mg X (625Å) and 7 km s^{-1} for O V (629Å). Yet *SKYLAB* spectra of a coronal hole on the disk (Doschek, Feldman, & Bohlin 1976) showed no velocity shift of O V (1218Å). Reanalysis suggests that both measurements are consistent with a static atmosphere (Brekke 1993; Dere 1994). The NRL High Resolution Telescope and Spectrograph (HRTS) observed C IV in a coronal hole on the solar disk (Dere et al. 1989) with a spatial resolution of ≈ 2 arcsec, and concluded downflows were present, although a greater proportion of outflows occurred in the hole (26%) as compared to the quiet Sun (7%). To date, direct observations of the solar *photosphere* show no signs of a solar wind. A consistent outward flow most likely to leading to mass loss from coronal holes is frequently not detectable until the chromosphere, transition region, and corona are reached (Warren et al. 1997; Dupree et al. 1996). Measurements above a coronal hole suggest the flow speed attains 200 km s^{-1} at $2R_{\odot}$ (Strachan et al. 1993) and the acceleration is almost complete by $10R_{\odot}$ (Grall et al. 1996). A hint of the heating processes occurring in holes may be found from the observed density fluctuations in polar coronal holes near $2R_{\odot}$ (Ofman et al. 1997) that could result from non-linear, high-amplitude compressional waves generated by Alfvén waves. A good summary of the current SOHO analysis and results of wind observations is contained in the useful Proceedings of the Fifth SOHO Workshop (ESA 1997).

However, the Sun has a wind, and the solar magnetic cycle is well known, thus providing a periodic change in atmospheric conditions; these facts fulfill at least two of the goals for this Workshop. The coronal effects of a magnetic activity cycle are dramatically demonstrated in the series of YOKHOH images in X-rays compiled over a solar cycle.[1] The measured proton flux from the Sun varies substantially depending on the solar latitude as recent results from ULYSSES have shown (Goldstein et al. 1996). High-speed flows are associated with high latitudes (± 30 degrees) confirming the coronal holes as the source of high-speed wind.

In order to address the question of variable mass loss, it is necessary to invoke semi-empirical models, since the total mass loss from the corona can not be

---

[1] Available from http://pore1.space.lockheed.com/SXT/homepage.html

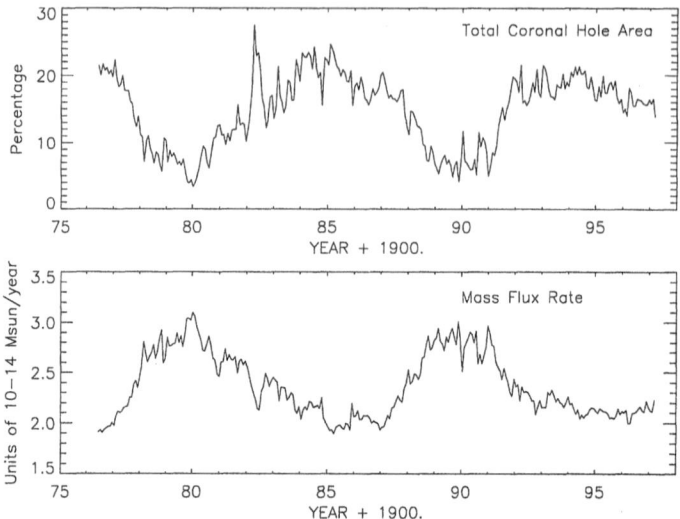

Fig. 1. Coronal-hole area and total mass flux from the Sun for the past two decades (adapted from Wang 1998).

measured. Starting from magnetograph observations, Wang (1998) has developed a semi-empirical model of the large-scale structure of the solar magnetic field and its cyclic variation. Of particular interest is the evolution of coronal holes which represent the low-density source regions of the solar wind where open magnetic field lines allow the coronal plasma to escape. Comparison (Wang et al. 1996) of the footpoints of the calculated field topology with the He I (λ10830) distribution (whose strength is related to the extreme ultraviolet coronal flux, and thus sensitive to its absence in coronal holes) indicates good agreement for the past two sunspot cycles. Additionally, because a simple empirical relationship exists between the solar wind speed and the coronal field geometry, the global patterns of the solar wind speed can be predicted. Over the last two sunspot cycles, the pattern is impressive (see Fig. 13 in Wang 1998): low speed wind ($v < 450$ km s^{-1}) dominates the low latitudes (latitude $\leq \pm 20°$), but the high latitudes cyclically vary between very high ($v > 750$ km s^{-1}) and low speed wind. The fast wind of the coronal holes dominates the poles at sunspot minimum and extends down to $\sim 20°$ latitude, but essentially vanishes at sunspot maximum. Since *in situ* measures show that the mass flux density and the wind speed are related, this fact is used to evaluate (Wang 1998) the mass flux from the whole surface of the Sun.

Figure 1 shows the coronal hole area and the total mass flux from the Sun for 2 solar cycles. The area of the coronal holes varies cyclically from 5 to 20 percent over the solar cycle with the maximum area occurring near solar minimum. The mass loss, however, varies inversely with the coronal hole area, with the maximum mass loss occurring at solar maximum. Thus even though the coronal holes carry fast (and hence easily detectable) wind, the densities are lower and the mass flux from the holes is less than in the slow wind features. The amplitude of the mass-flux variation is a factor of 1.5.

The implications of these results for stars are several. If the Sun, with its modest level of activity shows a cyclical modulation of mass flux, other dwarf stars may also have a varying mass flow. However, it will be hard (perhaps impossible) to find a wind signature in the photospheric spectrum of a single dwarf star. Chromospheric and coronal transitions offer the best chance for detection. The change of the wind speed with latitude over the solar surface, and the varying coverage of the high-speed wind during the solar cycle suggests that for other dwarf stars, it may help to "get lucky" and capture a spectrum at the right time when the hole features (containing the highest velocities) are prominent. This suggests that working in concert with Ca H&K programs that monitor stellar activity could prove fruitful in selection of targets and observing times. It also suggests that the "right" stars with low inclination (and the rotation axis pointed towards us!) may offer the optimum situation to detect a wind. However, the fast wind may give a lower limit to the mass-loss rate. Rapidly rotating dwarfs (certainly those in binary systems) appear to have a totally different magnetic structure with substantial spot activity at their poles (cf. Hatzes et al. 1996; Brickhouse & Dupree 1998).

3   A Hybrid Supergiant: α Aqr

The bright supergiant α Aqr (HD 209750, spectral type G2 Ib) belongs to a physically important group of stars denoted as "hybrid" objects. These represent a class of cool, luminous (Luminosity Class I, II, III) stars that were originally identified based on the appearance of their ultraviolet IUE spectra (Hartmann et al. 1980). C IV emission is present (signifying temperatures of at least 10^5 K), and asymmetric emission cores of Mg II are found, accompanied by absorption features at low and high velocities, indicating a massive stellar wind and circumstellar and/or interstellar material. The high-velocity absorption is variable as found in the earliest days of IUE (Dupree & Baliunas 1979) when changes in the line emission and absorption were first noted. These objects occur in a broad region of the HR diagram extending from luminous G supergiants ($M_V \approx -4$) to mid-K bright giants and giant stars ($M_V \approx 1$). Many members of this class have since been discovered (Reimers 1982; Hartmann et al. 1985; Judge et al. 1987). Because of their location in the color magnitude diagram, hybrids represent the physically important connection between solar-like stars (with coronas and fast winds of low mass-loss rate), and the cool supergiant stars (Alpha Ori-like) with cool outer atmospheres. Thus hybrid stars assume a pivotal role in the definition of atmospheric heating processes and mechanisms to drive winds of cool stars.

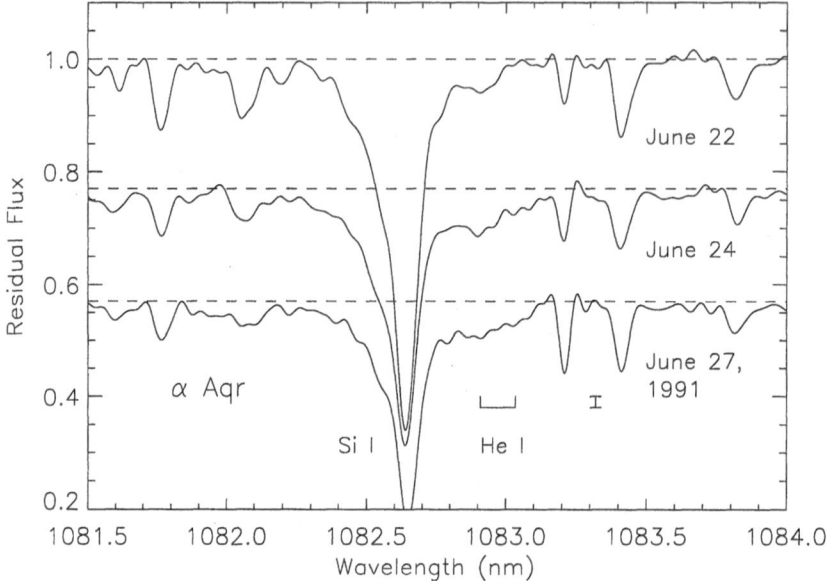

Fig. 2. FTS spectra obtained at the CFHT telescope showing the night to night variation of the chromospheric absorption line of He I at λ10830 in the hybrid star α Aqr. Note also the extension of the He absorption to short wavelengths beyond the photospheric Si I feature indicating an outflow of ∼ −150 km s⁻¹. The video shown at the Workshop of many consecutive spectra displays the fluttering characteristic of this profile (Dupree & Sasselov 1998).

The winds from hybrid stars have high terminal velocities (50 to 150 km s⁻¹) as measured by circumstellar Mg II absorption found in IUE spectra (see Fig. 5 in Dupree & Reimers 1989). And surprisingly, the wind from Alpha Aquarii, a hybrid supergiant, shows supersonic velocities deep in the chromosphere (Dupree et al. 1992). Spectra of the high excitation He I λ10830 transition (O'Brien & Lambert 1986) exhibit a P Cygni-type profile with absorption that extends to shorter wavelengths corresponding to −140 to −200 km s⁻¹ (see Fig. 2). These values are commensurate with the escape velocity from the surface of the star. Semi-empirical modeling of the He I line demonstrates that the observed high velocities *must* occur in the chromosphere. A rapid acceleration to ≈ 150 km s⁻¹ within 1.5 R⋆ appears necessary to produce the observed He I profile (Dupree et al. 1992). These values are supersonic at chromospheric temperatures, and the need to assume such a velocity profile may be an indication that shocks and transient events are present. The He I profiles themselves undergo significant night to night variation (see Fig. 2). In addition, these outflows *exceed* the velocities indicated by circumstellar absorption features suggesting that the wind undergoes both acceleration and deceleration.

An extended source of heating may be required to accelerate the wind at distances beyond the sonic point and achieve the velocities indicated by absorption troughs observed in the Mg II profiles of several hybrid stars (Hartmann et al. 1985; Judge et al. 1987). Some hint at the cause of this dynamic structure comes from recent photometric and spectroscopic monitoring.

Observations (Rao et al. 1993) of two hybrid supergiants, α and β Aqr, indicate both a short (about 80 days) and long (about 470 days) time scale of chromospheric Ca II emission modulation suggesting the presence of *both* magnetic activity (470 days from rotational modulation) and heating and non-magnetic phenomena in the atmosphere (80 days). And Butler (1998) finds a velocity variation in the photosphere of α Aqr, with amplitude ± 200 km s^{-1} and period also about 80 days. Pulsation thus appears to be the major culprit, driving substantial acceleration, extending the atmosphere, and leading to mass loss ($\dot{M} \approx 10^{-8}$ $M_\odot$ yr^{-1}).

High temperature material in hybrid objects as evidenced by X-ray detections (Reimers et al. 1996) is confined by loops (thus eliminating a thermally driven wind) and the absence of dust eliminates a dust-driven wind. The amplitude variation caused by pulsation in the chromosphere can damp out in the extended atmosphere thus explaining the lack of periodic modulation found by Brown et al. (1996) in the Mg II emission spectra. Now we believe the driving mechanisms for these atmospheres must include pulsation, and magnetically-driven waves may contribute too since there is evidence for magnetic activity from Ca II and X-ray observations.

4   A Cool Supergiant: Betelgeuse

Intensive study of the supergiant *Betelgeuse* (α Ori; M2Iab; HD 39801) began when the star reached a historic maximum in optical brightness in January of 1984. The first three years of synoptic ultraviolet observations revealed a distinct 420-day modulation of its ultraviolet continuum and chromospheric Mg II emission line flux (see Fig. 3); pulsation was suspected, and confirmed by the radial velocity measures of Smith et al. (1989). Continued measures with IUE through 1996 showed the weakening of the 420-day modulation. Effects of the pulsation traveling through the extended atmosphere are documented by the delay between continuum variations, and subsequently the Mg II h line and the Mg II k line (Dupree et al. 1987, 1998). The distinct behavior of the two Mg II resonance doublet lines is perhaps not surprising since the two form well-apart in the extended supergiant atmosphere as is demonstrated from the 0.2-year phase lag that is observed in the brightening of the (stronger) k line with respect to the h line intensity. Moreover, the line profiles of the h and k lines have opposite blue/red asymmetries, further evidence of the difference in dynamics between the two formation regions. IUE laid the foundation for studies of *Betelgeuse* with HST that is leading to the first direct look at the physics of stellar pulsation and mass loss.

In March 1995, the first direct images of the surface of a star other than

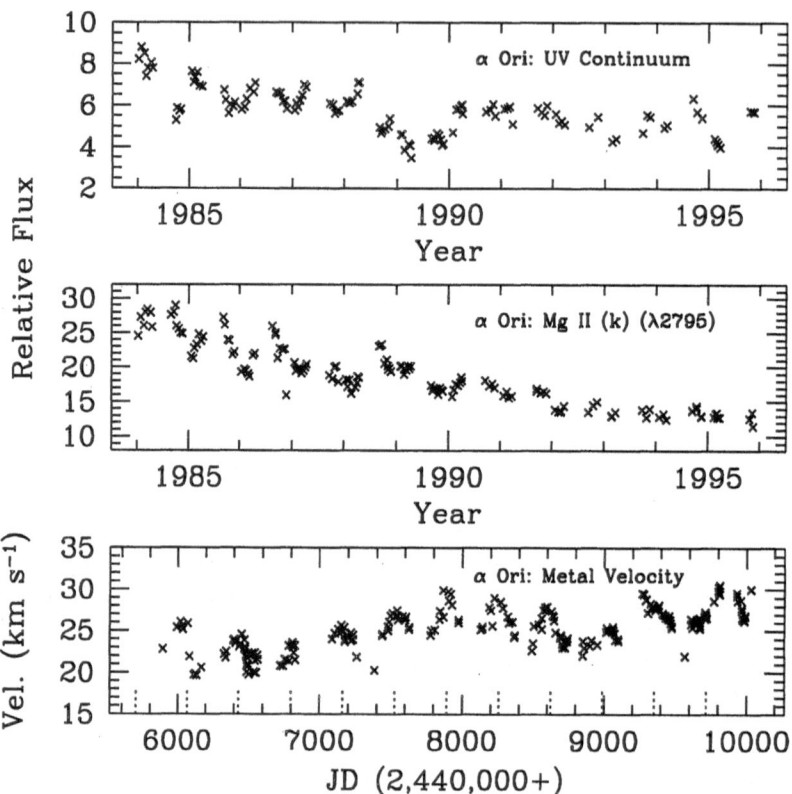

Fig. 3. The ultraviolet continuum flux and the Mg II k line flux obtained with IUE, and the photospheric metal velocities for α Ori measured at the US Kitt Peak National Observatory. Note that the pronounced periodic UV modulation (1984–1990) becomes less apparent from 1991–1996 although the photospheric velocity continues to exhibit short time scale variability (adapted from Dupree et al. 1987, Smith et al. 1989, Dupree et al. 1998).

the Sun were obtained (see Fig. 4, left hand panel). Using the Faint Object Camera on HST, *Betelgeuse* (M2 Iab) was imaged in two ultraviolet wavelength bands providing about 10 resolution elements over the stellar disk. Two important discoveries resulted (Gilliland & Dupree 1996). First, the apparent size of the star in the ultraviolet ($\sim\lambda 2550$) is much larger (by a factor of $\sim$2.2) than in the optical [$R_*(optical)$= 27.5 mas], demonstrating that the supergiant atmosphere extends far beyond the photosphere. Secondly, an unresolved bright feature appears prominently in the south-western quadrant of the disk. Its excess brightness indicates it is hotter by at least 200K than the rest of the surface.

The spatial resolution that is achieved by scanning the disk of *Betelgeuse* with the GHRS Small Science Aperture allows an estimate of the rotational velocity of the photospheric stellar disk in the two perpendicular scanning directions (Uitenbroek et al. 1998a, 1998b). The displacement of a photospheric line in the same spectral region shows that the north-west limb is approaching at approximately 5 km s^{-1} relative to the stellar radial velocity, and the south-east limb is receding at approximately 5 km s^{-1}.

 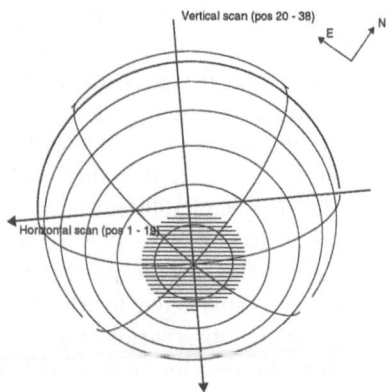

Fig. 4. UV image of *Betelgeuse* obtained in Mar. 1995 with the FOC (left panel) and schematic representation of orientation of the perpendicular series of offset pointings (vertical and horizontal scans) at which GHRS spectra were obtained (right panel). The shaded area indicates the position of the unresolved bright spot. The inclination angle of the rotation axis is ~ 22 deg to the line of sight (adapted from Gilliland & Dupree 1996 and Uitenbroek et al. 1998a, 1998b).

GHRS spectra, taken in the $\lambda 2800$ region with a series of different offset pointings each 27.5 mas apart and arranged in two perpendicular directions provide additional information on the nature of the spot. The excess spot emission is clearly visible in the Mg II h line ($\lambda 2802$), while it is hardly apparent at the wavelength of the k line ($\lambda 2795$). It appears that the perturbation causing the bright spot is present in the formation layers of h-line emission, but barely noticeable in the (higher altitude) layers forming the k-line. In even deeper layers than the h-forming region the spot seems more extended as is evidenced by the asymmetry of the intensity profile across the disk at continuum layers (Uitenbroek et al. 1998a, 1998b).

Based on the spectroscopic data we believe the bright spot on *Betelgeuse* is evidence for a non-spherically symmetric fundamental pulsation. The moderate rotation of the star would lead the shockwave associated with this pulsation in the higher levels of the atmosphere to break out initially at the poles giving rise to the brightening observed in the 1995 HST images (Gilliland & Dupree 1996), and some of the ground-based speckle imaging and interferometric studies (Klückers et al. 1997; Tuthill et al. 1997). The evidence from IUE-discovered chromospheric modulation (Dupree et al. 1987, 1998) and radial-velocity measures in the optical (Smith et al. 1989) shows that *Betelgeuse* is pulsating. Theoretical calculations demonstrate that rotation can affect the pulsational modes of a star causing amplitude amplification at the poles with respect to the equator for slow rotation rates (Clement 1981; Asida & Tuchman 1995). If we identify the unresolved bright spot with the pole of *Betelgeuse*, then the inclination $i = 20°$, and the rotational velocity is ~ 15 km s^{-1}. Attributing the radius to the ultraviolet continuum disk, and using the HIPPARCOS distance, we find a period of 17 years (Uitenbroek et al. 1998a, 1998b).

Betelgeuse has a dust shell, but long-baseline interferometry at 11μm reveals that the shell is of 1000 mas radius (Danchi et al. 1994) or at a distance of $36R_*$ with very little dust between the optical disk and the sharp edge of the dust shell. Thus, the chromospheric acceleration leading to a wind can not be caused by dust as is suspected for the coolest luminous stars.

Continued observations are now underway in a three-year program with HST to follow the development of the ultraviolet surface, and further define the atmospheric dynamics with STIS spectra. *Betelgeuse* is providing an unparalleled opportunity to track the appearance of bright spots, the response of the atmosphere to a pulsating photosphere, and the occurrence of mass flow directly from the stellar surface.

5   Concluding Remarks

It is clear from these examples that the variability of the outer atmospheres of cool stars is a common occurrence. The problem in documenting and characterizing this variability consists of finding the appropriate diagnostic, and acquiring sufficient continuous high-quality spectroscopic observations. Most single cool stars have longer rotation periods and longer activity cycles than the hot stars, thus making "campaigns" similar to the dedicated days of IUE rather difficult. One particularly good example for the rapidly rotating T Tauri star SU Aur is the synoptic observations of Giampapa et al. (1993) revealing the periodic modulation of the short-wavelength wing of the Hα line corresponding to the rotation period of the star. We need many more programs such as this to define the extent of the wind variability in a number of stars.

Results from the Sun suggest the mass flux is at a minimum when the coronal hole area is highest (solar minimum) but the mass flux ($\dot{M}$) is a maximum at solar maximum. Thus the slow wind carries more mass flux than the fast wind. The luminous cool stars (Luminosity Classes, I, II, and III) appear to have cooler winds than that in the Sun, and mass-loss rates larger by 6 orders of magnitude or more. Both of these facts make study of winds easier in the giants and supergiants. And, whenever there are sufficiently long measures of good quality, periodic or semi-periodic motions are found. For the hybrid stars, periodic photospheric velocity variations have been identified that may become a "fluttering" motion in the chromosphere (as observed in the He I $\lambda10830$ line), rapidly accelerate to supersonic speeds, and apparently decelerate at large distances from the star corresponding to the circumstellar absorption features. It is fair to suspect that the fluttering may also turn out to be semi-periodic when a sufficient sequence of measures is acquired. At present, while there are variations in the terminal velocities of hybrid stars, it is not clear whether a periodicity can be associated with the variations, and the mass-loss rate.

The supergiant star *Betelgeuse* presents a unique opportunity to sample a wind and its variability over the surface of a star other than the Sun. Knowing the orientation of the star, observations with STIS on HST, and its smaller slit and extended wavelength coverage than the GHRS should give indications of the two dimensional structure of the atmosphere, the wind, and its variation.

To return to the colorful poster of this Workshop, we can offer several answers to the question "What is pedaling the bicycle?" Cool stars possess a variety of culprits: hot coronas leading to thermally driven winds, magnetic fields and associated wave processes, and photospheric pulsations that can extend the atmosphere and drive winds as well. Dust does not appear to play a role in the start of the chromospheric acceleration that occurs in the Sun, the luminous hybrid stars, and in a supergiant such as *Betelgeuse*. Through observations such as those discussed here, we are now developing a comprehensive picture of the character of winds across the cool half of the color-magnitude diagram.

Acknowledgements

I appreciate the discussions and advice from Dimitar Sasselov and Yi-Ming Wang in preparing this manuscript. This research is supported in part by STScI grant GO-0649.01-95A to the Smithsonian Astrophysical Observatory.

References

Asida, S.M., Tuchman, Y. (1995): ApJ **455**, 286
Brekke, P. (1993): ApJ **408**, 735
Brickhouse, N.S., Dupree, A.K. (1998): ApJ, in press
Brown, A., Deeney, B.D., Ayres, T.R., et al. (1996): ApJS **107**, 263
Butler, R.P. (1998): ApJ **494**, 342
Clement, M.J. (1981): ApJ **249**, 746
Danchi, W.C., Bester, M., Degiacomi, G.G., et al. (1994): AJ **107**, 1469
Dere, K.P. (1994): Sp. Sci. Rev., **70**, 21
Dere, K.P., Bartoe, J.–D.F., Brueckner, G.E., Recely, F. (1989): ApJ **345**, L95
Doschek, G.A., Feldman, U., Bohlin, J.D. (1976): ApJ **205**, L177
Dupree, A.K., Baliunas, S.L. (1979): IAU Circ. **3435**, 1

Dupree, A.K., Reimers, D. (1989): in Exploring the Universe with the IUE Satellite, ed. Y. Kondo et al., (Kluwer, Boston), p. 321

Dupree, A.K., Sasselov, D.D. (1998): in preparation

Dupree, A.K., Baliunas, S.L., Hartmann, L., et al. (1987): ApJ **317**, L85

Dupree, A.K., Whitney, B.A., Avrett, E.H. (1992): in eds. M.G. Giampapa, J.A. Book-binder, ASP Conf. Ser. **26**, ASP, San Francisco, p. 525

Dupree, A.K., Penn, M.J., Jones, H.P. (1996): ApJ **467**, L121

Dupree, A.K., Guinan, E.F., Smith, M.J., Stefanik, R. (1998): in preparation

ESA, (1997): The Corona and Solar Wind Near Minimum Activity, ESA SP-404

Giampapa, M.S., Basri, G.S., Johns, C.M., Imhoff, C. (1993): ApJS **89**, 321

Gilliland, R.L., Dupree, A.K. (1996): ApJ **463**, L29

Goldstein, B.E., Neugebauer, M., Phillips, J. L., et al. (1996): A&A **316**, 296

Grall, R.W., Coles, W.A., Klinglesmith, M.T., et al. (1996): Nature **379**, 429

Hartmann, L., Dupree, A.K., Raymond, J.C. (1980): ApJ **236**, L143

Hartmann, L.W., Jordan, C., Brown, A., Dupree, A. K. (1985): ApJ **296**, 576

Hatzes, A.P., Vogt, S.S., Ramseyer, T.F., Misch, A. (1996): ApJ, **469**, 808

Judge, P.G., Jordan, C., Rowan-Robinson, M. (1987): MNRAS **224**, 93

Klückers, V.A., Edmunds, M.G., Morris, R.H., Wooder, N. (1997): MNRAS **284**, 71

O'Brien, G., Lambert, D. (1986): ApJ **253**, 716

Ofman, L., Romoli, M., Poletto, G., et al. (1997): ApJ **491**, L111

Rao, L., Baliunas, S.L., Robinson, C.R., et al. (1993): in Luminous High-Latitude Stars, ed. D.D. Sasselov, ASP Conf. Ser. **45**, (ASP, San Francisco), p. 300

Reimers, D. (1982): A&A **107**, 292

Reimers, D., Hünsch, M., Schmitt, J.H.M.M., Toussaint, F. (1996): A&A **310**, 813

Rottman, G.J., Orrall, F.Q., Klimchuk, J.A. (1982): ApJ **260**, 32

Smith, M.A., Patten, B.M., Goldberg, L. (1989): AJ **98**, 2233

Strachan, L., Kohl, J. L., Weiser, H., et al. (1993): ApJ **412**, 410

Tuthill, P.G., Haniff, C.A., Baldwin, J.E. (1997): MNRAS **285**, 529

Uitenbroek, H., Dupree, A.K., Gilliland, R.G. (1998a): in Cool Stars, Stellar Systems and the Sun, Proc. Tenth Cambridge Workshop, eds. R.A. Donahue, J. Bookbinder, ASP Conf. Series, (ASP, San Francisco), in press

Uitenbroek, H., Dupree, A.K., Gilliland, R.G. (1998b): AJ, submitted

Wang, Y.-M. (1998): in Cool Stars, Stellar Systems and the Sun, Proc. Tenth Cam-bridge Workshop, eds. R.A. Donahue, J. Bookbinder, ASP Conf. Series, (ASP, San Francisco), in press

Wang, Y.-M., Hawley, S.H., Sheeley, N.R., Jr. (1996): Science **271**, 464

Warren, H.P., Mariska, J.T., Wilhelm, K. (1997): ApJ **490**, L187

Gosling: I believe the mass-loss rate that you show for the Sun for the last two solar cycles is highly model dependent, being derived from photospheric magnetic-field measurements and potential field models, an assumed dependence between flux tube expansion and solar wind speed, and an average relationship between solar wind speed and density. The uncertainties are such that your derived mass-loss rate variation must be considered only as tentative rather than definitive.

Dupree: Actually, direct observations were used at every point where data are available. For instance, in converting global wind speeds into mass fluxes, actual *in situ* measures near Earth and the ULYSSES observations were incorporated. Unfortunately, global measures can not be made simultaneously (the Sun is not a distant point source like a star!), and so some modeling and spatial integration are needed. The total amplitude of the mass-loss rate variation is consistent with measures of individual features in the solar wind.

Linsky: α Ori has now been monitored at radio wavelengths for many years, and Stephen Drake and others find that the radio continuum flux is constant to within 25%, but with occasional dips. Yet your measures imply chromospheric variations are present. Why is that?

Dupree: Drake and his colleagues reported [Bookbinder, J.A. et al., 1987, in Cool Stars Stellar Systems and the Sun, Proc. Fifth Cambridge Workshop, eds. J.L. Linsky & R.E. Stencel, (Berlin: Springer-Verlag), p. 337] a variability at 6 cm on the 30-40% level based on 7 observational points over 7 months, so it would be hard to find convincing evidence of a 14 month period from those data. It is interesting that at the time of their observations, the ultraviolet continuum varied by only ~20 percent (Dupree et al. 1987; Dupree et al. 1998).

Hartmann: Recent VLA results by Lim & White suggest that the α Ori radio emission comes not from a compact 8000K chromosphere but from a much larger ~1000 K chromosphere. At these distances, wind/pulsation fluctuations will damp out, explaining why the radio fluxes don't vary much.

Dupree: Yes these observations were presented at the Tenth Cambridge Cool Stars Workshop in July 1997, and a paper is in press in *Nature* (Lim et al. 1998). They measure the temperature from 2 to $7R_*$ and find decreasing values from $3450\pm850K$ ($2R_*$) to $1370\pm330K$ ($7R_*$).

Willson: On the Mira models: violent pulsations deep in the atmosphere lead to a steady flow after a few stellar radii. So constant radio flux is what you expect if you sample on a larger scale.

Variations of Winds of Cool Giants in Hα

Joel A. Eaton

Tennessee State University, 330 10th Avenue North, Nashville, TN 37203-3401, USA

Abstract. I present results of monitoring Hα in eight cool supergiants and in a newer group of ∼ 100 cool giants and supergiants. Preliminary reductions of nine months of data for these latter stars indicate that about 40 % of them are variable in Hα, mostly the supergiants and bright giants.

1   Introduction

This is intended as a progress report for a program to monitor cool (types G-M) giants and supergiants (classes I-III) in Hα to detect changes in upper chromospheres caused by global variations of winds. I began the project several years ago as a way to detect outbursts in cool components of ζ Aurigae binaries (see Eaton & Bell (1995), sect. 3.4), observing 8 stars for roughly three years (Eaton & Henry 1996). The stars were all K or G supergiants, three of which (63 Cyg, ξ Cyg, and ε Peg) did not have bright B companions. Those observations showed that *all eight* stars vary in the blue wing of Hα (see Fig. 1). Changes in this part of the profile consisted of (1) slow, possibly cyclic, variations on timescales of 100 days to years and (2) intensification of the absorption on timescales of weeks (outbursts?). A typical example is given in Fig. 2. Smith & Dupree (1985) have obtained similar results for a few metal-poor giants, so we suspect all cool supergiants, and perhaps giants, would show the phenomenon. My new observations are testing this hypothesis with the hope of forcing theoreticians to look at the connection between winds and the interior structure of cool giants.

2   Why Hα?

Hα is a good line to use for detecting global changes in the outer part of the chromosphere where the wind is accelerating in the typical cool supergiant (see the discussion of Hα line formation in Cram & Mullan 1985). It is a very strong line with equivalent widths 1-1.5 Å in cool giants and supergiants (e.g., Eaton 1995), falling in a region of the spectrum where CCD detectors are easy to use. Furthermore, it is observed extensively for other purposes, which makes it even more convenient for monitoring programs. The huge equivalent width and low central intensity (of order 0.2) mean the line must be formed *globally* over the whole surface of the star and does not merely reflect exceedingly strong absorption in isolated chromospheric patches. Because of its excitation potential, its mere existence requires a chromosphere. The core of the line in models is formed at a depth of order 0.01 gm/cm², in the upper chromosphere but still

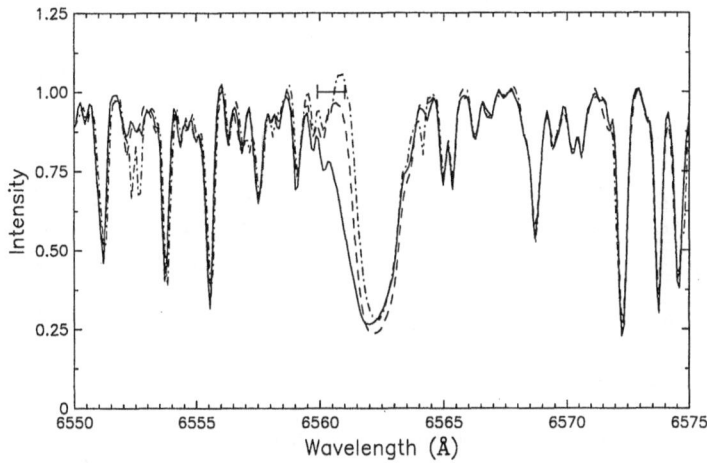

Fig. 1. The Hα profile of ξ Cyg. The shift of the core to shorter wavelength is usually taken to indicate formation at the base of the star's wind. Note the change in the blue wing, which actually sometimes goes into emission.

close to the surface of the star. Changes in the velocity structure of this layer give changes in the asymmetry of the line, while an increase in the temperature can cause the wing to go into emission.

In spite of these advantages, there are a couple of caveats. First, the line can be contaminated by telluric lines, especially during the summer. Second, changes in Hα profiles could reflect more complicated changes in chromospheric structure than wind variations, as Linsky has pointed out several times at this conference.

3   New Results

All the data I am discussing were taken with the stellar spectrograph of the McMath-Pierce Solar Telescope at the US National Solar Observatory. This decrepit instrument is probably the world's most effective for monitoring programs because it has a dedicated observer to make observations at regular intervals. This capability will be sorely missed if shut down before we can replace it with automatic spectroscopic telescopes.

I have nine months of data in hand for ∼ 100 stars, which have been observed for several days a month since January, 1997. Their properties are distributed as follows: spectral types – 22 % G, 50 % K, and 28 % M; luminosity type – 27 % I, 15 % II, and 58 % III. Since the program is still ongoing and may last for another year, I have only a preliminary analysis in which I reduced the spectra with IRAF and plotted them in groups on a computer screen to look for variations in the blue wing of Hα. Of the 100 stars, 40 % looked as though they are probably variable while 60 % probably are not. Most of the luminosity-

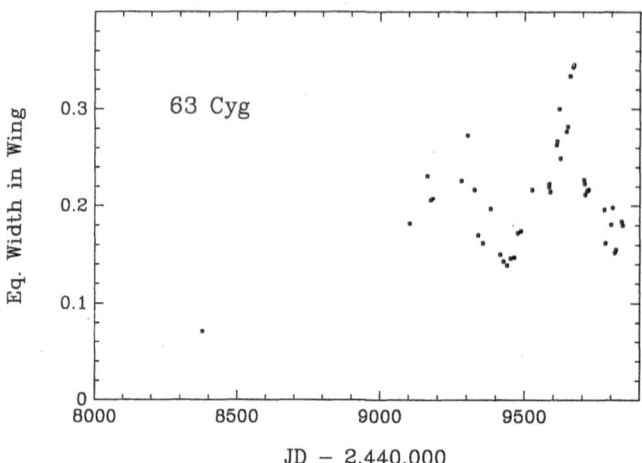

Fig. 2. Variation of the blue wing of Hα in 63 Cyg. The equivalent width plotted reflects absorption in the part of the profile marked by a horizontal bar in Fig. 1. Cyclic variations with a spike in the absorption are readily apparent.

class I and II stars are probably variable. Hardly any of the class III giants were variable enough for it to be detectable.

I gratefully acknowledge the dedicated work of Trudy Tilleman in obtaining the spectra of these many stars. This research is being supported by NSF grant HRD-9550561.

References

Cram, L.E., Mullan, D.J. (1985): ApJ, **294**, 626
Eaton, J.A. (1995): AJ, **109**, 1797
Eaton, J.A., Bell, C. (1995): AJ, **108**, 2276
Eaton, J.A., Henry, G.W. (1996): in Proc. IAU Symp. 176, ed. K.G. Strassmeier & J.L. Linsky (Kluwer, Dordrecht), 415
Smith, G.H., Dupree, A.K. (1985): AJ, **95**, 1547

Kubát: If the secondary is a B-type star, why do you neglect its radiation?

Eaton: You're referring to my observation of Hα in ζ Aur binaries. I'm trying to determine for these stars whether the B star has any direct effect on the chromosphere or wind. I don't think I have seen any effects of direct irradiation yet, since I've looked primarily at the sides of the K stars away from the B stars.

Little-Marenin: At what level were the Hα lines variable in the 40% of the stars on your observing program?

Eaton: At order 5-10% of the continuum.

Variability of the Circumstellar Shell of 89 Herculis

Janina Krempeć-Krygier[1], Miroslaw Schmidt[1], and Krzysztof Gesicki[2]

[1] NCAC, ul.Rabianska 8, PL-87-100 Torun, Poland
[2] Centrum Astronomii UMK, ul.Gagarina 11, PL-87-100 Torun, Poland

89 Herculis (F2Ia/Ib) is a pulsating binary (Smolinski et al. 1980; Waters et al. 1993) fulfilling the criteria for low-mass post-AGB stars.

The observational material consists of plates (2 and 6 Å mm^{-1} dispersion) made at the DAO (Victoria, Canada). The best model fits to hydrogen lines were obtained for $T_{\text{eff}} = 7250$ K and $\log g = 0.5$. Changes in the profiles of the Balmer lines, Hβ and Hδ, are shown in Fig. 1. For Hβ, the upper spectrum corresponds to 1975 May 29; the middle, two weeks later; the lower (dotted), six weeks later. For Hδ, the upper spectrum refers to 1975 June 6, while the lower spectrum was obtained one week later. A synthetic spectrum is plotted on top of the lower spectrum of Hδ to prevent accidental identification of photospheric absorption lines as shell features. An asymmetry (but no distinct circumstellar feature) is seen in the blue wing of Hδ in the upper spectrum, while one week later the circumstellar component is clearly visible. The circumstellar components are distinguished in all spectra of Hβ. Their radial velocities change from -90 to -120 km s^{-1} over two weeks. After six weeks an intermediate circumstellar component occurred at about -70 km s^{-1}. Hα profiles have P Cygni shapes that indicate outflow velocities of about 150 km s^{-1}. They exhibit only slight changes.

This work was supported by KBN grant No.2.P.03D.026.09.

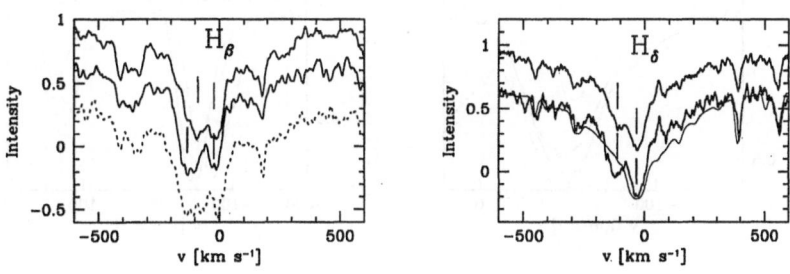

Fig. 1. Evolution of the 1975 shell episode seen in Balmer lines.

References

Smolinski, J., Climenhaga, J.L., Krempeć–Krygier, J., et al. (1980): in Proc. Fifth European Regional Meeting: Variability in Stars and Galaxies, Liège
Waters, L.B.F.M., Waelkens, C., Mayor, M., Trams, N.R. (1993): A&A 269, 242

Wind Variability in the Hypergiant HR 8752

Miroslaw Schmidt[1] and Krzysztof Gesicki[2]

[1] NCAC, ul.Rabianska 8, PL-87-100 Torun, Poland
[2] Centrum Astronomii UMK, ul.Gagarina 11, PL-87-100 Torun, Poland

The yellow hypergiant HR 8752 = HD 217476 (G0 Ia$^+$) probably belongs to the class of stars that cyclically "bounce" against the edge of the atmospheric instability region (Nieuwenhuijzen & de Jager 1995). One of the direct consequences of this instability seems to be brief periods of enhanced mass loss that occur when the star comes close to the edge of the instability region during its blueward evolution.

Our spectroscopic material consists of high-dispersion plates (6.5 Å mm^{-1}) taken at the DAO (Victoria, Canada) in 1969–70 and 1975 by Smolinski (Smolinski 1970). The left figure shows the blue wings of Fe I (multiplet 686) absorption lines in 1969, which show evidence for an outflow velocity of more than 50 km s^{-1}. Typical examples of the evolution of the absorption spectrum observed on all four plates are presented in the figure on the right for lines of Ba II and Ca I. There is a distinctive phase in 1970 when the lines are shifted to the blue and broadened. Most probably this is connected with the pulsation of the star, which may drive the outflow (which is also seen in Hα). These changes are accompanied by changes in the effective temperature: 5300 K (1969), below 5100 K (1970) and 5750 K (1975).

This work was supported by KBN grant No.2.P.03D.026.09.

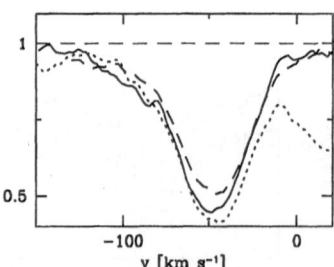 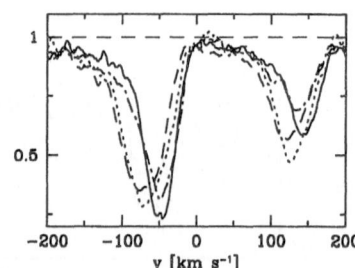

Fig. 1. Left: Fe I (686) lines as observed in 1969. Right: Evolution of Ba II λ5853 and Ca I λ5857 in 1969 (solid), 1970 (dotted and dashed), and 1975 (dash-dot).

References

Nieuwenhuijzen, H., de Jager, C. (1995): A&A 302, 811
Smolinski, J. (1971): in M. Hack (ed.), Colloquium on Supergiant Stars, Osserv. Astron. Trieste, Trieste, p. 68

Variable Dust Emission Observed in Miras

Irene R. Little-Marenin

CASA, CB 389, University of Colorado, Boulder, CO, 80309, USA

Mira variables pulsate with periods ranging from 100–2000 days and eject copious quantities of gas into space. Part of the gas condenses into dust grains when T < 1200 K in the shell. During their pulsational cycle Miras vary in luminosity, forcing the dust that has formed in the circumstellar shells alternately to evaporate and re-condense in the inner parts of their shells. Hence, emission from the dust will vary during the pulsational cycle. These variations have been observed in the low-resolution spectra (LRS) of Miras obtained by IRAS.

Comparison of the IRAS data to the optical light curves from the AAVSO for the Mira variables AU Cyg (P=439^d), R LMi (P=389^d), and R Cet (P=166^d) shows clear variations when their LRS spectra were obtained at different phases. However, spectra of Miras taken at similar phases (e.g., DL CMa, phase=0.98, 0.99) or of non-variable stars (e.g., π^1 Aur) are very similar. Spectra taken at different phases show variations in the total flux and in the contrast/strength of the 10 μm feature. However, the shape of the silicate emission feature is identical at the different phases.

Plots of the normalized fluxes of many stars as a function of phase show that the typical range of the 12 μm flux (from the IRAS point source catalog) is about 0.5–1.5 magnitudes. These same magnitude ranges are observed in the 2–3 μm region, where most of the stellar energy is emitted. Plotting the normalized contrast of many Miras versus phase shows that the contrast is largest near maximum light and smallest near minimum light. The contrast varies typically by about 10–20% between maximum and minimum. A few stars show only a small variation in contrast between maximum and minimum. For most of these Miras, the phases were extrapolated from epochs dating back 20 to 40 cycles. Small errors in period or epoch could have produced incorrect phases and resulted in misidentified times of maximum and minimum.

Since more dust is likely to form during minimum, when the condensation radius moves inward as the star cools, one would expect the contrast to be larger then, contrary to what was observed. There is no indication that the optical depth of the shell changed drastically between the two phases as is suggested by the similarity in shape of the emission feature. There is at present no clear-cut explanation for this effect. The observed variations may be related to different size distributions of the grains at maximum and minimum, possibly resulting from the presence of a significant population of small dust grains that is enhanced near maximum by the evaporation of larger grains. The variations may also be related to changes occurring in the spectrum shortward of 8 μm, which would lead to the 10 μm emission being underestimated near minimum. Data taken during a complete cycle, e.g., with ISO, are needed in order to understand the complex processes occurring during dust formation and destruction.

Variable Water Maser Emission from Circumstellar Disks

Irene R. Little-Marenin[1], Priscilla J. Benson[2]

[1] CASA, CB 389, University of Colorado, Boulder, CO, 80309, USA
[2] Whitin Observatory, Wellesley College, Wellesley, MA 02181, USA

Abstract. We observed the 22 GHz water maser emission in VX UMa (M8e, M=388^d) and V778 Cyg (C5,5J, SRb=300^d) from 1987-1992. The spectra of both stars show three distinct components. All three components vary periodically in intensity. We interpret these features as being produced by a rotating circumstellar disk

1 Introduction

We have monitored the 22 GHz water maser emission line in a dozen long period variable stars from 1987-1992 with the Haystack 37 m radio telescope. These stars are on the Asymptotic Giant Branch with typical masses of about 1-1.5 M$_\odot$, radii of about 1 AU and luminosities of about 10^4 L$_\odot$. The stars experience mass loss with typical rates of about 10^{-7} M$_\odot$ y^{-1}. All of our stars show highly variable maser emission with the same period as the optical period of the central star. But the maximum water maser emission appears to lag the optical maximum by about 0.3 phase. Since the period of variation of our stars is on the order of $100-600$ days, this phase lag corresponds to about $3 - 6$ months.

The 22 GHz spectrum for the Miras typically shows one strong component at the radial velocity of the central star. The line typically has a FWHM value of about 0.8 km s^{-1}. However, a few of our stars showed multiple components. Specifically the M star Mira variable VX UMa and the silicate-carbon star V778 Cyg showed three distinct components in their maser spectra.

2 Observations — VX UMa

The maser spectrum of VX UMa shows three distinct components at -48 (or -49), at -50.5 and at -52.2 km s^{-1}. Its LSR velocity is unknown and we predict it to be at -50.5 km s^{-1}. We observed water maser emission in VX UMa from 26 November 1989 through 22 December 1992 roughly every one or two months. A few additional observations were made in 1993 and 1994. The maser spectra vary in time with a period of about 390^d. Since the optical period of 215^d published in the GCVS (General Catalog of Variable Stars) did not fit our maser data, P. Benson re-observed the star and determined a new period of 388^d. The light curve is asymmetric with a fractional rise time of 30% of the total period. We find that the optical maximum lags the maser maximum by about 0.2 phase, similar to the phase lags observed for other Miras.

Not all three features of the maser spectra are present at all times. The central feature at -50.5 km s^{-1} is always present, both a red- and a blue-shifted feature are usually present at -49 km s^{-1} and -52.5 km s^{-1}. A strong feature at -48 km s^{-1} was present for a few months in 1989.

All three components vary in time roughly with an average period of 390^d during the four cycles of our observations, but not necessarily in phase with each other. Two distinct periods of flux variation are present. From November 1988 through February 1990, the flux variations of the two side-components at -52.5 and -48 km s^{-1} are anti- correlated. The -52.5 km s^{-1} feature becomes the strongest feature ever observed in VX UMa on 26 November 1989. With T=3.5 K (1 K $\simeq$ 10 Jy), it is about 3 times as strong as the flux from the central feature at that time. The -48 km s^{-1} feature is about as strong as the central feature when present. After June 1990, the -48 km s^{-1} feature is absent, but both a -49 and the -52.5 km s^{-1} feature are usually present. The fluxes of these side components vary in phase with each other as well as with the flux of the central component. However, the side components in general are only half as intense as the central feature.

We have only a few observations after November 1992, but by December 1992 we are again beginning to observe an anti-correlation of the side components. The blue-component is strengthening relative to the central peak and is about 40 % stronger than the central component in May of 1993. After that time it decreases in strength and by September of 1993 is roughly half as intense as the central component. In October 1993 the spectrum looks very similar to the ones observed in May or June of 1989. We have no observation from November 1993 through June 1994. If the pattern of anti-correlation is periodic, we suggest that the -48 km s^{-1} should have been very strong from November 1993 − February 1994 and that the blue component should have been strong during February − June 1994. Our last spectrum in July 1994 appears to show a reversion to correlated intensities. However, more observations would be needed to verify periodic switching from correlated to anti-correlated intensities.

2.1 Velocity Variations

The two side-components are shifted relative to the central component, but all three components show small velocity variations with time. The redshifted peak at -49.5 km s^{-1} is shifted by 1.65±0.15 km s^{-1} and the blue shifted one by 1.95± 0.1 km s^{-1} relative to the central peak. The central component varies periodically with an amplitude of 0.2 km s^{-1}. The limited data of the -52.5 km s^{-1} feature also shows periodic velocity variations of about 0.2 km s^{-1}. This velocity amplitude appears to diminish to about 0.1 km s^{-1} in 1992. The greatest flux approximately corresponds to the greatest blue shift of the line.

2.2 Model

The population inversion needed to emit the 22 GHz water maser line from circumstellar shells is produced by collisions of the water molecule with hydrogen

(Cooke & Elitzur 1985) at densities of $10^7 - 10^{13}$ cm^{-3}. At higher densities the inversion becomes quenched by collisional de-excitations and at lower densities not enough collisions occur to support the inversion. As suggested by others (Elmegreen & Morris 1978, Cesaroni 1990, Lekht et al. 1993), we propose that the triple maser spectrum is the result of a rotating disk. The blue- and red-shifted components are emitted by the lateral parts of the rotating disk seen edge-on, whereas the central component corresponds to the radial velocity of the star since no Doppler motion of the central part of the disk would be produced. For a symmetrically pumped maser, the models by Cesaroni (1990) give maser spectra with a stronger central peak and with weaker side features placed symmetrically about the central peak. This type of maser spectrum is observed in VX UMa from mid- 1990 until late 1992. In order to explain the anti-correlation of the side components from late 1988 until mid-1990, the models of Cesaroni (1990) and Lekht et al. (1993) suggest that the inversion is quenched in the receding side of disk in 1988, making the redshifted line unobservable. Because the diametrically opposite segment of the disk (i.e. the approaching side) is always in radiative contact with the receding side of the disk, the weakening of the red-shifted line will increase the emission of the blue-shifted line. In 1989 the situation reverses when quenching occurs in the approaching side of the disk. The quenching should be produced by inhomogeneities in the disk since a homogenous disk will produce symmetrically placed side components. Possibly the density structure and/or the velocity gradient changes, producing enough variations in the pumping rate to quench the maser. There are indications that the pattern of anti-correlated intensities to correlated intensities repeats after approximately four years.

3 Observations − V778 Cyg

V778 Cyg is another star with possible water maser emission from a circumstellar disk. V778 Cyg is one of only seven known carbon stars that are associated with oxygen-rich material in its circumstellar shell (Little−Marenin 1986). The oxygen-rich nature of the circumstellar material is identified by the presence of emission from silicate dust and water molecules. The most likely explanation is that we are observing a binary system, even though no widely accepted explanation exists as yet.

We have monitored the water maser emission from V778 Cyg since we detected it in 1987 until 1992 with a few additional observations in 1993 and 1994. We can identify three components at -20.5, -17 and -15 km s^{-1}. All three components vary in time. As in VX UMa, we observe periods of correlated flux variations of the side components (3/1987−6/1988) and periods of anti-correlated variations. However, the anti-correlated flux variation is one-sided. Only the receding (red-shifted component) at -15 km s^{-1} becomes very strong (1/1989−7/1991) while the approaching component becomes very weak and is effectively below our detection limit of about 0.15 K after April 1989. We do not observe a large increase in the flux of the approaching component with a corresponding decline in the flux from the receding side.

As in VX UMa, we suggest that a rotating disk produce the variations in the flux from V778 Cyg. However, the quenching appears to occur only in the approaching edge of the limb, unlike VX UMa where quenching appears to occur alternately in both edges. V778 Cyg is not the only silicate–carbon star to be surrounded by a rotating disk. Kahane et al. (1997) interpret the CO observations from BM Gem as being due to rotation of a disk.

References

Cesaroni, R. (1990): A&A, 233, 513
Cooke, B., Elitzur, M. (1985): ApJ, 295, 175
Elmegreen, B.G., Morris, M. (1979): ApJ, 229, 593
Kahane, C., Barnbaum, C., Uchida, K., Balm, S.P., Jura, M. (1997): (preprint)
Lekht, E.E., Likhackev, S.F., Sorochenko, R.L., Strelnitskij, V.S. (1993): AZh, 70, 731
    (Astron. Rep., 37(4))
Little–Marenin, I.R. (1986): ApJ, 307, L15

Rivinius: Can you estimate the radius of the maser emitting region from the separation of the disk emission peaks?

Little-Marenin: Yes, if you assume keplerian motion of the disk.

Systematic Variability in OB-Star Winds

Raman K. Prinja

Department of Physics and Astronomy, University College London, Gower Street, London WC1E 6BT, UK

Abstract. Spectroscopic results of OB stars, primarily from the *IUE* satellite, are reviewed which highlight the incidence of large-scale systematic wind structure. I summarize evidence for cyclic wind variability and rotational modulation, based on extended time-series UV data sets (including the '*IUE* MEGA Campaigns'). The constraints provided by these data indicate that the modulating 'clocks' apparent in some hot star winds must ultimately have an origin at the stellar photosphere. The precise mechanism connecting photospheric and wind changes (e.g. via nonradial pulsations and/or ordered weak magnetic fields) remains a crucial issue.

1 Preamble

Barely 4 or 5 years prior to this Workshop there were very few data sets available that were suitable for examining evidence for *repeatability* in the time-dependent characteristics of hot star winds. The situation has changed radically since then, however, to the extent that there are now demonstrable cases of *cyclic* (possibly periodic) modulations in the outflows of OB stars. We are consequently faced with a new and challenging multi-component picture of OB stars, where a continuously variable, highly structured stellar wind is shaped by inhomogeneities rooted at the stellar surface. Furthermore, it is exciting that current studies of the time-dependent nature of hot-star winds not only provide constraints on the mechanisms of mass loss via fast winds, but are also pertinent to stellar surface structure and the fundamental nature of massive stars. I will outline in this paper some of the key spectroscopic results that have provided the background, and led, to this new picture of cyclic variability in hot-star winds.

1.1 Large-scale Structure

It is necessary to distinguish between two principal forms of structure in the winds of early-type stars. Firstly, the winds are affected – but not grossly – by small-scale *stochastic* fluctuations. The origin of these is likely intrinsic to the winds, and related to the strong instability of the radiation-driven mechanism (e.g. Carlberg 1980; Owocki & Rybicki 1984, 1985; Owocki 1992). The result is a shocked outflow which qualitatively accounts for the observed X-ray emission (e.g. Collura et al. 1989; Berghofer et al. 1997; Feldmeier et al. 1998), Auger ionization (e.g. strong N V $\lambda\lambda1240$ UV resonance lines), 'black' troughs in saturated UV wind profiles (e.g. Puls et al 1993), and radio emission properties (Bieging et al. 1989).

However, it is the second form of wind structure in OB stars – i.e. large-scale, and diagnosed by systematic spectral variability – which is the most directly relevant to the theme of this Workshop. The large-scale structure is most probably induced by changes in the star itself and can substantially modify the overall wind. In this paper, the nature of large-scale structure in OB stars is discussed via properties of its principal tracer, namely, the incidence of short timescale (hourly) localised optical depth and velocity features in wind-formed spectral lines. Results from ultraviolet studies are emphasised here. (See e.g. Kaper; Kaufer, these proceedings, for an update on the optical wind properties of these stars.) There is no doubt that our present understanding of the nature and importance of variability in hot star winds owes much to the long-lived International Ultraviolet Explorer (*IUE*) satellite, which permitted observers to secure high-resolution time-series data for several OB stars. These results are discussed in the following sections.

2   UV Signatures of Stellar-wind Activity

More than two decades ago the first rocket and *Copernicus* satellite UV spectra were secured of OB stars (e.g. Underhill 1975; Morton 1976; Snow & Morton 1976), and they provided immediate indications that the extended absorption troughs of the UV resonance lines did not have a smooth 'textbook' morphology. 'Narrow absorption components' were discovered towards the shortward edges of the wind lines, indicating that the outflows of these stars were inhomogeneous (see Fig. 1). Gathier et al. (1981) and Lamers et al. (1982) carried out the first systematic investigations of the narrow absorption features. Prinja & Howarth (1986) and Howarth & Prinja (1989) carried out subsequent surveys based on *IUE* spectra. Localised optical-depth structure of this form is now established as almost ubiquitous in single 'snap shot' UV spectra of OB stars. Similar structure is also evident in other hot stars (see Fig. 1), though of course the physical origin may differ in some of these cases. The SMC example in Fig. 1, AV 388 (O4 V), illustrates an interesting case where the overall wind is considerably weaker compared to corresponding stars in the Milky Way (as expected for the lower-metallicity environment of the SMC), but evidence for wind structure (via narrow components) remains remarkably clear (see e.g. Walborn et al. 1995). UV time-series data sets (see below) revealed however that the narrow components in OB stars simply represent one aspect of *variability* across a very broad velocity range in the absorption troughs.

2.1   Patterns of Wind Variability

The stellar winds of OB stars are highly variable on characteristic timescales ranging from less than one hour to several days. Although *Copernicus* scans indicated wind fluctuations (e.g. Snow 1977; Gry et al. 1984), the nature of these changes was not even remotely understood until the launch of *IUE* in 1978. Time-series data sets of individual stars collected with the satellite have

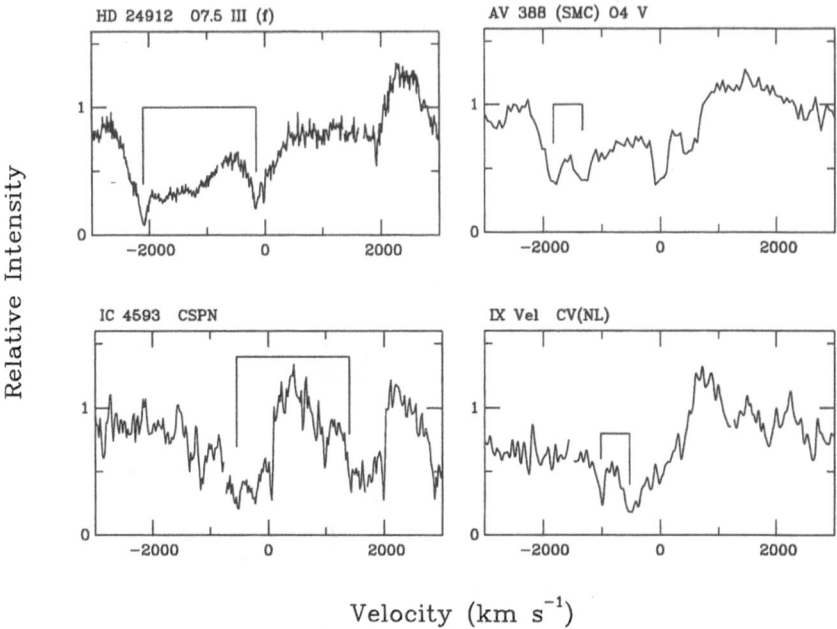

Fig. 1. Examples of 'narrow absorption components' (marked) in the UV wind profiles of hot stars. (Si IV; HD 24912, IC 4593 and C IV; AV 388 and IX Vel) IC 4593 is a planetary nebula central star and IX Vel is a nova-like cataclysmic variable.

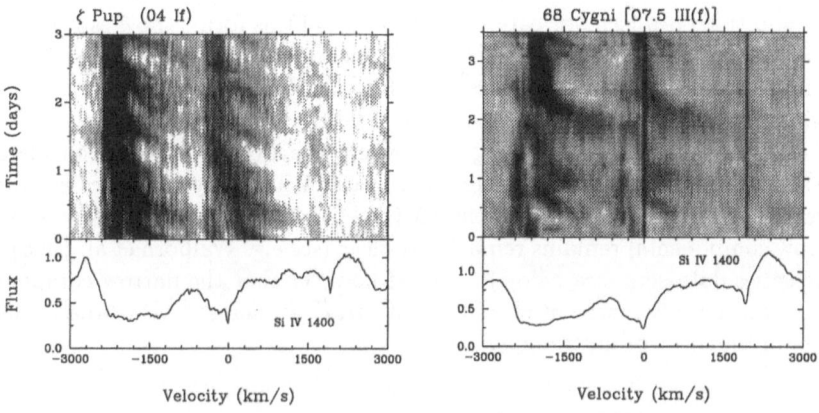

Fig. 2. Grey-scale representations of time variability in the UV stellar-wind profiles of two O stars. Darker shades represent greater absorption. Several episodes of blueward migrating 'discrete absorption components' are evident. These dynamic features diagnose large scale wind structures. (Mean Si IV profiles for the time-series are shown in the lower panels.)

been used to demonstrate that the wind absorption variability is not chaotic, but very systematic (e.g. Prinja et al. 1987; Prinja & Howarth 1988; Henrichs 1988; Kaper 1993).

The incidence of time-dependent wind activity in OB stars is diagnosed by the presence of blueward migrating 'discrete absorption components' (DACs) in the UV absorption troughs. Some examples of this temporal variability are shown in Fig. 2 as grey-scale image representations. The progressive DACs trace substantial wind perturbations. Between about 1980 and 1994, several IUE time-series data sets of individual stars were secured, spanning $\sim 2 - 6$ days. They provided the following key constraints (briefly):

1. Individual DACs usually migrate bluewards over a substantial velocity range, i.e. from less than 0.3 v_∞ to ≥ 0.9 v_∞. Their full-widths decrease from $\sim$0.3 v_∞ to ≤ 0.1 v_∞ during the blueward progression.
2. The structures are substantial, with DACs accounting for $\sim$ 20–50% of the observed absorption, and line central optical depths regularly exceeding 0.5. (Typical Si^{3+} DAC ionic column densities are $\sim 1 \times 10^{14}$ cm^{-2}, or 1.7 $\times$ 10^{21} cm^{-2} relative to hydrogen).
3. The DACs mostly have rather slow observed accelerations, typically varying from $\sim 5 \times 10^{-2}$ km s^{-2} (at lowest central velocities) to less than 1 $\times$ 10^{-3} km s^{-2} near the shortward profile edge. The DAC accelerations are no more than $\sim 50\%$ of the values predicted by steady-state wind models, suggesting that the observed features are due to perturbations in the wind, through which the ambient wind material flows (i.e. we are not tracking the migration of mass-conserving features such as blobs or shells, which are 'riding' with the wind).
4. The UV profile variability is dominated by absorption-trough changes, with the P Cygni emission components remaining relatively stable (this also remains true of HST GHRS data, e.g. Heap 1994). Spherical symmetry is therefore unlikely to be the correct geometry for the wind structures.
5. The $v_e\sin(i)$ connection: A key constraint from these short IUE time-series data sets came from evidence that the characteristic timescale of variability in OB star winds appeared to be related to the rotation period of the stars. The observed accelerations and recurrence timescales of DACs show trends as a function of projected rotation velocity ($v_e\sin(i)$), such that faster-developing, more-frequent DACs are apparent in stars with higher $v_e\sin(i)$ (e.g. Prinja, 1988, 1992; Henrichs et al. 1988; Kaper 1993). These results provided the inspiration and background for the 'IUE MEGA Campaigns' discussed in Section 3.

The optical waveband contains important diagnostics of activity in the inner (denser) regions of hot star winds, including the He I λ5876 and Balmer lines. Time-series results presented elsewhere in these proceedings (see also Henrichs et al. 1994; Fullerton et al. 1992; Prinja et al. 1995a; Reid & Howarth 1996) suggest that; (i) the inner stellar wind (i.e. between $1 - 2R_*$; $v \leq 0.15$ v_∞) is also unstable and affected by coherent evolving structures, and (ii) the growth rate of these perturbations remains large close to the photosphere.

3   The '*IUE* MEGA Campaigns'

The UV results quoted in Section 2 relied on data sets which mostly extended over about 3 – 4 days; they are therefore not suitable for a rigorous investigation of wind activity on rotational timescales. Recognising this limitation, Derck Massa proposed in 1993 August (at the 'Workshop on Instability and Variability in Hot Stars; Moffat et al. 1994) that there should be a major collaborative effort to secure intensive *IUE* time-series data sets for selected stars, extending over several rotational periods. Two major efforts – known as the '*IUE* MEGA Campaigns' – were born out of this initiative, and the principal results from these rich data sets are outlined here. The key aim of these projects was to address whether stellar rotation merely provided a characteristic timescale for wind activity, or it was instead *causally* related to the initiation of wind structure. [Note that, in addition, an independently acquired *IUE* data set of ξ Per is also available in the archives, which also continuously spans several rotation periods of the star; these data are described by Henrichs et al.; these proceedings.]

The two '*IUE* MEGA Campaigns' involved:

1) Observations of HD 66811 (ζ Pup; O4 If(n); $v_e\sin(i) \sim 219$ km s^{-1}), HD 64760 (B0.5 Ib; $v_e\sin(i) \sim 216$ km s^{-1}) and HD 50896 (EZ CMa; WN 5) carried out between 1995 January 13–29, accumulating about 150 high-resolution spectra of each star. The OB star results are summarised here, and have also been reported on by Massa et al. (1995), Prinja et al. (1995b), Howarth et al. (1995), Fullerton et al. (1997), and Howarth et al. (1998).

2) A natural selection effect in the first 'Campaign' was the inclusion of rapid rotators only. The approach of the second program was to monitor winds of stars with more typical $v_e\sin(i)$'s. Observations of HD 91969 (B0 Ia; $v_e\sin(i) \sim 83$ km s^{-1}), HD 96248 (BC1.5 Iab; $v_e\sin(i) \sim 83$ km s^{-1}), and HD 93843 (O5 III, $v_e\sin(i) \sim 95$ km s^{-1}) were secured between 1996 May 3 – June 2. (The run finished barely four months before the termination of spacecraft operations!). Typical sampling intervals were $\sim 8.5 - 10.5$ hours, with about 70 spectra per star. (The B stars from this run are also discussed by Massa; these proceedings.)

3.1   Repeatable Wind Structure

The time-variability of the wind lines in the 'MEGA Campaign' stars ζ Pup, HD 64760, HD 93843 and HD 91969 is shown as dynamic spectra in Fig. 3. (Individual spectra are normalised to the mean spectrum for each time-series in these displays). Note that in each case some degree of repeatability may be identified in the wind activity, but that the characteristics of the profile changes differ greatly from star to star, even within a small range in spectral type. In all these cases two-dimensional Fourier analysis (presenting information on temporal frequency versus spatial velocity) reveals evidence for cyclic or periodic behaviour. The individual spectra for each star, phased on the dominant periods identified, are also displayed in Fig. 3 (uppermost panels), Some of the key results and physical constraints derived from these data sets are summarised below (see also references in Section 3 above).

ζ **Pup; O4 If(n)** - the *IUE* data set of ζ Pup is dominated by two periods; (i) $\sim$ 19.2 hours, which is the recurrence time of the DACs (Fig. 3), and (ii) $\sim$ 5.2 days, which is the same as the estimated stellar rotation period, and reflects a modulation of the entire wind, out to v_∞ (and including v_{edge}). The UV data do not provide any clear trace of the 8.5 hour optical photospheric absorption period (e.g. Baade 1991; Reid & Howarth 1996), which likely arises from nonradial pulsation (NRP). The fact that there is significant power in the wind lines *at* the rotation period of ζ Pup, and that the modulations span over almost the entire wind velocity range, indicates that the wind is influenced on large angular and radial scales once per rotation period. These results have thus led to a resurrection of the 'oblique magnetic rotator' model proposed by Moffat & Michaud (1991) for ζ Pup, invoking a low-order dipole field, off set from the rotation axis (e.g. Howarth et al. 1995).

HD 64760; B0.5 Ib - a striking 1.2-day wind modulation is apparent in the 'MEGA Campaign' data of this star (e.g. Fig. 3, and references in Section 3). The estimated maximum rotation period of HD 64760 is $\sim$ 6.0 days. Individual modulations cover a large velocity range almost instantaneously, occasional exhibiting a bow-shaped morphology in the grey-scales such that blueward and redward extensions are apparent at a give time. In combination with the detection of weak co-variations in the emission components of the P Cygni profiles, these results suggest that the wind structures are azimuthally extended (spiral-like) in HD 64760, and not radial.

The *IUE* observations of HD 64760 may be compared to predictions from models of 'corotating interaction regions' (CIRs) in hot stars (e.g. Mullan 1986; Cranmer & Owocki 1996). The simulations of Cranmer & Owocki (1996) in particular, invoke localised stellar surface regions which may affect the radiative force, leading to high-density, low-speed gas streams. These streams then interact with the faster ambient wind material, yielding a spiral pattern as the star rotates. Interestingly these calculations predict enhanced absorption from extended regions of near uniform velocity, that mark the initial response of the unperturbed wind to the slower moving CIR ahead.

Recently, Howarth et al. (1998) have stressed that the structures in the wind of HD 64760 do not have to corotate with the surface. Interpretations of corotating stellar wind features in this star are essentially based on plausibility arguments only. The ratio of wind modulation and rotation periods, for example, offers no firm information. Using a cross-correlation technique they report on the discovery of a 1.2 day UV *photospheric* period (extracted from the *IUE* MEGA data), which they argue is consistent with a low order ($l = 2$) NRP mode. The wind modulations – seen only once every 1.2 days – may then arise from the leakage of pulsation energy.

HD 93843; O5 III - two-dimensional Fourier analysis provides support for the repeatability of the wind structure in HD 93843 on a 7.1 day timescale (Fig. 3; Prinja et al., in preparation). Power at this frequency is only evident from intermediate to high velocities in Si IV and N V, and in the blue wings of the saturated C IV profile. The rotation, photometric, pulsation and magnetic

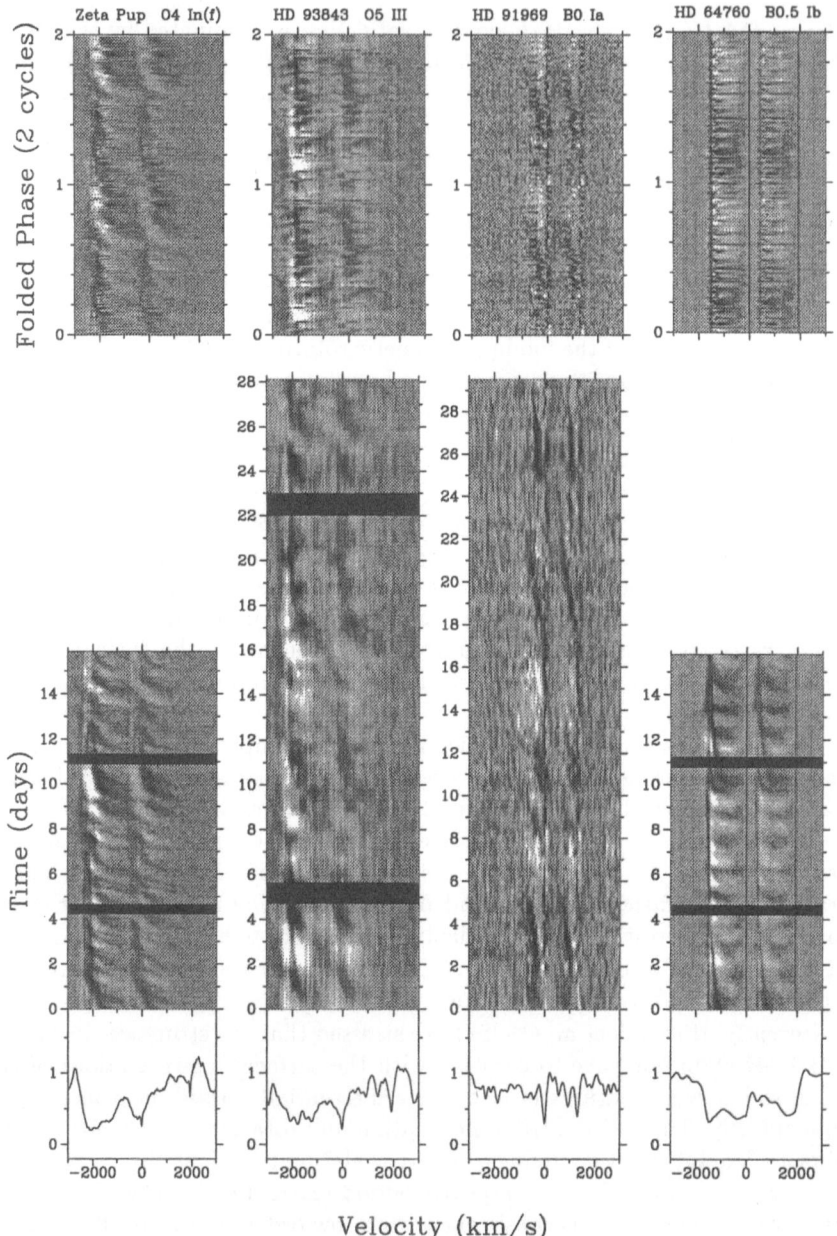

Fig. 3. Grey-scale representations of cyclic variability in the UV wind lines of (from left to right): ζ Pup (Si IV), HD 93843 (Si IV), HD 91969 (Al III), and HD 64760 (Si IV). In each case the uppermost images show the individual spectra phased on the dominant modulation period (see text and Table 1).

properties of HD 93843 are either poorly constrained or unknown! An estimated radius of 14 $R_\odot$ leads to a maximum rotation period $\sim$ 7.5 days. If the 7.1 UV period was in fact the rotation period then a single perturbation or 'feature' at the stellar surface would have to be invoked to produce power primarily at the rotation frequency. (This potential similarity to ζ Pup is also apparent from Fig. 3.)

HD 91969; B0 Ia - Massa (these proceedings) describes in detail the properties of this star, monitored during the second 'MEGA Campaign'. Fourier analysis reveals a 7.8 day period in the wind lines of HD 91969 (Fig. 3); the estimated maximum rotation period is $\sim$ 18 days. The basic result of cyclic variability being present in the wind, may be combined with the following additional points to support the notion that variability in HD 91969 is rooted at the stellar surface: (i) the observed temporal coherency is strongest in the low ion species (e.g. Al III) which likely form in the deeper regions of the wind, (ii) periodic modulations of the wind lines are seen down to very low velocities ($\leq$ 0.05 v_∞), and (iii) variations are present in the 'photospheric' lines (Si III, Fe III, Fe IV) which are broadly in phase with the low-velocity absorption changes in the winds lines. These 'photospheric' changes may then be the tracers of localised surface inhomogeneities responsible for the wind activity.

4 Final Remarks

A summary of the cyclic wind behaviour evident in the 'MEGA Campaign' stars, and in ξ Per (see e.g. Henrichs et al. these proceedings) is given in Table 1. Note that these five OB stars probably have the most extended *IUE* time-series data sets available, and in *each case* some degree of *cyclical* wind modulation is present. The detailed spectroscopic properties differ however. Only in the case of ξ Per, and possibly HD 91969, can the modulations be ascribed to 'classical' DACs (e.g. Section 2.1), with enhancements migrating bluewards to the profile edge. In ζ Pup and HD 64760 the obvious cyclic behaviour is due to other phenomena (global wind changes, or modulations about a well-defined mean state, respectively), although DACs *are* also present in the wind profiles of these stars. In fact, all the OB stars in Table 1 show instances of wind structure co-existing on separate timescales, but their relationship (if any) is unknown.

The principal periods associated with the wind modulations in Table 1 range from $\sim$ 1.2 – 7.8 days, i.e. grossly exceeding the characteristic radial flow times of the winds ($\sim R_*/v_\infty \sim$ hours to 1 day). It is unlikely then that the wind structures are entirely due to physical processes which are intrinsic to the wind. The modulating 'clock' on these timescales must relate to inhomogeneities at the stellar surface. The spectral properties of individual stars discussed in Section 3.1 led to scenarios invoking weak ordered magnetic fields (not yet firmly detected in any normal OB star) or nonradial pulsations or a combination of both processes. Clearly challenging questions remain concerning the precise coupling of variability in the photosphere to the generation of structure in the stellar winds. It seems that given an appropriate and extended enough data set, cyclic wind variations

Table 1. Cyclic UV variability in OB stars winds

Star	Sp.type	Max. rotation period (d)	Wind modulation period (d)	Modulation due to DACs?	Photos. abs. period
ζ Pup	O4 If(n)	5.1	5.2	No	8.54 hr
HD 93843	O5 III	7.5	7.1	?	–
ξ Per	O7.5 III ((f))	2.6	2.1	Yes	16 hr?
HD 91969	B0 Ia	18	7.8	Yes	–
HD 64760	B0.5 Ib	6.0	1.2	No	1.2 d, 9 hr?

(with modulation timescales exceeding several days) are readily detected. If this is indeed the case, then whatever the nature of any 'associated' surface phenomenon, its occurrence would need to be commonplace in OB stars. From an observational prospective, and without the services of the *IUE* satellite, interesting constraints may come from high-quality spectroscopic (and photometric) monitoring of stellar surfaces, continued efforts to measure the magnetic properties of normal hot stars, and the potential offered by 'whole-Earth' campaigns like MUSICOS to monitor optical wind diagnostics over rotational timescales.

Acknowledgements

I am grateful for the support of the Royal Society.

References

Baade, D. (1991): in *ESO Workshop on Rapid Variability of OB-Stars: Nature and Diagnostic Value* (ESO:Munich), 21

Berghofer, T.W., Schmitt, J.H.M.M., Danner, R., Cassinelli, J.P. (1997): A&A, 322, 167

Bieging, J.H., Abbott, D.C., Churchwell, E.B. (1989): ApJ **340**, 518

Carlberg, R.G. (1980): ApJ **241**, 1131

Collura, A., Sciortino, A., Serio, S., et al. (1989): ApJ **338**, 296

Cranmer, S.R., Owocki, S.P. (1996): ApJ **462**, 469

Feldmeier, A., Pauldrach, A., Puls, J. (1998): in *Boulder-Munich II: Properties of Hot, Luminous Stars* (ASP Conf Ser: San Francisco), 278

Fullerton, A.W., Massa, D.L., Prinja, R.K., et al. (1997): A&A 327, 699

Fullerton, A.W., Gies, D.R., Bolton, C.T. (1992): ApJ, **390**, 650

Gathier, R., Lamers, H.J.G.L.M., Snow, T.P. (1981): ApJ, **247**, 173

Gry, C., Lamers, H.J.G.L.M., Vidal-Madjar, A. (1984): A&A, 137, 29

Heap, S.R. (1994): in *Instability and Variability of Hot-Star Winds* (Kluwer: Dordrecht), 87

Henrichs, H.F. (1988): in *O Stars and Wolf-Rayet Stars* (NASA: Washington) 199

Systematic Variability in OB-star Winds   101

Henrichs, H.F., Kaper, L., Nichols, J.S. (1994): in *Pulsation, Rotation and Mass Loss in Early-Type Stars* (Kluwer: Dordrecht), 517

Henrichs, H.F., Kaper, L., Zwarthoed, G.A.A. (1988): in *A Decade of UV Astronomy with the IUE satellite* (ESA: Noordwijk), 145

Howarth, I.D., Townsend, R.H.D., Clayton, M.J., et al. (1998): MNRAS, in press

Howarth, I.D., Prinja, R.K. (1989): ApJS **69**, 527

Howarth, I.D., Prinja, R.K., Massa, D. (1995): ApJL **452**, 65

Kaper, L. (1993): *Wind Variability in Early-type Stars* (PhD thesis, Univ. of Amsterdam)

Lamers, H.J.G.L.M., Gathier, R., Snow, T.P. (1982): ApJ **258**, 186

Massa, D. et al. (1995): ApJL **452**, 53

Moffat, A.F.J., Owocki, S.P., Fullerton, A.W., St-Louis, N. (1994): in *Instability and Variability of Hot-Star Winds* (Kluwer: Dordrecht)

Moffat, A.F.J., Michaud, G. (1991): ApJ **251**, 133

Morton, D.C. (1976): ApJ **203**, 386

Mullan, D.J. (1986): A&A **165**, 157

Owocki, S.P. (1992): in *Atmospheres of Early-Type Stars* (Springer: Berlin), 393

Owocki, S.P., Rybicki, G.B. (1984): ApJ **284**, 337

Owocki, S.P., Rybicki, G.B. (1985): ApJ **299**, 265

Prinja, R.K. (1988): MNRAS **231**, 21P

Prinja, R.K. (1992): in *Nonisotropic and Variable Outflows from Stars* (ASP Conf Ser: San Francisco), 167

Prinja, R.K., Howarth, I.D. (1986): ApJS **61**, 357

Prinja, R.K., Howarth, I.D., Henrichs, H.F. (1987): ApJ **317**, 389

Prinja, R.K., Howarth, I.D. (1988): MNRAS **233**, 123

Prinja, R.K., Fullerton, A.W., Crowther, P.A. (1995a): A&A, **311**, 264

Prinja, R.K., Massa, D., Fullerton, A.W. (1995b): ApJL **452**, 61

Puls, J., Owocki, S.P., Fullerton, A.W. (1993): A&A **279**, 457

Reid, A.H.N., Howarth, I.D. (1996): A&A, **311**, 616

Snow, T.P. (1977): Ap J **217**, 760

Snow, T.P., Morton, D.C. (1976): ApJS **32**, 429

Underhill, A.B. (1975): ApJ **199**, 691

Walborn, N.R., Lennon, D.J., Haser, S.M., et al. (1995): PASP **107**, 104

Linsky: Does the evidence for line profile changes (discrete absorption components etc.) indicate significant changes in the mass-loss rates or structures in a grossly constant mass-loss wind?

Prinja: The DACs etc. are really tracing variable wind structure in a relatively constant (within a factor of 2) mass-loss rate wind. The success of the line-driven wind theory in reproducing the steady-state properties is, I think, a confirmation of this fact.

Ignace: You made reference to the result that DAC recurrence times show a correlation with $v_{rot} \sin i$. Is the dataset sufficient to consider separately (i.e. statistically) the dependence of t_{rec} with v_{rot} and $\sin i$ to obtain some information about the latitudinal properties of the DAC phenomena?

Prinja: The $v_{rot} \sin i$ versus DAC recurrence timescale relation is I think statistically significant in highlighting the role of stellar rotation in hot-star winds. I doubt, however, that the sample (of about a dozen stars) is substantial enough to reliably constrain the v_{rot} and $\sin i$ information.

Kubát: The idea of corotating structures was originally introduced by Harmanec in 1989 for the case of Be stars.

Prinja: It is true that Harmanec discussed the application of CIRs to Be stars in 1989, but I believe that Mullan first discussed CIRs in the context of hot stars in 1986.

Owocki: Regarding to the inferred variations in mass loss, I think it is important to distinguish between the global mass-loss rate and the local mass-flux density. As I understood it, the mass-flux density in, e.g. the solar wind, is remarkably constant between high and low-speed wind. On the other hand, the overall solar wind mass-loss rate variation referred to by Dupree earlier, depends on variations of the fraction of open field during the solar cycle. For hot stars, it seems that the DACs that evolve only very slowly may represent a dense ejection of mass towards the observer, accelerated more slowly by the radiative driving. In contrast, the periodic modulations discovered in the IUE MEGA campaign show a nearly sinusoidal variation with both *decreases* and *increases* in absorption. Here you can imagine that the mean mass loss be really quite constant.

Henrichs: (1) In how many stars you know of one observes "phase bowing" in the UV wind lines? (2) How would a pulsation period show up in the wind line behavior, as you mentioned as a possible explanation for the B supergiant HD64760?

Prinja: (1) I know of 4 stars (all B supergiants) which show clear evidence of "bowed" structures in the UV wind lines. With hindsight, a re-examination of the many short O-star time-series in the IUE archives may reveal further examples; (2) Howarth et al. (1998, MNRAS) report a 1.2d UV photospheric period in HD64760 which is of course the same as the wind modulation period. Their notion is that the wind modulation may then result from leakage of pulsation energy into the supersonic outflow.

Kaper: The phase bowing corresponding to the CIR modulations has also been detected in the O7.5 III star ξ Per (see poster De Jong et al.). However, in the case of ξ Per the phase bowing seems to be connected to the evolution of the DACs itself, contrary to the B0.5 Ib star HD64760 where the DACs do not seem to be related to the CIR modulation. It would be interesting to find out whether this effect is different in O-type stars.

Cyclical Variability in O-Star Winds (The Hα Project)

Lex Kaper[1], Alex Fullerton[2,3], Dietrich Baade[1], Jeroen de Jong[4],
Huib Henrichs[4], Peer Zaal[4]

[1] European Southern Observatory, Karl-Schwarzschild-Str. 2,
   D-85748 Garching bei München, Germany
[2] Universitäts Sternwarte München, Scheinerstr. 1, D-81679 München, Germany
[3] Center for Astrophysical Sciences, Dept. of Physics & Astronomy,
   The Johns Hopkins University, 3400 N. Charles Street, Baltimore, MD 21218, USA
[4] Astronomical Institute, University of Amsterdam, Kruislaan 403,
   1098 SJ Amsterdam, The Netherlands

Abstract. In order to test the hypothesis that hot-star winds are modulated by the rotation of the underlying star, we monitored a large sample of bright O stars for Hα line-profile variations. Most (75%) of the O stars we observed intensively (excluding the known short-period spectroscopic binaries) exhibited line profile variations in Hα. With a few interesting exceptions, the dwarfs in our sample did not show any variability that we could detect, probably because of their low wind densities. In contrast, the supergiants displayed the largest amplitude variations. A period could be determined for 41% of the single Hα-variables, which was in all cases consistent with (an integer fraction of) the expected stellar rotation period.

1 Introduction

In the past decade, extensive monitoring of the UV resonance lines of a dozen bright O stars with the *International Ultraviolet Explorer* has shown that the variability of their stellar winds is not chaotic, but systematic. The variable and blue-shifted discrete absorption components (DACs) are the most prominent features of wind variability and represent a fundamental property of hot-star winds; for recent reviews, see Prinja (1998; this volume) and Kaper (1998). A key point is their regular appearance: it turns out that their (apparent) acceleration and recurrence is faster when the projected rotational velocity of the star is higher. This suggests that stellar rotation determines the cyclical behaviour of the wind variability, and that its origin has to be found close to or at the stellar surface. It must be stressed, however, that the detailed "pattern" of variability is often very complicated. Subsequent DACs can reach different (sometimes alternating) terminal velocities, vary in strength, and in some cases two different, intermingled patterns seem to be present simultaneously. The pattern of variability is, however, repeatable for a given star (e.g. Kaper et al. 1996, 1998a).

The potent instability of radiation-driven winds very likely explains the observed X-ray flux and the extended regions of saturation exhibited by some UV P Cygni lines, but does not seem to be the main cause for the development and evolution of DACs. A more promising explanation for DACs is in terms of "co-rotating interaction regions" (CIRs; Cranmer & Owocki 1996, see also Owocki 1998, this volume). The CIR model, which was originally developed to explain features in the solar wind, invokes fast and slow wind streams that originate at different longitudes on the stellar surface. Streams coming from one surface location are curved due to the rotation of the star, but by an amount that depends on whether the location is a source of "fast" (less curvature) or "slow" (more curvature) wind. The result is that fast wind material collides with slower material emitted earlier from an adjacent longitude, and a shock is formed at the interaction region. The shock "pattern" in the wind is determined by the boundary conditions at the base of the wind and corotates (or nearly corotates) with the star. The physical origin of the local changes in the radiative driving force is not specified in the CIR model, but temperature and velocity fields at the stellar surface due to non-radial pulsations (NRP) or weak magnetic fields (which, however, have up to now remained undetected) are implicated.

Since the demise of IUE, it is practically impossible to obtain continuous time series of UV spectra to study cyclical wind variability for a larger sample of O stars. This is a very important task, since the confirmation of rotational modulation as the cause of the observed variability (instead of an origin solely related to the instability of radiation-driven winds) would represent an important paradigm shift in our understanding of hot-star winds, and would imply that there is an underlying piece of physics (e.g. magnetic fields? NRP?) that still needs to be incorporated in models of hot-star atmospheres.

It has been known for a long time that strong subordinate lines in the optical spectrum (like Hα and He II 4686 Å) are also variable. In a survey of a dozen OB supergiants, Ebbets (1982) found dramatic changes in the shape and strength of Hα, but the time sampling of his study was too irregular to permit the time scales (1 – 10 days) to be estimated reliably. On the basis of coordinated UV and Hα observations of a small number of O stars, Kaper et al. (1997) demonstrated that the variability in the Hα line is also cyclical and directly related to the regular appearance of DACs in the UV resonance lines. Thus, the Hα line provides a good diagnostic of (cyclical) wind variability. Furthermore, Hα preferentially samples the denser regions of the wind close to the photosphere, since the strength of Hα wind emission scales as the square of the local density; it is accessible from the ground; and can be observed for stars in other galaxies.

2   O-Star Hα Survey

Here we report on the results of a large campaign that was organised to obtain time-resolved Hα profiles for a large sample of bright O stars (Kaper et al. 1998b). In total, 71 O stars were observed from ESO, OHP, and La Palma; of these, 38 were monitored intensively. The high signal-to-noise (≥ 200) and spectral resolution ($R \geq 35,000$) permit very small variations in Hα to be detected. As an example, we show in Fig. 1 two time series of Hα spectra. The left panel shows the time series for HD 57682 (O9 IV), which – together with θ^1 Ori C and

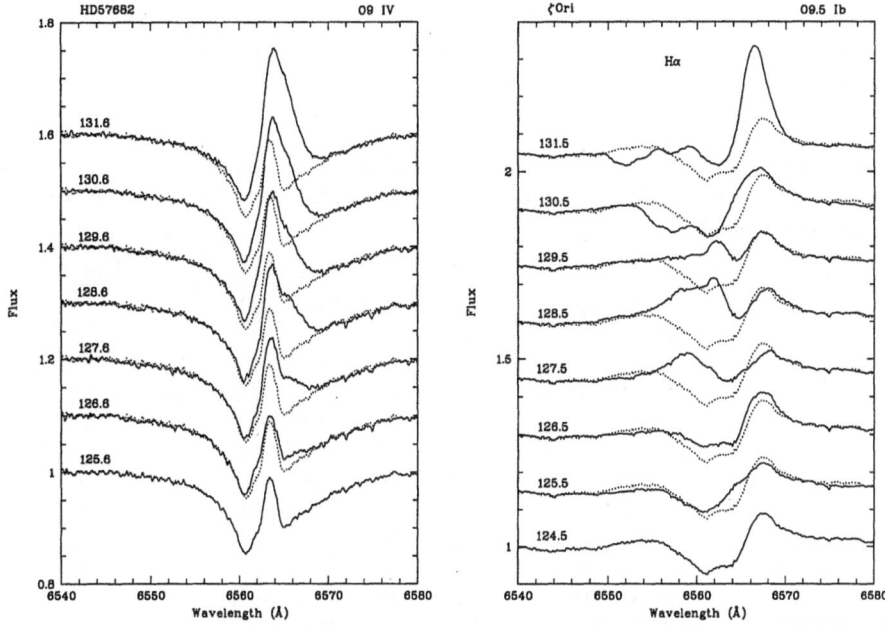

Fig. 1. Left: Spectroscopic time series showing Hα line profile variations for the O9 IV star HD 57682. Time runs upwards with JD-2450000 indicated at the left. The dotted line represents the first spectrum in the time series and is repeated for comparison. Note the strong increase in Hα emission. **Right:** Same for the O9.5 Ib star ζ Ori. Dramatic emission and absorption events are evident.

ζ Oph – belongs to a group of exceptional "unevolved" stars that exhibit Hα variability. The variations consist of steady growth in the emission component and progressive "filling in" of the blue part of the absorption line. These changes occur over a time scale that is longer than the observing run, and may or may not be cyclical.

The right panel shows the dramatic absorption and emission variability exhibited by the Hα profile of ζ Ori (O9.5 Ib). Blue-shifted emission appears at day 127, and slowly evolves to the red. At day 130, blue-shifted absorption is observed again (as at day 124), while the evolution of the emission event to the red continues, resulting in the strongly increased red emission peak at day 131. Fourier analysis of the complete time series (only a portion of which is illustrated in Fig. 1) suggests a period of ∼6 days, which we interpret as half of the full rotation period associated with a high-density region in the stellar wind. This region first appears as blue-shifted emission at the approaching limb of the star and subsequently becomes visible in absorption as it moves across the line of sight to the receding limb of the star. Similar behaviour has been observed in other O supergiants (e.g. α Cam O9.5 Ia; λ Cep O6 I(n)fp).

Table 1. Measured "wind" periods in O-star spectra compared to the estimated maximum rotation period (based on the stellar parameters listed by Howarth & Prinja [1989, ApJS 69, 527] and references therein). Runaway stars are indicated by (R). References: 1) Kaper et al. 1998a; 2) Prinja et al. 1992, ApJ 390, 266; 3) Howarth et al. 1993, ApJ 417, 338; 4) Fullerton et al. 1992, ApJ 390, 650: from He I 5876 Å; 5) Prinja 1988, MNRAS 231, 21P; 6) Stahl et al. 1996, A&A 312, 539; 7) Kaper et al. 1997; 8) Kaper et al. 1998b.

Name	Sp. Type (Walborn)	$v \sin i$ (km s^{-1})	R ($R_\odot$)	$P_{\max}$ (days)	P_{wind} (days) UV	Hα	Reference
ζ Oph (R)	O9.5 V	351	8	1.2	0.9		3
68 Cyg (R)	O7.5 III:n((f))	274	14	2.6	1.4	1.3	1,7
ξ Per (R)	O7.5 III(n)((f))	200	11	2.8	2.0	2.0	1,7
λ Cep (R)	O6 I(n)fp	214	19	4.5	1.3	1.2	1,7
ζ Pup (R)	O4 I(n)f	208	19	4.6	0.8	0.9	2,8
HD 34656	O7 II(f)	106	10	4.8	1.1		1
15 Mon	O7 V((f))	63	10	8.0	> 4.5		1
63 Oph	O7.5 II((f))	80	16	10.2	> 3		5
HD 135591	O7.5 III((f))	65	14	10.9		3.1	8
λ Ori A	O8 III((f))	53	12	11.5	~ 4	2.0:	1
θ¹ Ori C	O6-O4 var	50:	12:	12:	"15.4"	15.4	6
19 Cep	O9.5 Ib	75	18	12.1	4.5	~ 5	1,7
μ Nor	O9.7 Iab	85	21	12.5		6.0	8
α Cam (R)	O9.5 Ia	85	22	13.2		5.6	8
ζ Ori A	O9.7 Ib	110	31	14.3	~ 6	6	1,8
10 Lac	O9 V	32	9	15.3	~ 7		1
HD 112244 (R)	O8.5 Ib(f)	70	26	18.8		6.2	8
HD 57682 (R)	O9 IV	17	10	29.8		> 6	8
HD 151804	O8 Iaf	50	35	35.4		2-3,7.3	4,8

For each time series a temporal variance spectrum (Fullerton et al. 1996) was computed in order to assess the extent and distribution of statistically significant profile variability. The spectroscopic binaries show a characteristic double-peaked profile. Some of the binaries show evidence for colliding winds (cf. Thaller 1997), so that the level of intrinsic wind variability is hard to determine. All 9 of the stars that did not exhibit variability with time scales and amplitudes that we could detect are dwarfs, several of which are known to have variable DACs in their UV resonance lines. This is probably a selection effect due to the low density of the winds of main-sequence stars, which drastically reduces the contribution of the wind to the observed Hα feature and causes any variations about an average state to remain below our detection threshold. Excluding the spectroscopic binaries (15) and the stars that we observed infrequently (18), we are left with 29 (out of 38) that are intrinsically variable. We were able to estimate a "wind" period (or a lower limit) for 12 of these objects. These estimates are listed in Table 1, where they are compared to the period derived from UV resonance lines and the estimated maximum rotation period. The UV and Hα periods are consistent with each other, and are generally close to an integer fraction of the estimated maximum rotation period. These results support the hypothesis that the winds of O stars are rotationally modulated.

References

Cranmer, S.R., Owocki, S.P. (1996): ApJ 462, 469
Ebbets, D. (1982): ApJS 48, 399
Fullerton, A.W., Gies, D.R., Bolton, C.T. (1996): ApJS 103, 475
Kaper, L., Henrichs, H.F., Nichols, J.S., et al. (1996): A&A Supp. Ser. 116, 1
Kaper, L., Henrichs, H.F., Fullerton, A.W., et al. (1997): A&A 327, 281
Kaper, L., Henrichs, H.F., Nichols, J.S., Telting, J.H. (1998a): A&A, submitted
Kaper, L., Fullerton, A.W., Baade, D., et al. (1998b): in prep.
Kaper (1998): in Proc. "Ultraviolet astrophysics beyond the IUE final archive", ESA-SP 413, Eds. Gonzalez-Riestra, Wamsteker, Harris, in press
Thaller, M. (1997): ApJ 487, 380

Prinja: Regarding the nature of the Hα changes: do you see evidence for any localised optical depth features progressing bluewards, perhaps like the ones Alex Fullerton and I detected in He I 5876 Å P Cygni profiles of O8 stars?

Kaper: The only stars for which we detect distinct blue-shifted absorption features in Hα are the supergiants. Our time coverage (typically a few spectra per night) is probably not sufficient to resolve their evolution. However, the migrating features in these profiles are often in emission (see e.g. Fig. 1).

Chochol: What is the range of photometric variability of the stars in your sample?

Kaper: Balona (1992, MNRAS 254, 404) conducted a photometric survey of 16 bright O-type stars. He shows that microvariability (~ 0.01 mag) with a timescale of the order of days is present in all the O supergiants and in some dwarfs.

Cyclic Polarization Variability of Bright O Stars

D. McDavid[1,2]

[1] Limber Observatory, P.O. Box 63599, Pipe Creek TX 78063-3599
[2] Guest Observer, McDonald Observatory, University of Texas at Austin

Several bright O stars were monitored by broadband optical linear polari-metry at the University of Texas McDonald Observatory in conjunction with ultraviolet and optical spectroscopy during a series of international multiwave-length campaigns from 1986 through 1992. With a typical instrumental uncer-tainty of about 0.03% for a single observation, no polarization variability was detected at the 3σ level for any of the program stars. However, two of the stars observed during the 1991 October campaign have small-amplitude periodicities in polarization that are significant because they match those found by Kaper et al. (1997) in the simultaneous optical and ultraviolet spectra.

A CLEANed Fourier periodogram of the polarization of 68 Cygni (Spectral Type O7.5 III:n((f)); $v \sin i = 274 \, \mathrm{km \, s^{-1}}$) shows maximum power at $0.75 \pm 0.03 \, \mathrm{d^{-1}}$ (period $1\overset{d}{.}33$), which is equal to the frequency of both the Hα and Si IV equivalent width (EW) variations. The amplitude of this variation is about 0.045%. Interestingly, the polarization is in antiphase with the Hα EW, exactly as expected in the corotating interaction region (CIR) model as applied to O-star winds (Cranmer & Owocki 1996; Cranmer 1996). Assuming a radius of 14 R$_\odot$ and an inclination of 90°, the rotation period of 68 Cyg is $2\overset{d}{.}59$. This gives two cycles per rotation, also in agreement with the CIR model with two diametrically opposite equatorial bright areas.

The case of ξ Persei (Spectral Type: O7.5 III(n)((f)); $v \sin i = 200 \, \mathrm{km \, s^{-1}}$) is more complicated. Kaper et al. (1997) found that the Hα and Si IV EWs vary with a frequency of $0.50 \pm 0.10 \, \mathrm{d^{-1}}$ (period $2\overset{d}{.}0$) and also noted the presence of another Hα frequency at $1.12 \pm 0.03 \, \mathrm{d^{-1}}$ (period $0\overset{d}{.}89$), which they interpreted as a harmonic. The simultaneous polarimetry shows exactly this "harmonic" frequency, with an amplitude of about 0.025%. Assuming a radius of 11 R$_\odot$, the rotation period of ξ Per is less than $\sim2\overset{d}{.}78$, depending on the inclination. It is possible, then, that ξ Per had only one photospheric bright area and associ-ated CIR structure. This would result in one EW cycle per rotation, but two polarization cycles.

References

Kaper, L., Henrichs, H.F., Fullerton, A.W., et al. (1997): A&A 327, 281
Cranmer, S.R., Owocki, S.P. (1996): ApJ 462, 469
Cranmer, S.R. (1996): Ph.D. thesis, University of Delaware

Confirmation of a 2.3-Day Periodicity in the Wolf-Rayet Star WR 134: A Twin of EZ CMa?

Thierry Morel[1], Sergey V. Marchenko[1], Philippe R.J. Eenens[2],
Anthony F.J. Moffat[1], Gloria Koenigsberger[3], Thomas Eversberg[1],
Gaghik H. Tovmassian[4], Grant M. Hill[5], Octavio Cardona[6], and Nicole St-Louis[1]

[1] Université de Montréal, Canada
[2] Universidad de Guanajuato, México
[3] Instituto de Astronomía, UNAM, México
[4] Instituto de Astronomía, Ensenada, México
[5] McDonald Observatory, Fort Davis, TX, USA
[6] Instituto Nacional de Astrofísica, Optica y Electronica, Puebla, México

We confirm the existence of the $2\overset{d}{.}25 \pm 0\overset{d}{.}05$ periodicity in the line-profile changes of the apparently single Wolf-Rayet star WR 134 that was first proposed by McCandliss et al. (1994). This period dominates the skewness, centroid, and FWHM variations of He II $\lambda4686$ in spectra obtained in 1992 and 1993. Furthermore, the line-profile changes demonstrate a coherent, although complex, pattern of variability when phased with this period. Loss of coherency on a ~monthly time scale is also observed.

Although definitive statements regarding the nature of the variability must await a more detailed analysis of the data, two mechanisms can be proposed in order to account for the periodic nature of the variability:

1. the presence of a collapsed companion (neutron star or black hole) orbiting in the Wolf-Rayet wind (Antokhin & Cherepashchuk 1984).
2. the existence of large-scale wind structures that persist for ~one month in a rotating single star, as is likely for the peculiar WN 5 star EZ CMa (Morel et al. 1997; 1998).

References

Antokhin, I.I., Cherepashchuk, A.M. (1984): Sov. Astron. Lett. 10, 155
McCandliss, S.R., Bohannan, B., Robert, C., Moffat, A.F.J. (1994): Ap&SS 221, 155
Morel, T., St-Louis, N., Marchenko, S.V. (1997): ApJ 482, 470
Morel, T., St-Louis, N., Moffat, A.F.J., et al. (1998): ApJ, in press

No Need for a White & Becker Disk-Like Wind in WR 140

A.M.T. Pollock

Computer & Scientific Co. Ltd., 230 Graham Road, Sheffield S10 3GS, England

Fig. 1 shows the beautiful observations obtained by White & Becker (1995, ApJ 451, 352) of the non-thermal radio emission of the colliding WC7+O4.5 winds of WR 140 through its 7.94-year orbit. The radio peak near phase $\phi = 0.8$ and the high expected free-free absorption led them to argue that the conventional models of spherically symmetric winds of Stevens et al. (1992, ApJ 386, 265; hereafter SBP) do not apply and that, instead, most of the Wolf-Rayet mass loss is confined to a plane. Consider, however, the details of SBP's model for WR 140. Because the stars are far apart, the shocked interaction region between the winds is both wide and hot. The radio emission peaks when the line of sight lies along the cone of shocked, hot gas, where the free-free absorption is much reduced and any radio sources near the stagnation point can be seen. Separate Wolf-Rayet and O-star components seem to be visible. There are eight different potential opportunities to observe near binary-phase, ϕ, along the length of shocked !WR(ϕ)! or !O(ϕ)! gas (where the notation "! !" designates shocked gas) moving left or right, towards or away, along the sides of the interaction cone. The periastron/apastron wind density ratio of about 150 ensures very heavy modulation of the free-free absorption. !WR(0.75)! was seen at 2 cm and 6 cm but not at 20 cm because of high absorption; !O(0.84)! was seen at 2 cm, 6 cm, and 20 cm; neither !WR(1.)! or !O(1.)! were seen due to high absorption and rapid variability; !O(0.02)! and !WR(0.09)! just made an appearance at 2 cm because the absorption falls as the stars retreat after periastron. Far from denying spherical symmetry, the radio data appear to support SBP's models. A full account of this work is in preparation by Pollock et al. (1998).

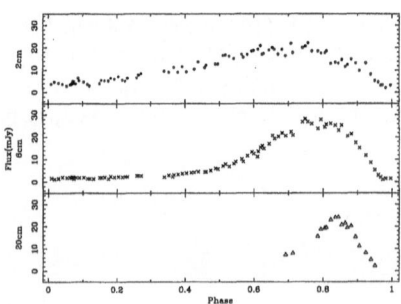

Fig. 1. White & Becker's (1995) radio measurements of WR 140.

Phase Shifts Between Cyclical Spectral Parameters of P Cygni

Indrek Kolka

Tartu Observatory, EE2444 Tõravere, Estonia

The results of measurements of three typical spectral lines of hydrogen covering two time-intervals (1981–83 and 1989–91) are presented in Fig. 1.

The right-hand graph gives an example of the variability in the selected lines. The measured spectral line parameters (the blue-shifted velocity of the absorption maximum or absorption core, V_r; the residual flux in the absorption core, R_c; and the flux at the emission peak, F_e) are marked in boldface. The variability of profile shapes is indicated by the separation of three different profile types.

In the left-hand graph the main result of our investigation is illustrated. The obvious time delay between cycles of the different parameters is indicated by thick lines connecting the critical phases (maxima or minima) in these cycles.

The explanation of this phenomenon could be the migration of a distinct zone in the stellar wind from the inner part at lower expansion velocities towards the outer layers at higher velocities. The zone should become observable cyclically every hundred or more days, and should be characterized by enhanced hydrogen opacity.

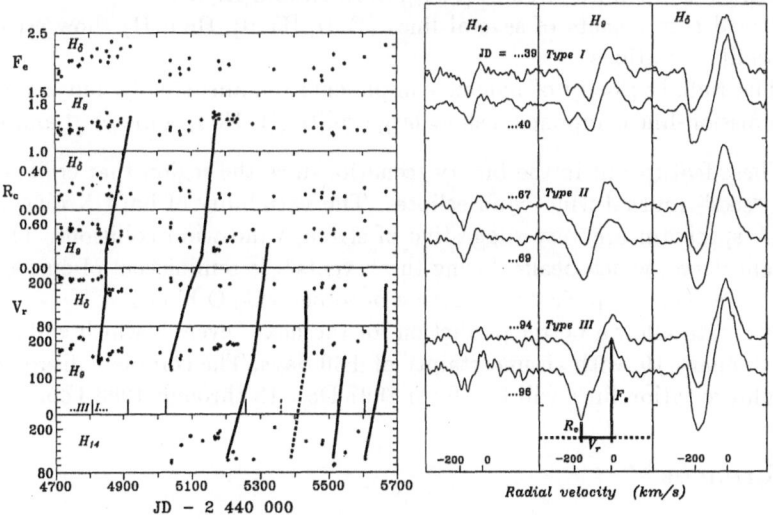

Fig. 1. Spectral parameters of P Cygni in 1981–83 and 1989–91.

Long-Term Spectroscopic Variability of η Carinae

Otmar Stahl[1] and Augusto Damineli[2,3]

[1] Landessternwarte Heidelberg Königstuhl, Germany
[2] JILA, University of Colorado and NIST, Boulder, CO, USA
[3] Inst. Astr. e Geof. da Universidade de São Paulo, São Paulo, Brazil

Damineli (1996) and Damineli et al. (1997) recently discovered a possible strict periodicity of 5.52 years in the intensity of the high-excitation lines of η Car and the radial velocity variations associated with these cycles. These observations suggest that η Car is a long-period binary.

We analyzed a series of high-dispersion spectra to examine the features of this "spectroscopic event", which are:

1. High-excitation lines disappear completely ([S III], [Ar III], [Ne III], [Fe III], [N II]). The same occurs for some lines excited by fluorescence (e.g., Mn II).
2. Narrow-line components become fainter approximately in proportion to the level of excitation, ranging from complete disappearance (H, He I, Si II, Fe II) to a small decrease (some Fe II, [Fe II]).
3. Some lines become *stronger* during the event, like the broad component of some Fe II and [Fe II] lines, and lines of Ti II.
4. Some emission lines remain almost unchanged, e.g., Na I and some narrow line components.
5. Strong, violet-displaced absorption components (P Cygni components?) appear in some lines of Fe II, [Fe II], Si II, H, and He I.
6. Broad components of several lines (Fe II, [Fe II], He I, H) show strong line profile variations.
7. The radial velocity of narrow components remains steady, but some broad emission-line components show large shifts (H, He I, some Fe II and [Fe II]).

These features fit in the binary scenario, since the region that emits nebular lines "sees" mutual wind-wind eclipses. The variability of hard X-rays and the ASCA spectrum are very suggestive of strong wind-wind collisions. The near-IR light-curve, which peaks during the "events", is reminiscent of episodic dust formation near the periastron passage of some WC+O binaries.

Our data allow a better prediction for the next "event", which will occur on 1998 January 19, with an uncertainty of ±50 days. The complete disappearance of high-excitation lines will last from 1997 Dec. 18 through 1998 Feb. 1.

References

Damineli, A. (1996): ApJ 460, L49
Damineli, A., Conti, P.S., Lopes, D.F. (1997): New Astronomy 2, 107

A Kinematic Study of the LBV Nebula around η Carinae

Kerstin Weis[1,2], Wolfgang J. Duschl[1,3], and You-Hua Chu[2]

[1] Institut für Theoretische Astrophysik, Tiergartenstr. 15, 69121 Heidelberg, Germany
[2] Astronomy Dept., University of Illinois, 1002 W. Green Street, Urbana, IL 61801, USA
[3] Max-Planck-Institut für Radioastronomie, Auf dem Hügel 69, 53121 Bonn, Germany

The most massive stars, those with ZAMS masses above $50\,M_\odot$, evolve into Luminous Blue Variables (LBVs) within 3×10^6 years (Langer et al. 1994) when they reach the empirical upper boundary in the Hertzsprung-Russell diagram (which is known as the Humphreys-Davidson limit; Humphreys & Davidson 1994). Winds with high mass-loss rates ($\sim10^{-4}\,M_\odot\mathrm{yr}^{-1}$) and giant eruptions during the short (25,000 years) LBV phase lead to the formation of nebulae around these objects, the so-called LBV-nebulae (LBVN).

The star η Carinae is one of the best known LBVs, and it was found to be surrounded by a nebula, historically called "the Homunculus". The Homunculus nebula is the result of an unstable LBV phase of η Car about 150 years ago, when part of the star's envelope was peeled off and formed the LBVN. Hubble Space Telescope images reveal the almost perfectly bipolar shape of the Homunculus. The first deep images in Hα (F656N) show an overwhelming number of tiny filamentary structures outside the bipolar lobes.

We obtained high-resolution spectra (FWHM 14 $\mathrm{km\,s}^{-1}$) with a long-slit echelle spectrograph and identified spectroscopic features with the knots and structures seen in deep HST images. By comparing morphology and kinematics we obtained our first results:

1. about *200 individual features* were identified in the spectra;
2. many features have kinematical and morphological sub-structures;
3. radial velocities range from -1217 $\mathrm{km\,s}^{-1}$ to $+1963$ $\mathrm{km\,s}^{-1}$;
4. most features have radial velocities between -400 $\mathrm{km\,s}^{-1}$ and $+600$ $\mathrm{km\,s}^{-1}$;
5. the largest *velocity spread* of a single coherent feature is 1250 $\mathrm{km\,s}^{-1}$;
6. the features follow the *bipolar symmetry of the Homunculus*: the south-eastern features are approaching us, while the north-western features are receding.

References

Langer, N., Hamann, W.-R., Lennon, M., et al. (1994): A&A 290, 819
Humphreys, R.M., Davidson, K. (1994): PASP 106, 1025

Cyclic Variability in BA-Type Supergiants

Andreas Kaufer

Landessternwarte Heidelberg, Königstuhl 12, D–69117 Heidelberg, Germany

Abstract. From extended spectroscopic monitoring programs with high resolution in wavelength and time a new picture of the circumstellar environment and the photospheric velocity fields of late B- and early A-type supergiants is developed. Four major conclusions result from time-series analysis of the dramatic and to date unknown profile variations of the wind-sensitive Hα line and the complex variations of photospheric line profiles:

a) the extended envelopes show strong deviations from spherical symmetry as seen in the characteristic cyclical V/R variations superimposed on the wind line profiles, b) rotational modulation plays an important rôle in explaining the observed Hα variability and therefore, rotating magnetic surface structures modulating the lower wind region are proposed, c) the envelopes temporarily display extraordinarily large and extended rotating circumstellar structures observable as suddenly appearing, highly blue-shifted, and strong absorption features. In one case it was possible to derive the true corotation period of the structure to predict and observe its reappearance after four rotational cycles, d) the profile variations of the numerous photospheric lines indicate multiperiodic radial and non-radial pulsations in the photospheres with periods clearly distinct from the time scales observed in the circumstellar envelope. Therefore, no obvious causal connection between the photospheric and the circumstellar variations could be established.

1 Introduction

The late B- and early A-type supergiants are evolved massive ($\approx 20\,\mathrm{M_\odot}$) luminous ($\approx 10^5 \mathrm{L_\odot}$) stars at the hot end of the transition zone between hot and cool supergiant stars. UV observations of resonance lines from low-ionized species indicate accelerating outflows with terminal velocities of several hundred $\mathrm{km\,s^{-1}}$. These outflows are interpreted as due to spherically symmetric mass loss via a radiation-driven wind with a typical mass-loss rate of $10^{-8}\,\mathrm{M_\odot yr^{-1}}$. The BA supergiants are well known to be photometrically and spectroscopically variable showing semiperiodic brightness and radial-velocity variations on time scales from days to several months (e.g. Burki 1978, Rosendhal 1970). In addition, peculiar variability of the wind-sensitive Hα line (unresolved in time) was observed which was early interpreted as variable mass loss and deviations of the spherical symmetry of the envelope (Wolf & Sterken 1976).

2 Observations

Due to the long time scales involved, no homogeneous spectroscopic data set with adequate sampling and coverage in wavelength and time was available for a detailed examination of the structure and variability of the BA supergiant's extended atmospheres. To overcome this lack of observational material, extended monitoring campaigns on a small sample of bright BA supergiants (Tab. 1) were carried out with a fiber-linked echelle spectrograph (HEROS) mainly mounted at the ESO 50-cm telescope on La Silla, Chile. The HEROS spectrograph provides a resolving power of $R = 20,000$ and covers the wavelength range from 3500 to 8700 Å in one exposure. In every year from 1993 to 1996 about 100 spectra per object were obtained in observing runs of up to 120 nights length. A detailed description of the instruments and the campaigns can be found in Kaufer (1998).

Table 1. Program stars of the BA supergiants monitoring campaigns

Object	spectral type
HD 91619	B7 Ia
HD 34085 (= β Ori = Rigel)	B8 Ia
HD 96919	B9 Ia
HD 92207	A0 Ia
HD 100262	A2 Ia
HD 197345 (= α Cyg = Deneb)	A2 Ia

3   Circumstellar Variability

Figure 1 shows dynamical spectra of the wind sensitive Hα line for two typical representatives of the BA supergiants: HD 92207 (right) which in the time average shows a distinct P Cygni-type wind profile indicating a strong radially accelerated outflow and β Ori (left) where no clear P-Cyg profile is present but a photospheric absorption profile filled in by wind emission, roughly up to the continuum level. The crucial finding regarding the complex variability patterns in *all* examined program stars is that – independent of the average shape of the Hα profile – the variability is concentrated symmetrically around the system velocity with the maximum power of the variations just beyond $\pm v \sin i$ (Kaufer et al. 1996a). The variations are mainly due to additional violet (V) and red (R) shifted emission components superimposed on the otherwise constant underlying wind or even photospheric profiles. This is clearly seen in the case of HD 92207 (Fig. 1, right) where the red P Cyg emission peak is modulated in height and the blue shifted P Cyg absorption is filled in with a blue-shifted emission peak of variable strength. The same type of V and R emission peak modulation is seen in β Ori (Fig. 1, left). The amplitudes of the modulations are measured for all program stars in the corresponding "temporal variance spectra" (TVS) and give equal amplitudes for the V and R peaks with values between 5% and 20% which are characteristic for the individual objects. This characteristic V/R

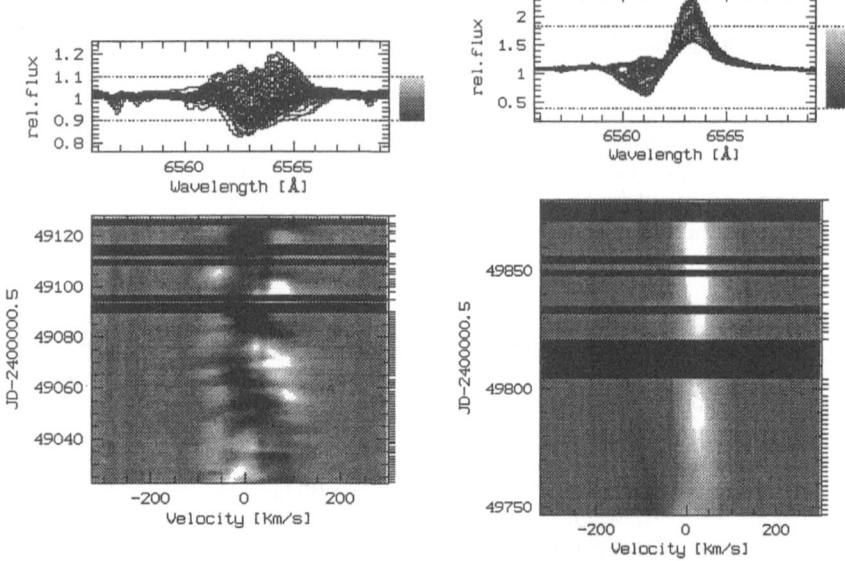

Fig. 1. Dynamical Hα spectra of β Ori in 1993 (left) and HD 92207 in 1995 (right) showing the typical V/R variability caused by rotational modulation of the envelope.

variability is indicative of deviations from spherical symmetry in the circumstellar envelopes of BA supergiants. Further – in analogy with the Be stars – this double-peaked emission variability suggests an equatorial concentration of the emitting circumstellar material.

The measured equivalent width curves of the variable "excess" emission directly show the observed cyclical variability. Time-series analyses of the equivalent-width curves reveal for all examined stars basically *one* dominant period which is in all cases close to the rotational period estimated from $v \sin i$. Therefore, rotational modulation of the lower circumstellar envelope by stellar surface structures is proposed as the major source of the observed Hα variability. An efficient coupling of the stellar rotation into the lower wind region could be provided by weak magnetic surface structures.

Further insight into the circumstellar structures in BA supergiants was obtained by the occasional observations of so-called "high-velocity absorptions" (HVA) which were for the first time identified and described in detail by Kaufer et al. (1996b). In Fig. 2 the two most prominent and so far best documented HVAs are shown as observed in the dynamical Hα spectra of β Ori (left) and HD 96919 (right). HVAs are described as suddenly appearing, extraordinarily deep, and highly blue- but also beyond $+v \sin i$ red-shifted absorptions in the circumstellar envelope. The appearance of the HVAs is announced either by V or $V + R$ emission peaks as known from the "normal" Hα variability. The sudden

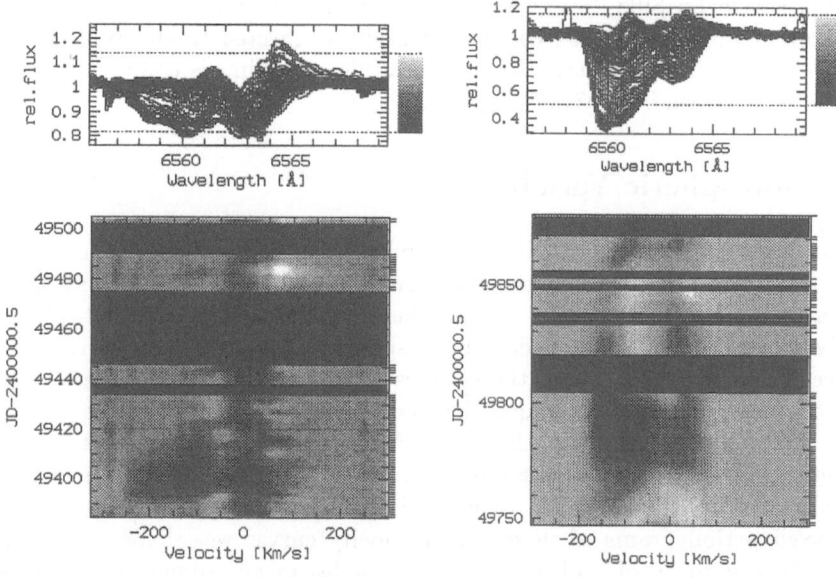

Fig. 2. Dynamical Hα spectra of β Ori in 1994 (left) and HD 96919 in 1995 (right) showing the two best observed HVA events.

appearance within a few days of the blue shifted absorptions in the velocity range $v_{sys} \to v_{\infty}$ is not compatible with the typical wind-flow times and, therefore, the HVAs *cannot* be explained as discrete mass-loss events at the base of the wind with a subsequent β-type acceleration of the "blob" within the ambient wind. In addition, the extreme depth of the absorptions suggests that the HVAs are caused by extended regions in the envelope which consist of dense and therefore probably recombined circumstellar structures. As a model to describe the observed HVAs, a rotating region of enhanced mass loss localized at the base of the wind near the stellar equator was proposed. The continuously injected material is radially accelerated according to a slow β-type velocity law and, due to the conservation of angular momentum, builds up an extended rotating streak line in the equatorial plane of the envelope. This simple model primarily accounts for the sudden appearance of the HVA because the streak line rotates into the line of sight of the observer and the complete velocity field $v_{sys} \to v_{\infty}$ becomes visible as extended and – since the observer looks along a large column of enhanced density – also deep absorption against the stellar disk. Assuming that the large streak-line structures are stable for at least a few stellar rotation cycles, this model predicts the reappearance of the HVA events after integer multiples of the stellar rotation period. For the case of the extreme HVA event in HD 96919 in 1995 the reappearance dates of the maximum blue absorption were predicted with the derived stellar rotation period of 93 days. And indeed, an HVA event with a strikingly similar time-velocity structure was observed during the 1996

HEROS run in La Silla – exactly after four rotational cycles. In this sense the HVA event in β Ori in 1994 could be causally connected to the even stronger HVA event observed 108 days before[1] since this cycle time is comparable to the estimated $P_{\rm rot}/\sin i$ of 107 days for β Ori.

4   Photospheric Variability

Apart from the few optical lines that probe the circumstellar envelope, the HEROS wavelength range contains numerous metallic absorption lines of different strength and·formation depth. Therefore, these lines are well-suited to study the variability of the deeper photosphere but also of the transition zone between the photosphere and the circumstellar envelope. By the use of an efficient cross-correlation technique the photospheric variability of the program stars was studied in subsets of the metallic lines with increasing formation depth (Kaufer et al. 1997). For all line groups, complex multiperiodic variations of the radial velocities with a typical velocity dispersion of $3\,{\rm km\,s}^{-1}$ were found. The CLEANed periodograms of these radial-velocity curves reveal for all program stars the excitation of multiple pulsation modes in accordance with the results found by Lucy (1976) for α Cyg. The periods detected in the photospheric spectra are found to be clearly different from the much longer stellar rotation periods, which raises the possibility to distinguish pulsation from rotation as the source of circumstellar variability. The fact that the power spectra do not exactly reproduce from year to year heavily complicates any identification of the pulsation modes. The finding of periods longer and shorter than the estimated radial fundamental pulsation periods at least suggests the excitation of non-radial and radial overtone modes. A closer inspection of the photospheric "line-profile variations" (LPVs) of all studied stars shows occasionally prograde travelling features with crossing times over the profile considerably shorter than the estimated stellar rotation periods. These features are identified as azimuthally running non-radial pulsation patterns, most probably g-modes of low order ($l = |m| \leq 5$). Figure 3 illustrates the described photospheric variability showing the LPVs of the Si IIλ6347 line in HD 92207. The comparison of this photospheric variability pattern with the circumstellar variations in Fig. 1 (right panel) demonstrates the above mentioned separation of the respective time scales of the line-profile variability.

5   Discussion

From extended spectroscopic monitoring programs with high resolution in wavelength and time a new and refined picture of the circumstellar environment and the photospheric velocity fields of late B- and early A-type supergiants has been obtained. At the current state of the evaluation of this unique data set the

---

[1] For a more detailed discussion of this HVA event cf. the contribution by G. Israelian, this volume.

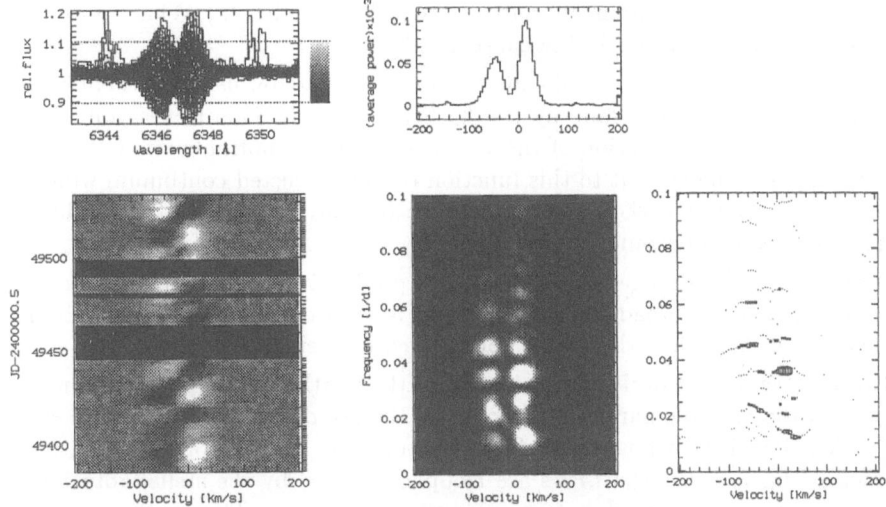

Fig. 3. Left: Dynamical spectrum of the photospheric Si IIλ6347 line (quotient with the mean spectrum); middle: corresponding power spectrum; right: CLEANed periodogram, the estimated frequency of the radial fundamental pulsation is 0.05 days^{-1}.

emerging scenario is a bipartite one: on one hand, the interaction of mass loss, stellar rotation, and magnetic surface structures is found to be the major source of a highly variable lower circumstellar envelope. On the other hand, multi-mode radial and non-radial pulsations are found to cause complex photospheric velocity fields. To date, no causal connection of this photospheric variability to the envelope's variability could be established mainly due to the fact that both parts of the atmosphere show variability on very different time scales. Therefore, also the possibility of a non-interacting coexistence of the observed phenomena should be considered in the current discussion on the cyclical wind variability of luminous hot stars.

References

Burki G. (1978): A&A **65**, 357
Kaufer A., Stahl O., Wolf B., et al. (1996a): A&A **305**, 887
Kaufer A., Stahl O., Wolf B., et al. (1996b): A&A **314**, 599
Kaufer A., Stahl O., Wolf B., et al. (1997): A&A **320**, 273
Kaufer A. (1998): In: *Reviews in Modern Astronomy* **11**
Lucy L.B. (1976): ApJ **206**, 499
Rosendhal J.D. (1970): ApJ **159**, 107
Wolf B., Sterken C. (1976): A&A **53**, 355

Kubát: (1) How can you guarantee that the continuum level in the Hα line is the same for different spectra? According to my experience it is very difficult to find a proper continuum for such a broad line as Hα; (2) What causes the several percent variations in the continuum?

Kaufer: For a fiber-fed echelle spectrograph as used by us, the echelle-ripple correction is accurately done by the flatfielding alone. This leaves you with a very slowly varying function of the wavelength for the normalization. The normalization is done by a fit to this function through selected continuum windows (resulting accuracy 1-5%). Apart from this: the variable part of the Hα line in these objects is only some 15 Å wide.

Lucy (by meaning): The photospheric line profiles in A supergiants are known to show additional broadening mechanisms apart from rotation. How does this macroturbulence affect the values you derived for $v \sin i$?

Kaufer: We do not derive the $v \sin i$ from the width of the lines but from the width of the variable part of the profile by the use of "temporal variance spectra". The idea is that possible surface structures like spots or pulsation patterns which probably cause the LPVs are Doppler-mapped by the stellar rotation to a maximum velocity which is to be identified with $v \sin i$. This method assumes that an azimuthal velocity component of a surface structures is small compared to the value of $v \sin i$.

Prinja: Did you also detect *infall* in the case of β Ori? And how does this fit in with your CIR picture for late B supergiants?

Kaufer: Together with the strengthening of the blue-shifted absorption during a HVA event, we indeed see absorption red-shifted well beyond $+v \sin i$. This cannot be easily understood with the stream-line model. I refer to the contribution by Israelian (this proceedings) for further explanation.

Owocki: Do you mean to imply that the recurring Hα variation you see is to be identified with stream material far from the star? If so, I don't see how this can be so strong in a ρ^2 diagnostic like Hα, since material from the star should expand and so have a low density.

Kaufer: We don't claim to see the same material reappearing after one revolution which would have expanded and diluted - as you just said. We think that we see a streak line filled with denser material rotating into the line of sight.

Puls: Hα in A-supergiant winds behaves differently compared to O-stars, since here the second level becomes the effective ground state. In consequence, it behaves more or less like a resonance line.

Hao: How do you distinguish pulsations from rotational modulation?

Kaufer: We have two arguments: (i) the periods found in the Hα lines are about a factor of 5 larger than the periods found in the variation of the photospheric radial velocity; (ii) the photospheric line-profile variations show prograde travelling features with crossing times from $-v \sin i \rightarrow +v \sin i$ of the order of the break-up rotation period. Therefore, we identify these features as NRP modes and not as rotating surface features.

Rotational Modulation of B Supergiant Winds

D. Massa[1], A.W. Fullerton[2], and R.K. Prinja[3]

[1] Hughes STX, GSFC, Code 631, Greenbelt, MD 20771
[2] Universitäts Sternwarte München, Scheinerstraße 1, D–81679 München, Germany
[3] Dept. of Physics & Astronomy, University College London, Gower Street,
London WC1E 6BT, UK

Abstract. We present a 30 day *IUE* time series of the B0 Ia star HD 91969, a member of the Carina open cluster NGC 3293. We show that wind lines which probe more deeply into the wind vary more regularly and that the lowest stages of ionization and the Si III $\lambda\lambda1300$ triplets have a dominant period of ~ 7.9 days. The photospheric lines (primarily Fe IV) also vary regularly, except with a period of 3.95 – half the dominant wind line period. Further, the photospheric lines *do not* vary at 7.9 days. While these results show a clear relationship between the wind and photospheric variability, the physical nature of the connection remains elusive.

1 Introduction

The link between the incidence and strength of wind activity and apparent stellar rotational velocity, $v \sin i$, has been known for some time (Prinja 1988). This link was demonstrated to be causal by the *IUE* MEGA campaign (Massa et al. 1995) in which the rapidly rotating B supergiant HD 64760 was observed for 15 consecutive days (Prinja et al. 1995). At about the same time, Kaufer et al. (1996) published their study of luminous late B and early A supergiants which also display rotational modulation at Hα.

One problem with the MEGA campaign results is that they apply to rapidly rotating stars, leaving open the question of whether the observed rotational modulation is typical of all B supergiants or only rapidly rotating stars. To test this question, we observed a B0 Ia with a typical $v_{eq} \sin i$ during the final episode of *IUE*.

2 The Program Star

We observed the B0 Ia HD 91969 in the Carina open cluster NGC 3293. This star is a bit more luminous than optimal for wind studies, but it was our only option given the severe operating constraints during the final months of *IUE*. Because of its cluster membership, its M_v is known from main sequence fitting of the cluster B stars. Curiously, even though the star is classified B0 Ia by Walborn (1976), its absolute magnitude is closer to a B0 Iab. Table 1 lists the properties of HD 91969, along with references. A range for the derived quantities, $\log L/L_\odot$, $R/R_\odot$ and P_{max}, is given. The uncertainty results from adopting either the Humphreys & McElroy (1984) or the de Jager & Nieuwenhuijzen

(1987) temperature calibration. Because HD 91969 has a typical $v_{eq} \sin i$, its $\sin i$ and, hence, its actual rotation period are poorly constrained. The maximum period is between 14.0 and 18.4 days, and if we assume that the star would be spectroscopically peculiar if it had a v_{eq} much greater than 200 km s^{-1}, a minimum period of 4.6 – 7.9 days results.

Table 1

Sp Ty	B0 Ia	Walborn 1976
$v_{eq} \sin i$	83 km s^{-1}	Howarth et al. 1996
M_v	-6.3 mag	Turner et al. 1980, Feinstien &
Distance	2.6 kpc	Maraco 1980, Shobbrook 1983
$\log L/L_\odot$	5.5 dex	Humphreys & McElroy 1984,
$R/R_\odot$	22.9–31.6	de Jager & Nieuwenhuijzen,
P_{max}	14.0-18.4 d	1987

3   Results

We obtained 82 spectra over 29.6 days in May-June 1996 with a mean sampling of 8.6 hours and no major gaps in the time series. This time series spans at least 2 stellar rotation cycles. The spectra were reduced using the IUEDR package described by Giddings & Rees (1989). Figure 1 shows selected lines from the series as dynamic spectra normalized by the mean profile for the series.

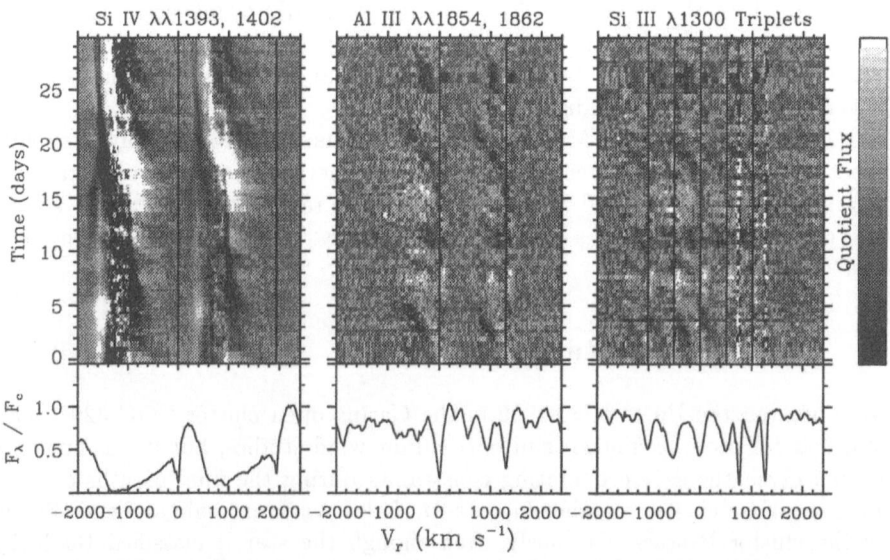

Fig. 1. Dynamic spectra of selected lines in HD 91969. Notice that periodicity is not obvious in the saturated Si IV doublet, but is progressively more apparent in the weak Al III resonance doublet and the Si III triplets.

Several aspects of the time series shown in Figure 1 are immediately apparent. First, variability extends to $v \sim -1800$ km s^{-1} in the strongest wind lines; second, there is activity to $v = 0$ km s^{-1} in Al III and the Si III triplets, and; third, as is often the case, two distinct, simultaneous forms of variability are present – strong, long period (~ 20 days) variability at high v in the high ions and regular variations at low v in the low ions.

In addition to the wind variability, we also noted activity in the strongest Fe IV photospheric lines. This was analyzed by cross-correlating the individual spectra with a template of 201 strong Fe IV lines lying between 1415 and 1845Å (omitting the wind line regions). The result is shown as a dynamic spectrum of the dispersions about the normalized mean profile in Figure 2. Notice that the mean profile has the expected position and width, but that the variability is primarily on the red portion of the line, centered at $v \sim +30$ km s^{-1}. Furthermore, some features can be seen to move from blue-to-red, as expected for surface features.

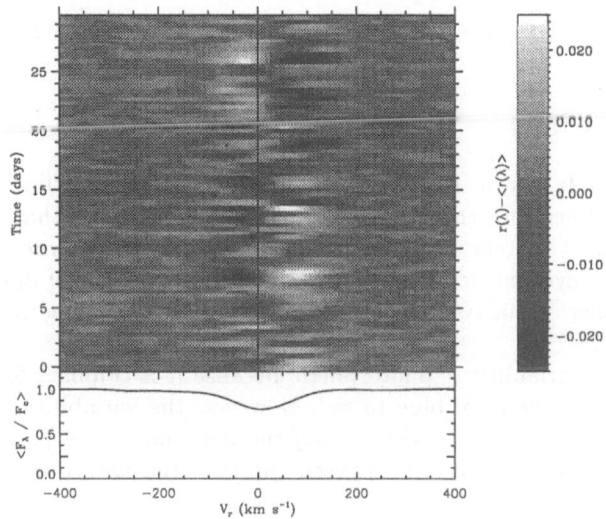

Fig. 2. Dynamic spectrum of the cross-correlation of the spectra and a template of Fe IV lines. The mean line is shown at the bottom.

4   Periodic Behavior

Figure 3 displays the amplitudes of the cleaned Fourier transforms, FTs, of the wind lines (see, Fullerton et al. 1997). There is a strong peak at ~ 7.9 days in Al III (also visible in the raw data in Figure 1), C II and the Si III triplets. The high ions do not repeat exactly, but all do have a peak in their power at ~ 8 days. Because $\sin i$ is so poorly constrained and since P_{rot} is so poorly determined, it is difficult to decide whether the 8 day period in the wind corresponds to a disturbance which occurs once or twice per revolution.

Figure 4 shows the amplitudes of the cleaned FTs of the Fe IV data. In this case there is a distinctive peak at 3.95 days, but nothing near the 7.9 day period favored by the wind lines. Notice that the dominant peak occurs at roughly half of the period of the dominant peak in the wind lines.

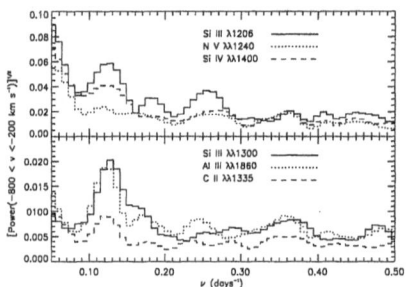

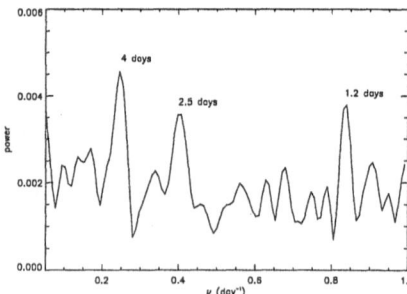

Fig. 3. Cleaned FT power in the wind lines.

Fig. 4. Cleaned FT power in Fe IV.

5   Conclusions

A typical B0 Ia shows rotational modulation of its wind lines increasing in strength for lines formed deepest in the wind – implying that the coherency is introduced at or very near the surface and weakens as structures propagate outward. The cyclical period of the wind structures is $\sim$ 7.9 days which could represent either one or two fixed surface features or a traveling wave with several crests.

The Fe IV variability is photospheric because it is confined to $\pm v_{eq} \sin i$ and some features move from blue to red. However, the variability is not centered on the line. Its strongest period is *half* the dominant wind period. While this suggests a distinct connection between the two, the nature of this connection remains unclear.

References

Fullerton, A.W., Massa, D., Prinja, R.K., et al. (1997): A&A, 327, 720
Giddings, J.R., Rees, P.C.T. (1989): SERC Starlink User Note 37
Howarth, I.D., Siebert, K.W., Hussain, G.A.J., Prinja, R.K. (1997): MNRAS, 284, 265
Humpherys, R.M., McElory, D.B. (1984): ApJ, 284, 565
de Jager, C., Nieuwenhuijzen, H. (1987): A&A, 177, 217
Kaufer, A., Stahl, O., Wolf, B., et al. (1996): A&A, 305, 887
Massa, D., et al. (1995): ApJ, 452, L53
Prinja, R.K. (1988): MNRAS, 231, 21p
Prinja, R.K., Massa, D., Fullerton, A.W. (1995): ApJ, 452, L61
Walborn, N.R. (1976): ApJ, 205, 419

Expression of Turbulence and the Variability of Stellar Winds in Resonance Line Profiles

Arved Sapar and Lili Sapar

Tartu Observatory, Tõravere EE2444, Tartumaa, Estonia

Observations suggest that supersonic turbulent velocities are generated in stellar winds by instabilities. We treat these velocities as thermal velocity distributions (i.e., Gaussian functions) with higher equivalent temperatures in the radial direction. We derived a general frequency redistribution function assuming coherent photon scattering in the rest frame of each particle for turbulence and partially coherent thermal scattering of photons. In such a scattering layer, the source function grows with optical depth. We found the source function, the emergent scattered radiation intensity at the outer and inner surfaces of the layer, and the increased radiative acceleration in the stellar wind due to the transfer of photon momentum by the prevalence of backscattering (i.e., due to the recoil effect). This effect explains the wide, dark plateaus seen in resonance line profiles and approximately doubles the momentum transfer to the stellar wind. Collisions of high-speed clumps produce superionization and large high-speed clumps are seen as narrow absorption components (NACs).

With the above picture in mind, we have analyzed the peculiarities and variability of the resonance line (RL) profiles in IUE spectra of the B supergiants κ Cas, 24 CMa, and ϵ Ori. For κ Cas and 24 CMa we found long-term variations in the blue-edge slopes of the RL-doublets of Si IV, C IV, Al III, and C II, but did not find short-term variations. The RL-profiles of these lines have wide, dark plateaus produced by the backscattering of radiation in optically thick, supersonically turbulent layers. The slope of the broad blue wing indicates the presence of high-speed turbulence in the optically thin outer layers of the stellar wind. The RL-profiles of Al III and C II exhibit NACs that are variable in strength and radial velocity.

The RL-profiles of C IV and Si IV in spectra of ϵ Ori also have wide, dark plateaus; strongly changing violet slopes; and NACs (sometimes several) that migrate blueward with time. The available time series exhibit rapid changes in NACs over about a week and indicate the presence of long-term variations that are about twice as large as the daily variations. Variations in the unsaturated RL-profile of N V correlate with variations in the RL of Si IV and C IV except in SWP 6450, when two strong absorption components appeared at -1200 km s^{-1} in both components of the N V doublet. The unsaturated Si III singlet also has a NAC, but its variations are confined to different regions and do not correlate with the behaviour of the other resonance lines due its origin in other layers.

If the stellar wind variability is connected with stellar rotation, then it must stem from the stellar surface, and is most probably connected with a stellar magnetic field. Details of this work will be published elsewhere.

Understanding A-Type Supergiants:
Visible and Ultraviolet Spectral Variability

Eva Verdugo[1], Antonio Talavera[2], and Ana I. Gómez de Castro[3]

[1] ISO Science Operations Centre, P.O. Box 50727, E-28080 Madrid, Spain
[2] LAEFF/INTA, P.O. Box 50727, E-28080 Madrid, Spain
[3] Instituto de Astronomía y Geodesia (CSIC-UCM), Facultad de CC Matemáticas,
    Universidad Complutense de Madrid, Avda. de la Complutense s/n,
    E-28040 Madrid, Spain

We have performed an extensive spectroscopic analysis of high-resolution visible (Balmer lines, Na I D lines, Ca II H and K lines, and Mg II λ4481) and ultraviolet (IUE LWP spectra) observations for 43 A-type supergiants. Our study shows that A-type supergiants can be divided into two groups that depend on the strength of mass-loss indicators, which are correlated with stellar luminosity. Group I contains the A supergiants with luminosity class Ib, which only show weak signs of stellar winds in the Mg II resonance lines, while Group II contains the most luminous stars, which show strong evidence for winds and mass loss in the visible and ultraviolet spectral ranges. Moreover, the analysis of spectral variability shows:

1. **In the visible:** We have only found significant variations in the Hα line profiles for most of the stars showing asymmetric emission profiles. Some of these stars show double-peaked emission profiles at Hα and exhibit V/R variability on time scales of days, which is consistent with the rotational periods of these stars.

2. **In the ultraviolet:** The ultraviolet spectra of A-type supergiants are characterized by the presence of variable, blue-shifted discrete absorption components in the resonance lines of Mg II and Fe II. These components are usually observed only in spectra of the most luminous stars. However, we have observed similar components in less luminous stars (luminosity class Ib) that exhibit spectacular variations. We observed the appearance and evolution of shortward-shifted components superimposed on the profile of the Mg II lines. Moreover, the observed life times of these components are compatible with the rotational periods of these stars. Similar components have not been detected in the Fe II lines. In the luminous A-supergiants we have observed variations in the components of the Mg II and Fe II lines, but the time scales of variation are much longer than in the less luminous stars. It seems that the features indicating mass outflow are stronger and more steady in the most luminous A-supergiants, while the less luminous stars exhibit these features in a smoother but more variable way.

Cyclic Variability in Be Star Winds: The Observations

Geraldine J. Peters

Space Sciences Center, University of Southern California,
University Park, Los Angeles, CA 90089-1341, USA

Abstract. The winds in Be stars vary cyclically on both short (≤ 1.5 days) and long (months-years) time scales. *IUE* data on 15 Be stars from 12 multiwavelength campaigns reveal three classes of behavior for the stars that display short-term photometric/spectroscopic variability: 1) wind strength and FUV flux correlated, period less than the star's rotational period, and ($\ell=2$) NRP modes identified from ground-based spectroscopy with a hot crest in front when the star is bright, 2) wind strength cyclically varies but does not correlate with light variations, and 3) FUV flux varies cyclically but wind variability is minimal and does not correlate with the flux. For the latter two cases, the period is close to the expected rotation period. The light variations appear to be caused by a modulation of the photosphere temperature. Analysis of long-term wind variations in λ Eri and 66 Oph confirms that most of the variability is in the DACs. Prior to an episode of wind enhancement, the DACs show the largest outflow velocities (< -1000 km s^{-1}) and this activity precedes the strengthening of the Balmer emission.

1 Introduction

The hallmark of a Be star is its spectroscopic/photometric/polarimetric variability on time scales of $\lesssim 1^{\rm d}$ to years. This variability can be cyclical, episodic, or transient. Current and historical information on the nature of this diverse activity can be found in the recent conference proceedings (Balona et al. 1994, Baade 1991, Slettebak & Snow 1987, Jaschek & Groth 1982). One of the most important characteristics of a Be star is its conspicuous wind and the phenomenal long lifetime of the *IUE* satellite has allowed us to study the variability in these winds on both short and long time scales. Some results derived from the 17 years of FUV observations of Be stars with *IUE* are discussed in this paper.

2 Short-Term Variability in the Winds

Although there was sporadic evidence of cyclic short-term photometric and spectroscopic variability in Be stars before the early 1980s (cf. Percy 1987), it seemed to be the reports on rapid variability in λ Eri and 28 (ω) CMa by Bolton (1982) and Baade (1982), respectively, at IAU Symposium No. 98 held in München in 1981 that caught the attention of the community. In the decade to follow two models were proposed to explain the activity, nonradial pulsations (NRP) and

Table 1. Campaign Chronology and Periods Found from UV and Optical Observations

Star	Sp. Type[a]	$v \sin i$ [a] km s^{-1}	Duration (hr)	Start Date of *IUE* Obs.	FUV Flux/Wind Period (d)	Optical NRP[b] Period (d)
μ Cen	B2 IVe	155	16	1985 Jun 28	0.50±0.10(w)	0.505±0.005
28 Cyg	B3 IVe	320	16	1985 Sep 29	...	0.69±0.02
o And	B6 IIIe-sh	260	24	1987 Aug 3	1.57:(f)	1.48±0.33
λ Eri	B2 IVe	310	40	1987 Nov 5	0.70±.02(f)	0.71±0.12
ω Ori	B2 IIIe	160	40	1987 Nov 5	1.50±0.17(f)	...
ε Cap	B3 IIIe-sh	250	40	1988 Sep 30	0.98±0.02(f,w)	...
28 Cyg	B3 IVe	320	56	1989 Sep 20	0.69±0.01(f,w)	0.64±0.06
η Cen	B2 IVe	350	64	1991 Mar 29	0.63±0.01(f,w)	0.61±0.04
48 Lib	B3 IVe-sh	400	64	1991 Mar 29	0.8:(f)	...
ζ Tau	B1 IVe-sh	220	56	1991 Oct 7	0.81±0.05(f)	0.80±0.06
ψ Per	B5 IIIe-sh	280	56	1991 Oct 7	1.04±0.02(w)	...
2 Vul	B0.5 IV	330	24	1992 Sep 13	...	1.27±0.15
120 Tau	B1.5 IVe	210	16	1993 Feb 24	0.5:(f,w)	...
EW Lac	B3 IVe-sh	300	24	1993 Sep 6	0.60:(w)	...
α Eri	B4 Ve	225	64	1995 Sep 7	1.28±0.02(f)	...
λ Eri	B2 IVe	310	64	1995 Sep 7	1.0±0.1(w)	...
DU Eri	B1 Ve	230	64	1995 Sep 7	1.2±0.1(w)	...
ω Ori	B2 IIIe	160	72	1996 Feb 2	1.20±0.10(w)	...

[a]From Slettebak (1982)
[b] From Hahula & Gies (1994), except for μ Cen (Baade 1984) and 28 Cyg, first obs. (Peters & Penrod 1988)

rotational modulation due to the presence of spots (RM). To date no consensus has been reached, and both models are still debated (cf. Baade & Balona 1994).

To determine the cause for short-term photometric and spectroscopic variability in Be stars, and in particular look for evidence of a wind/photosphere interaction, we carried through 12 multiwavelength campaigns. Typically a campaign would consist of a 24–72 hour interval of uninterrupted, repeated *IUE* observations that were supported with simultaneous optical photometry, high resolution spectroscopy, and polarimetry and sometimes FUV photometry from the *Voyager UVS*. The campaign chronology is summarized in Table 1.

Several of the program stars showed clear evidence of correlated cyclic variability in their wind absorption and FUV flux. Of the stars for which we achieved good simultaneous ground-based coverage, five (λ Eri, 28 Cyg, η Cen, ζ Tau, and

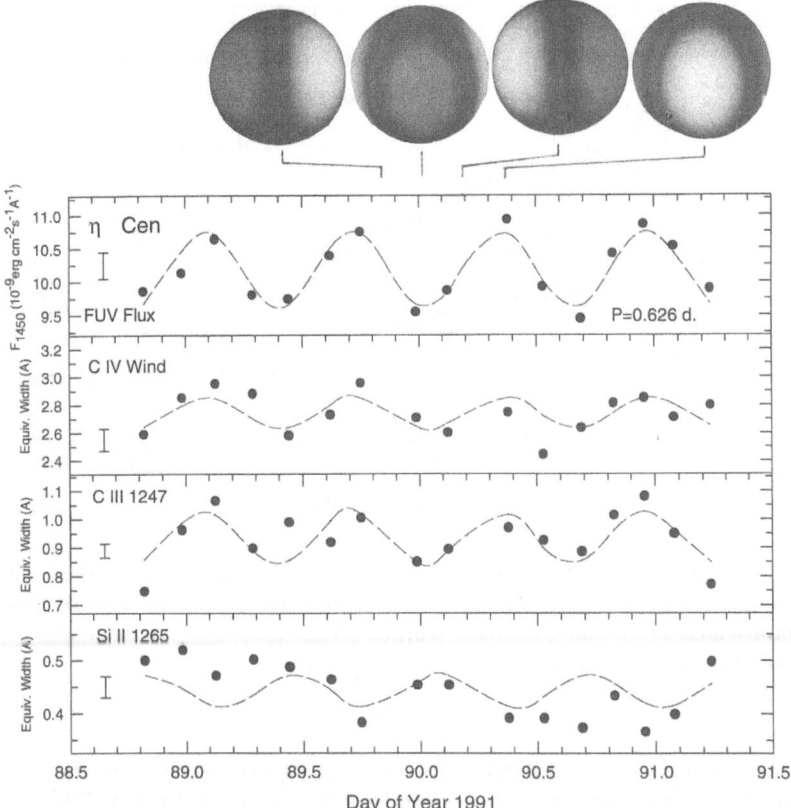

Fig. 1. The photometric, wind, and photospheric behavior of η Cen during the 1991 March campaign. On top are illustrations of the photosphere showing the orientation of crests (light gray)/troughs (dark gray) of the $\ell = 2$ mode based upon our modeling of optical line profile data. The *dashed* lines are sine curve fits to the 1450 Å flux and the EWs of the C IV wind and C III/Si II photospheric lines.

2 Vul) displayed NRP with sectorial, $\ell = -m = 2$ modes (Hahula & Gies 1994) and there was good agreement between the periods derived from the NRP analysis and FUV flux/wind (cf. Table 1). Modeling of the optical line profile data confirms that in η Cen and 28 Cyg a hot crest of the $\ell=2$ NRP mode crossed the star's meridian oriented along our line-of-sight when the star was brightest in the FUV. These two objects also showed the strongest correlation between the wind strength and the FUV flux.

The wind/flux behavior shown by η Cen is representative of this group of Be stars. Key results are summarized in Fig. 1, which can be compared with those for 28 Cyg and ϵ Cap in Peters (1991). During the 64 hr of the *IUE* run the FUV flux in η Cen showed a clear sinusoidal oscillation throughout the four cycles that were covered. A sine curve fit to the data (*dashed* line in Fig. 1)

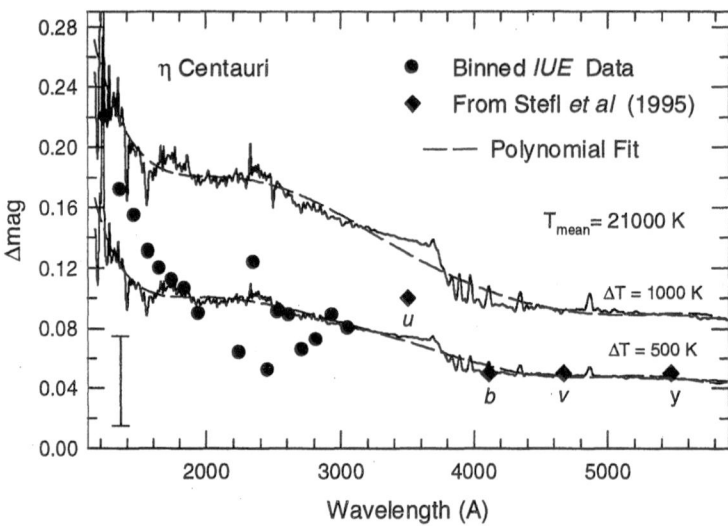

Fig. 2. The amplitude of the light curve for η Cen versus wavelength. *Filled* circles are from binned (10Å) *IUE* data; *filled* diamonds are from Štefl et al. (1995). The *solid* and *dashed* lines are predictions from Kurucz (1994) models if the color effect is due to a modulation of the star's photospheric temperature.

suggests a period of 0.626 ± 0.010 d (in excellent agreement with Štefl et al. 1995 and our NRP analysis) and an amplitude of 0.12 mag at 1450 Å. The wind in η Cen displayed a striking cyclic variation in strength (cf. Peters 1994), and from the sine curve fit to the C IV EWs (Fig. 1) it becomes apparent that there is a strong correlation between the wind strength and the FUV flux. Models of the photosphere derived from the NRP analysis are shown at the top of Fig. 1. A hot crest/cool trough faced us alternately at light maximum/minimum.

The photometric behavior of η Cen in the FUV is typical of the Be stars we have investigated (Peters 1991, 1994) in that the amplitude of the light curve increases with decreasing wavelength. Results for η Cen are shown in Fig. 2. Amplitudes from the light curves derived from binned high dispersion SWP & LWP images are compared with predictions from Kurucz (1994) models. Combining the *IUE* data with ground-based photometry from Štefl et al. (1995), one can see that the observations suggest that the photospheric temperature of η Cen is modulated by 500-750 K over the photometric period. Since this is an average over the entire facing hemisphere, local temperature deviations are much larger. The slope of the observed amplitude curve is steeper at the shortest wavelengths, and the NUV data fall conspicuously below the model predictions, but considering the observational errors and the fact that we expect emission from the asymmetric wind to be more prominent at photometric minimum (decreasing the photometric amplitude in spectral regions with numerous Fe group lines) the data do not provide compelling evidence for any other interpretation of the variability than a temperature modulation. Nevertheless to support this explanation, I have looked for cyclic effects in the strengths of key photospheric lines that vary strongly with temperature around 21,000 K, the approximate T_{eff} of η Cen. Results for C III 1247 and Si II 1265 are seen in the lower two panels in Fig. 1. Independent sine curve fits to the EW data reveal cyclic behavior with the above-mentioned photometric/wind period (within observational errors) and C III maxima and Si II minima when the star is at a photometric maximum.

For a few stars, cyclic wind variability was apparent but there did not appear to be a strong correlation with the FUV flux level. Showing a C IV wind line that varies by 75% in strength with a period of 1.04 d, the Be-shell star ψ Per is the most striking example. Other stars include ω Ori (cf. Peters 1996), DU Eri, and EW Lac. The nature of the wind behavior in ψ Per is seen in Peters (1994). As is typical for Be-shell stars, the absolute strength of the emission component to the P Cyg C IV wind line is strongly *anticorrelated* with the EW of the absorption part (cf. Peters 1991, 1994). If there are two regions on opposite hemispheres where the wind is enhanced (e.g. over the hot crests of an $\ell = 2$ NRP pattern), C IV absorption would be a maximum when these areas crossed our line-of-sight but the emission from the same region would appear enhanced when the denser part of the wind was viewed at the limb.

Other Be stars showed clear evidence of cyclic variability in the continuum but weak correlation with wind strength. Stars in this group are α Eri, ζ Tau, 48 Lib, and o And. Most but not all are Be-shell stars. Notable is α Eri in which the FUV flux displays a cyclic (P=1.28^d) 7% peak-to-peak variation at 1450 Å (30% at 1050 Å Holberg et al. 1998) that suggests a ΔT of ~500 K.

The periods for the cyclic wind/flux variations in the latter two groups of stars tend to be longer than those found in Be stars that display correlated wind strength and flux and are close to the expected rotation period of the star. For some (e.g. ζ Tau) NRP modes have been identified, but for others the ground-based data did not confirm the presence of NRP. Typically, all of the Be stars that showed evidence of FUV flux variability (cyclic or not) displayed light curves with amplitudes that increased with decreasing wavelength. The bulk of the observations seem to imply that NRP is not only responsible for the light variability, but also for the modulation of the wind.

3   Long-term Wind Variability

During the 17 years of the *IUE* project we were able to monitor the long-term wind behavior in λ Eri and 66 Oph. Both objects displayed cyclic epochs of strong winds flanked by brief periods of minimal or no wind presence. λ Eri's wind episodes recur at intervals of 1.30±.06 yr and last 1–5 mo. Typical behavior is illustrated in Fig. 3. It is apparent that most of the strength increase in C IV comes from the DAC that strengthens when the "outburst" begins. As in 66 Oph (Peters 1988) the DACs show the largest outflow velocities prior to an episode of wind enhancement, and this activity precedes the increase in Balmer emission.

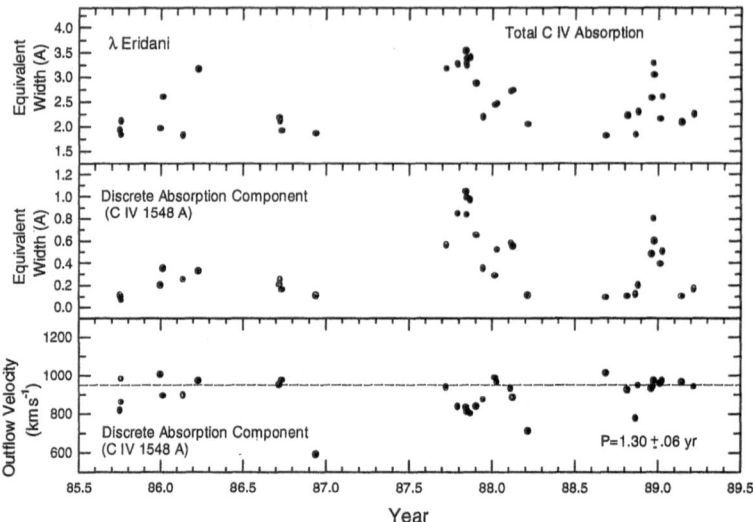

Fig. 3. Long-term wind behavior in λ Eri during the latter 1980s. The onset of an enhanced wind episode is accompanied by a strengthening DAC with an outflow velocity < -1000 km s^{-1}. As in 66 Oph, the DAC velocity increases as the active phase progresses then wanes.

This research would not have been possible without the faithful support of many colleagues, especially D. Gies, H. Henrichs, D. McDavid, & J. Percy, who participated in the multiwavelength campaigns. I appreciate financial support from NASA grants NSG-5422, NAG5-1296, & NAG5-2313.

References

Baade, D. (1982): in *Be Stars*, eds. Jaschek, M., Groth, H.-G. (Reidel, Dordrecht, Boston, London), 167

Baade, D. (1984): A&A, **135**, 101

Baade, D., ed. (1991):*Rapid Variability of OB-Stars: Nature and Diagnostic Value*, ESO Conf. & Workshop Proc. No. 36

Baade, D., Balona, L.A. (1994): in *Pulsation, Rotation, and Mass Loss in Early-Type Stars*, eds. L.A. Balona, H.F. Henrichs, J.M. Le Contel (Kluwer, Dordrecht, Boston, London), 311

Balona, L.A., Henrichs, H.F., Le Contel, J.M., eds. (1994): *Pulsation, Rotation, and Mass Loss in Early-Type Stars* (Kluwer, Dordrecht, Boston, London)

Bolton, C.T. (1982): in *Be Stars*, eds. Jaschek, M., Groth, H.-G. (Reidel, Dordrecht, Boston, London), 181

Hahula, M.E., Gies, D.R. (1994): in *Pulsation, Rotation, and Mass Loss in Early-Type Stars*, eds. L.A. Balona, H.F. Henrichs, J.M. LeContel, (Kluwer, Dordrecht, Boston, London), 100

Holberg, J.B., Sandel, B., Drake, V, Peters, G.J. (1998): in preparation.

Jaschek, M., Groth, H.-G., eds. (1982): *Be Stars* (Reidel, Dordrecht, Boston, London)

Kurucz, R.L. (1994): SAO Kurucz CD-ROM, No. 19

Percy, J.R. (1987): in *Physics of Be Stars*, eds. A. Slettebak & T.P. Snow (Cambridge Univ. Press, Cambridge, New York, Melbourne), 49

Peters, G.J. (1988): PASP, **100**, 207

Peters, G.J. (1991): in *Rapid Variability of OB-Stars: Nature and Diagnostic Value*, ed. D. Baade, ESO Conf. & Workshop Proc. No. 36, 171

Peters, G.J. (1994): in *Pulsation, Rotation, and Mass Loss in Early-Type Stars*, eds. L.A. Balona, H.F. Henrichs, J.M. Le Contel (Kluwer, Dordrecht, Boston, London), 284

Peters, G.J. (1996): *Be Star Newsletter*, No. 31, 17

Peters, G.J., Penrod, G.D. (1988): in *A Decade of UV Astronomy with the IUE Satellite*, ESA SP-281, Vol 2., 117

Slettebak, A, Snow, T.P., eds. (1987): *Physics of Be Stars* (Cambridge Univ. Press, Cambridge, New York, Melbourne)

Štefl, S., Baade, D., Harmanec, P., Balona, L.A. (1995): A&A, **294**, 135

Hao: Is the period of 0.626 d of η Cen pulsational or rotational?

Peters: Both. It reflects the sum of the prograde rotational motion of the star and the slower retrograde motion of the NRP pattern. Bright patches (hot crests) appear on opposite hemispheres and light maximum occurs as they sweep past the meridian oriented along our line of sight to the star.

Kubát: Do the variations in the C IV line really come from a variable wind or are they caused by some rotating structure?

Peters: The "observed" variations must be due to azimuthal asymmetry in the wind that we sample as the star+wind rotates with respect to our line of sight. If the period is < 1 d, then it appears that NRP in a $\ell=2$ mode moving in a retrograde direction is modulating the wind; for a star with a period of $\gtrsim 1$ d one cannot say with certainty that the azimuthal wind pattern moves with respect to the star.

Stellar Wind Variability of Be Star FY CMa

Huilai Cao[1,2], Joachim Dachs[2]

[1] Beijing Astronomical Observatory, Beijing, China
[2] Astronomisches Institut der Ruhr-Universität, D-44780, Bochum, Germany

Abstract. Archival IUE spectra including the ultraviolet resonance lines and optical spectra of the Hα and the He I λ5876 Å line of the Be star FY CMa show inverse P Cygni profiles, indicating an unusual activity in February 1987. The edge velocity measured in the blue-shifted absorption wings profile reached up to ~ -950 km s^{-1}. We estimate the mass loss rate $\dot{M}$ to be about $4.1 \times 10^{-9} M_\odot$ yr^{-1} (assuming spherical symmetry). If we assume instead a conical wind shape, for a cone half angle of $\sim 30°$, $\dot{M} \approx 3.4 \times 10^{-10} M_\odot$ yr^{-1}, or $\dot{M} \approx 2.2 \times 10^{15}$ gm s^{-1}. We attempt to give a qualitative explanation of the abrupt activity in FY CMA in terms of the infall of material.

1 Introduction

Since 1949, the spectrum of the Be star FY CMa (HR 2855, HD 58978, MWC 179, B0.5 IVe, $v \sin i \sim 280$ km s^{-1}) has changed. Slettebak (1949) noted that the hydrogen emission lines in the photographic region had disappeared. Burbidge et al. (1954) reported that this object showed double emission lines at Hβ and Hγ with R/V > 1, the Balmer lines from H9 onward appeared in absorption only. In 1987, ultraviolet and optical observations showed that this star exhibited remarkable variations in its mass-loss properties. The high resolution IUE spectra obtained between 1981 to 1987 in the range $1150 - -1950$ Å (SWP), were supplemented by almost simultaneous optical spectra of the Hα line on Feb. 07/08, and the He I λ5876 Å line on Feb. 11/12, 1987. The resonance lines of N V and C IV, and the optical Hα and He I λ5876 lines displayed striking inverse P Cygni profiles (Fig. 3). This activity sustained in the period from Jan. 28, 1987 to Feb. 26, 1987.

2 Obtained Data and Reduction

The high-resolution ($\lambda/\delta\lambda \sim 10^4$) ultraviolet spectra from the International Ultraviolet Explorer (IUE) spacecraft (Boggess et al. 1978) were obtained from ESA Satellite Tracking Station at Villafrance del Castillo, Spain, or provided by the archives of the ESA Vilspa Data Center.

The high-quality optical spectra of Hα and He I λ5876 were obtained with the ESO Coudé Echelle Spectrometer fed by the 1.4m Coudé Auxiliary Telescope at La Silla, Chile. The spectral resolution was measured to be 76 mÅ at either wavelength, by means of IHAP and MDAS image processing software systems provided by ESO.

The resonance doublet lines N v $\lambda\lambda1240$ and C iv $\lambda\lambda1550$ show in 81263 a very flat bottom absorption structure, but in 87028-87057 they appear with a strong discrete component absorption core (Fig. 1) (Grady et al. 19 87). The doublet lines of Si iv $\lambda\lambda1400$ in the active period from Jan. 28, 1987 to Feb. 26, 1987 do not exhibit obvious variability, only the equivalent width (EW) of Si iv

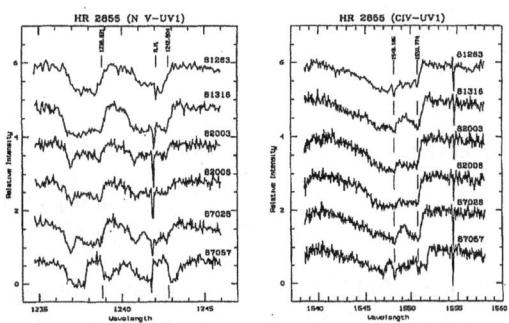

Fig. 1. The spectral variations of FY CMa in the resonance doublets of N v and C iv; high-resolution spectra obtained with IUE.

$\lambda1393.755$ Å decreased from 3.5 to 3.0 Å, and the EW of Si iv $\lambda1402.770$Å changed from 3.0 to 2.5 Å, respectively (Fig. 2).

3   Mass-loss Rate of the Stellar Wind

Theoretical P Cygni profiles have been calculated by Castor & Lamers (1979) assuming resonance scattering in a spherically symmetric wind. The model is specified by the radial optical depth as a function of the flow velocity, $\tau_{rad}(v)$, and the velocity law of the envelope, $v(r)$, where r is the distance from the center of the star. Assuming that the lines are formed in the outflowing material, the edge velocities of the absorption wings represent a lower limit to the terminal velocities of the wind, v_∞, where $v_\infty \sim v_{edge}$ (Cassinelli & Abbott 1981). The mass-loss rate is given by,

$$\dot{M} = 4\pi r^2 \rho v \,, \tag{1}$$

where ρ and v are the density and velocity, respectively.
The radial optical depth in an expanding envelope is given by

$$\tau_{rad} = \frac{\pi e^2}{mc} f \lambda_0 n_i \left(\frac{dv}{dr}\right)^{-1}, \tag{2}$$

where f is the absorption oscillation strength, λ_0 (in cm) is the rest wavelength, n_i (in cm^{-3}) is the number density, $(dv/dr)^{-1}$ (in sec) is the inverse velocity gradient.
(a)   Velocity law:

$$w = w_0 + (1 - w_0)\left(1 - \frac{1}{x}\right)^\beta \tag{3}$$

β is a variable parameter having the value: $\beta = 0.5$ or $\beta = 1$.
(b)   Optical depth law (Olson 1982):

$$\tau_{rad}(w) = T(\gamma + 1)(1 - w_0)^{-1-\gamma}(1 - w)^\gamma, \quad \gamma \geq 0 \tag{4}$$

T : the total optical depth

$$T = \int_{w_{photo}}^{1} \tau_{rad}(w)dw = \frac{\pi e^2}{mc} \frac{f\lambda_0}{v_\infty} N_i .$$ (5)

N_i is the column density (cm^{-3}) of the absorption ion in the envelope.

$$\dot{M} = \tau_{rad}(w) \frac{4mcm_H v_\infty^2 R_*}{e^2 f\lambda_0 g A_E} \left(x^2 w \frac{dv}{dr} \right) ,$$ (6)

where $\tau_{rad}(w)$ is computed from equation (4), m_H is the atomic mass of hydrogen (Snow et al. 1981), g is the ionization fraction for the observed species, and A_E is the abundance of the observed element with respect to hydrogen. For $[x^2 w(dv/dr)] \approx 2$ at $w = 0.5$,

$$\dot{M} = 1.74 \times 10^{-18} \tau_{rad}(w) \frac{v_\infty^2 R_*}{f\lambda_0 g A_E}$$ (7)

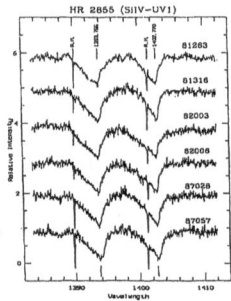

Fig. 2. The Si IV doublet lines at λλ1393.755,1402.770Å did not show any obvious variations during Jan.28 – Feb.26, 1987.

$\tau_{rad}(w)$ has to be evaluated for $w = 0.5$, and $\gamma = 2$, for comparison of the profile with the theoretical profile calculated by Castor & Lamers (1979), v_∞ is in units of km s^{-1}, R$_*$ in solar radii. The abundance of Si IV/H I = 3.5 × 10^{-5} , the f–value is 0.528 for λ1393.755 Å and 0.262 for λ1402.770 Å, respectively. For $v_\infty \sim v_{edge}$ around –900 km s^{-1}, we can estimate the mass-loss rate of FY CMa: about 4.1 × 10^{-9}M$_\odot$ yr^{-1} (if the wind were spherically symmetric). If we instead adopt a conical shape for the wind, $\dot{M}$ is reduced by a factor $\Omega_w/4\pi$, where Ω_w is the solid angle into which the wind flows.

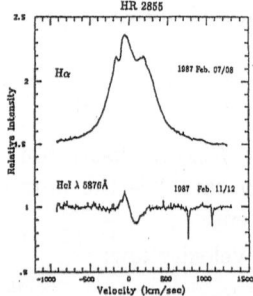

Fig. 3. The abrupt activity in the Hα and He Iλ5876 Å lines showed up as inverse P Cygni features.

For a cone half angle of $\sim 30°$ and assuming that the wind emanates from above and below the envelope, $\dot{M} \approx 3.4 \times 10^{-10}M_\odot$ yr^{-1}, or $\dot{M} \approx 2.2 \times 10^{15}$ gm s^{-1} (Cordova & Mason 1982). The momentum rate of the wind, $\dot{M} v_\infty$, is $\sim 3 \times 10^{24}$ ($\Omega_w/0.3\pi$) erg cm^{-1}. The momentum rate of the radiation, L/c, is 1× 10^{23} (L/L$_\odot$) erg cm^{-1}. If the luminosity of FY CMa is $\sim 100 L_\odot$, the radiation pressure may be sufficient to account for the observed wind. The ratio of the wind momentum to radiation momentum is of order unity.

This implies the stellar wind is driven mainly by radiation pressure. If the wind momentum exceeds the radiation momentum, a source of mechanical energy (e.g. rotation or magnetic field) may be required to initiate the wind. Radiation pressure in the lines could then complete the acceleration of the wind to high velocities (Hanuschik et al. 1993).

4 Discussion

The ultraviolet and optical observations clearly show that an abrupt change occured in FY CMa, involving both infall and outflow of material (Peters 1988). The collected data show that this activity persisted for at least 30 days. Combining with Peters observation it can be estimated that this activity extended over more than 3 months. Such sudden spectral variations have been seen in pre-main sequence objects in which accretion of matter is occurring, but such bursts of activity have rarely been observed in classical Be stars.

A more plausible explanation of the abrupt activity is probably a sudden accretion event in the circumstellar envelope, as evidence for the infall of matter is clearly seen in several species of moderate ionization. Such spectral changes could be due to magnetic activity, i.e. magnetic flares or loops. FY CMa perhaps has large magnetic loops or jets with a flow of material that originates near the pole but fall down upon the star in line of sight, in this case the line profile presents a strong blueshift with an inverse P Cygnis feature (Baade et al. 1989).

References

Baade, D., Dachs, J., Weygaert, R. van de, Steemann, F. (1989): A&A 198,211
Boggess, A. et al. (1978): Nature 275, 372
Burbidge E.M., Burbidge G.R. (1954): ApJ 119, 496
Cassinelli, J.P., Abbott, D.C. (1981): in The Universe at Ultraviolet Wavelengths: The First Two Yrs. of IUE (Editor: R.D. Chapman), p127
Castor, J.I., Lamers, H.J.G.L.M. (1979): ApJS 39, 481
Cordova, F.A., Mason, K.O. (1982): ApJ 260, 716
Grady, C.A., Bjorkman, K.S., Snow, T.P., et al. (1987): ApJ 320, 376
Hanuschik, R.W., Dachs, J., Baudzus, M., Thimm, G. (1993): A&A 274, 356
Peters, G.J. (1988): ApJ 331, L33
Olson, G.L. (1992): ApJ 255, 267
Slettebak, A.: cf. P.W. Merrill, C.G. Burwell (1949): ApJ 110, 387
Snow, T.P. (1981): ApJ 251, 139

Prinja: How did you account for contributions due to the photospheric absorption (in the Si IV region) in your wind model profile fitting?

Cao: I used a Kurucz model for the same spectral type to fit the photospheric absorption in the Si IV region, to separate the photospheric absorption contribution from the stellar wind line profiles.

Fundamental Parameters for Seven Be Stars

Yves Frémat

Université de Mons-Hainaut, 20, Place du Parc, B-7000 Mons, Belgium

The fundamental parameters (distance, luminosity, temperature, mass) of seven Be stars were determined from a study combining ultraviolet fluxes, Balmer discontinuity parameters, near-infrared observations, and trigonometric parallaxes measured by the Hipparcos satellite.

We determined the $E(B-V)$ colour excess by studying the 2200 Å interstellar band according to the method used by Beeckmans & Hubert-Delplace (1980). In order to find the effective temperature of the programme stars, we used the temperature determinations made by Zorec & Briot (1991) for a large sample of B and Be stars as a function of the Balmer discontinuity. From the ratio between the equivalent width of the P14 and P12 Paschen lines we computed the surface gravity (column 4). For each star, we determined the angular diameter by combining the Hipparcos V magnitudes and the theoretical fluxes integrated over the V filter passband. The same procedure was applied to the TD1 UV fluxes. Then, from the trigonometric parallax, we computed the stellar radius and luminosity and located the objects in a theoretical H-R diagram. We could then estimate their mass and derive a second estimate of their surface gravity (column 5).

HD	Sp. Type (This work)	$\log(T_{\mathrm{eff}})$	$\log(g)$ (c.g.s.)	$\log(g)$ (c.g.s)	$M/M_\odot$	$R/R_\odot$	$\log(L/L_\odot)$
12882	O9-B0 III	4.471±0.045	...	4.42	12.2	3.56±3.64	3.64±2.23
21650	B5 V	4.162±0.015	4.23	4.10	4.3	3.03±0.81	2.57±0.59
33604	B2 IV	4.328±0.039	...	3.91	8.4	5.29	3.71
40978	B2 III	4.311±0.044	...	3.90	7.8	5.51±4.64	3.62±1.98
89884	B5 IV	4.182±0.027	4.22	4.24	4.3	2.58±0.71	2.50±0.66
175863	B5 III	4.182±0.034	3.52	3.70	5.4	5.48±2.01	3.16±0.87
189689	B8 III	4.074±0.032	...	3.61	4.0	5.17±1.94	2.68±0.88

References

Beeckmans, F., Hubert-Delplace, A.-M. (1980): A&A 86, 72
Zorec, J., Briot, D. (1991): A&A 245, 150

Wind and Jets in the Be Star 4 Herculis

J. Kubát[1], P. Koubský[1], P. Harmanec[1], A.M. Hubert[2], and P. Hadrava[1]

[1] Astronomický ústav, Akademie věd ČR, 251 65 Ondřejov, Czech Republic
[2] Observatoire de Paris, Section d'Astrophysique, F-92195 Meudon, France

4 Her is a binary star with a period of $46\overset{d}{.}192 \pm 0\overset{d}{.}004$ that is located at a distance of 150 pc (as determined by Hipparcos). It consists of a Be primary with $T_{\mathrm{eff}} = 12500\,\mathrm{K}$ and $\log g = 4.0$ and an unseen secondary that does not fill its Roche lobe (Koubský et al. 1997). The behaviour of its Hα line has been monitored for a long time. Recently, the star has entered the third emission-line episode detected since the 1920's. The two periods when it exhibited a normal B spectrum are well documented, but their lengths are very different: 13 to 15 years vs. 3 to 5 years. Episodes of activity of 4 Her are characterized by a gradual development of emission in Hα and the appearance of metallic shell lines. The effective temperature during the emission phase decreases to $T_{\mathrm{eff}} = 8500\,\mathrm{K}$ and the gravity decreases to $\log g = 2.5$, but the spectrum is different from that of an A3 supergiant. Future systematic observations of this star may help to determine the nature of the physical mechanisms that cause the changes between normal and shell spectra.

Koubský et al. (1997) discovered a fascinating absorption feature in the violet peak of the Hα emission line. It is detectable at most orbital phases. It moves from blue to red across the emission peak until it reaches a radial velocity (RV) of about $-90\ \mathrm{km\,s^{-1}}$ and then reappears again at a RV of $-190\ \mathrm{km\,s^{-1}}$. It is observable only when the Hα emission strength exceeds $\sim$1.5 of the continuum level. It is somewhat reminiscent of a feature reported recently by Pogodin (1997) for HD 50138 which, however, is not known to be a periodic RV variable. In the case of 4 Her, this phenomenon – which has so far been monitored for more than 1300 days – repeats strictly with the $46\overset{d}{.}192$ orbital clock. Another similarity can be found in the UV spectrum of HD 77581 (Vela X-1; Sadakane et al. 1985).

The cause of this absorption feature is not completely clear. We tested several hypotheses, but none of them is completely consistent. In order to determine whether this feature comes from Hα or the blended He II 6560 Å line, we suggest that the program of continuous observations of this star be extended to include the region of He II 4686 Å.

The full version of this study is published elsewhere (Koubský et al. 1997).

References

Koubský, P., Harmanec, P., Kubát, J., et al. (1997): A&A 328, 551
Pogodin, M.A. (1997): A&A 317, 185
Sadakane, K., Hirata, R., Jugaku, J., et al. (1985): ApJ 288, 284

Variability of Be Stars
Using Hipparcos Photometry

Anne Marie Hubert, Michèle Floquet, Ana Gómez, and Danièle Morin

URA 335, Dasgal, Observatoire de Paris-Meudon, 92195 Meudon Cedex, France

Rapid, mid-term, and long-term variability for bright Be stars ($V \leq 7.1$) was sought in Hipparcos photometry, which was obtained between mid-1989 and mid-1993. The Hipparcos magnitude, designated as Hp, is defined by a broad pass-band that ranges from 340 to 850 nm. This system yields magnitudes close to the visual magnitude, V. The accuracy of individual measurements depends on the magnitude of the star. A sample of 261 Be stars was considered; the number of measurements per star was between 100 and 200. We found:

1. The degree of variability of Be stars decreases with effective temperature.
2. Short-term variability ($\leq 3^{d}.5$) is present in 86% of early Be stars, but only in 18% of late Be stars. For many Be stars exhibiting both short-term and long-term variability, we have been able to make a determination or an estimation of short periods after subtracting the long-term variations. The full amplitude of the short-term light curves can be as large as 0.1 magnitude in some cases, which gives some evidence for multiperiodicity.
3. Outbursts of variable strength and duration (short-lived outbursts, outbursts followed by a slow decrease, and outbursts linked to a Be phase) are observed more frequently in early-Be stars. They are preferentially seen in stars with a low- to moderate- $v \sin i$.
4. Fadings are also observed: short-lived fading events, which could be seen preferentially in Be stars with higher $v \sin i$; and fadings preceding a large increase of brightness or quasi-periodic oscillations.
5. Quasi-periodic oscillations (QPOs) with periods $\leq$ 100–200 days were observed in about 20 stars.
6. Quasi-cyclical changes over several years are conspicuous in several Be stars, but the evidence is restricted by the life time of Hipparcos. In shell stars, they are linked to spectroscopic V/R and/or radial velocity cycles.
7. Many Be stars exhibit slow variations. In intermediate and late types, they are generally not disturbed by short-lived events (outbursts or fadings), but are only affected (in some cases) by a short-period modulation.

The Hipparcos photometry illustrates the presence of superimposed periodic, quasi-periodic, and aperiodic phenomena of different time scales in the light curves of a large percentage of early Be stars. Some B0-B1e stars are characterized by irregular, weak but very numerous short-lived events. The light curves of 35% of B0-B3e stars are ruled by outbursts and/or fading events and/or mid-term quasi-periodic oscillations. About 50% of late Be stars have constant light, and the other 50% are essentially affected by rather smooth, long-term variations.

The Interrelation Between Long-Term Cycles and Rapid Variations in Be Stars: A Photometric View

K. Pavlovski[1], Ž. Ružić[2], and D. Dominis[1]

[1] Faculty of Geodesy, University of Zagreb, Kačićeva 26, HR–10000 Zagreb, Croatia
[2] Ružmarinka 17, HR–10000 Zagreb, Croatia

Between 1972 and 1990, in the course of the Hvar-Ondřejov photometric monitoring of bright northern Be stars, about 14 000 UBV measurements were secured for 76 objects (Pavlovski et al. 1997). We used the Hvar archival data (Harmanec et al. 1997) to study the incidence and characteristics of variability patterns for 48 Be stars with good time coverage.

Are there common photometric characteristics among those Be stars that have shown long-term cycles in the light variations? The variability patterns are very complex and we summarize some of our findings as follows:

1. Both long-term and rapid variations were simultaneously present in 25% of the sample.
2. There is no linear relationship between the light amplitudes of the long-term variations and the rapid variations. Moreover, we have found that in three cases (o And, EW Lac, and 88 Her) the light amplitudes of the rapid changes strongly depend on the phase of the long-term cycle.
3. Typical time scales for the long-term light variations were found to be around 2,400 days for early-type Be stars and about 3,000 days for late-type Be stars.
4. The largest amplitudes were found for stars with moderate $v \sin i$, with no dependence on luminosity and spectral types.
5. Colour variations are only minute in $B - V$, but complicated and quite pronounced in $U - B$. Their paths in colour vs. magnitude and colour vs. colour diagrams do not repeat strictly from cycle to cycle.

References

Pavlovski, K., Harmanec, P., Božić, H., et al. (1997): A&AS 125, 75
Harmanec, P., Pavlovski, K., Božić, H., et al. (1997): J. Astron. Data (in press)

Long-Term Cyclical Circumstellar Changes and He I *lpv* in the Be Star EW Lacertae

M. Floquet[1], A.M. Hubert[1], H. Cao[2], and the 1993 Multi-Site Campaign Team

[1] URA 335, Dasgal, Observatoire de Paris-Meudon, 92195 Meudon Cedex, France
[2] Beijing and Xinglong Observatories, Beijing, China

The well-known Be shell-star EW Lac entered an active phase in 1977–1978. Since then, it has exhibited long-term quasi-cyclical variations in the V/R ratio of Balmer emission lines, the radial velocities of shell lines, and light. It has also exhibited short-term, multiperiodic variations in light and in the photospheric line profiles of He I λ6678, which are interpreted in terms of non-radial pulsations.

Several observational facts led us to search for a link between short-term phenomena and the behaviour of long-term, quasi-cyclical variations.

1. Time series analysis of He I λ6678 (which was intensively monitored in 1989 at the Haute Provence Observatory and in 1993 as part of an international campaign) has shown that: a) the "mean variance" was larger in 1993; b) the same frequencies were detected in 1989 and 1993 but with different amplitudes; c) the mean equivalent width was larger in 1993 than in 1989 and its variation modulated by a period of 10–15 days.

2. The V/R ratio and the equivalent width (W_λ) of the Hα emission line were rapidly and highly fluctuating in 1983 (Xinglong data); a discrete blue-shifted component of shell Balmer lines and abrupt light changes occurred more or less at the same epoch. Analogous abrupt light changes (or fading events) were quite conspicuous in 1990 and 1991 in Hipparcos photometry.

3. The time scales of the radial velocity (RV) variations of the higher Balmer lines (formed in the inner parts of the circumstellar envelope) are about 2600 days, and are consistent with the time scales of the slow variations seen in photometry from Hvar and Hipparcos. RV minima for H10 occurred near the brightness maxima. The abrupt, episodic variations described above seem to occur at the same phases of the long-term light curve, on the ascending branch of the RV curve.

4. The behaviour of the V/R ratio of Hα is not closely correlated with the long-term light variations. Its amplitude began to increase again early in 1990 in response to strong, new perturbations in the inner part of the envelope.

5. Similarly, the W_λ of Hα does not follow the V/R curve. Though highly variable in 1983, the W_λ decreased toward a minimum in 1995 December. Observations at Xinglong in 1997 show that the W_λ has increased again, which suggests that a new episode of mass loss has begun in EW Lac.

The strengthening of He I λ6678 modulated by a 10–15 day period; the higher amplitude in rapid line-profile oscillations that occur when the V/R ratio of Hα started a new cycle: are these signs of higher activity levels or an enhanced stellar wind?

Wind Variability in PMS-Stars

Mikhail Pogodin

Pulkovo Observatory, Pulkovo, 196140 Saint Petersburg, Russia

Abstract. The available amount of observational data on non-stable phenomena in gaseous envelopes around pre-main sequence (PMS) T Tau and Herbig Ae/Be stars are analysed here. The analysis allows to conclude that stellar wind as well as matter infall are present in the envelopes of both classes of pre-main sequence objects, but the balance of wind/accretion processes in these objects is not the same. The role of accretion in the formation of envelopes around the Herbig Ae/Be stars is unlikely to be so important as in the case of T Tau stars. Cyclic variability of observational parameters with different periods is discovered in the Herbig Ae/Be stars. It seems to be connected with spatial and kinematic stratification of circumstellar media containing azimuthal inhomogeneities.

1 Wind and Infall Phenomena in the Envelopes of PMS-Stars

Two main classes of PMS objects are well-known now. The first of them are the low-mass T Tau stars (TTSs). A review specially devoted to these objects is presented in this volume (Johns-Krull 1997). Therefore, I accentuate results concerning presumably the class of PMSs of intermediate mass – so called Herbig Ae/Be stars (HAEBEs, Herbig 1960). The data on the TTSs are considered here only briefly for the sake of comparison.

1.1 Accretion Magnetospheric Model

All the PMS objects show observational evidence for gaseous envelopes of rather complex structure, where the wind is present only as one component.

For the classical TTS (CTTSs) which seem to be surrounded by accretion disks, the magnetospheric envelope model has been developed. Different versions of this model have been presented by a number of authors (Camenzind 1990, Königl 1991, Calvet & Hartmann 1992, Hartmann et al. 1994, Shu et al. 1994). According to this model, the interaction of the stellar magnetic field with an accretion disk results in a quasi-stationary magnetic configuration. Inside the corotation radius the gas is falling onto the star along closed lines of force. Matter having diffused into the open field lines of the star and the disk is accelerated outward. At large distances from the star the azimuthal pinch stresses collimate the field and the wind parallel to the rotation axis.

This model can explain many observational features of the TTSs (Herbst 1995, Fernandes & Eiroa 1996, Gullbring et al. 1996, Petrov et al. 1996), namely:

– rapid (and sometimes periodic) photometric and spectral variability connected with rotation of bright spots on the stellar surface tracing a funnel matter infall;
– the presence of low-velocity and high-velocity components of the disk winds displayed in forbidden lines;
– prominent jets and outflows seen at large distances from the star.

It is clear that emission lines originating in the envelope of such a complicated structure, in general should not be considered as wind indicators, since the bulk of a line probably arises in the magnetospheric infalling flows joining the disk and the star and does not trace the outflowing material.

Hence, the TTS winds are likely to be powered by disk accretion. Nevertheless, their non-stable behaviour can be directly studied by investigation of:

– variability of blueshifted absorption components on the profiles of permitted lines;
– structural features in the forbidden lines;
– morphological structures of large-scale jets and outflows reflecting not only the interaction between the wind and inhomogeneous circumstellar media but also real wind variability.

1.2 Envelopes of the Herbig Ae/Be Stars

The situation with the HAEBEs seems to be less clear. It has recently become obvious that they possess more complex and diverse sources of activity than their low-mass counterparts. There is a systematic deficiency of our knowledge on:

– the sources of the observed activity;
– the origin of high-temperature zones of circumstellar gas;
– the driving mechanisms for the stellar wind.

The envelope model commonly accepted for the TTSs can hardly be applied to the HAEBEs. According to the conventional models of stellar evolution, classical dynamo-mechanisms for the generation of magnetic fields cannot work in such early-type stars.

The questions regarding the presence of accretion disks and the balance of outflow/infall processes in the formation of gaseous circumstellar envelopes of the HAEBEs also remain a matter of debate.

A way to determine similarities and distinctions between properties of circumstellar gas in the TTSs and HAEBEs can be based on a comparative analysis of observational features originating in different parts of their envelopes.

The most prominent of them is the Hα line, which forms throughout the envelope. Its profile contains information on structural and kinematical peculiarities of circumstellar gas in a significant part of the envelope.

A histogram of the Hα profiles with the signs of outflow/infall for large samples of CTTSs (Reipurth et al. 1996) and HAEBEs (from the catalogue by Finkenzeller & Mundt 1984) is presented in Fig. 1.

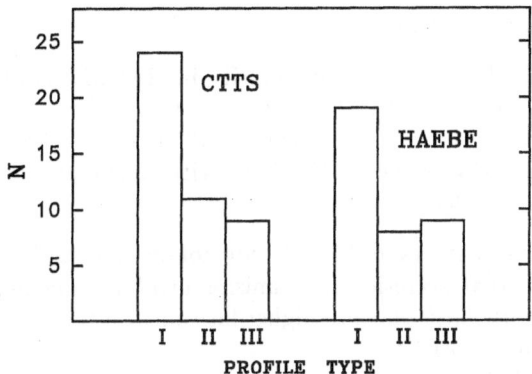

Fig. 1. Distribution of the Hα profile types for the CTTSs and the HAEBEs. The profile type classification is as follows: I – profiles with wind signatures; II – single or symmetric double-peaked profiles; III – profiles with signatures of matter infall.

One can see that the TTSs and the HAEBEs do not demonstrate differences in these distributions. The stars of both classes display profiles with signatures of outflow and infall, but the objects with outflowing circumstellar gas are observed more often.

This speaks in favour of some similarity in the kinematical structure of the gaseous envelopes in CTTSs and HAEBEs. However, T. Böhm and C. Catala in a number of papers presented evidence against the presence of prominent disk accretion onto Herbig stars. Their arguments are based on the absence of noticeable optical veiling in the majority of these objects (Böhm & Catala 1993) and on the apparent symmetry of the forbidden line profiles in their spectra (Böhm & Catala 1994, Böhm & Hirth 1997).

However, a number of stars exhibit developed manifestations of disk-like structures. High-resolution spectroscopy of the candidate HAEBE HD 36112 and the related object HD 50138 near the sodium doublet Na I D has been carried out at ESO and the Crimean Astrophysical Observatory (Beskrovnaya et al. 1998b, Pogodin 1997a). These results have shown that the profiles are very asymmetric with the blue part being more developed. This phenomenon can be connected with the existence of an optically thick disk in a remote part of the envelope which screens the receding flow.

A survey of high-resolution UV and optical spectra of HAEBEs and related objects was published by Grady et al. (1996). Signatures of accreting circumstellar gas were found in 36% of objects and signs of outflows in 48%. Analysis of the optical data revealed signs of matter infall in only 21% of the objects. This value is rather small in comparison with 85% for the CTTSs.

In general, the HAEBEs with accretion signatures:

– have higher projected velocities $V \sin i$ than the objects with signatures of outflow;
– tend to exhibit larger amplitudes of optical light variability;

– demonstrate, as a rule, polarimetric variability.

These data imply that accretion activity in HAEBEs is predominantly observed when the line-of-sight transits the equatorial disk. A similar conclusion has been made by Grinin & Rostopchina (1996) on the basis of a statistical investigation of the Hα profiles in HAEBEs with strong photometric variability.

In contrast to the TTSs:

1. The accretion signatures in HAEBEs are not observed at high polar latitudes. This suggests that accretion mechanisms involving magnetic channeling of material from the disk plane to higher latitudes on the star are probably not effective in the HAEBEs.
2. Analysis of infall features shows that accretion toward the HAEBEs is clumpy rather than continuous (Graham 1992).
3. The presence of accretion material with free-fall velocities implies that the density of the infalling gas is lower in comparison with TTSs and that viscosity is unimportant in slowing the accretion gas down to values expected for the inner disk of a TTS.

These conclusions are confirmed by results based on the He I 5876 line in the PMS stars.

The analysis of the He I emission line observed in CTTSs (Edwards 1997) shows that it is characterized by a two-component structure consisting of a broad component (BC) and a narrow one (NC). The emission peak of the BC is not so intense in comparison with the NC and its profile has sometimes the inverse P Cyg-type. This feature is likely to arise in the magnetospheric accretion flow. The NC forms possibly in the postshock region at the base of the accretion flow.

The He I 5876 line profiles of a sample of HAEBEs have been investigated by Böhm & Catala (1995). They found that the line is of pure photospheric origin in hotter stars and signs of activity become noticeable only in objects later than B8. In general, the line consists of two components. The first one is a strong absorption centered at the rest velocity of the star, the second component (if present) is blueshifted and is seen in emission.

Later on, this line in the prototype Herbig Ae star, AB Aur, was monitored in the framework of the MUSICOS campaign (Böhm et al. 1996). The authors analysed the behaviour of these two line components and suggested the following interpretation:

– The "basic" component is a broad and rather symmetric emission line. It is generated in the increasing temperature zone of the chromosphere.
– The periodically variable absorption is formed in hypothetical streams between the stellar surface and the high-temperature region. These streams are corotating with the star due to the local magnetic fields.

The authors considered different kinematics inside the streams: from ejected jets to magnetic tubes containing accretion material.

Rapid variability of the He I 5876 line profile in several Herbig Ae stars was investigated in the framework of the Pulkovo observing programme using the

1.4m CAT at ESO and the 2.6m Shajn telescope of the Crimean Astrophysical
Observatory. Fig. 2 illustrates different types of profile variability in two objects
on timescales from hours to days.

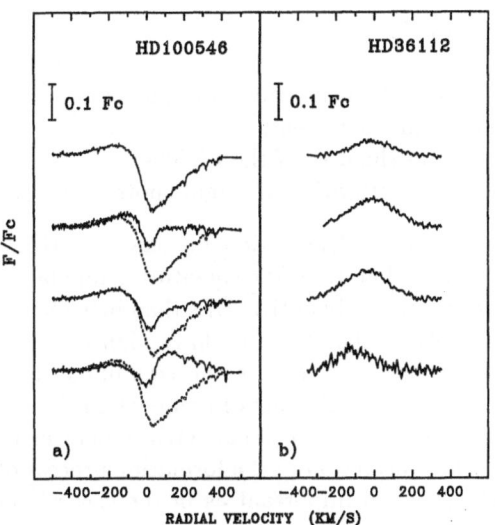

Fig. 2. The He I 5876 line profiles typical for Herbig Ae stars.
a) Two-component profile observed in HD 100546 (Pogodin & Vieira 1997). Similar profiles were observed in AB Aur, HD 163296 (Beskrovnaya et al. 1998a), and HD 50138
(Pogodin 1997a)
b) Profile observed in HD 36112, which contains only a blueshifted emission peak (Beskrovnaya et al. 1998b)

Studies have shown that a similar type of variability is detected in the majority of the Herbig Ae stars, with strong changes being observed only in the central
and red parts of the profile in contrast to the blue emission peak which is rather
stable (Fig. 2a). Such behaviour resembles the variability observed in AB Aur.
Only one object under investigation – HD 36112 – demonstrates the He I 5876
line profile to display only a blueshifted emission peak (Fig.2b). Therefore, the
inner envelope of a typical Herbig Ae star is likely to be kinematically stratified.
It contains the wind as the "basic" component combined with zones of discrete
non-stationary accretion.

In contrast to the CTTSs, the He I 5876 line profiles of HAEBEs do not
exhibit so-called NCs, which are thought to be connected with bright spots on
the stellar surface, the latter being in accordance with photometric data. As it
follows from the review by Herbst (1995), the photometric variability of HAEBEs
cannot be caused by either rotation of bright spots or by variations in the veiling
continuum.

1.3 Wind/Accretion Balance in the Envelopes of CTTSs and HAEBEs

Analysis of observational data allows to conclude that a stellar wind is present in the envelopes of HAEBEs along with matter infall, but their nature is not the same as in TTSs.

Accretion in the HAEBEs:

- is strongly concentrated to the equatorial plane;
- is discrete rather than continuous;
- is not so intense as in the case of the CTTSs;
- does not result in the formation of bright spots on the stellar surface.

Different character of wind/accretion processes in HAEBEs and in TTSs is confirmed by the fact that the disk signatures, bipolar jets and large-scale outflows are observed more rarely in HAEBEs than in TTSs. According to Mundt & Ray (1995), only 15% of the HAEBEs have highly collimated winds, which could be produced in the same way as the collimated TTS winds. All these distinctions can be caused by differences in magnetic field configurations on the stellar surface and/or in the circumstellar environment of these two classes of PMS objects. The role of accretion in the formation process of envelopes around HAEBEs is unlikely to be so important as in the case of the TTSs, and, as a consequence, the stellar wind is the dominant component in the envelopes of intermediate-mass PMS stars.

2 Wind Variability in the Herbig Ae/Be Stars

A more detailed description of circumstellar peculiarities in the HAEBEs can be derived from the analysis of their variability. It is known to occur on various timescales from hours to years and may be connected either with real variations in the envelope parameters or with the modulation of observed parameters by azimuthal inhomogeneities rotating around the star. In the latter case one can observe cyclic variations if the life time of these inhomogeneities is longer than a few rotation periods.

In the late eighties this type of spectral variability was discovered in three HAEBEs by Praderie et al. (1986), and Catala et al. (1986, 1989, 1991). They found periodic variations in the blueshifted absorption components of a number of envelope lines. It was also established that the periods are longer for the lines originating further away from the star: AB Aur – 32 hours for the Ca II k and He I 5876 lines, in comparison with 47 hours for the UV Mg II doublet; HD 163296 – 35 hours for the Ca II k line, and 50 hours for the Mg II doublet; HD 250550 – 41 hours for the Ca II k line. This fact was explained assuming the existence of stream-like inhomogeneities corotating with the star close to the stellar surface and rotating with smaller angular velocities at larger distances from the star. The origin of inhomogeneities has been suspected to be connected with local magnetic fields on the stellar surface powering the wind near the star.

In fact, one more important conclusion can be derived from this observational phenomenon. It is not difficult to show that the rotation period of a long-lived stream ejected from the stellar surface should be exactly equal to the stellar spin period at any distance from the star no matter how low the rotation velocity is. The period at any distance is determined by the phase rotation velocity of the stream rather than by its physical velocity. Therefore, longer periods for lines originating at larger distances should be connected with long-lived streams forming not on the stellar surface but at some distance from the star, where the angular velocity of rotation is smaller.

This type of cyclic variability in HAEBEs was investigated in detail by means of spectral and polarimetric observations which were carried out in the framework of the Pulkovo research programme (Beskrovnaya 1998; Beskrovnaya et al. 1994, 1995, 1998a,b; Pogodin 1994, 1997, 1998; Pogodin & Vieira 1997; Vieira et al. 1998). Cyclic variations of the blue edge of the Hα absorption component and of the Stokes parameters of linear polarization ($P = 32$ hours) have been found in AB Aur (Beskrovnaya 1998; Beskrovnaya et al. 1995; Pogodin 1998). Different periods have been discovered in the variations of different observational parameters in HD 36112 (Beskrovnaya 1998, Pogodin 1998): 42 hours (the Stokes parameters), 48 and 55 hours (some features on the Hα profile forming close to the region of maximum wind velocity and in the outer envelope respectively).

The following types of circumstellar inhomogeneities have been revealed:

- Very small short-lived condensations ("bullets") with a life time of 2–3 hours are sometimes observed in the HAEBE star HD 163296 (Pogodin 1994).
- Short-lived azimuthal inhomogeneities are present in the gaseous envelopes of HAEBEs. They manifest themselves as "standing waves" on the residual spectra of the Hα line in all the objects of the programme.
- Long-lived jet-like inhomogeneities can appear episodically in the winds of HAEBEs. Their rotation results in a cyclic variability of observed envelope parameters with periods $P = 1 - 5$ days. The period of variability is correlated with the distance from the region where this parameter originates.
- Longer periods of variability (P of the order of tens of days) are also observed in some HAEBEs. They are probably connected with the existence of dense condensations rotating around the star in the remote envelope (Beskrovnaya 1998, Beskrovnaya et al. 1998a).

3 A General Picture of Phenomena in the Envelope Around a HAEBE Star

Taking into account all the observational data available so far, the conclusion can be drawn that the wind and accretion are likely to be independent processes in the envelopes of HAEBEs.

Appearance of the wind in the high-temperature chromospheric region is confirmed by the detection of a blue-shifted component in the He I 5876 line. Similar character of variability of the He I and of some photospheric lines observed in the typical HAEBE star AB Aur (Catala at al. 1998) favours that the stellar wind in HAEBEs probably originates close to the photosphere (in contrast to winds of TTSs). The physical mechanism of wind generation is unknown for the time being, but it is possibly connected with local surface magnetic fields. In any case, the kinematical stratification of the wind may result in the interaction between outflowing streams with different velocities and, as a consequence, in spatial stratification of the envelope. Shocks near the boundaries of interacting flows may lead to high-temperature regions in the envelope in the form of either the chromosphere near the stellar surface or very hot local inhomogeneities in the wind which can be traced by emission in the N v lines (Bouret et al. 1998).

The accretion onto a HAEBE star is likely to be fed by an external envelope which may contain a protostellar disk with inner boundary at rather large distance from the star (in comparison with the TTSs) and, additionally, of a "secondary shell" supported by the low-velocity wind. This portion of the wind with initial velocities smaller than the escape velocity cannot leave the system and stimulates formation of a secondary disk-like gaseous envelope. Interaction of the external dense envelope with the faster wind can result in the loss of angular momentum by some portion of the circumstellar gas rotating around the star, leading to its subsequent infall onto the stellar surface. This hypothetical scenario can explain the origin of discrete accretion components and other phenomena observed in the HAEBEs.

As it is commonly recognized, the class of HAEBEs is very inhomogeneous. By now only a limited subgroup of HAEBEs with P Cyg-type profiles of the Balmer lines and with spectral classes ranging from B8 to A2 is studied in detail. Active phenomena and circumstellar peculiarities in other objects belonging to the class of HAEBEs are poorly studied so far. Reconstructing a complete picture of active phenomena in the envelopes of HAEBEs requires long continuous series of high-resolution spectral, photometric and polarimetric observations, which should be incorporated in future studies.

References

Beskrovnaya, N., Pogodin, M., Shcherbakov, A., Tarasov, A. (1994): A&A **287**, 564

Beskrovnaya, N., Najdenov, I., Pogodin, M., Romanyuk, I. (1995): A&A **298**, 585

Beskrovnaya, N. (1998): this volume

Beskrovnaya, N., Pogodin, M., Yudin, R., et al. (1998a): A&AS **127**, 1

Beskrovnaya, N., Miroshnichenko, A., Pogodin, M., et al. (1998b): A&A, submitted

Böhm, T., Catala, C., (1993): A&AS **101**, 629

Böhm, T., Catala, C., (1994): A&A **290**, 167

Böhm, T., Catala, C. (1995): A&A **301**, 155

Böhm, T., Catala, C., Donati, J.-F., and the MUSICOS collaboration (1996): A&AS **120**, 431

Böhm, T., Hirth, G. (1997): A&A **324**, 177

Bouret, J.-C., Catala, C., Simon, T. (1998): this volume

Calvet, N., Hartmann, L., (1992): ApJ **386**, 239

Camenzind, M. (1990): *Reviews in Modern Astronomy 3: Accretion and Winds* (Springer, Berlin), 234

Catala, C., Felenbok, P., Czarny, J., et al. (1986): ApJ **308**, 791

Catala, C., Simon, T., Praderie, F., et al. (1989): A&A **221**, 273

Catala, C., Czarny, J., Felenbok, P., et al. (1991): A&A **244**, 166

Catala, C., Donati, J.-F., Böhm, T., and the MUSICOS collaboration (1998): this volume

Edwards, S., (1997): in *Herbig-Haro Flows and the Birth of Low Mass Stars* (Kluwer Academic Publishers), 433

Fernandes, M., Eiroa, C. (1996): A&A **310**, 143

Finkenzeller, U., Mundt, R. (1984): A&AS **55**, 109

Grady, C., Pérez, M., Talavera, A., et al. (1996): A&AS **120**, 157

Graham, J. (1992): PASP **104**, 479

Grinin, V., Rostopchina, A. (1996): Astron. Reports **40**, 171

Gullbring, E., Petrov, P., Ilyin, I., et al. (1996): A&A **314**, 835

Hartmann, L., Hewett, R., Calvet, N. (1994): ApJ **426**, 669

Herbig, J. (1960): ApJS **4**, 337

Herbst, W. (1994): PASP **62**, 35

Königl, A. (1991): ApJ **370**, L39

Mundt, R., Ray, T. (1994): PASP **62**, 237

Petrov, P., Gullbring, E., Ilyin, I., et al. (1996): A&A **314**, 821

Pogodin, M. (1994): A&A **282**, 141

Pogodin, M. (1997): A&A **317**, 185

Pogodin, M. (1998): this volume

Pogodin, M., Vieira, S. (1997): in *Low Mass Formation–from Infall to Outflow* (BP 53, 38041 Grenoble, France), 244

Praderie, F., Simon, T., Catala, C., Boesgaard, A. (1986): ApJ **303**, 311

Reipurth, B., Pedrosa, A., Lago, M. (1996): A&AS **120**, 229

Shu, F., Najita, J., Ostriker, E., et al. (1994): ApJ **429**, 781

Vieira, S., Pogodin, M., Franco, G. (1998): A&A, to be submitted

Kundt: I disagree with the conclusion that an inverse-P-Cygni profile implies difficulties. Inverse-P-Cygni profiles can form in the receding hemisphere of a windzone.

Pogodin: Inverse P-Cygni profiles can form in the receding hemisphere of a wind only in the case of very unrealistic conditions when the emissivity of the CS gas is greater than the surface brightness of the star.

Owocki: I disagree with the previous comment regarding the impossibility of inflow. If material is driven away from a star's surface, but not maintained to escape speed, it can readily fall back down directly onto the star. Indeed, in multidimensional simulations, you can get wind outflow from some parts and inflow in others. The objection regarding angular momentum only applies to material originating from outside, which generally must pass through an accretion disk, preventing direct infall. Material of stellar origin does not suffer this problem.

Kaper: In the beginning of your talk you suggested that a possible difference between CTTSs and HAEBEs might be that HAEBEs cannot support a dynamo mechanism. Is a dynamo mechanism really necessary to generate a magnetic field or would a fossil magnetic field be sufficient to explain the observations?

Pogodin: The existence of magnetic fields in HAEBEs remains a matter of discussion. In any case they (if they exist) are smaller than in CTTs. It is possible that the observed distinctions in the outflow/infall balance between CTTs and HAEBEs are a result of the different strength and configuration of their large-scale magnetic fields. Maybe only a fossil magnetic field exists in HS. Really, a role of magnetism as a trigger of HAEBE-activity is rather indefinite (in contrast to the CTTs).

Ikhsanov: The interpretation of the non-symmetric single outflows observed in the Ae/Be Herbig stars requires a relatively high strength of the magnetic field and, that is the most important, the local, non-dipole configuration. The magnetic field of this type is rather difficult to achieve within the model of a collapsed protostellar cocoon. On the other hand, the effective amplification of the magnetic field in a local region is just the case of the dynamo process. In this context, the magnetic field amplification in the envelope or at the star seems quite reasonable.

Kakouris: The accretion disks of Ae/Be stars are not Keplerian as in TTs. Would you like to comment a little more on disk thickness and structure?

Pogodin: According to observational data, the main distinctions between the disks around CTTs and HAEBEs are the following:
1. The inner boundary of the HAEBE's disk is not as close to the star as in the case of TT disks.
2. The gas density is smaller in the case of the HAEBE disk.
3. Matter infall toward the HAEBE is discrete rather than continuous.

Wind and Accretion Variability in T Tauri Stars

Christopher M. Johns–Krull

University of California, Space Sciences Laboratory, Berkeley, CA 94720, USA

Abstract. The winds and accretion flows of classical T Tauri stars are highly variable on a variety of timescales. While much of this variability appears random, or chaotic, there are a few instances of periodic variations in the wind/accretion signatures on some stars. The periods detected are equal to either the stellar rotation period or in one instance equal to the orbital period in a close binary system. We discuss these results in terms of the popular magnetocentrifugally driven flow models for classical T Tauri stars.

1 Introduction

T Tauri stars (TTS) are newly formed, roughly solar mass stars, which have recently emerged from their parent molecular cloud cores to become optically visible. The classical TTS (CTTS) are those stars which are still surrounded by a dusty, active accretion disk (see review by Basri & Bertout 1993). The naked TTS (NTTS) are those stars which are no longer surrounded by an accretion disk. They appear to derive their activity from a solar-like dynamo operating in these typically very fastly rotating stars (Feigelson, Giampapa, & Vrba 1991).

Strong stellar winds flowing away from CTTS have been known since the earliest high-resolution spectra were made of the Balmer lines, particularly Hα (Kuhi 1964). It was quickly realized that the mass loss rates implied by the observations were much too large to be accounted for by a solar-like, thermally driven wind (DeCampli 1981), and thus began the long quest for the origin of the winds in these stars. The contribution by L. Hartmann in this volume reviews some of the current theories regarding wind acceleration from young stars. A strong connection between the accretion disk and the winds in CTTS was suggested by Walter et al. (1988) who noted that the NTTS lack both the evidence for disks and winds while the CTTS exhibit both phenomena simultaneously. This connection between the wind and accretion processes in CTTS is further supported by observations which show a positive correlation between the strength of the two flows for CTTS as a whole and individually with time (Cabrit et al. 1990; Hartigan et al. 1995; Johns–Krull & Basri 1997). Here, we present the results of a detailed analysis of the wind and accretion variability in a modest sample of CTTS.

2 SU Aur

SU Aurigae is a G2 CTTS. Herbst et al. (1987) found a tentative photometric rotation period of either 2.73 or 1.55 days, but their sampling was not dense enough to distinguish between these two periods. The high $v\sin i$ (~ 60 km s^{-1} – Johns–Krull 1996) combined with the rotation period imply that we are viewing the star nearly edge-on. We have analyzed over 100 spectra of SU Aur. The observations and data reduction are described fully in Giampapa et al. (1993) and Johns & Basri (1995a).

2.1   Periodogram Analysis of Line Profile Variations

Hα: The longest observing run in which nightly or better sampling was made of the spectra of SU Aur occurred over 14 nights in February 1988. The Hα line profiles are shown in Figure 1. This surface plot of the profile variations appears to show periodic intensity modulations in the blue wing of the line profile. Indeed, when these data are subjected to power spectrum analysis, significant power ($>$ 99% confidence) is found at a period of ~ 3 days in the blue wing where a variable blue-shifted absorption component appears (Johns et al. 1992; Giampapa et al. 1993). This is illustrated in Figure 2.

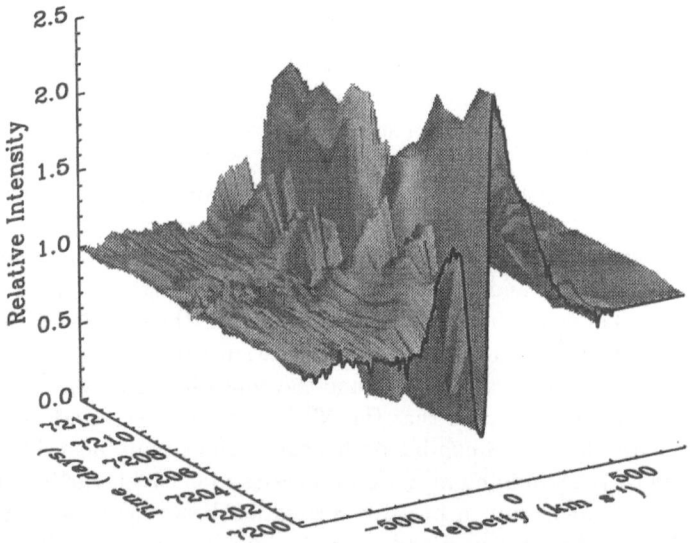

Fig. 1. A three-dimensional surface plot of the Hα line profile variations in SU Aur from February 1988. Velocity (wavelength) runs on the x-axis, time on the y-axis, and intensity on the z-axis. Note the apparent periodic intensity variations in the blue wing of the line profiles.

Figure 2 shows a strong peak at a period of ~ 3 days; equal to the suspected rotation period to within the uncertainty of the period determinations. This Figure also shows that the power spectrum estimate is very nearly equal to the

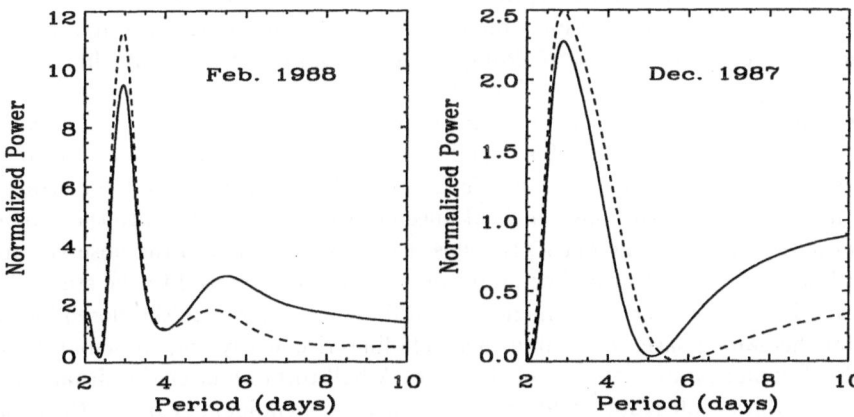

Fig. 2. The solid line in the left panel shows the power spectrum in the blue wing for the Hα profile variations from February 1988. The dashed line shows the power spectrum of the best-fit sine wave to this data. The right panel shows the same for the December 1987 run.

power spectrum of a pure sine wave sampled at the same times as the observations. As stressed by Horne & Baliunas (1986), with such sampling, power spectrum peaks should not be extremely narrow or strong. Figure 2 also shows the power spectrum estimate from the second longest run of nightly sampled spectra in our data: a 6 night run from December 1997. Again, there is a peak near 3 days and the power spectrum of the data is nearly identical to that of a pure sine wave sampled at these times. However, due to the very small number of observations, the confidence level for the data (and the pure sine wave) is too low to conclude there is periodicity present based on the December observations alone. Though not shown, the third longest continuous run (5 nights) also shows the same behavior. Taken together, we believe these power spectra do indicate that the periodicity is usually present in the blue wing of the Hα profile.

Given that the periodicity in the blue wing of the Hα profile appears to be present most of the time, the analysis of the entire time series taken together should produce a very strong, sharp feature near 3 days; however, the data shows only a broad, noisy feature slightly stronger than the strongest peak seen in Figure 2. This can be explained if there is a loss of coherence in the periodic signal at random times in the data span. This could correspond to the emergence and subsequent decay of an "active longitude" in the wind modulation which is replaced by a different "active longitude" later in time. This emergence and decay would then need to continue throughout the total time series.

Hβ: Johns & Basri (1995b) examined the behavior of many of the other line profiles in SU Aur. Most notably, they found the same periodicity in the blue wing of Hβ as is seen in Hα. In addition, Hβ shows a 3 day period in the intensity

variations of a red-shifted absorption component tracing mass accretion onto the
star. Together, the Balmer lines indicate simultaneous rotationally modulated
wind and accretion flows surrounding SU Aur. Comparing sine wave fits to the
wind and accretion signatures show that the variations in these two flows are
180° out of phase. This can be understood in terms of the Shu et al. (1994)
magnetocentrifugally driven flow model if the stellar dipole magnetic field is
tilted with respect to the rotation axis. The situation is schematically illustrated
in Figure 3. The closed magnetic field lines funnel disk material onto the star in
an accretion flow. At one phase, the closed field lines visible to the observer are
tilted nearly parallel to the disk, allowing easy flow of material to the star. The
wind is driven along the open magnetic field lines just exterior to the closed loops,
but at this same phase, the visible open field lines rise nearly perpendicularly out
of the disk and inhibit the flow of the wind. A half rotation later the situation is
reversed: the visible closed field lines rise nearly perpendicularly out of the disk,
making accretion difficult; while the open field lines are more nearly parallel to
the disk which results in an efficient centrifugally driven wind.

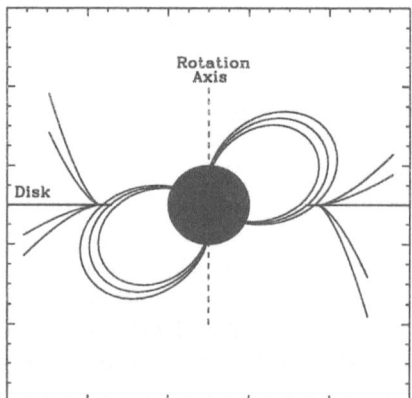

Fig. 3. A schematic diagram of the Shu et al. (1994) model with the magnetic axis
tilted with respect to the rotation axis. Disk material accretes onto the star via the
closed magnetic field lines and a wind is launched on the exterior open magnetic field
lines threading the disk. The observer is assumed to be to the far right, slightly above
the disk plane.

3   Sz 68

Johns–Krull & Hatzes (1997) have examined the spectroscopic variations of the
CTTS Sz 68, a K2 IV star with a photometric rotation period of 3.9 days. Sz 68
is a less extreme CTTS than SU Aur. The wind component of the star is hardly
noticeable, but Hα often displays a red-shifted absorption component tracing

the accretion of material onto the star. Two weeks of observations were made of this star in June/July 1995 at McDonald Observatory. The primary goal of these observations was to perform Doppler imaging on the star which revealed a decentered polar spot. Periodogram analysis of the Hα profile variations revealed periodicity in the red-shifted absorption component of Hα with the same period as the photometric rotation period. Sz 68 appears to be the second example of rotationally modulated accretion found in CTTS.

4   DQ Tau

DQ Tau is a newly discovered CTTS spectroscopic binary system (Mathieu et al. 1997; Basri et al. 1997). It is of particular interest because its highly eccentric orbit brings the stars so close to one another that their magnetospheres probably overlap. It is highly unlikely that circumstellar accretion disks can exist in this situation; however, the circumbinary disk appears to be continually feeding material to the stars. Accretion onto the stellar surfaces, as traced by the continuum veiling emission produced as this material shocks on the stellar surface, is present throughout the orbit, though it is strongly enhanced during periastron passage. DQ Tau's accretion is definitely cyclic, but with a period equal to the orbital period.

Typically, the wind in DQ Tau as traced by obvious blue-shifted absorption in the Hα line profile is undetectable for most of the orbit. However, at periastron passage, the wind strengthens along with the accretion in this system as is shown in Figure 4. At this point, the stars are furthest from the circumbinary disk and it is unlikely that their magnetospheres are interacting with the disk. Since there are no circumstellar disks, it therefore appears that CTTS can launch very strong outflows *without* the interaction of the star and a proper disk, as long as there is material near the star and its magnetosphere.

5   Conclusions

Some CTTS display cyclical variations in both their wind and accretion flows on rotational and orbital timescales. The work discussed here presents evidence for periodicity detected in 3 CTTS. It should be noted that Petrov et al. (1996) confirm the results for SU Aur from an entirely different data set. Hessman & Guenther (1997) present quasi–periods for two additional CTTS, but they rightly caution that their time series are too short to verify periodicity in these targets. One of their stars, DR Tau, was also analyzed by Johns–Krull & Basri (1995b) where extensive Hα time series for 5 additional CTTS were analyzed with no periodicities found. Taking all these facts together, we feel that something like the Shu et al. (1994) model shown in Figure 3 is at work in CTTS, but that the magnetic geometry on the stellar surface is probably evolving very rapidly, so that a true steady state is never achieved. The result is generally chaotic line profile variations which occasionally show hints of the underlying periodicity caused by the stellar rotation.

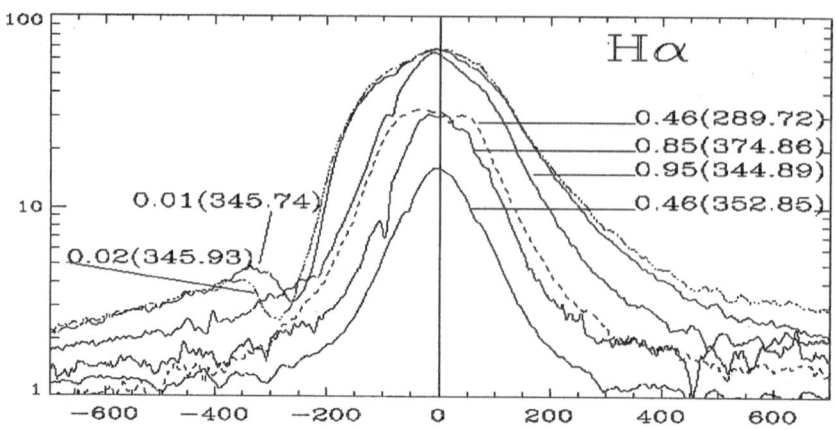

Fig. 4. Shown are a collection of Hα profiles during the orbit of the DQ Tau binary system. Each curve is labeled by orbital phase (with periastron at 0) as well as with a modified Julian date in parenthesis.

References

Basri, G., Bertout, C. (1993): in *Protostars and Planets III* (Univ. Arizona, Tucson), 543–566

Basri, G., Johns–Krull, C., Mathieu, R. (1997): AJ **114**, 781

Cabrit, S., Edwards, S., Strom, S., Strom, K. (1990): ApJ **354**, 687

DeCampli, W. (1981): ApJ **244**, 124

Feigelson, E. D., Giampapa, M. S., Vrba, F. J. (1991): in *The Sun in Time* (Univ. Arizona, Tucson), 658

Giampapa, M., Basri, G., Johns, C., Imhoff, C. (1993): ApJS **89**, 321

Hartigan, P., Edwards, S., Ghandour, L. (1995): ApJ **452**, 736

Herbst, W., et al. (1987): AJ **94**, 137

Hessman, F., Guenther, E. (1997): A&A **321**, 497

Horne, J., Baliunas, S. (1986): ApJ **302**, 757

Johns, C., Basri, G. (1995a): ApJ **449**, 341

Johns, C., Basri, G. (1995b): AJ **109**, 2800

Johns, C., Basri, G., Giampapa, M., DeFonso, E. (1992): in *Cool Stars, Stellar Systems, and the Sun: Seventh Cambridge Workshop* (ASP, San Francisco), 441

Johns–Krull, C. (1996): A&A **306**, 803

Johns–Krull, C., Basri, G. (1997): ApJ **474**, 433

Johns–Krull, C., Hatzes, A. (1997): ApJ **487**, 896

Kuhi, L. (1964): ApJ **140**, 1409

Mathieu, R., Stassun, K., Basri, G., et al. (1997): AJ **113**, 1841

Petrov, P., Gullbring, E., Ilyin, I., et al. (1996): A&A **314**, 821-834

Shu, F., Najita, J., Ostriker, E., et al. (1994): ApJ **429**, 781

Walter, F., Brown, A., Mathieu, R., et al. (1988): AJ **96**, 297

Kundt: I disagree with your infall interpretation. From the excitation conditions and the equivalent width of your lines, you get constraints on the separation and extent of the line-forming region. In my estimates, this region is much farther from the star than shown in Frank Shu's cartoon.

Johns-Krull: I really think the infall indicators are well founded. In numerous cases we see red-shifted absorption below the stellar continuum. I think this is only possible if the absorbing material is between the star and the observer. If this is the case, the material must be in infall.

Foing: Could you track the signature of matter infall from the disk to the surface along funnels, from your data sets? What is the expected duration for this infall?

Johns-Krull: Since the vast majority of the data I have taken is on a nightly time scale, we really do not have the sampling to observe this effect. According to the Shu et al. models, the time for material to fall from the disk to the star is the rotation period divided by 2π. For SU Aur this is about 0.5 days. For SU Aur, rotation has moved the initial material out of the star-observer line of sight by the time it reaches the star, so it is a difficult observation.

Ikhsanov: Do you really need an assumption about the open magnetic field lines for the interpretation of your data? Have you observed a collimated outflow?

Johns-Krull: Since these winds are magnetocentrifugally driven (at least in the models I discussed), the field lines must open for the wind to flow. However, it is my impression from the models, that it is the wind itself, once it gets beyond the Alfven radius, that forces the field lines open. In SU Aur, we have not actually observed a collimated outflow; however, line profile calculations show the winds cannot be spherically symmetric.

Hot Spots on Classical T Tauri Stars with High Mass-Loss Rates

Matilde Fernández[1], Carlos Eiroa[2] and William Schuster[3]

[1] MPI für Astronomie, Königstuhl 17, D–69117 Heidelberg, Germany
[2] Dpto. Física Teórica, C-XI, Universidad Autónoma de Madrid,
Cantoblanco, E–28049 Madrid, Spain
[3] Observatorio Astronómico Nacional, UNAM,
Apartado Postal 877, C.P. 22800 Ensenada, B.C., Mexico

Abstract. UBV(RI)$_c$ and Strömgren photometry has been carried out on a sample of 29 classical T Tauri stars for which a wide range of mass-loss rates has been reported in the literature. We measured their brightness variability in the 3400-8000 Å and 3400-5600 Å wavelength ranges during observing runs between 7 and 14 days long. To this sample we added other stars for which this information was taken from the literature. We found a correlation between the amplitude of the brightness variability and the strength of the mass-loss process.

Since high mass-accretion rates have been observed for those CTTSs that drive strong winds, we discuss our results in the framework of the magnetospheric accretion model. In this model the inner parts of the disk are disrupted and the matter from the disk is channelled along the magnetic field lines to high latitudes on the star, where an accretion shock (hot spot) develops. We estimated the temperature of the regions responsible for the variability and, although the derived relation suggests that variability could be related to the accretion process, hot spots were found less frequently than expected from this model. We explain this discrepancy as due to the effect of the blue non-photospheric continuum (veiling).

1 Introduction

T Tauri stars are young, low-mass stars ($\leq$ 2 M$_\odot$) still contracting towards the main sequence (Bertout 1994). This communication deals only with one of the two subgroups of these stars: the classical T Tauri stars (CTTSs), those in which a more or less strong accretion of mass still goes on. Nowadays, the model that best explains this accretion process is the magnetospheric accretion model, which has already been introduced (see Gullbring, this volume). To explain it very briefly: in this model the matter is accreted from a disk that surrounds the star (the so-called accretion disk). The inner parts of this disk are disrupted by the stellar magnetic field, and the matter that is to be accreted is channelled along the magnetic field lines to high latitudes on the star; at the bottom of the accretion column, an accretion shock (hot spot) develops. These hot spots are regions a few thousand degrees hotter than the unaltered photosphere.

Simultaneous with this mass-accretion process, mass loss has been observed in the CTTSs, the strength of both processes being strongly correlated (Hartigan et al. 1995).

In our project, winds were not the goal but the starting point of the study. We chose a sample of CTTSs with a wide range of mass-loss rates. At the time of the selection of the first sample, 1988, reliable mass-loss rates had only been estimated for a few T Tauri stars. We used as indicators of this mass loss the presence of P Cygni profiles of optical lines, wide wings of CO and/or optical jets. Broad- and intermediate-band photometry for 29 CTTSs has been carried out and the observations have been fit by simple stellar hot- and cold-spot models. As a result of this analysis, hot spots were found less frequently than expected from the magnetospheric accretion model. The discussion of this *lack of hot spots* is the subject of this communication.

2   Observations and First Analysis

A photometric monitoring program has been carried out on two, partially over-lapping, samples of CTTSs. For the first one (1988-92; Fernández 1995) we used the $UBV(RI)_c$ system and for the second one (1994-96) the Strömgren uvby system. The light curves (brightness versus time) of the stars almost never showed a periodic behaviour; thus, the amplitudes were computed as the difference between maximum and minimum levels of brightness. We found that the amplitude of the variations was larger for the stars with higher mass-loss rates. Due to the correlation between the mass-accretion and mass-loss processes found in CTTSs, the observed variability could be related to the accretion process.

We tried to fit the amplitudes at the different bands by a simple model of stellar spots. For both the cold[1] and the hot spot case the amplitude of the variability increases as the wavelength decreases, the rate of the change being much larger for the hot than for the cold spot. Nevertheless, the comparison between the observations and the model was not straightforward because the observed amplitudes presented an anomalous behaviour at short wavelengths, that is, below 4500 Å. As an example, the observed amplitudes of some of the stars of the two samples are shown in Figs. 1 and 2. For some stars the amplitude in the U band is smaller than that in the B band and for a few of them, larger. In Fig.1 for $LkH_\alpha 264$, for example, if we constrain the comparison with the stellar spot model to the $BV(RI)_c$ bands, no cold spot can reproduce the rate at which the amplitude decreases as wavelength increases, while hot spots in the range 6000 to 6400 K reproduce it within the errors of the measurements (Fernández & Eiroa 1996), the nonuniqueness of the solution being due to these errors.

In order to study this anomalous behaviour in more detail, Strömgren photometric observations were carried out. We chose this system because of its narrower band passes and because it covers the blue side of the spectrum (3400-5600 Å). Figure 2 shows the results of our uvby observations. The amplitudes observed for $LkH_\alpha 264$ are very well matched by hot spots with temperatures between 6300 and 6700 K.

---

[1] Cold spots have also been reported for the T Tauri stars; they are supposed to be stellar analogues of sunspots. They fit very well in the frame of the enhanced solar-type activity that has been proposed for the non-accreting, weak-line T Tauri stars.

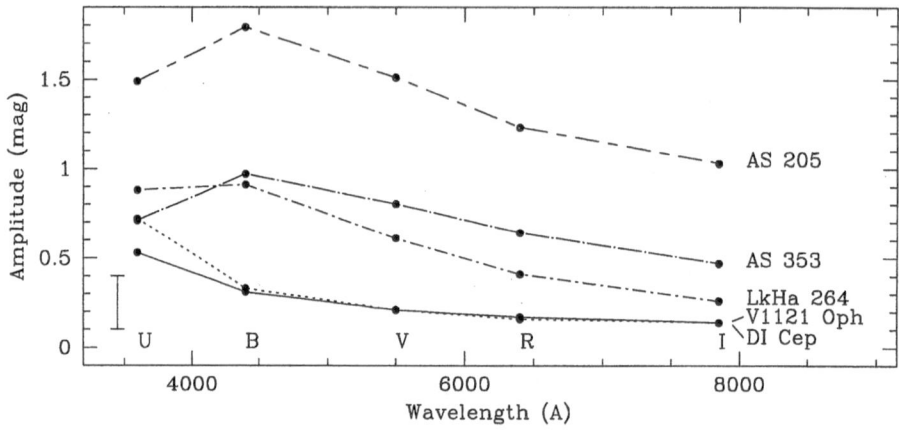

Fig. 1. The amplitude of the brightness variability versus wavelength for some of the stars observed during 1988-92. For each photometric band the position of its letter indicates its central wavelength, the error bar shows the average error of our measurements, and for the U band the error is sometimes a bit larger than indicated. ($\cdots$ corresponds to V1121 Oph).

3   Discussion

Data from the literature show that there is not a very strong correlation (at least not so strong as we had expected from the magnetospheric accreting model) between the amplitude of the brightness variability and the mass-loss (or mass-accretion) rate. At this point it is important to realize that if the star does not look like the black-body photosphere that has been assumed in the star-spot model, but there is something else superimposed on its continuum (as, in fact, has been observed for many CTTSs in the ultraviolet and infrared regions), the same amount of energy added by a hot spot will correspond to a smaller amplitude. This effect has already been described (Vogt, 1981; Bouvier et al. 1993): companions can change the rate at which the amplitude increases as wavelength decreases, e.g., a cold companion that adds flux at long wavelengths (I band) can make the amplitude at that band smaller and, thus, the spot to look hotter than what it really is. On the other, an extra source of flux at short wavelengths reduces the amplitude at the U band, making the spot appear colder.

Thus, the anomalous behaviour observed at short wavelengths suggests that something alters the blue continuum of some stars in our sample; maybe hot regions that are always visible, like the annular regions that can be generated by the matter channelled along the magnetic field lines. We have taken one of these annular regions into account in our model and the observed amplitudes are very well matched (although a 4-parameter model for a 5-amplitude set of observational data makes the results less convincing than a 2-parameter one). This hypothesis is supported for V1121 Oph, for example, by the fact that during

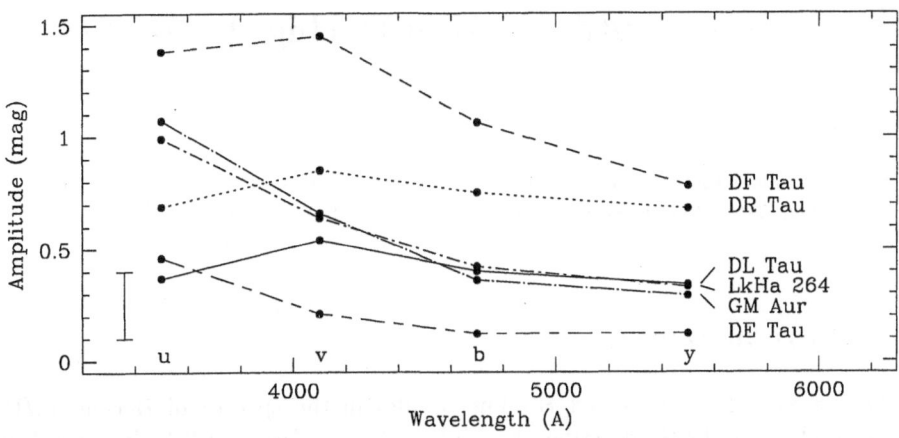

Fig. 2. The amplitude of the brightness variability versus wavelength for some of the stars observed in 1994. For each photometric band the position of its letter indicates its central wavelength, the error bar shows the average error of our measurements, and for the U band the error is sometimes a bit larger than indicated. (— corresponds to DL Tau).

the observing runs in which the amplitude at the U band was smaller than expected from the stellar spot model, the star was also bluer. For the stars with an amplitude in the U band much larger than what is expected from the stellar-spot model, a second, small and very hot spot (~30,000 K) can explain the observations.

From this analysis it can be concluded that it is dangerous to decide, from only optical photometry, whether the variability of CTTSs is due to a hot or a cold spot.

References

Bertout, C. (1994): in *Star formation and techniques in infrared and mm-wave astronomy*, eds. T.P. Ray and S.V.W. Beckwith, Springer-Verlag p. 49
Bouvier, J., Cabrit, S., Fernández, M., et al. (1993): A&A 272, 176
Fernández, M. (1995): A&AS 113, 473
Fernández, M., Eiroa, C. (1996): A&A 310, 143
Hartigan, P., Edwards, S., Ghandour, L. (1995): ApJ 452, 736
Vogt, S.S. (1981): ApJ 250, 327

Katsova: At what time scales do you observe the optical continuum variations?

Fernández: At time scales between a few days and two or three weeks.

Grinin: The cyclic brightness variability of T Tauri stars due to the hot spots is complicated by the variable circumstellar extinction. This is the reason why the hot-spot variability is clearly observed only in very few cases.

Anisotropic Outflows from Herbig Ae/Be Stars

Vladimir Grinin[1,2]

[1] Astronomical Institute of St. Petersburg University, Russia
[2] Crimean Astrophysical Observatory, Crimea, Nauchny, Ukraine

1 Introduction

The variety of the observed Hα line profiles in the spectra of Herbig Ae/Be (HAEBE) stars firstly described by Finkenzeller & Mundt (1984) demanded, as believed, the same or almost the same variety of kinematical models for their interpretation (Catala 1994). Traditionally, the outflow of matter is considered as the main physical process in HAEBE stars and the star AB Aur which exhibits strong P Cygni profiles in many emission lines is frequently considered as the prototype of HAEBE stars.

In this paper we argue that the real situation is, probably, opposite: the main physical process in the neighbourhood of HAEBE stars is the disk accretion and the outflow of matter is a secondary phenomenon.

2 UX Ori Stars: their Variability and Orientation in Space

The observational evidence supporting this point of view stems from the study of the most photometrically active HAEBE stars (so-called UXORs) which prototype is the star UX Ori. In many respects these interesting objects are the key for understanding the physics of HAEBE stars. Their large-amplitude Algol-type variability is caused by variable circumstellar (CS) extinction and the role of the other mechanisms of variability (such as the surface magnetic activity) is negligible. The observational evidence of this suggestion was found in the course of the multi-year photopolarimetric monitoring of UXORs started in 1986 at the Crimean Astrophysical Observatory (Grinin et al. 1991; Grinin 1994). These observations have shown that as a rule the linear polarization of UXORs anti-correlates with the brightness changes. Such an anti-correlation is a natural property of young stars surrounded by protoplanetary disks in which the opaque dust fragments (clouds) cross from time to time the line of sight. When screening a star from an observer, they act as a natural coronograph. As a result the weak scattered radiation of the CS disk begins to dominate when the star fades.

The high linear polarization (up to 5-8%) systematically observed in deep minima led us to the suggestion (Grinin 1992) that the UX Ori type stars are those young objects whose disk-like dust envelopes are oriented edge-on relatively to the observer. This suggestion was supported in a recent paper by Natta et al.

(1997) in which the masses of CS dust envelopes of the UXORs were estimated from millimeter observations. It was shown that the amount of dust is the same as in non-active HAEBE stars of similar spectral type.

3   Variability of HAEBE Stars and Statistics of the $H\alpha$ Line Profiles

The dependence of the photometric activity of HAEBEs on the orientation of their CS disks relatively to an observer provides an interesting possibility for studying the geometry and kinematics of the emission regions around these stars. Taking this into account we have analyzed the $H\alpha$ emission line in the spectra of UXORs and found that almost all of them have two-component profiles typical for rotating disks (Grinin 1992). This result was recently confirmed by a more detailed analysis (Grinin & Rostopchina 1996). Fig. 1 shows the statistical dependence of $H\alpha$ profiles on the level of photometric activity of HAEBE stars: the majority of single and P Cygni profiles are observed in the photometrically quiet stars (the pole-on objects); the two-component profiles are more frequently observed in the photometrically active stars (i.e. UXORs).

In several UXORs the ratio $V/R > 1$ in the $H\alpha$ line indicates a combination of rotation and radial motion to the star. Numerous spectroscopic signatures of the gas infall onto UX Ori type stars were observed over the last years in the optical and ultraviolet region of the spectrum (Pérez et al. 1993; Grady et al. 1995; Grinin et al. 1994, 1995; de Winter 1995; Kozlova et al. 1996). Summarizing these data we conclude (Grinin & Rostopchina 1996) that *the accretion phenomenon is observed in those HAEBE stars whose equatorial planes are close to the line-of sight; the opposite motion (outflow) is more typical for the pole-on or intermediate orientation of CS disks*. A similar conclusion is drawn in the paper by Grady et al. (1996) in which the spectroscopic signatures of the accretion and stellar winds in HAEBE stars were also considered in the context of their photopolarimetric activity.

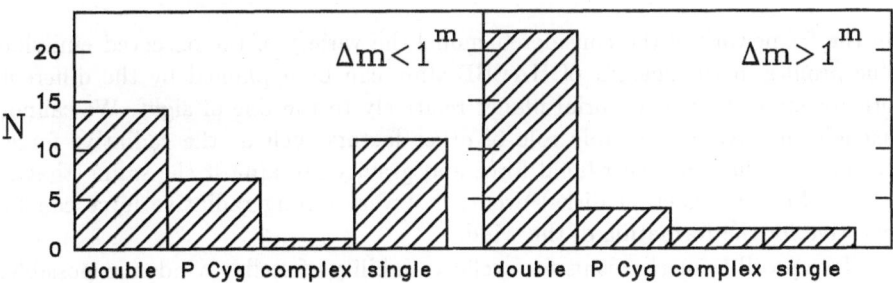

Fig. 1. Statistics of $H\alpha$ profiles in the spectra of photometrically quiet and active Herbig Ae/Be stars (from Grinin & Rostopchina 1996).

4   Discussion

The complex kinematical picture discussed above resembles the gas motion expected in the magnetospheric accretion models (Camenzind 1990; Königl 1991; Shu et al. 1994). These models were successfully adopted by Hartmann et al. (1994) to the classical T Tauri stars. However, their application to HAEBE stars faced several problems (Böhm & Catala 1994), such as: i) the absence of veiling in the spectra of HAEBEs (see also Ghandour et al. 1994), ii) the absence (or weakness) of asymmetry in the forbidden lines (see also Corcoran & Ray 1997), iii) the large optical thickness of the accretion disks (Hartmann et al. 1993). These difficulties gave rise to the suggestion that the CS disks are not a common property of these young stars but are only a special feature of the low-mass young objects (Hartmann et al. 1993).

In reality, the absence of veiling and asymmetry of the forbidden lines is compatible with the accretion models if the mass accretion rate $\dot{M}_a \leq 10^{-7} M_\odot/\mathrm{yr}$. The calculations show (Tambovtseva & Grinin 1997) that the hydrogen and helium lines in the spectra of UXORs can be explained even with smaller $\dot{M}_a$. The main problem of the low accretion rates is connected with the explanation of the large IR luminosities of HAEBE stars in the framework of the classical model of accretion disks (Hartmann et al. 1993). A possible way to solve this problem is by modification of this model. It has to take into account that the CS disks around young stars are in reality more complex objects than predicted by the models of viscous geometrically thin accretion disks. In the case of UXORs we observe the young star immediately through its CS disk and we see that the disk is a highly heterogeneous object consisting of a lot of gas-dust fragments of various sizes. The model calculations by Mitskevich (1996) show that the inhomogeneous structure of the CS dust envelopes is important in modeling their IR radiation and leads to a quite realistic spectral energy distribution.

5   Conclusion

In the framework of the considered model the variety of the observed emission line profiles in the spectra of HAEBE stars can be explained by the different orientation of their equatorial planes relatively to the line of sight. We cannot exclude, however, a possible role of other factors such as the radiative force. In the case that the centrifugal force and gravity are almost the same (that is typical for the accretion disks) even a not very strong radiative force can be effective for the formation of the wind.

Two possible mechanisms for cyclic variability of stellar winds are possible: a) the rotational modulation of the CS gas envelopes (in the case of inclined magnetic dipole) and b) the periodic variations of the mass-accretion rate in the case of binary systems. Both these mechanisms are discussed in the context of T Tauri stars (see e.g. Giampapa et al. 1993; Gullbring et al. 1996; Mathieu et al. 1997) and can also be applied to HAEBE stars.

References

Böhm, T., Catala, C. (1994): PASPC, **62**, 26

Camenzind, P. (1990): Rev. in Modern Astronomy, **3**, 233

Catala, C. et al. (1986): ApJ, **308**, 791

Catala, C. (1994): PASPC, **62**, 91

de Winter, D. (1995): *Observational Aspects of Herbig Ae/Be stars*, Dissertation, Univ. of Amsterdam

Finkenzeller, U., Mundt, R. (1984): A&AS, **55**, 109

Ghandour, L. et al. (1994): PASPC, **62**, 223

Giampapa, M. et al. (1993): ApJS, **89**, 321

Grady, C. et al. (1995): A&A, **302**, 472

Grady, C. et al. (1996): A&ASS, **120**, 157

Grinin, V.P., Kiselev, N.N., Minikhulov, N.Kh. et al. (1991): ApSS, **186**, 283

Grinin, V.P. (1992): Astron. Astrophys. Tran. **3**, 17

Grinin, V.P. (1994): PASPC, **62**, 63

Grinin, V.P., Thé, P.S., de Winter, D., et al. (1994): A&A, **292**, 165

Grinin, V.P., Kozlova, O.V., Thé, P.S., Rostopchina, A.N. (1996): A&A, **309**, 474

Grinin, V.P., Rostopchina, A.N. (1996): Astron. Rep., **40**, 171

Gullbring, E. et al. (1997): A&A, **314**, 835

Hartmann, L et al. (1993): ApJ, **407**, 219

Hartmann, L. et al. (1994): ApJ, **426**, 669

Kozlova, O.V. et al. (1996): Astron. Astrophys. Trans. **8**, 18

Könadvl, A. (1991): ApJ, **370**, L39

Mathieu, R.D. et al. (1997): AJ, **113**, 1841

Mitskevich, A.S. (1996): A&A, **298**, 231

Mundt, R., Ray, T.P. (1994): PASPC, **62**, 237

Natta, A., et al. (1997): ApJ, accepted

Perez, M., Grady, C., Thé, P.S. (1993): A&A, **274**, 381

Shu, F. et al. (1994): ApJ, **429**, 797

Kundt: How do you support your proposed thick accretion disks? How hot are they?

Grinin: Herbig Ae/Be stars are more massive objects than T Tauri stars and I suppose that their CS disks are geometrically thicker due to their larger "turbulent" velocity connected with their more rapid Keplerian rotation. The higher turbulence can also be caused partially by formation of planetesimal bodies in the protoplanetary disks of HAEBE stars.

The maximal temperature in the part of the disk closest to the star should be enough (about $20000\,K$) to explain the appearance of the He I absorption line at $5876\,\text{Å}$ in the spectra of UXORs.

Variability in the Wind of the FU Ori Object Z Canis Majoris

D. Chochol[1], M. Teodorani[2], F. Strafella[3], L. Errico[2], and A.A. Vittone[2]

[1] Astronomical Institute, Slovak Academy of Sciences, Tatranská Lomnica,
The Slovak Republic
[2] Osservatorio Astronomico di Capodimonte, Via Moiariello 16, I-80131 Napoli, Italy
[3] Dipartimento di Fisica, Universitá di Lecce, I-73100 Lecce, Italy

We present high-resolution, time-resolved spectroscopy of the pre-main sequence FU Ori variable Z CMa obtained with the 1.82 m telescope of the Asiago Observatory between 1997 January 14–17. The time evolution of Hα and Hβ is shown in Fig. 1. Large night-to-night and hour-to-hour variations in the P Cygni profiles were observed. Variations in the red wings of hydrogen emission lines were also detected. The radial velocity (V_r) of the broad Hα P Cygni absorption decreased from -590 km s^{-1} (Jan. 14) to -450 km s^{-1} (Jan. 17). The absorption became narrow, while the broad emission bump with a peak at $V_r = -900$ km s^{-1} appeared on Jan. 17. Hα absorption is formed in the wind along the line of sight to the star, so the changes in the absorption are caused either by variations of the mass outflow or changes of the angle of collimated outflow to the line of sight. The wind could be collimated and accelerated from the underlying star/accretion disk system by a magnetic field. The acceleration of the wind is supported by the velocity gradient within the line-forming region (the blue edge of the absorption component is at a higher velocity in Hα than in the Na I D lines). A magnetically driven bipolar wind from an inclined rotator provides the best explanation of the observed spectroscopic changes. In this case we expect cyclical variability in the wind.

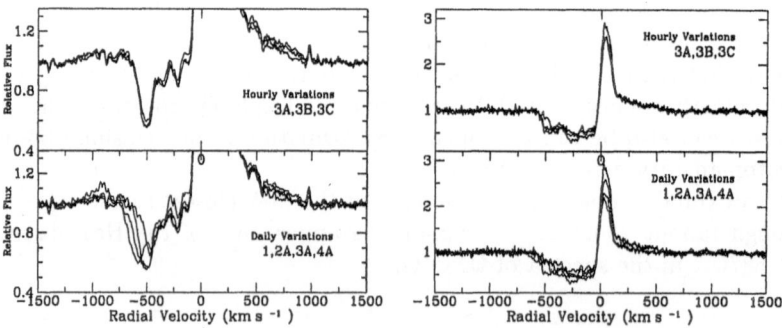

Fig. 1. Night-to-night and hour-to-hour variations of Hα (left) and Hβ (right) over four consecutive nights from 1997 Jan. 14 to 17.

A Variability Study of High-Resolution Profiles of DL Tauri

Jorge F. Gameiro and M. Teresa V.T. Lago

Centro de Astrofísica and Departamento de Matemática Aplicada da Faculdade de
Ciências da Universidade do Porto, Rua do Campo Alegre, 823, 4150 Porto, Portugal

We report on the variability of the profiles of Hα, He I λ5876, and the Na I
D lines (λλ5890, 5896) for the T Tauri star DL Tau. The data were obtained at
the INT (La Palma Observatory) with the Intermediate Dispersion Spectrograph
between 1990 October 1 and 7, and includes repeated observations on time scales
of one day.

Hα is the line that displays the largest variations during the week. The profile
broadly maintains its shape with a strong blue shifted absorption. The variability
in the equivalent width results mainly from variations in the intensity of the line
as a whole. The lines of He I and Na I are much weaker, but also vary in intensity
as a whole without varying in shape. The Na I D lines show an inverse P Cygni
profile that disappears during one night but appears again on the following
nights.

From our set of data it is not clear if there exist any time scale for the
variations.

We conclude that the observed variations probably result from variations in
the continuum, namely fluctuations in the veiling.

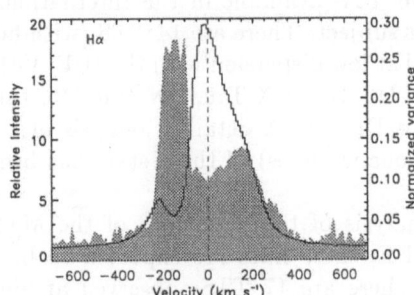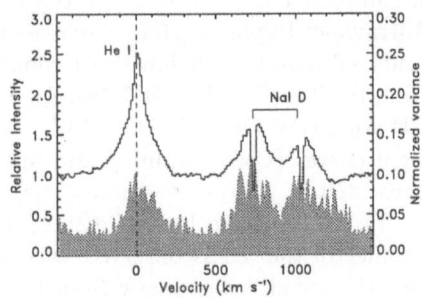

Fig. 1. Average profile and normalized variance of Hα, He I λ5876, and the Na I D
lines during 1990 October 1–7.

The UV Variability of T Tauri Stars

Ana I. Gómez de Castro[1], Merche Franqueira[1], Nuria Huélamo[1],
and Eva Verdugo[2]

[1] Instituto de Astronomía y Geodesia (CSIC-UCM), Facultad de CC Matemáticas,
Universidad Complutense de Madrid, Avda. de la Complutense s/n,
E-28040 Madrid, Spain
[2] ISO Science Operations Centre, P.O. Box 50727, E-28080 Madrid, Spain

The T Tauri stars (TTS) are low mass pre-main-sequence stars that show an enhanced continuum and line emission when compared with main sequence stars of similar spectral types. This excess is especially noticeable at ultraviolet wavelengths, where a major fraction of the gravitational binding energy released during the accretion of the circumstellar material is radiated and the stellar continuum emission fades dramatically. The chromosphere and transition region may also contribute significantly to the UV spectrum. Disentangling both contributions and understanding whether and how they may be related is a major problem in the physics of star formation and early stellar evolution. We have approached it by studying the overall UV properties of the TTS and their variability in detail.

Some classical TTS exhibit periodic $UBVRI$ photometric variability with periods between 2 and 15 days. In some sources these variations are produced by the presence of hot spots on the stellar surface and result from the modulation of the observed flux by rotation. The hot spots likely indicate the location of accretion shocks on the stellar surface. In this contribution we have presented a complete compilation of all the data on TTS available in the International Ultraviolet Explorer (IUE) Archive on this subject. There are 14 TTS (with hot and cool spots) which have been monitored at low dispersion with the IUE: V410 Tau, BP Tau, RY Tau, HD 283572, T Tau, DF Tau, UX Tau, DN Tau, DR Tau, SU Aur, GW Ori, AB Dor, TW Hya, and DI Cep. A detailed analysis of the light curves (UV continuum and main resonance lines) of these stars has been presented.

In addition, we have presented an analysis of the variations of the Mg II profiles of the TTSs observed with the IUE; these lines are expected to be a major coolant in magnetic funnel flows. There are 17 TTSs observed at high dispersion with the IUE and 12 of them have been observed more than once. We show that significant variations are found in the profiles of BP Tau, RY Tau, DG Tau, and SU Aur.

Acknowledgements: We thank the IUE Observatory VILSPA staff for their assistance. This research was financed by the Ministerio de Educación y Cultura of Spain through Research Grant PB93-0491.

Spectroscopic Investigations of Cyclical Variability in the Gaseous Envelopes of Early-Type Emission-Line Stars

Mikhail Pogodin

Pulkovo Observatory, Pulkovo, 196140 Saint Petersburg, Russia

This contribution presents some results of high-resolution spectroscopy for three Herbig Ae stars. The investigation was performed within the framework of a large programme to study the structural and kinematical peculiarities of the gaseous envelopes around Ae/Be stars of different types. More than 300 high-resolution spectra of three Herbig Ae stars (Herbig 1960) have been obtained at ESO (Chile, 1.4 m CAT/CES), the CrAO (Crimea, 2.6 m Shajn telescope), and the LNA (Brazil, 1.6 m telescope). Analysis of the observational data shows that these objects are likely to contain azimuthal inhomogeneities rotating around the star. In particular:

1. The anti-correlated behaviour of variations in the blue- and red- emission wings of $H\alpha$ is clearly seen in HD 163296 and HD 36112 (Beskrovnaya et al. 1998a,b).
2. In 1991 July, the central peaks of the $H\alpha$ and $H\beta$ emission lines in the spectrum of HD 163296 exhibited sinusoidal-like positional variability with a period close to 4.5 days (Beskrovnaya et al. 1998a).
3. Periodic variations of localized features in the $H\alpha$ line profile have been found in AB Aur ($P = 32^h$) and in HD 36112, with periods from 42 to 55 hours for features originating in different regions of the envelope (Beskrovnaya et al. 1995, 1998b).

References

Beskrovnaya, N., Najdenov, I., Pogodin, M., Romanyuk, I. (1995): A&A 298, 585
Beskrovnaya, N., Pogodin, M., Yudin, R., et al. (1998a): A&AS 127, 243
Beskrovnaya, N., Miroshnichenko, A., Pogodin, M., et al. (1998b): A&A, submitted
Herbig, G. (1960): ApJS 4, 337

Polarimetric Studies of Cyclic Phenomena in the Circumstellar Envelopes of the Young Herbig Ae/Be Stars

Nina Beskrovnaya

Pulkovo Observatory, 196140 St. Petersburg, Russia

Analysis of line profiles allows the geometric and kinematic characteristics of gaseous inhomogeneities to be determined. In addition, multi-colour polarimetry makes it possible to study the structure of the circumstellar medium close to the star, and to diagnose the magnetic field inside a moving inhomogeneity. The influence of the magnetic field on the polarization parameters was investigated in the framework of a simple model of a gaseous jet rotating in the equatorial plane of the star (Beskrovnaya & Pogodin 1998), with allowance for the Faraday rotation of the polarization plane due to electron scattering in the magnetized media (Gnedin & Silant'ev 1984). This method was used to study circumstellar peculiarities of a well-known Herbig Ae star, AB Aur. We investigated variations of the Stokes parameters observed on 1994 January 9, when the rotating jet was travelling close to the line-of-sight (as determined from analysis of the spectroscopic data; for details see Beskrovnaya et al. 1995). The observed locus of the Stokes parameters on the (q, u)-plane is in good agreement with the results of calculations for $i = 60°$ and $B = 0$. Another type of cyclic phenomena can be noticed in the polarimetric variations of HD 163296 (for details see Beskrovnaya et al. 1998a). Sinusoid-like curves with a period of $7\overset{d}{.}5$ are clearly seen in both Stokes parameters in the V-band, while in the I-band sinusoidal variations are seen only in the q-parameter. This effect can be explained by assuming that the rotating gaseous condensation is magnetized, since the influence of the magnetic field is stronger at longer wavelengths. A multi-component picture of circumstellar inhomogeneities moving with different periods at different distances from the star has also been revealed from spectral and polarimetric analysis of HD 36112 (Beskrovnaya et al. 1998b).

References

Beskrovnaya, N.G., Pogodin, M.A., Najdenov, I.D., Romanyuk, I.I 1995, A&A 298, 585

Beskrovnaya, N.G., Pogodin, M.A. 1998, Astrofizika, in preparation

Beskrovnaya, N.G., Pogodin, M.A., Yudin, R.V., et al. 1998a, A&AS 127, 243

Beskrovnaya, N.G., Pogodin, M.A., Miroshnichenko, A.S., et al. 1998b, A&A, submitted

Gnedin, Yu. & Silant'ev, N. 1984, Ap&SS 102, 375

Comparing Observations of Cyclical Variability in Hot- and Cool-Star Winds

D.J. Mullan

Bartol Research Institute, University of Delaware, Newark 19716, USA

Abstract. In hot-star winds, mass-loss events propagate through the wind on time-scales of order days: this makes it possible for a campaign lasting a few weeks (e.g. IUE MEGA) to follow the evolution of a number of events in their entirety. Here, we present the first results of an archival analysis of wind lines in two isolated cool stars, one a giant, the other a supergiant: the analysis is based on a set of Fe II lines with a broad range of intrinsic strengths. The results suggest that discrete mass-loss events can be identified as they move through the wind: the time-scale for propagation of an event through the wind seems to be 2–3 years for the giant and 4–8 years for the supergiant. Although the conditions for applying the Sobolev approximation may be only marginally satisfied in cool star winds, we find that nevertheless the SEI code is useful in fitting wind line profiles observed in cool stars.

1 Introduction

Hot stars have been the subject of many observational campaigns (e.g. IUE MEGA) in order to study the evolution of events in the winds. Here we report on a study of cool star winds from a similar perspective.

2 Studies Using the h and k Lines of Mg II

The first exploration of temporal variability in the winds from cool stars was based on a study of the asymmetry of the Mg II h and k lines (Mullan & Stencel 1982): the ratio S/L of the maximum flux in the shortward (S) and longward (L) emission peaks is variable in some stars. The variations are especially pronounced in the K0 IIp star 56 Peg: in this star, S/L varies from values less than unity (suggesting mass outflow) to greater than unity. Time-series analysis of S/L variability in a sample of hybrid stars over an interval of a few years (Brosius et al. 1985) suggested periods in the wind which in some cases are consistent with rotational modulation (as predicted from evolutionary models). These rotational estimates led to the CIR hypothesis (Mullan 1984, 1986). The ratio S/L is sensitive to variations *in the wind*, while the sum $S + L$ is a measure of *chromospheric* behavior. The fact that time-series analysis of the *total flux* in Mg II lines (i.e. $S + L$) over a 15 year interval shows no significant periods in some hybrid stars (Brown et al. 1996) proves that the chromospheric heating in these stars is not coherently periodic over 15-year time-scales. But there may be short-lived active regions which come and go at random longitudes on much shorter time-scales: effects of these would be easier to identify in shorter time-series (e.g. Brosius et al. 1985). Moreover, to study the occurrence of periodicities *in the wind*, it is better to use S/L data (which is sensitive to changes on the blue side of the line, rather than total flux). Thus, the claim of Brown et al. that the periodicities identified by Brosius et al. (1985) are spurious may be premature.

Specific cases of variations in the winds of two hybrid stars on time-scales of 5-6 months have been reported by Brown et al. in their study of Mg II h and k lines. In one case (α Aqr: G2Ib), the h and k lines changed in a 6-month interval in a way that suggested a change in terminal velocity of $\sim$25 km/sec.

3   Studies Using Fe II Lines

A disadvantage of using Mg II lines is their great strength: they typically probe conditions only in the outermost parts of the wind. (Ly-α might probe even farther out [J. Eaton, priv. comm.]) The limitation of this is obvious when we consider the solar wind: if we were to take a snapshot of the wind at any instant, different processes might be at work at different radial distances. For example, there might be a recently launched coronal mass ejection (CME) in the inner wind, one or more CIR's in the intermediate wind, merged interaction regions farther out, and maybe the remains of an older CME in the outermost wind.

To probe a broad range of radial distances in the wind, we use a set of lines of varying strength: Fe II lines serve well here, because of the broad range on intrinsic strengths of the lines which exist in the near UV. As a measure of the strength of a line, we use a relative optical-depth parameter $\tau_{rel} = \tau_o$ where τ_o is defined by Carpenter et al. (1995) and Carpenter & Robinson (1997): the lines in their lists have τ_o values which range over some 4 orders of magnitude. By choosing a set of lines from these lists, we can probe conditions in the wind at a broad range of radial distances. Moreover, the Fe II wind lines are known to exhibit large variations from one epoch to another (Mullan et al. 1998).

4   Method of Analysis of Wind Effects in the Line Profiles

We use GHRS spectra and archival IUE data for two isolated cool stars: λ Vel (a K-supergiant) and γ Cru (an M giant). We define an empirical optical depth $\tau_{emp}(v)$ of any line by comparing the intensity on the blue side $I_b(|v|)$ at velocity -v to the intensity on the red side $I_r(|v|)$ at velocity +v relative to the photosphere: $\tau_{emp}(v) = \ln[I_r(|v|)/I_b(|v|)]$. The relation between τ_{emp} and the physical optical depth of the wind at velocity $|v|$ is complicated (Mullan et al. 1998), and we regard τ_{emp} as a purely empirical quantity to study variations in time in various lines.

We refer to the area under the τ_{emp} versus v curve to as the "Integrated optical depth". And the velocity at which τ_{emp} first passes through zero is referred to as the "terminal velocity": this is actually a combination of the physical terminal velocity of the wind and a turbulent broadening contribution which increases with relative line strength.

We display measurements of these quantities, and their variations with time, in two different formats: (i) as a function of relative line strength τ_{rel} at a single epoch, and (ii) as a function of time for individual lines.

5 Results for λ Vel

In Fig. 1, we show how the integrated empirical optical depth of 11 lines varies with τ_{rel}. The solid curve refers to data taken in 1994: note the monotonic trend towards larger empirical optical depth in the stronger lines. The relation between empirical optical depth in the wind and physical optical depth in the wind is clearly not a simple linear one: thus, whereas the relative line strengths span a range of some 3 orders of magnitude, the values of empirical optical depth span a range of only 1.5 orders of magnitude. Nevertheless, the existence of a monotonic curve in 1994 is a consistency check that the empirical optical depth parameter is a physically meaningful quantity. In 1982, when the curve in Fig. 1 is also close to monotonic, the values of wind optical depth τ_{emp} are smaller than in 1994 by factors of ∼2 in all parts of the wind except the inner regions. In 1978, the exposures did not allow us to extract data for all lines: the curve in Fig. 1 is therefore too sparse to allow much comment.

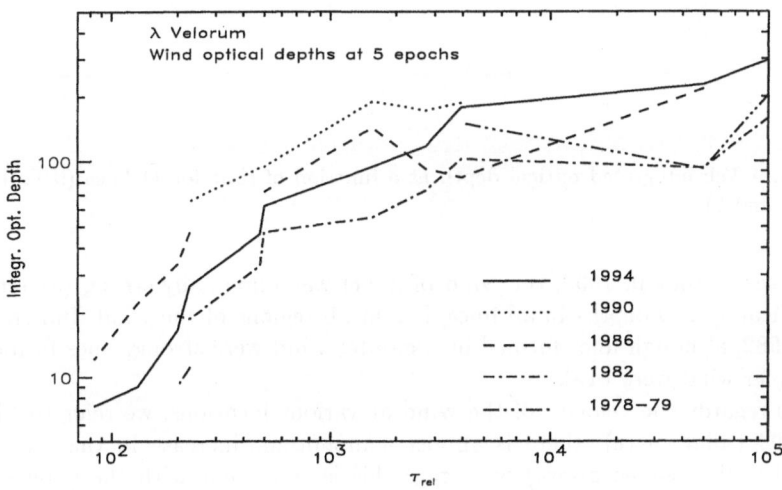

Fig. 1. λ Vel: integrated optical depth as a function of the relative line strength τ_{rel} at 5 epochs.

A clear departure from monotonicity appears in 1986: lines with $\tau_{rel} \leq 1000$ are stronger (by ∼1.5-2) than the same lines in 1994, whereas lines with τ_{rel}

around 4000 are weaker (by ~2) than in 1994. In 1990, the lines with τ_{rel} around 1000 are stronger than in 1986 at all τ_{rel} for which measurements exist.

To show how the individual lines vary in strength as a function of time, we plot in Fig. 2 the integrated τ_{emp} for each line with a different symbol. Many of the 11 lines can be measured reliably at only some of the 5 epochs: at other epochs, the spectra are either over- or under-exposed at the wavelength of that line. Only three lines are measurable at all 5 epochs: these are plotted with the largest filled symbols in Fig. 2. The weaker two of these 3 lines show a trend of increasing in strength from 1978 to 1990, and then decreasing in 1994; these two lines have $\tau_{rel} = 200$ and 500. The strongest of the three lines (with $\tau_{rel} = 4000$) also has maximum strength in 1990, but it was also strong in 1978 and 1982.

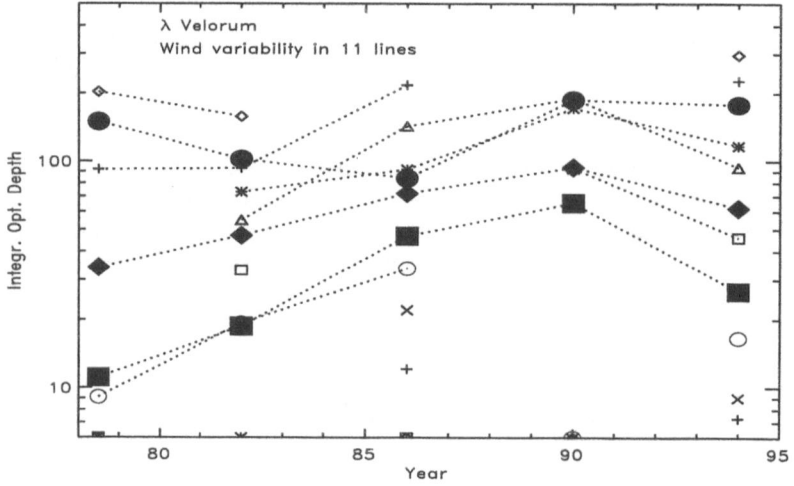

Fig. 2. λ Vel: integrated optical depth as a function of time for 11 lines (9 Fe II lines plus h and k).

It seems that in 1990, the wind of λ Vel was maximally strong (as regards empirical optical depth) in all lines, i.e., at all regions of the wind. But in 1978 and 1982, although lines formed in the outer wind were strong, lines formed in the inner wind were weak.

As regards the velocity of the wind at various locations, we refer to Fig. 3. In 1994 (solid curve), there is almost a monotonic increase in the "terminal velocity" (as defined above) with τ_{rel}: this is consistent with the expectation that the stronger lines sample the outer wind which has accelerated to higher velocities. And again in 1982, the results are also close to monotonic. But in 1986 and 1990, there is a large enhancement in velocity for lines with τ_{rel} around 1000: velocities are 100-120 km/sec in 1986 and 1990, compared to 50-70 km/sec in 1994. These are precisely the lines which appeared in Figs. 1 and 2 as unusually strong in 1986 and 1990. Thus, these lines are not only more absorbing than usual: they are also moving faster than usual. Results for individual lines are

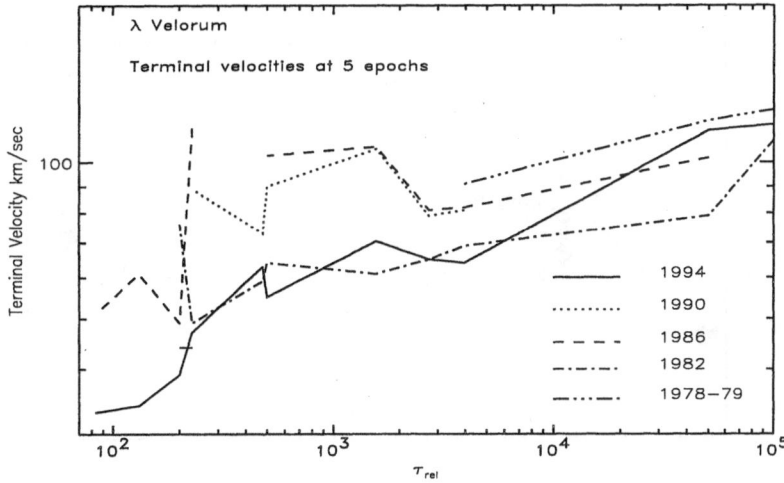

Fig. 3. λ Vel: terminal velocity as a function of relative line strength τ_{rel} for 11 lines (9 Fe II lines plus h and k).

presented in Fig. 4. The years 1990 and 1986 are clearly epochs in which the wind material is on average faster than in 1994 or in 1982.

6   Results for γ Cru

Analogous results for γ Cru are presented in Figs. 5–8. In this case, the wind was thickest in 1989, and thinnest in 1984. The inner wind also appeared relatively thick in 1978 and 1986, and relatively thin in 1992.

7   Discussion of Variability in Cool Star Winds

The results in Figs. 1–8 indicate the advantage of using a set of Fe II lines to probe conditions at different radial distances in the wind.

For both stars, epochs of maximally thick wind coincide also with epochs of fastest wind. This is an interesting conclusion as regards the mechanism of wind driving: whatever is doing the driving is able to push denser material to higher speeds. This behavior is not consistent with the general behavior in the solar wind: there, the density and velocity are *anti-correlated* such that nv^2 remains almost invariant (Mullan 1983). What we are seeing in λ Vel and γ Cru is exactly the opposite behavior: n and v both increase together. Such behavior is more typical of exceptional periods in the solar wind when a transient (such as a CME) is present. Moreover, the finite extent of the CME means that only a finite region of the solar wind is affected by the CME at any instant of time.

The results in Figs. 1–4 are consistent with the possibility that a CME is moving through the intermediate wind of λ Vel in 1986 and 1990, and has reached

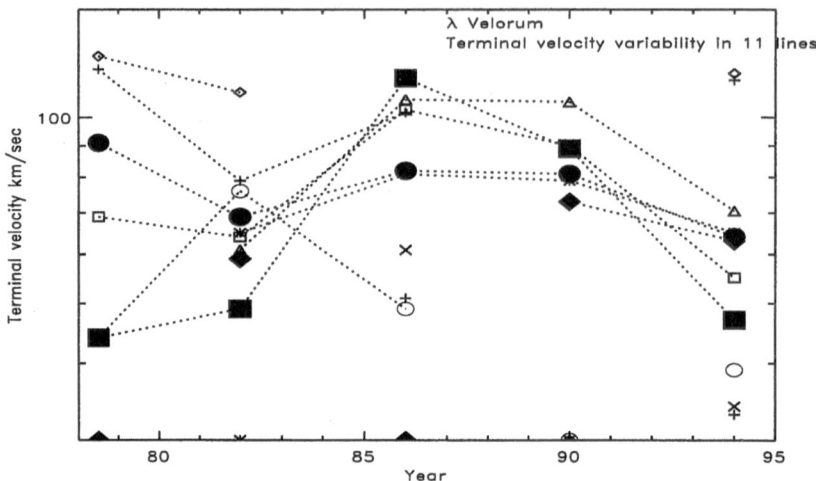

Fig. 4. λ Vel: terminal velocity as a function of time for 11 lines (9 Fe II lines plus h and k).

the outer wind in 1994. Also in 1978, the outer wind may still contain material left over from an earlier CME. If this interpretation is correct, it seems that, in this supergiant star, time-scales of 4-8 years are required before the effects of a CME are "flushed" through the wind.

As for γ Cru, although the data are noisier, it seems that there is fast and dense material in the inner wind in 1978, but this has disappeared by 1984. Then another dense and fast event is apparent in 1986 and 1989. By 1992, this material is fading away from the inner wind. Thus, in this giant star, time-scales of 2-3 years may suffice for the "flushing". (In reply to a question of J. Gosling, other IUE data indicate no significant variations on shorter time-scales.) The fact that these are somewhat shorter than in λ Vel is consistent with the smaller radial extent of the giant compared to the supergiant.

Note the logarithmic scales in the figures: in 1992, the empirical optical depth of the wind in γ Cru, as measured by lines with $\tau_{rel} \approx 20$, is 10 times weaker than in 1989. Thus, the wind perturbations are by no means trivial.

Are the perturbations cyclic? Even though we have used essentially the entire IUE data base, plus more recent spectra obtained with HST, there are not yet enough data to decide.

8 Sobolev Approximation in Cool Star Winds

The Sobolev approximation has been very useful in extracting quantitative information on the properties of winds in hot stars. We have explored the possibility that it might also be useful in cool-star winds. The criterion for the Sobolev approximation to be valid is that the velocity gradient dv/dr must exceed the

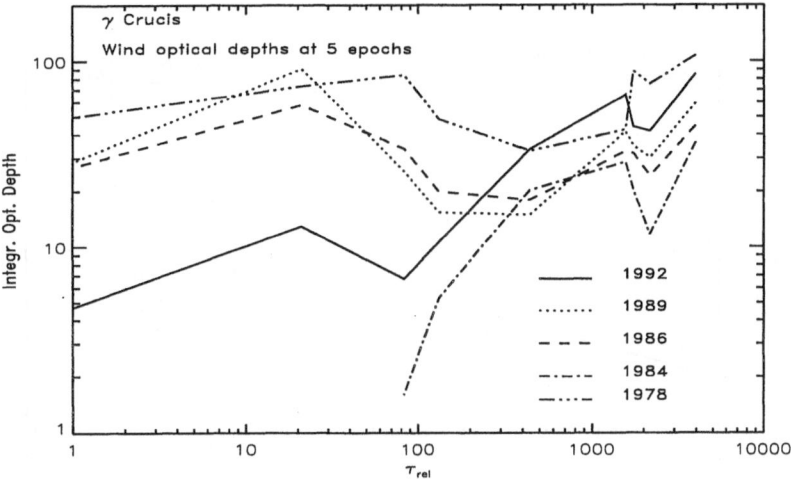

Fig. 5. γ Cru: integrated optical depth as a function of the relative line strength τ_{rel} at 5 epochs.

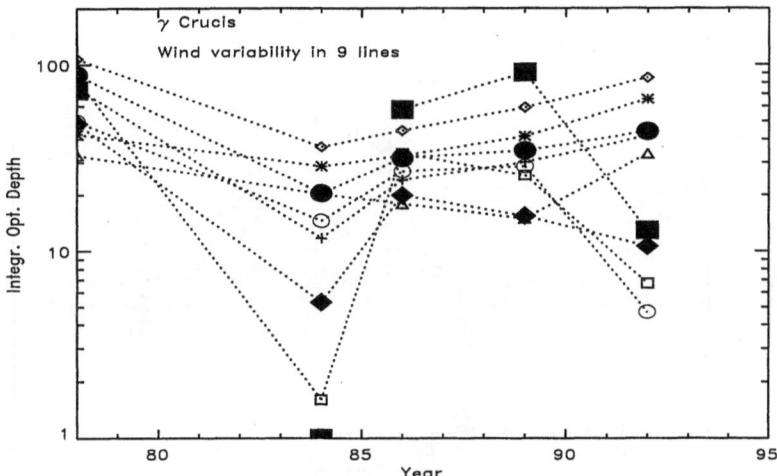

Fig. 6. γ Cru: integrated optical depth as a function of time for 9 Fe II lines.

ratio of v_t (the turbulent broadening of the line) to L (the length scale over which density varies in the wind). In a wind which accelerates to terminal speed v_∞ in radial distances of order R_*, the Sobolev ratio $R_s = (dv/dr)/(v_t/L)$ is of order $(v_\infty/v_t) \times (L/R_*)$. In hot stars, where $v_\infty \geq 1000$ km/sec, and $v_t \sim 100$ km/sec, R_s is of order 10 if $L \sim R_*$. Thus, the Sobolev approximation is well justified in hot stars. When we turn to cool stars, the criterion is almost certainly not as well satisfied because v_∞ is much smaller. Nevertheless, in order

180   D.J. Mullan

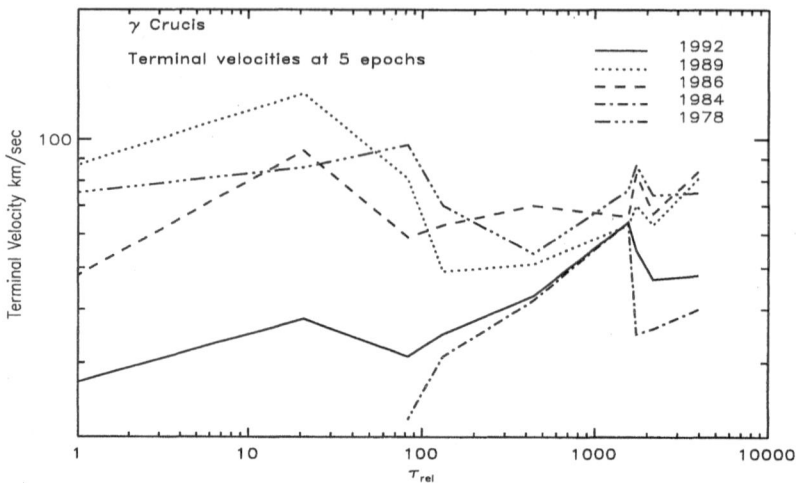

Fig. 7. γ Cru: terminal velocity as a function of relative line strength τ_{rel} for 9 Fe II lines.

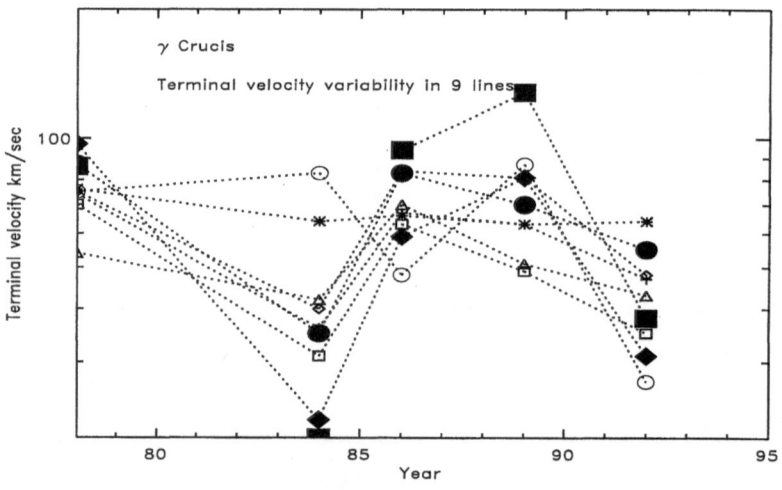

Fig. 8. γ Cru: terminal velocity as a function of time for 9 Fe II lines.

to determine some of the parameters of winds in cool giants, we have applied the SEI code (Lamers et al. 1987) to the line profiles in cool giants.

In using the SEI code, we vary five wind parameters in order to fit an observed profile: (i) β and (ii) v_∞ in the velocity law ($v(r) = v_\infty(1 - R_*/r)^\beta$); (iii) α in the optical depth law ($\tau(v) \sim v^\alpha$); (iv) total optical depth τ_{tot} in the line; and (v) turbulent velocity $w_{turb} = v_t/v_\infty$. We also need to input a line profile at the base of the wind: to do this, we use the red wing of the line we wish to fit

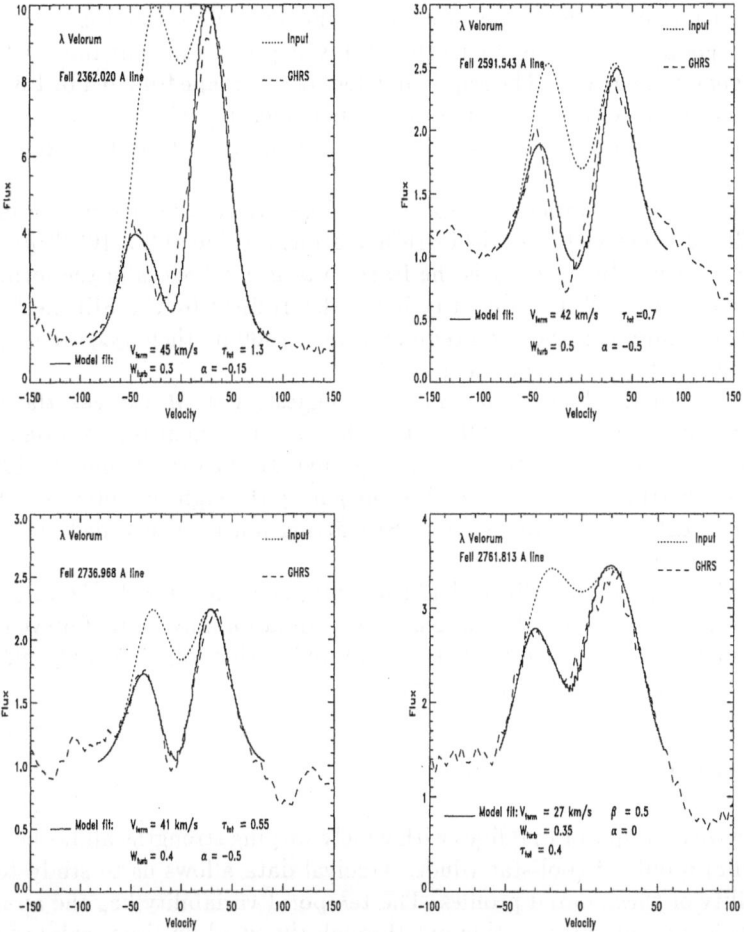

Fig. 9. λ Vel: SEI fits to observed lines. Solid: SEI. Dashed: observed profile. Dotted: input profile at the base of the wind.

and construct a symmetric profile which fits both the far red wing and the far blue wing. In our calculations, we assume the case of conservative scattering: the wind is so cold that no photons are generated in the wind. The assumption of a symmetric profile at the base of the wind is probably acceptable for the weaker lines in our study. However, for the strongest lines, asymmetries introduced by velocity fields in the chromosphere are expected to become more pronounced. This is certainly true for the case of the Mg II k line (A. Dupree, priv. comm.), and it may also affect some of the strongest Fe II lines in our sample.

In Fig. 9, we show preliminary results of fitting 4 lines in the 1994 GHRS spectrum of λ Vel. (Abscissa is velocity in km/s relative to the photosphere: ordinate is flux in arbitrary units.) We have not attempted an exhaustive study of parameter space: the figures are merely meant to illustrate that the SEI code yields a good zeroth-order fit to the observed profiles of four lines, whose τ_{rel} values span a range of 6. The required values of τ_{tot} range from 0.4 to 1.3, increasing monotonically (but, as noted above, not linearly) with τ_{rel}. It is especially noteworthy that the fits require a large turbulent velocity in the wind, as large as $0.5v_\infty$.

Using the total optical depths of each line, we can determine the mass-loss rate: all four lines yield $\dot{M}$ within 10% of a mean value of $2 \times 10^{-10}/g_{FeII}$ solar masses per year. Here, g_{FeII} is the fraction of iron which is in the form of Fe II in the wind of λ Vel: this quantity is not determined by the SEI method. But, by way of example, if the wind temperature is 7000 K, then $g_{FeII} \approx 0.1$, and $\dot{M} \approx 2 \times 10^{-9}$ solar masses per year.

SEI fits to the line profiles in 1990 suggest that $\dot{M}$ exceeds the value in 1994 by a factor of about 4. Of course, the SEI treatment is based on spherical symmetry, and on a monotonic velocity profile. To the extent that the 1990 wind of λ Vel is perturbed by an "event" propagating through the intermediate wind, the numerical values obtained from SEI fitting must be regarded with caution (H. Lamers, priv. comm.).

The fact that all four lines yield consistent estimates for the $\dot{M}$ value (at the 10% level) suggests that the Sobolev approximation may be useful even though the formal criterion suggests that the approximation might be only marginally applicable (e.g. Drake & Linsky 1983).

9 Summary

Analysis of a sample of Fe II lines with widely varying strengths allows us to probe the radial profile of cool-star winds. Archival data allows us to study temporal variability of these radial profiles. The temporal variability has the appearance of mass loss events propagating out through the wind on time-scales of years.

References

Brosius, J., et al. (1985): ApJ **288**, 310
Brown, A. et al. (1996): ApJS **107**, 263
Carpenter, K.G. et al. (1995): ApJ **444**, 424
Carpenter, K.G., Robinson, R.D. (1997): ApJ **479**, 970
Drake, S.A., Linsky, J.L. (1983): ApJ **273**, 299
Lamers, H.J.G.L.M. et al. (1987): ApJ **314**, 726
Mullan, D.J. (1983): ApJ **272**, 325
Mullan, D.J. (1984): ApJ **283**, 303
Mullan, D.J. (1986): A&A **165**, 157
Mullan, D.J., Stencel, R.E. (1982): ApJ **253**, 716
Mullan, D.J. et al. (1998): ApJ **495**, (March 10 issue, in press)

Nitrogen V in the Wind of the Pre-main Sequence Herbig Ae Star AB Aurigae

J.-C. Bouret[1], C. Catala[1], and T. Simon[2]

[1] Laboratoire d'Astrophysique de Toulouse, Observatoire Midi-Pyrénées,
14 avenue Edouard Belin, F-31400 Toulouse, France
[2] Institute for Astronomy, University of Hawaii,
2680 Woodlawn Drive, Honolulu, Hawaii 96822, USA

The Herbig Ae star AB Aur was observed at intermediate spectral resolution with the GHRS aboard the HST in the spectral region surrounding the N V doublet at 1238.8, 1242.8 Å. We identified the emission feature detected at $\lambda \sim 1238.8$ Å as one component of the 2s–2p resonance doublet of N V.

The presence of N V lines lead us to introduce an additional high-temperature zone to the current model for the wind of AB Aur (Catala & Kunasz 1987). Spherically symmetric models produce features that are not observed, but which could be made undetectable by lowering the filling factor of the hot zone. We reduced this filling factor by assuming the presence of corotating interaction regions (CIRs), which are generated by a fast and slow wind stream structure controlled by the surface magnetic field. This assumption is justified because azimuthal structures are known to exist in the wind, as witnessed by the rotational modulation detected in some lines of AB Aur. Since the temperature in the CIRs is expected to rise far above that of the unperturbed stellar wind, we assumed that the N V lines can originate in part of the CIRs, while X-ray emission is produced in the hottest parts of them. The observed N V line is fairly well reproduced by such a model, assuming a filling factor of 7×10^{-3} for the CIR and temperatures on the order of 1.4×10^4 K. If we assume that higher temperatures, in the range of $1-2 \times 10^6$ K, are also present in other regions within the CIRs, then the same filling factor is consistent with the observed X-ray flux from AB Aur. By evaluating the amount of non-radiative energy deposited in the wind, we have found that the presence of CIRs keeps the global energy balance of the wind of AB Aur almost unchanged. These results suggest that the wind of AB Aur is very inhomogeneous and includes azimuthal structures that cause CIRs.

References

Catala, C., Kunasz, P.B. (1987): A&A 174, 158
Catala, C., Praderie, F., Felenbok, P. (1987): A&A 182, 115
Mullan, D. (1984): ApJ 283, 303
Zinnecker, H., Preibisch, Th. (1994): A&A 292, 152

Part 3

PROCESSES AFFECTING
THE EMERGENCE
OF THE STELLAR WIND

Line-Profile Variability as a Diagnostic for Non-radial Pulsation Mode Identification

John Telting[1] and Coen Schrijvers[2]

[1] Isaac Newton Group of Telescopes, ASTRON, Santa Cruz de La Palma, Spain
[2] Anton Pannekoek Instituut, University of Amsterdam, Netherlands

Abstract. We use a model of a rotating and adiabatically pulsating star to investigate the observable spectroscopic characteristics of non-radial pulsations. We calculate time series of absorption line profiles in a carefully chosen domain of parameter space.

We find that the intensity variations in time series of theoretical spectra, at each position in the line profile, cannot be described by a single sinusoid: at least one harmonic sinusoid needs to be included. Across the line profile the relative amplitudes and phases of these sinusoids vary independently.

The blue-to-red phase difference found at the main pulsation frequency turns out to be an indicator of the degree ℓ, rather than the azimuthal order $|m|$; the phase difference of the variations with the *first harmonic* frequency is an indicator of $|m|$.

We present linear relations between observable phase differences and the parameters ℓ and $|m|$. These relations can be used to identify pulsation modes. This method works for spheroidal and toroidal, sectoral and tesseral modes. The method is also applicable to multi-periodic multi-mode pulsations.

We apply the method to new data of three rotating β Cephei stars.

1 Introduction

The study of pulsations in stars provides direct tests for the validity of stellar evolution models. With observed pulsation frequencies one can constrain these models, provided that the pulsation modes can be identified. For non-radial pulsations the relevant parameters are the degree ℓ and azimuthal order m, which specify the tangential shape of the pulsation mode. The intrinsic pulsation frequency is physically linked to the degree of the pulsation (e.g. Dziembowski & Pamyatnykh 1993, Gautschy & Saio 1993). For rotating stars the azimuthal order affects the apparent frequency, since the modal pattern is rotating with the star. Hence, for asteroseismological purposes one needs accurately determined values of ℓ, m and the observed frequency. In this paper we focus on the possibility to spectroscopically determine ℓ and m in rotating early-type stars; in many cases, our results are applicable to δ Scuti stars as well.

Other work relating to spectroscopic mode identification is presented by e.g. Vogt & Penrod (1983), Campos & Smith (1980), Kambe & Osaki (1988), Gies & Kullavanijaya (1988), Kennelly et al. (1992, 1998 in press), Aerts et al. (1992), Aerts (1996), Townsend (1998 in preparation), Hao et al. (1998 in press).

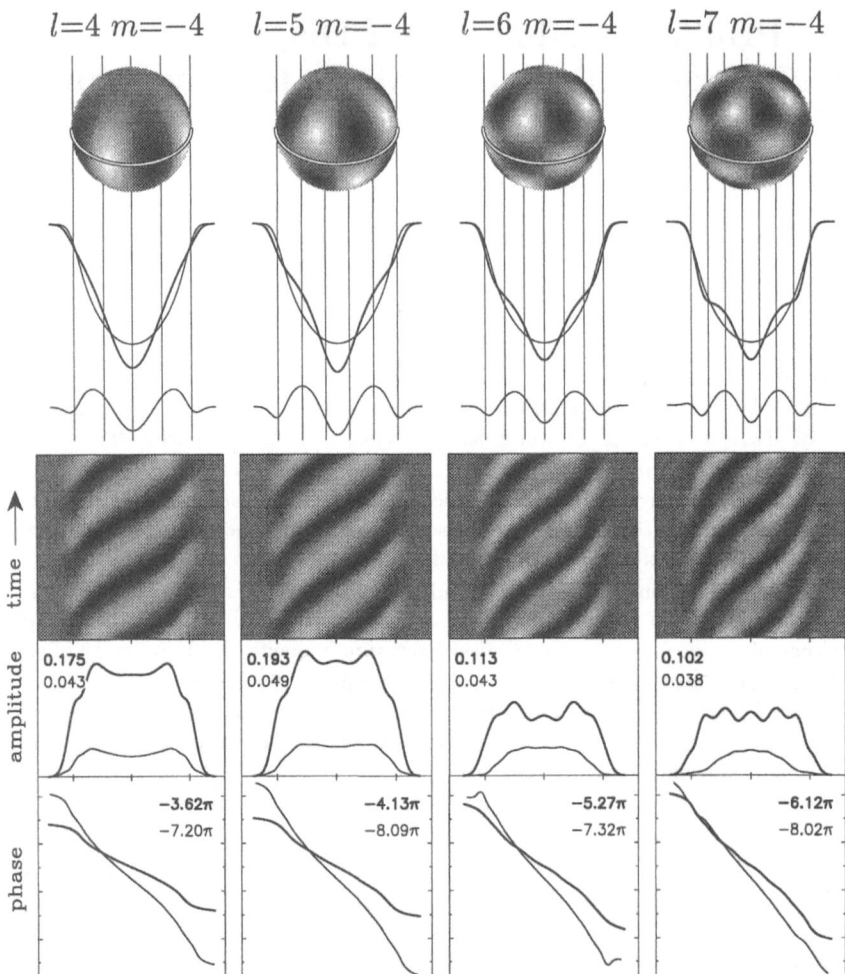

Fig. 1. Time series of generated spectra (lower grey-scale panel) are Fourier transformed to give amplitude and phase diagrams. For the pulsation frequency (thick lines) and its first harmonic (thin lines), the amplitudes and phases are plotted as a function of position in the line profile (bottom two panels). The phase difference from blue to red in the profile, of the main phase diagram, is a measure of pulsation parameter ℓ. That of the *harmonic* phase diagram gives $|m|$.

2   Model and Analysis

We model line-profile variability caused by adiabatic non-radial pulsations of a rotating star, with the rotation axis as the symmetry axis of the pulsation. Our model of the surface-velocity field is essentially the same as the one described by Aerts & Waelkens (1993), with a few improvements which have been discussed in Schrijvers et al. (1997). We account for local brightness and equivalent-width changes, which are induced by the oscillatory temperature variations (Buta & Smith 1979, Lee et al. 1992). A complete and detailed description of our model, the parameters, and the analysis of time series of spectra is given by Schrijvers et al. (1997), Telting & Schrijvers (1997) and Schrijvers & Telting (1998, in preparation).

We treat the unknown pulsational velocity amplitudes V_{max}, $k^{(0)}$, the degree ℓ and order m, and the rotation parameter $\Omega/\omega^{(0)}$ as free and independent parameters. Together with the inclination i of the star, the width W of the intrinsic profile and the three temperature related parameters $(\delta T/T)_{max}$, ϕ_{lag}, and $\alpha_{W_{\dot{E}}}$, they form a set of 10 parameters that determine the variation of the line profiles.

To investigate the effects of the relevant parameters on observable character-istics of line-profile variations (Figure 1), we generate time series of absorption line profiles and analyse these with a technique equivalent to that proposed by Gies & Kullavanijaya (1988). For each wavelength bin in the line profile, the in-tensity variations are Fourier analysed. The variational power and phase at the pulsation frequency and its first harmonic can then be plotted for each position in the line profile.

The line-profile variations are usually not strictly sinusoidal (see e.g. Gies 1991, Reid & Aerts 1993); we find that in general one harmonic frequency needs to be added to properly describe the variability. Throughout the line profile the phases and amplitudes of the main pulsation frequency and its first harmonic show independent behavior, which is a clear indication that the inclusion of a harmonic term provides additional information in the analysis of the line-profile variability.

The blue-to-red phase differences $\Delta\Psi_0$ and $\Delta\Psi_1$ are obtained by reading off the *maximum phase change* between the outermost wavelength positions in the profile (Figure 1, bottom panel). Note that we do not require a priori knowledge of a value of $V_e \sin i$ to read off the blue-to-red phase differences.

3   Relation Between the Observable Blue-to-red Phase Differences and Pulsation Parameters ℓ and m

For diagnostic purposes we quantify the relation between ℓ and $|\Delta\Psi_0|$, and that between $|m|$ and $|\Delta\Psi_1|$; here we restrict ourselves to spheroidal modes.

Table 1. Parameter ranges used in our Monte-Carlo simulations

ℓ	$0 - 15$	$\log k^{(0)}$	$-2.0 - 0.5$
m	$-\ell - \ell$	$\Omega/\omega^{(0)}$	$0.025 - 0.35$
i	$25° - 90°$	$(\delta T/T)_{max}$	$0.0 - 0.05$
$W/V_e \sin i$	$0.05 - 0.1$	ϕ_{lag}	$-15° - 15°$
$V_{max}/V_e \sin i$	$0.05 - 0.2$	α_{W_E}	$-1.5 - 1.5$

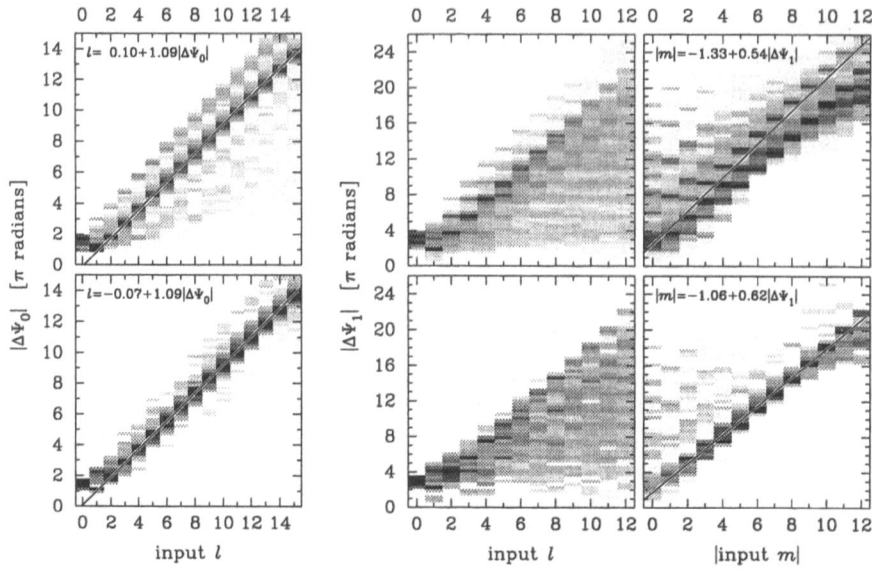

Fig. 2. Results of Monte-Carlo calculations. We plot the blue-to-red phase difference $|\Delta\Psi_0|$ against ℓ, and $|\Delta\Psi_1|$ against ℓ and $|m|$, with the number of occurrences on a grey scale. For each vertical bin in the plots the grey scale is normalized to the total number of computed modes in that bin. **Top:** All 15360 modes, with stellar and pulsation parameters as specified in Table 1. **Bottom:** Selection of all modes, with $k<0.3$ (mainly p modes), $i>45°$ and without slope reversals in $\Psi_0(\lambda)$.

We compute 60 time series of line profiles for each combination of ℓ and m with $\ell\leq15$. The other relevant parameters (i, W, V_{max}, $\Omega/\omega^{(0)}$, $k^{(0)}$, $(\delta T/T)_{max}$, ϕ_{lag}, α_{W_E}) are chosen at random within the ranges specified in Table 1. Values for the inclination are drawn according to the probability $p(i) = \sin i$. Values for $\log(k^{(0)})$ are drawn such that the combination of $k^{(0)}$ and $\Omega/\omega^{(0)}$ corresponds to an equatorial velocity of less than 50% of break-up. All other parameters are drawn from a flat distribution. For these 15360 time series we derive the absolute blue-to-red phase differences $|\Delta\Psi_0|$ and $|\Delta\Psi_1|$.

In Figure 2 we present the results of these calculations. We plot $|\Delta\Psi_0|$ against ℓ, and $|\Delta\Psi_1|$ against both ℓ and $|m|$, with the number of occurrences displayed as a grey value. From our calculations, we conclude that the assumption that the phase difference $\Delta\Psi_0$ is a measure for $|m|$ (Gies & Kullavanijaya 1988) is only correct for sectoral modes, which have $|m|=\ell$. In general the phase difference $\Delta\Psi_0$ is a measure of ℓ (rather than $|m|$), while the phase difference $\Delta\Psi_1$, if detectable, is a reasonable estimator of the value of $|m|$.

To quantify the relation between the observable phase differences and ℓ and $|m|$ we perform a least-squares fit of a straight line to the data in Figure 2, with ℓ as a function of $|\Delta\Psi_0|$ and $|m|$ as a function of $|\Delta\Psi_1|$. We use an iterative rejection algorithm to discard the outlying points. We find that throughout parameter space the fitted coefficients of the lines are remarkably stable, and that in general the phase differences relate to the pulsation parameters as

$$\ell \approx 0.10 + 1.09|\Delta\Psi_0|/\pi, \qquad |m| \approx -1.33 + 0.54|\Delta\Psi_1|/\pi . \qquad (1)$$

More fitted coefficients, for various subsets of parameter space, as well as error estimates for the retrieval of ℓ or $|m|$, are provided by Telting & Schrijvers (1997).

4   Application

In April 1996 we took time series of high-resolution spectra of three rotating β Cephei stars: ω^1 Sco (B1 V, not previously known to be variable), ϵ Cen (B1 III, a multi-periodic photometric variable, Heynderickx 1992) and δ Sco (B0.3 V e, a known non-radial pulsator, Smith 1986). Detailed description of these data and our analysis can be found in Telting & Schrijvers (1998, in preparation).

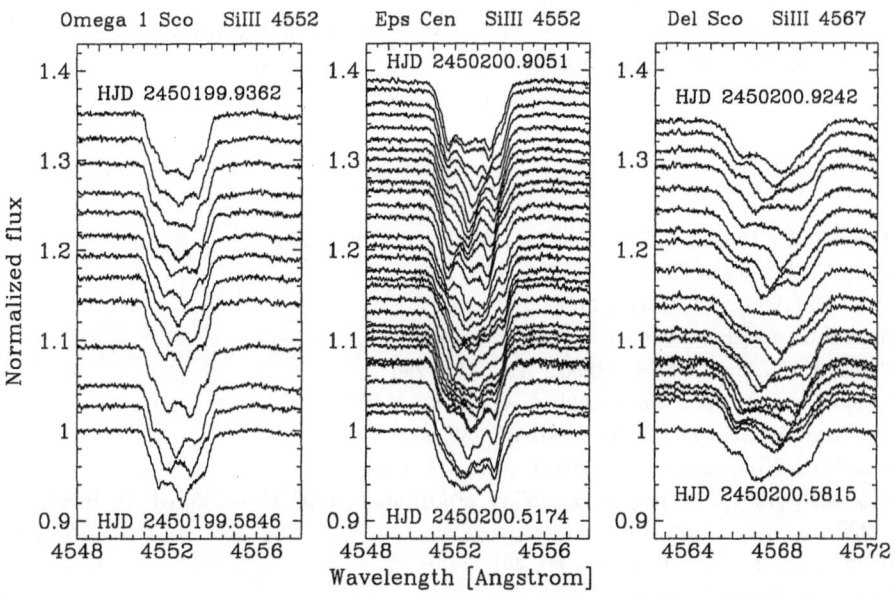

Fig. 3. One night of data taken with ESO CAT/CES in April 96

Following the analysis steps as depicted in Figure 1, and using Equation 1, we find that ω^1 Sco is a single-mode pulsator with $\ell=9\pm1$ and apparent pulsation frequency of 15 cycles/day. For δ Sco we find a dominant mode with apparent pulsation frequency 10.3 cycles/day. The phase diagram at this frequency implies $\ell=5\pm1$. For both stars the temporal sampling of the data does not allow a determination of m of the modes.

For the multiperiodic star ϵ Cen the 6-day dataset proved not to be sufficient to resolve the pulsation frequencies; more data, taken over a longer time base, are needed.

5   Conclusions

We find that for stars with $V_e \sin i$ larger than approximately five times the half-width (HWHM) of the intrinsic profile, one can derive both the degree ℓ and the order $|m|$ from the phase diagrams of the line-profile variations. It is possible to derive values of $\ell \lesssim 20$ and values of $|m| \lesssim 10$.

Reasonable uncertainty estimates for a derivation of ℓ or $|m|$ with the method described above are ± 1 and ± 2, respectively. For p modes values of $|m|$ with accuracy ± 1 can be derived. If stellar and pulsational parameters are constrained, better accuracies can be achieved. The values of ℓ and $|m|$, as derived from the phase diagrams, can then be used as initial guesses in more detailed modelling of the line-profile variability.

We applied our method to derive the ℓ values of the dominant pulsation modes in ω^1 Sco ($\ell=9\pm1$) and δ Sco ($\ell=5\pm1$).

References

Aerts C., De Pauw M., Waelkens C. (1992): A&A, 266, 294

Aerts C., Waelkens C. (1993): A&A, 273, 135

Aerts C. (1996): A&A, 314, 115

Buta R.J., Smith M.A. (1979): ApJ, 232, 213

Campos A.J., Smith M.A. (1980): ApJ, 238, 250

Dziembowski W.A., Pamyatnykh A.A. (1993): MNRAS, 262, 204

Gautschy A., Saio H. (1993): MNRAS, 262, 213

Gies D.R., Kullavanijaya A. (1988): ApJ, 326, 813

Gies D.R. (1991): in: Rapid Variability of OB stars, ESO Proc. 36, ed. D. Baade, p. 299

Heynderickx D, (1992): A&AS, 96, 207

Kambe E., Osaki Y. (1988): PASJ, 40, 313

Kennelly E.J., Walker G.A.H., Merryfield W.J. (1992): ApJ, 400, L71

Lee U., Jeffery C.S., Saio H. (1992): MNRAS, 254, 185

Reid A.H.N., Aerts C. (1993): A&A, 279, L25

Schrijvers C., Telting J.H., Aerts C., et al. (1997): A&A, 121, 343

Smith M. (1986): ApJ, 304, 728

Telting J.H., Schrijvers C. (1997): A&A, 317, 723

Vogt S.S., Penrod G.D. (1983): ApJ, 275, 661

Prinja: It seems that low-order pulsation modes may have potentially interesting links to cyclical wind variability (on days timescales) in hot stars (e.g. HD64760 Howarth et al. 1998, MNRAS submitted). Do you think that rotation effects are now properly treated in NRP modes to enable reliable identification of these low-order modes?

Telting: If the rotation frequency is comparable to the pulsation frequency in the corotating frame, then it is not meaningful to assign a single ℓ-value to an oscillation mode. Instead, a series of spherical harmonic functions is necessary to describe one eigenfunction. A proper identification should therefore be based on a model that fully accounts for these effects.

We refer the interested reader to: Saio 1981, ApJ 244, 299; Martens & Smeyers 1982, A&A 106, 317; Lee & Saio 1990, MNRAS 254, 185; Papaloizou et al. 1995, MNRAS 277, 471.

Hubert: In the case of tesseral modes, the width of the amplitude curve across the line profile is narrower in the case of the 1st harmonic than in the case of the fundamental frequency. Could you deduce information on the latitudinal node position from that?

Telting: In general, the answer to your question is, unfortunately, negative. Indeed, a subset of the tesseral modes shows the property you describe. But we find this for a subset of the sectoral modes as well. So far, we could not determine to which property of the pulsation mode this effect is related. Therefore, we do not see how to deduce conclusive information on the latitudinal node position from this property.

Kaper: My question relates to the time sampling of the data; in deriving the NRP modes of massive stars, what is more important: the time span of the dataset or the "density" of the dataset.

Telting: The time span determines the accuracy at which the frequencies can be measured: the longer the dataset the better. Additionally, a longer time span will automatically lead to an increased sampling of the phase of the line-profile variation (i.e. a higher "density" in phase coverage). So for a coherent signal it is better to have a long time-base of observations, than a shorter one with higher density.

Nevertheless, a high density is needed to track short-period variability, and is needed if you are not sure that the cyclic variability is coherent.

When identifying periods, it is essential to have a dataset which is relatively free of aliasing. This means that periodic gaps in the data should be avoided; in many cases multi-site observations are a must.

The length of the exposure time will affect the observed line-profile variability, since each observation is averaged over the exposure time. We are currently investigating the effects that this might have on the diagnostics of line-profile variability due to NRP.

Linsky: You have presented evidence that β Cephei stars are non-radial pulsators with power primarily in one or two modes, whereas the sun pulsates in a very large number of modes. 1) Is the identification of only a few modes an observational artifact or real evidence that there is little power in these other modes? 2) If the latter is true, why is the oscillation power mostly in only very few modes?

Telting: 1) For ω^1 Sco, for example, we find only one dominant mode with $\ell=9$. From yet to be presented modelling we find that this mode is probably sectoral. If other sectoral modes with similar ℓ were excited we would certainly see these, provided that the pulsational amplitude would be similar. So although there is an observational bias towards modes with a small value of $\ell-|m|$, we can definitely say that this star is very selective in which modes to excite with large amplitudes and which not. Other rotating β Cephei stars show similar behaviour.

2) Your question nails down the problem that we have to solve. Theoretically we expect a wide range of modes to be excited, but for rotating β Cephei stars we find only few. We feel that the effects of rotation on the mode selection should be investigated.

NRP in Light and Colour: Diagnostic Value

D. Vinković[1] and K. Pavlovski[2]

[1] Dept. of Physics & Astronomy, University of Kentucky, Lexington, KY 40506, U.S.A.
[2] Faculty of Geodesy, University of Zagreb, Kačićeva 26, HR–10000 Zagreb, Croatia

Nonradial pulsations (NRP) seem to be common among the stars in the upper-left part of the H-R diagram. Apart from the work of Buta & Smith (1979), little has been done to model light and colour changes in NRP variable stars.

We have developed a new code to model the light and colour variations of nonradially pulsating, rapidly rotating stars. Here, we demonstrate the effectiveness of the new code for diagnosing NRP modes by deriving pulsational modes for 53 Persei from the observed photometric variations.

We used the new code to calculate light amplitude ratios for different mixtures of modes and compared the results to UBV (Huang et al. 1994) and *Voyager 2* FUV measurements (Smith & Huang 1994). We determined that the observed variations are due to a combination of $(l, m_1, m_2)=(2, -2, -1)$ or $(2, -1, 0)$ modes; see Table 1. This finding is consistent with recent spectroscopic results obtained from analysis of line-profile variations.

Table 1. Calculated and observed light-amplitude ratios for 53 Per

(l, m_1, m_2):	$(1, -1, 0)$	$(2, -1, 0)$	$(2, -2, -1)$	Obs.
B/V	1.09	1.10	1.10	1.2±0.1
U/V	1.68	1.78	1.76	1.7±0.2
UV/V	2.7	3.2	3.2	3.0±1.0
FUV/V	4.0	5.1	5.1	4.6±1.0

References

Buta, R.J., Smith, M.A. (1979): ApJ 232, 213
Huang, L., Guo, Z., Hao, J., et al. (1994): ApJ 431, 850
Smith, M.A., Huang, L. (1994): in L. Balona et al. (eds), Proc. IAU Symp. 162, Pulsation, Rotation and Mass Loss in Early-Type Stars, Kluwer, Dordrecht, p. 37

Nonradial Pulsations in Relation to Wind Variability in Early-Type Stars

Dietrich Baade

European Southern Observatory
Karl-Schwarzschild-Str. 2, D-85748 Garching b. München, Germany

Abstract. The observational findings are briefly reviewed that variously led to the conjecture that nonradial pulsation could influence the mass-loss processes from early-type stars. Starting from this historical background, a systematic attempt is made to characterize the pulsational properties of the known types of pulsating OB-stars in connection with their mass-loss properties. Two groups of stars on the fringe of the domain in the HRD, where the strong radiative forces dominate over everything else, emerge as exhibiting the highest potential for conclusive observational tests: (i) At the low-temperature limit, these are the B/A supergiants. The available observational facts indicate that positive evidence is weak at best. By contrast, there are numerous cases of rotational modulation of the wind. Contrary to this initial expectation, the pulsation connection is rescued by the textbook example of stellar winds, ζ Pup. It is probable that at least the azimuthal velocity and density structure of the wind is co-shaped by low-order nonradial pulsation. (ii) At the low-luminosity limit, an intriguing picture develops in which the line-emission outbursts of some Be stars arise from the (periodic) interaction between several very-long-period low-degree g-modes.

1 Introduction

The title to this talk, proposed by the Workshop organizers, is a diplomatic master piece. It neither suggests that in early-type stars nonradial pulsations (NRP's) and wind variability have anything at all to do with one another nor does it indicate that such claims have, in fact, been made and alternately been met with euphoria and severe reservations (if not criticism). It even remains relatively ambiguous about the directionality of any such putative link. One phase of the vacillation between uncritical acceptance and overcautious rejection is highlighted by the title of the workshop *On the Connection* (emphasis added) *between Nonradial Pulsation and Mass Loss in Massive Stars* the proceedings of which (Abbott et al. 1986) document many of the historical roots of the subject.

The discovery of NRP's in Be stars (Baade 1982; Vogt and Penrod 1983) was at a time particularly effective in preparing the ground for the belief that NRP could play an active role in triggering mass loss from early-type stars. Be stars are puzzling in that they have a wind at luminosities and effective temperatures, where winds are not generally observed. The formation out of the wind or through other processes of the variable and presumably Keplerian disk, from where the emission lines arise, is another unsolved problem.

The β Cephei stars (with spectral types between B0.5 and B2) in some cases experience strong shocks caused by large-amplitude radial pulsation (e.g.,

Mathias and Gillet 1993). The possible analogy to pulsation-driven atmospheric shocks and associated mass loss in Mira stars (e.g., Willson, these proceedings) seemed to provide further circumstantial support for the hypothesis. However, the idea was at variance with the more pertinent observational fact that the mass-loss rate of β Cephei does not strongly vary with phase of the pulsation (Blomme and Hensberge 1985; but there is evidence of a regular modulation of the profiles of resonance lines.) The case became more diffuse when it was realized (cf. Baade 1998 for a review) that NRP is ubiquitous in the region of the HRD where radiatively driven winds prevail. Since in the vast majority of these stars effects other than radiation pressure are not needed to explain the time-averaged properties of their mass loss, the general impact of NRP could obviously only be rather subtle at most.

When it was recognized that the momentum in the wind of Wolf Rayet stars exceeds the single-scattering limit by up to an order of magnitude, the tentative detection of rapid periodic light variations (cf. Bratschi and Blecha 1996) seemed to provide another NRP-wind connection. However, since multiple scattering can largely eliminate the problem (cf. Lucy, these proceedings) whereas observations have not so far established rapid periodic variability as a common property of Wolf Rayet stars, these conjectures have not been pursued much further.

The discrete absorption components (DAC's) of UV wind lines are a time-dependent phenomenon that is beyond the scope of stationary stellar-wind theory. The intrinsic instability of the winds and the ensuing shocks provide the basis for very successful numerical simulations of this behaviour (Owocki, these proceedings). The high susceptibility of the wind flow to external triggers of instabilities offers a new motivation for dealing with the subject of this paper. However, since there is evidence that in many luminous OB stars the primary timescale of the DAC variability is set by the stellar rotation rate, the need to invoke NRP would again be reduced.

Returning to the Be stars, one should complement the above by adding that at least in some of them discrete mass-loss events (outbursts) seem to have a significant share in the total mass loss. Any involvement of NRP's in this activity would be very different from the first two scenarios, namely (i) the direct driving of a wind and (ii) the injection of seeds of instability into the wind.

It is worthwhile to recall that many of the above speculations emerged well before more accurate opacity calculations finally established the availability of the κ-mechanism as a driver of NRP's in early-type stars. Although this long-lasting very basic problem is now solved, it is important to keep in mind that continuous significant energy leakage into the wind will damp, if not quench, the oscillations. This is a strong practical constraint which would make the first of the three pictures, namely direct driving of a significant wind by NRP, the least probable. Therefore, it will be omitted from the rest of this paper.

Given the frequently demonstrated potential of the subject for confusion, it ought to be useful to start this summary of the current observational knowledge with a more differentiated view of the properties of NRP's in perspective with mass loss in various groups of pulsating OB stars. This is attempted in Sect. 2.

On this basis, hypotheses are developed that lead to the design of two 'virtual observing experiments' which exploit the existing observational facts. The first one (Sect. 4.1) addresses the second of the three possible NRP / mass-loss interactions listed above and searches for evidence of NRP-triggered wind peculiarities in supergiants. ζ Pup provides an illustrative but not necessarily representative example (Sect. 4.2). The second experiment focuses on the discrete mass-loss events from Be stars (Sect. 5.1); it is supplemented by a case study of μ Cen (Sect. 5.2). The main conclusions are summarized in Sect. 6.

2   Observational Evidence of NRP's in OB Stars

For nonradial pulsations of high-mass stars there is no obvious observational upper limit in effective temperature. To date, the hottest star with detected NRP's is ζ Pup at 42,000 K (Baade 1991, Reid and Howarth 1996). Towards lower temperatures, the incidence and amplitude of NRP's drop steeply around spectral types B8/B9, at least close to the main sequence (Baade 1989). Within this so delineated domain a variety of groups of pulsating stars can be observationally distinguished. Their pulsational properties have been reviewed only recently (Baade 1998). The following can therefore focus on the synopsis with mass-loss properties:

β **Cephei stars** are restricted to the range B0.5 to B2 in spectral type and only pulsate in short-period (2-6 h) radial and/or nonradial modes.

Even in large-amplitude *radial* pulsators there does not seem to be a significant periodic modulation of the mass-loss *rate* (Blomme and Hensberge 1985; but there are periodic variations of UV resonance line *profiles*).

In *nonradial* pulsators, the variation of the effective temperature throughout the pulsation cycle seems to be low. For the second proto-typical β Cephei star, β CMa, Cassinelli et al. (1996) derive from EUV observations a peak-to-peak amplitude of only ~200 K even for the largest-amplitude mode.

Slowly Pulsating B Stars (SPB's) (cf. Waelkens et al. 1998) have much longer pulsation periods of 1-5 days which are, therefore, due to low-degree nonradial g-modes. They also seem to be restricted to later spectral subclasses, namely B2-B9. Broad-lined stars with $v \sin i$ significantly above 200 km/s are strongly underrepresented among the SPB's.

Nothing is presently known about any unusual mass loss from SPB's.

53 Per stars may only be some mixture of SPB's and ϵ Per stars. They would surround the β Cephei stars in the HRD. All known stars of this type were selected on the basis of their low $v \sin i$ values. They pulsate in low-degree g- and/or p-modes.

The 53 Per stars are not known for conspicuous mass-loss properties.

ϵ **Per stars** in the HRD connect to the SPB's towards higher temperatures, up to about O9 in spectral type. The average rotational line broadening is also higher. Low-degree modes ($\ell \leq 3$-4) are not usually diagnosed in these stars whereas the upper limit in ℓ may be set by the line broadening and the temporal resolution of the observations.

ϵ Per stars have not been noted for special mass-loss characteristics.

Supergiants hardly ever show genuine periodic variability in their photometry (or are strongly multiperiodic, possibly with time-dependent amplitudes). A slightly sharper picture emerges spectroscopically (cf. Sect. 4.1). The timescales are in the range attributable to low-order g- or strange modes but also extend into the domain of the rotation periods. Only in broad-lined supergiants, almost all of which have spectral types O or early B, have line-profile variations reminiscent of intermediate- and higher-degree p-modes ('moving bump' phenomenon) been found.

There are some hints that some print-through of the photospheric variability is still detectable in the wind.

Bn stars are rapidly rotating ($v \sin i$ above ~ 200 km/s) B stars of any spectral subclass but without Hα line emission. Moving bumps are quite commonly seen in their line profiles. It could be indicative of higher-degree p-modes but this still needs confirmation.

Bn stars do not show evidence of unusual mass loss.

Be stars are equally frequent as Bn stars and have a similar distribution of $v \sin i$ values. However, in addition to the moving-bump phenomenon, the vast majority of them shows also low-order g-modes (very late B stars may be the only major exception) with periods close to the typical rotation period of one day.

In addition to these differences with respect to the Bn stars, the Be stars have a number of mass-loss characteristics not common to the Bn stars:

- Be stars have a rotationally flattened circumstellar disk from where their emission lines arise.
- The V/R ratio of double-peaked emission lines often varies with the period of the main low-degree mode of the underlying star.
- In η Cen, the Balmer discontinuity of the *disk* continuum emission has been seen to vary with the stellar period (Štefl et al. 1995).
- Some Be stars experience numerous line-emission outbursts (Baade et al. 1988, Hanuschik et al. 1993).
- During an outburst, there are additional *cyclic* V/R variations due to the ejected matter. Part of the ejecta may also contribute to replenishing the disk (Rivinius et al. 1998).
- Be stars have a fast wind with variable DAC's (e.g., Doazan 1982).

3 Test Target Selection Strategies

From Sect. 2 one may tentatively infer that if it makes any sense at all to think about NRP and mass loss simultaneously, supergiants and Be stars should be the targets to be studied first. The most immediate reason is that there is hardly any other group of stars which observationally proved interesting pulsation- as well as mass-loss-wise. However, two other considerations suggest the same selection.

Firstly, one may wish to stay clear of those regions in the HRD where radiation pressure dwarfs all other forces. Accordingly, one would explore the low-temperature and low-luminosity limits of the radiatively-driven wind domain. The stars that have the lowest luminosity but still undergo significant mass loss are the Be stars. Favouring lower temperatures would mean to prefer among the very luminous stars the cooler B- and maybe even A-type supergiants.

Secondly, NRP modes with short spatial wavelengths, i.e. with ℓ in excess of about 4, and/or periods that are much shorter than the wind flow timescale are less likely to have an easily detectable effect on the large-scale properties of the wind. Most Be stars and - presumably - many B/A supergiants would therefore qualify as test targets on account of their low-degree long-period g-modes. A potential complication to be kept in mind is that in these objects rotation and pulsation periods may/do overlap. SPB's are the only other OB stars with known or suspected g-modes. But since there is no reason to assume that they are conspicuous also with regard to mass loss, SPB's are less promising for inclusion into a first survey.

It is attractive that in the two groups the objectives for their study are different. The supergiants may offer the opportunity to learn about the effect, if any, of NRP's on a steady wind. By contrast, in the Be stars the emphasis is probably more on discrete mass-loss events. Sects. 4.1 and 5.1, respectively, present and discuss the general observational database, each followed by a specific case study (Sects. 4.2 and 5.2).

4   Cyclic Modulation of Stellar Winds

4.1   General Observational Evidence

The most comprehensive spectroscopic studies to date of the variability in the optical of O/B and B/A supergiants have been presented by Fullerton et al. (1996) and Kaufer et al. (1996, 1997), respectively, in the PhD thesis works of the first authors. The two investigations are complementary to each other in so far as Fullerton observed many stars only a few times each whereas Kaufer monitored few stars for a long time. They arrive at very similar conclusions:

o In some stars, multiple g- or strange modes are probably excited. However, the phase coherence becomes quite uncertain already after few cycles.

o The main wind variability takes place on a rotational time scale.

o There is no *obvious* link between the two variabilities even though the results of all plausibility checks have been positive.

Kaufer et al. report also that within the entire optical spectrum the velocity amplitudes do not depend on the depth of formation of the spectral lines. Fullerton et al. observed a much shorter range in wavelength but, similarly, found a significant correlation between C IV $\lambda\lambda 5801, 5812$ and He I $\lambda 5876$. The incidence and amplitude of variability decreases towards the main sequence (Fullerton et al.).

That the variability of the wind is coupled to the rotational timescale was more firmly concluded in a number of major continuous IUE monitoring campaigns dedicated to only very few objects each (cf. Prinja, these proceedings; see also Sect. 5.2). In all cases a cyclically repeating pattern in the development of DAC's was observed but already within the few cycles covered by the observations their phase coherence was relatively uncertain. The timescales are sometimes much longer than the one of the wind flow so that the variability is probably not intrinsic to the wind itself but somehow triggered by one or more other processes.

The rotation periods usually exceed the repetition timescale of the DAC by a factor of a few. Unfortunately, the uncertainties in both quantities do not permit to conclude whether this ratio is an integer number. If it were, a fixed multipole surface magnetic field would undoubtedly be the strongest contender for the explanation. On the other hand, in the case of a definitive non-integer ratio, the traveling-wave character of some NRP modes might be favoured. Both pictures have in common that they do not presently include a physical model of the coupling mechanism.

4.2 ζ Puppis - a Case Study

The wind of this O4 If star is the textbook example of radiatively driven mass loss. In addition, ζ Pup has a considerable record of variability. The most persistent timescale of 5.1 d was detected in the Hα emission (Moffat and Michaud 1981, Berghöefer et al. 1996), optical photometry (Balona 1992), and the DAC's of UV wind lines (Howarth et al. 1995), but not in X-rays (Berghöfer et al. 1996). A timescale so much longer than the wind flow time can hardly be intrinsic to the wind and, therefore, almost certainly is the rotation period. Accordingly, ζ Pup rather convincingly confirms the conclusion derived for other stars that there is rotational modulation of the wind.

ζ Pup is also a nonradially pulsating star (Reid and Howarth 1996 and references therein). A low-degree mode ($|m| \leq 4$) with a period of 8.54 hours has been observed in at least two different seasons. Other modes may also be excited but this is less clear. The important point is that in the blue wing of the Hα emission Reid and Howarth (1996) find a modulation with the same 8.5-hr period. The features they detect migrate from red to blue, i.e. in the direction opposite to the photospheric components, and at velocities well beyond the stellar $v \sin i$. This is the first clear evidence of an NRP-induced perturbation propagating at least into the low-velocity regime of the wind.

A third period, namely some 19 hours, was detected in both the DAC's of UV wind lines (Howarth et al. 1995) and the blue Hα emission component (Reid and Howarth 1996). This shows that a structure imposed on the wind at low velocity can persist up to its terminal velocity. Since the IUE and the ground-based observations were obtained more than 4 years after one another, this variability, too, is of a long-term nature. On the other hand, the repetition timescale for the appearance of UV DAC's is not always the same. Prinja et al. (1992) measured a value of only about 15 hours. These two numbers bracket the timescale of ~17 hours observed by Berghöfer et al. (1996) contemporaneously in the Hα and the X-ray emission between 0.9 and 2.0 keV. The simultaneity of these variations underlines again the cohesion between variations at low and high velocities.

Each of the data sets, in which one of the 15-, 17- and 19-hr timescales was found, appears irreconcilable with the respective other two candidate periods.

Attempts to trace them back to photospheric variability is at this moment limited to the numerical coincidence of $17\,\mathrm{hr} = 2 \times 8.5\,\mathrm{hr}$ (the primary NRP period). However, ζ Pup very probably at least occasionally pulsates in more than one mode. Published ground-based observations are insufficient to shed further light on this. But the variable interaction between several NRP modes would at present seem to be the simplest hypothesis for the 15-19-hour bandwidth in the timescales of the wind variability.

5   Mass-loss Events

5.1   General Observational Evidence

Be stars are at the same time among the most rapid rotators and the slowest pulsators. In fact, they rotate so rapidly and pulsate so slowly that the two timescales are quite similar to one another in the inertial system and in the co-rotating frame the pulsation periods are extremely long. It is this combination which makes Be stars unique among the OB stars. If only these two parameters are admitted as extra constituents of the mass-loss mechanism of Be stars, most of the possible combinations have little potential for generalization:
  – Rotation alone (Struve's model of the 1930's) does not make a B star a Be star. (The equally rapidly Bn stars do not undergo outbursts.)
  – Pulsation alone does not seem to make any B star lose significant amounts of mass. (Otherwise, the large-amplitude β Cephei stars should have rather pronounced Be characteristics.)
  – The combination of two or more pulsation modes is not probably any more effective. (Multi-mode β Cephei stars provide again the counter example.)
  – Rotation in combination with pulsation is not sufficient, either. (There are rapidly rotating β Cephei stars, and the moving bump phenomenon in Bn stars may be due to higher-degree NRP.)
  – So far, there is little observational evidence for a temporary equatorial spin-up to the break-up velocity which has been considered as the result of angular momentum transport by NRP waves from the stellar core to the surface (Saio 1994 and references therein).
At first glance, this would imply that not even by appealing to the principle of the minimum number of assumptions one can avoid the necessity of introducing other parameters such as magnetic fields.

On the other hand, the above list does not cover what really seems to make Be stars unique, namely the length of the co-rotating pulsation periods.

5.2 μ Centauri - a Case Study

This 3rd-magnitude Be star has a rich observational history of numerous line-emission outbursts (cf. Rivinius et al. 1998). A particularly long and homogeneous series of echelle spectra was obtained by Rivinius et al. (1998; see also Rivinius et al., these proceedings) during the initial stages of the formation of a new Hα-emitting disk. They permit the following phases to be distinguished which, however, in their appearance depend sensitively on the state of development of the disk and also differ somewhat between different lines:

Precursor: Two major symptoms appear within a day or two: (i) For about 5-15 days, the *peak height* of all emission lines drops. This is unlikely to be mimicked by an increase in the continuum flux. (ii) The *wings* of all emission lines grow in width. Weak emission lines not previously present flare up with a base width close to what may be the rotational break-up velocity.

Since the drop in peak height is larger when the disk is well developed, the implied reduced number of recombinations is probably not due to the sudden disappearance of ionized matter but would more plausibly be attributable to a decrease in the ionizing flux reaching the disk. Since the enhanced wings indicate the presence of additional material close to the star, it is possible that this gas partly shields the disk against the stellar irradiation. (Given only the above description, one might also suspect that for some reason matter from the disk moves temporarily closer to the star. But this is unlikely on the basis of outbursts with broad-winged emission lines at times when there is virtually no persistent emission from the disk.)

Outburst proper: The peak heights recover their previous value while the wings, after having reached their maximum, gradually decay. Therefore, the total emission strength reaches a shallow maximum. Variations in the violet-to-red (V/R) peak height ratio, which may have set in already during the precursor phase, now attain their maximum amplitude. This phase lasts about as long as the precursor phase.

Relaxation: This phase begins when the V/R variations have ceased. Over 5-15 days, the emission wings slowly vanish, and the emission strength asymptotically approaches its pre-outburst level.

Relative quiescence: For 20-30 days, the line emission is dominated by the steady-state disk and may even marginally decrease. This phase is rarely observed in purity as it is often already contaminated by the next outburst.

At the photospheric level, the star's activity is characterized by 4 closely spaced periods around 0.5 d and another 2 near 0.28 d (Rivinius et al., these proceedings; Rivinius et al., in preparation). All 6 periods are due to low-order NRP modes but of different azimuthal order, $|m|$, for the two groups. After careful time-series analyses it turned out that the line emission outbursts repeat periodically with the beat periods between the half-day period with the largest amplitude, A_1 and one each of the two half-day periods with next lower amplitudes, A_2 and A_3. These are the two combinations with the largest combined amplitude so that there seems to be an amplitude threshold for the occurrence of an outburst. The distribution of pulsation amplitudes is very fortunate because with more combinations exceeding the outburst threshold, the recognition of a pattern in a quasi-continuous outburst would have been much more difficult. Conversely, if only one pair of modes would satisfy the outburst condition, the inference of an amplitude threshold would be much more uncertain.

From the discussion in Sect. 5.1 it appears unlikely that the existence of such a velocity threshold means that matter is ejected just because it is moving at this (presumably still subsonic) velocity. Furthermore, optically thin emission lines show that some of the gas lost from the star reaches velocities that are larger than the rotational velocity by several times the combined pulsational velocities.

One other possibility might be that, when one NRP travelling-wave pattern is overtaking another one, additional turbulence develops which adds to the kick needed to eject matter. But, although one of the 0.28 d periods has an amplitude between A_2 and A_3, there are no recognizable events associated with the times when wave crests of the two patterns are superimposed. (Note, however, that this argument is somewhat weakened by the fact that the two groups of modes differ in their m-values which means that only one wavecrest of each of the two patterns will have the same stellar longitude at any one time.)

A further possibility is provided by the fact already mentioned in Sect. 5.1. In the corotating frame, the periods are very long. Therefore, it may well be that not the instantaneous velocity is the critical quantity but rather that it is the displacement which results from matter moving with this velocity *for a long time*. Balona (Baade and Balona 1994) has argued that Be stars cannot be nonradial pulsators because the near-identity of rotation and inertial-frame NRP periods leads to arbitrarily long co-rotating periods and correspondingly arbitrarily large mass displacements that would eventually disrupt the star which is not observed.

This claim lacks observational confirmation because the high similarity in Be stars of the distribution of rotation and inertial-frame periods does not imply that this is true also for every individual star. However, it is well conceivable that the periodic positive interference between two velocity fields does lead to outbursts if, for given pulsation amplitudes, this phase just lasts long enough.

6   Conclusions

It appears safe to conclude that NRP *can* modulate the large-scale structure of stellar winds as well as that NRP *can* lead to the ejection of matter in single events. However, the governing physical processes are not yet known so that it not possible to predict under what *general* conditions these two effects take place. Yet, these conditions are not too likely to be very restrictive: ζ Pup is *the* proto-typical OB star with strong radiation-driven mass loss, and μ Cen is the first Be star for which an observational database has been accumulated that permits the claimed relation between NRP and outbursts to be diagnosed. In fact, in a recent paper Howarth et al. (1998; see also Prinja, these proceedings) report for the B0.5 Ib star HR 3090 results that are very similar to those of ζ Pup. Finally, the β Cephei stars clearly deserve to be re-examined (Massa 1997, private communication). It could be interesting to check whether the picture developed for the Be star μ Cen could also be applicable to some of the longer timescales in the winds of supergiants.

Observationally, progress is most likely to result from (i) very long time series, (ii) multi-longitude observations, (iii) Doppler imaging (which, however, introduces a bias towards rapid rotators), (iv) coverage of numerous spectral lines forming under different physical conditions, and (v) high S/N. Polarimetry is also extremely useful as it is the observing technique most complementary to all others. It is sensitive to circumstellar matter very close to the star, i.e. samples the space-UV wind regime at ground-based cost and detects events very early, and to non-axisymmetric gas distributions.

References

Abbott, D.C., Garmany, C.D., Hansen, C.J., Henrichs, H.F., and Pesnell, W.D. (eds.) (1986): *Connection between NRP and Mass Loss in Massive Stars*, PASP **98**, 29

Baade, D. (1982): A&A **105**, 65

Baade, D., Dachs, J., v.d. Weygaert, R., Steeman, F. (1988): A&A **198**, 211

Baade, D. (1989): A&A **222**, 200

Baade, D. (1991): in *ESO Workshop on Rapid Variability of OB-Stars: Nature and Diagnostic Value*, ed. D. Baade, ESO Proc. No. 36, p. 21

Baade, D. (1998): in Proc. IAU Symp. No. 185, eds. F.-L. Deubner and D. Kurtz, in press (= ESO Scient. Prepr. No. 1251)

Baade, D., Balona, L.A. (1994): in Proc. IAU Symp. No. 162, eds. L.A. Balona, H.F. Henrichs and J.M. Le Contel, Kluwer, Dordrecht, p. 311

Balona, L.A. (1992): MNRAS **254**, 404

Berghöfer, T.W., Baade, D., Schmitt, H.J.M.M., et al. (1996): A&A **306**, 899

Blomme, R., Hensberge, H. (1985): A&A **148**, 97

Bratschi, P., Blecha, A. (1996): A&A **313**, 537

Cassinelli, J.P., Cohen, D.H., MacFarlane, J.J., et al. (1996): ApJ **460**, 949

Doazan, V, (1982): in *B Stars with and without Emission Lines*, eds. A. Underhill and V. Doazan, NASA SP-456, p. 279

Fullerton, A.W., Gies, D.R., Bolton, C.T. (1996): ApJS **103**, 475

Hanuschik, R.W., Dachs, J., Baudzus, M., Thimm, G. (1993): A&A **274**, 356

Howarth, I.D., Prinja, R.K., Massa, D. (1995): ApJ **452**, L65

Howarth, I.D., Townsend, R.H.D., Clayton, M.J., et al. (1998): MNRAS, in press

Kaufer, A., Stahl, O., Wolf, B., et al. (1996): A&A **314**, 599

Kaufer, A., Stahl, O., Wolf, B., et al. (1997): A&A **320**, 273

Mathias, P., Gillet, D. (1993): A&A **278**, 511

Moffat, A.F.J., Michaud, G. (1981): ApJ **251**, 133

Prinja, R.K., Balona, L.A., Bolton, C.T., et al. (1992): ApJ **390**, 266

Reid, A.H.N., Howarth, I.D. (1996): A&A **311**, 616

Rivinius, Th., Baade, D., Štefl, S., et al. (1998): A&A, in press

Saio, H. (1994): in Proc. IAU Symp. No. 162, eds. L.A. Balona, H.F. Henrichs and J.M. Le Contel, Kluwer, Dordrecht, p. 287

Štefl, Baade, D., Cuypers, J. (1995): in Proc. IAU Coll. 155, eds. R.S. Stobie and P.A. Whitelock, PASPC **83**, p. 303

Vogt, S.S., Penrod, G.D. (1983): ApJ **275**, 661

Waelkens, C., Aerts, C., Kestens, E., et al. (1998): A&A **330**, 215

Peters: μ Cen is known to display some rather spectacular transient mass-loss events. Can you comment on how these might be explained with your scenario for the star's recurrent outburst activity?

Baade: You are probably referring to some of your own observations (1986: ApJ **301**, L61; 1995: IAU Symp. 176, p. 212). Using our ephemeris for the outbursts (Rivinius et al., these proceedings), we find that the blue-shifted high-velocity components and the sudden increase in Hα emission strength seen by you in 1994 occured at exactly the time predicted by the ephemeris. The 1985 event is still within the uncertainty of the match between the ephemeris and the 1987 observations of Hanuschik et al. (1993: A&A **274**, 356). Moreover, in some of the stronger outbursts we detect such features also in our data.

Savonije: I was intrigued by the outburst data you showed of the Be star μ Cen. Did you mention there was evidence for the outflowing matter to be (almost) ejected in Keplerian orbit?

Baade: From optically thin emission lines we can be confident that right at the beginning of an outburst there is some material with velocities in the vicinity of the critical velocity (Rivinius et al., A&A, submitted). However, we have not yet attempted to estimate the fraction of the total ejecta that reaches this velocity or to derive from our much longer time series whether some of the ejected gas eventually merges with the disk. The latter would be a plausible guess.

Savonije: It is interesting that the modes seen around outburst seem almost co-rotating with the star. That would indicate that these modes are rotational (Rossby-type) modes. Such quasi-toroidal modes have displacement velocities strongly confined to the horizontal direction ($v_r \approx 0$) which could be of dynamical interest in relation to the mass ejection observed.

Baade: In fact, during some outbursts we find around $2/3 \, v \sin i$ towards either line wing features that wax and wane at almost constant wavelength (Rivinius et al., to be submitted to A&A) as described by Smith (1985: ApJ **297**, 224) for Spica and compared by him to Rossby waves. We cannot presently say whether such Rossby wave-like velocity fields further the mass ejection and whether they arise from one g-mode wave pattern slowly passing the other one.

Predicting the Outbursts of the Be Star μ Cen

Th. Rivinius[1], D. Baade[2], S. Štefl[3], O. Stahl[1], B. Wolf[1], and A. Kaufer[1]

[1] Landessternwarte Königstuhl, D–69117 Heidelberg, Germany
[2] European Southern Observatory, Karl-Schwarzschild-Str. 2,
    D–85748 Garching bei München, Germany
[3] Astronomical Institute, Academy of Sciences, CZ–251 65 Ondřejov, Czech Republic

Abstract. By analyzing 409 high-resolution spectra covering the whole Paschen continuum of the Be star μ Cen taken from 1992 to 1997, we found six genuine periods in the photospheric radial-velocity variations. The properties of the variability, sorted into two period groups of around 0.28 and 0.5 days, point to nonradial pulsation in low order g-modes as mechanism. Whenever the modes of the 0.5 d group come into positive, strong interference, line-emission outbursts occur. The outbursts are indicative for mass transfer to the circumstellar environment. Calculating the phase differences and overall amplitude, outbursts in archival datasets taken up to ten years ago are reconstructed, but also outbursts in early 1997 were predicted correctly.

1 Introduction

Between 1992 and 1997 we observed μ Cen (=HR5193, B2IV-Ve, V=3.47 mag) with the HEROS spectrograph, described e.g. by Rivinius et al. (1998, Paper I). μ Cen has a well documented history of frequent line-emission outbursts (e.g. Hanuschik et al. 1993). As most other Be stars the star is also known to exhibit low-order line-profile variability with a suspected period of 0.505 days (Baade 1984). This combination of variability, low-order lpv of relatively long periods and outbursts, is observed for other Be stars, too, as e.g. λ Eri (Smith, 1989). However, no mechanism triggering outbursts could be positively identified so far in this or in any other Be star.

2 Definition of the Beginning of an Outburst

Since the properties of an individual outburst depend strongly on the mean emission strength (Paper I), we had to find the most invariant quantity for our purpose, to define the beginning of an outburst homogeneously. In older datasets, with basically zero mean emission level (e.g. Hanuschik et al. 1993), the Hα equivalent width works well. From 1995 on Hα was presumably already emerging from an area too large to be sensitive to small scale changes close to the stellar surface, but the higher Paschen lines still showed considerable variability. In 1997, however, even the Paschen lines became less sensitive to smaller bursts. The only quantity working equally well in all our data subsets are broad emission wings appearing in the higher Balmer lines at the times of outbursts (Fig.1).

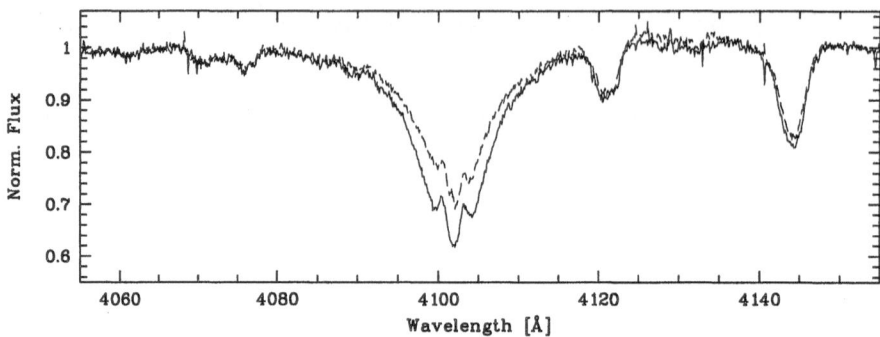

Fig. 1. The Hδ line of μ Cen shortly before (full line) and during an outburst (dashed line). We measure the strength of the wings by means of fitting a Gaussian to the complete line profile. Although the absolute numbers of the equivalent width measured by this procedure have obviously systematic errors and should therefore not be taken at face value, the quantity is far more sensitive to changes of the broad emission wings than to variations of the emission peaks. Since we found the wings to be the most general indicator for outbursts, this quantity can be used to monitor the activity.

3   Photospheric Periodicity

We used the radial velocities (RVs) of photospheric lines least affected by emission to perform a search for periodicity following the method described by Kaufer et al. (1996), which is based on Scargle's algorithm (1982). By using all data from 1992 to 1997 we found six periods sorted in two groups. One group contains four periods near 0.505 d, the other two periods close to 0.28 d (Table 1). However, a 1D-analysis technique does not make use of the full information contained in high-resolution spectroscopy. Thus we applied also a 2D-technique (Kaufer et al. 1996) on the normalized flux over the line profile in 5 km s^{-1} wide bins to obtain phase and amplitude distribution over the spectral line (Fig. 2). This can be used to obtain information on the nature of the variability, for which we favour low-order g-modes of equal angular indices l and m within each group, but high radial orders n differing by only a few to explain the splitting (Rivinius et al. 1997a, Dziembowski et al. 1993). Detailed modelling is in progress.

	Period $\mathcal{P}$ [days]	Phase ϕ at MJD 50 000	Amplitude A [km s^{-1}]
$\mathcal{P}_1$	0.502925 ± 06	0.897 ± 10	14.3 ± 1.0
$\mathcal{P}_2$	0.507519 ± 09	0.150 ± 16	8.4 ± 0.9
$\mathcal{P}_3$	0.494523 ± 11	0.646 ± 19	5.8 ± 0.7
$\mathcal{P}_4$	0.516358 ± 15	0.922 ± 22	4.8 ± 0.7
$\mathcal{P}_5$	0.281405 ± 05	0.850 ± 17	7.6 ± 0.7
$\mathcal{P}_6$	0.279137 ± 08	0.425 ± 32	3.3 ± 0.7

Table 1. Parameters for the sine fits of the radial velocity variation in the line cores of He I λλ4121, 4168, 4438. The epoch is given for Modified Julian Date 50 000, i.e. MJD ≡ JD - 2 400 000.5. Note that the amplitudes given are *not* the physical pulsation amplitudes.

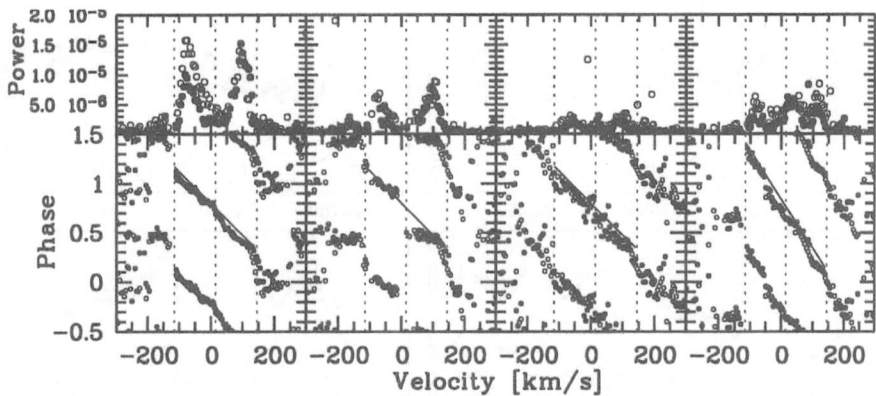

Fig. 2. The phase diagrams of $\mathcal{P}_1$, $\mathcal{P}_2$, $\mathcal{P}_3$, and $\mathcal{P}_5$ (from left to right) for He I $\lambda4121$ ($\bullet$) and He I $\lambda4713$ (o). The dotted vertical lines mark the systemic velocity of $v_{sys} = 14.5\,\mathrm{km\,s^{-1}}$, which was derived from the mean radial velocity of the Si III $\lambda\lambda4553,4568$ lines, and $v_{sys} \pm v\sin i$. The solid lines in the left three panels are the linear regression of the phase with velocity computed for $\mathcal{P}_1$ only (in order to stress the similar behaviour of these three periods), whereas in the rightmost panel the $\mathcal{P}_5$ data are used. Note the steeper slope for $\mathcal{P}_5$ compared to $\mathcal{P}_1$, $\mathcal{P}_2$, and $\mathcal{P}_3$.

4 An Ephemeris for Outbursts

Most interesting in relation to the circumstellar variability is the coincidence of positive mode interference of the 0.505 d group with line-emission outbursts. In Fig. 3 we show the times of equal phases compared to the circumstellar activity as defined above. The agreement is striking for 1996. The behaviour is more complex for 1995, but still well visible. We tested this correlation with an older dataset taken by Hanuschik et al. (1993) in 1987. Although the times are shifted due to period uncertainties, the temporal behaviour of major outbursts (circles and squares together) and intervening smaller bursts (just circles or squares) can be well recognized (Fig. 3, upper left panel). The positive results encouraged us to submit a prediction to the Be Star Newsletter (Rivinius et al. 1997b) in Feb. 1997 (MJD 50 500). The observations thereafter confirmed the prediction well (Fig. 3, lower right panel, rightmost datastring).

5 Conclusions

The linear superposition of the pulsation alone cannot eject the material, not to speak of the angular momentum transfer necessary to reach orbit. So our model remains on an entirely phenomenological base, but for the first time observations of the purely photospheric variability could be connected to the circumstellar emission behaviour of a Be star.

Although the physics of line-emission outbursts, which presumably are mass ejections to the circumstellar environment, remains unknown, our results point

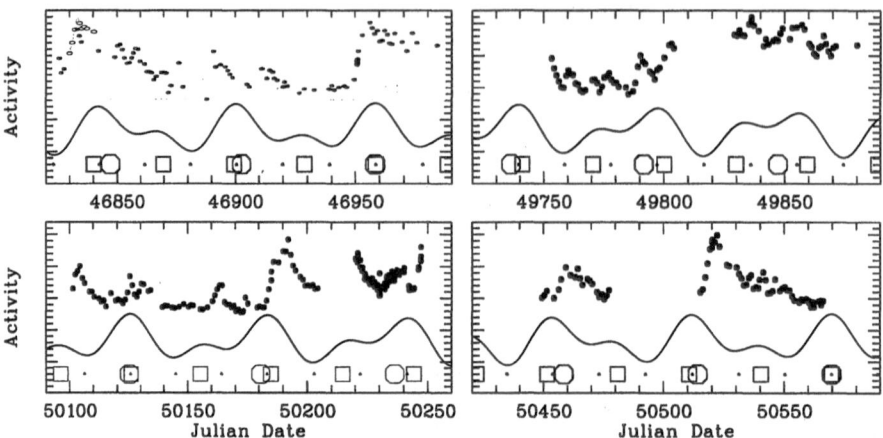

Fig. 3. The observed emission activity (upper dots), the overall reconstructed amplitude (solid lines) and the times of zero phase difference for $\Delta\phi_{1-2}(\circ)$, $\Delta\phi_{1-3}(\square)$, $\Delta\phi_{2-3}$(dots) (bottom part of the panels). The sizes of the symbols denote the relative importance. It can be well seen that $\mathcal{P}_2 + \mathcal{P}_3$ causes only minor additional effects, if at all. The observed activity was derived from Hα for Hanuschik's data (upper left), while for 1995 (upper right), 1996 (lower left), and 1997 (lower right) the Hδ data (see Fig. 1 and Sect. 2) is shown. The uncertainty of the periods may cause shifts in the order of ten days, but the temporal pattern of the outbursts can still be recognized.

to multiperiodic nonradial pulsation as the trigger for the episodic mass-loss events, and possibly even as key ingredient for the Be mechanism acting in μ Cen besides its fast rotation.

Acknowledgements

We are grateful to Conny Aerts, Martin Kürster, and Michelle Thaller for having taken some additional spectra. This work was supported by the Deutsche Forschungsgemeinschaft (Wo 296/20-1, 436 TSE 113/18) and Academy of Sciences and Grant Agency of the Czech Republic (436 TSE 113/18, 202/97/0326).

References

Baade, D. (1984): A&A **135**, 101
Dziembowski, W.A., Moskalik, P., Pamyatnykh, A.A. (1993): MNRAS **265**, 588
Hanuschik, R.W., Dachs, J., Baudzus, M., Thimm, G. (1993): A&A **274**, 356
Kaufer, A., Stahl, O., Wolf, B., et al. (1996): A&A **305**, 887
Rivinius, Th., Baade, D., Štefl, S., et al. (1997a): In: A Half Century of Stellar Pulsation Interpretation, Bradley, P.A., Guzik, J.A. (eds.), PASPC, in press
Rivinius, Th., Štefl, S., Baade, D., et al. (1997b): Be Star Newsletter **32**, 14
Rivinius, Th., Baade, D., Štefl, S., et al. (1998): submitted to A&A (Paper I)
Scargle, J.D. (1982): ApJ **263**, 835
Smith, M.A. (1989): ApJS **71**, 357

Kundt: (by meaning) What is the order of magnitude for your Fig. 3? You give no scale for the axes.

Rivinius: The measured Hδ equivalent width is about 3.5 Å in the inactive state with a maximal amplitude of about 1 Å. However as shown in Fig. 1 the values are not to be taken at face value.

Foing: What are the observed velocities (and derived vertical velocities) of the individual modes, and how do they compare with the soundspeed? Do you then explain the outbursts as the nonlinear combination of these velocities leading to a shock front?

Rivinius: These numbers can be given only after detailed pulsation modelling, which is not finished yet. However the velocities most probably do not exceed sound speed. Even the combination of all modes should exceed sound speed only marginally, if at all. Also shocks might not be enough to eject the material, since μ Cen may rotate 'only' in the order of 50% of its critical velocity of about $450\,\mathrm{km\,s^{-1}}$. There surely is a nonlinear mechanism, but it is probably more than just shocks.

Owocki: (by meaning) A first idea for this nonlinearity may be something like Rossby waves. They have high horizontal velocities.

Rivinius: Yes, actually we observe some typical pattern, so-called spikes, that is also present in another star, Spica, and has been suggested to be caused by somehow modified Rossby waves by Smith (1985, ApJ, 297, 224).

Grinin: (by meaning) How do you exclude rotational modulation as the process responsible for the variability?

Rivinius: The distribution of the observed periods in two groups cannot be explained by rotation. Furthermore all periods are already below the critical value for this star's rotation, even if we would see it equator on.

Kubat: The half day period you found may be a result of aliasing. How can you exclude this possibility?

Rivinius: Part of our observations cover a whole night with ten to twenty spectra. In these data we are able to identify the times of maximal asymmetry on both the blue and red side from the shape of the line profile. For $\mathcal{P} = 0.5\,\mathrm{d}$ these times are separated by 0.5 cycles, while for $\mathcal{P} = 1\,\mathrm{d}$ (the only alias we could have confused it with) this separation would only be 0.25 cycles. The latter can be excluded, since the RV-curves are of sinusoidal shape folded both with $\mathcal{P} = 0.5\,\mathrm{d}$ and its one-day alias.

Mullan: It was once suggested that mass ejections in Be stars are caused by mode switching. Do you think that this old suggestion is no longer valid?

Rivinius: I am quite confident that the case of μ Cen has nothing to do with mode switching. However, I would not dare to say that there is only one possibility for a B star to become a Be star.

Surface Magnetic Fields of Non-degenerate Stars

Jean-François Donati

Lab. d'Astrophysique, Observatoire Midi-Pyrénées, F–31400 Toulouse, France

Abstract. In this paper, I review the various existing techniques for measuring magnetic fields at the surface of non-degenerate stars. I also give a quick summary of the main results obtained to date for the few classes of objects on which fields have been unambiguously detected (chemically peculiar and active late-type stars). As a conclusion, I identify a few classes of stars for which potential field measurements could improve our understanding of various physical processes related to mass loss.

1   Introduction

Magnetic field is very often a crucial parameter in physical problems. It indeed plays a key role basically everywhere in the universe, at all astrophysical time and spatial scales, and in particular in stars, during their formation, their evolution and the last stages of their lives.

The first concern is of course to understand the origin of these stellar magnetic fields, whether they come from a tiny seed galactic field imprisoned in the parent protostellar cloud and amplified during the gravitational collapse of the cloud (which should result in a field structure stable on evolution time scales typically, like that of chemically peculiar stars for instance), or whether they are self-generated fields resulting from turbulent motions in a partly ionised plasma within the stellar envelope (which should then be variable on relatively short time scales, like that of our Sun for instance). Studying stellar magnetic structures can also provide a wealth of information on various physical processes in which magnetic fields play a major role (e.g. dynamos, mass loss). A good understanding of such mechanisms is indeed of direct interest for us in the context of our Sun, of its impact on our solar system and on the Earth in particular. Better input physics also implies better understanding of stellar evolution in general, and of angular-momentum evolution in particular (which affects chemical evolution through transport processes such as meridional circulation), or pre-main sequence evolution (through magnetic interaction with accretion discs).

I will first briefly recall the various existing techniques (polarimetry, spectroscopy and spectropolarimetry) for detecting and modelling stellar surface magnetic fields in Sect. 2. They all use the fact that magnetic fields, through the Zeeman effect, affect the shape and polarisation of spectral line profiles. As numerous extensive papers have already been published on this subject (e.g. Babcock 1958; Landstreet 1980, 1992), I only concentrate on the methods that are still in use now. I do not mention either the possibility of using the Hanle effect (Faurobert-Scholl 1993), as this method has never been used to estimate magnetic fields at the surfaces of stars other than the Sun. Similarly, I do not discuss indirect methods based on radio or X-ray measurements, which only suggest the presence of a field in the extended atmosphere or wind, but cannot be used to demonstrate unambiguously that stellar photospheres do host magnetic fields. I will then summarise in Sect. 3 the main results obtained to date with these different techniques, and will try to outline in Sect. 4 (as a conclusion) the most promising directions' of research in this field as well as results one can reasonably expect in the near future.

2   Measuring Magnetic Fields

2.1   Photopolarimetric Methods

One possibility consists in measuring *circular polarisation* in the wings of a broad spectral line (usually Hβ) with a narrow-band photopolarimeter. The observed circular polarisation (Stokes V) is translated into an estimate of the line-of-sight (longitudinal) component of the field vector averaged over the visible hemisphere. One can easily identify circular polarisation from magnetic fields, as it is expected to switch *sign* from one spectral-line wing to the other. This method (developed by Angel & Landstreet 1970) has provided us with very reliable measurements of longitudinal magnetic fields (e.g. Borra & Landstreet 1980; Bohlender et al. 1987, 1993). A variation of this technique (similar in principle though more complex in its instrumental implementation) was recently proposed by Bedford et al. (1995) for solar-type stars, using an infrared K I line.

Another possibility consists in measuring *broadband linear polarisation* in stellar spectra. What one aims at detecting in this case is the net residual linear polarisation (Stokes Q and U) over the whole profile of one (or several) spectral lines, resulting from the *differential saturation* of π and σ Zeeman components. The broadband filter is usually centred on a spectral domain which maximises the density of spectral lines for the stars of interest. The observed linear-polarisation rates give access to the disc-integrated *transverse* component of the field vector (perpendicular to the line of sight), as well as its *azimuth* with respect to the instrument angular reference (usually north-south). Broadband linear polarisation from surface magnetic fields can be distinguished from continuum polarisation by its specific colour dependence and by its expected modulation with the stellar rotation period. This method was first used by Leroy (1962) for investigating solar magnetic fields, then by Kemp & Wolstencroft (1974) for stellar studies. It is now used by different groups over the world (e.g. Leroy 1995; Tinbergen & Zwaan 1980; Huovelin et al. 1985).

The obvious advantage of these two polarimetric techniques is that they only necessitate a small and relatively cheap device (a photometer rather than a spectrograph). Moreover, they are basically insensitive to stellar rotation (Landstreet 1980), as opposed to some of the methods presented in the following section. However, as they provide no more than a disc-integrated estimate of the longitudinal or transverse field vectors, they can only reveal (through the rotational modulation of these vectors) the large scale structure of a magnetic topology (e.g. Borra & Landstreet 1980; Leroy et al. 1994), and are only poorly sensitive (especially Stokes V polarimetry) to small scale features in this distribution. Both methods yield essentially null results for complex field topologies (e.g. Bedford et al. 1995; Leroy & Leborgne 1989).

2.2   Stokes I Spectroscopy

For these techniques, magnetic information is extracted from the shape of spectral lines, recorded at high spectral resolution and signal to noise ratio.

A first method consists in measuring the *differential broadening* between lines that are highly and weakly sensitive to magnetic fields (Preston 1971; Robinson 1980). The output is an estimate of the disc area covered with magnetic fields (called filling factor), as well as an average field strength value within these regions. This method requires the observed star to spin slowly enough (with $v \sin i < 8$ km/s) to avoid rotational drowning of potential magnetic information contained in line profiles (Saar 1988). Simulations indicate that systematic errors on the derived magnetic flux (i.e. the field strength times the filling factor) can potentially be quite large (Basri et al. 1990, Saar & Solanki 1992). Several multiline approaches were proposed to extend the potential of this technique and decrease systematic errors (Mathys & Solanki 1989; Valenti et al. 1995; Babel et al. 1995a; Rüedi et al. 1997). In particular, these methods have established that magnetic-flux values from earlier studies (using only one line pair) can be overestimated by as much as a factor of 3.

When the magnetic field is strong enough, however, spectral lines can be split into individual Zeeman components, allowing the *magnetic-field modulus* to be measured accurately. This technique has been used to measure disc-averaged field strengths in about 40 objects (e.g. Mathys et al. 1997).

Another method aims at detecting the *equivalent width increase* some spectral lines can be subject to in presence of a magnetic field, depending on the actual Zeeman pattern of the selected line and orientation of field lines (Babcock 1949). However, the dependence of this effect on parameters such as the field orientation for instance, is sufficiently complex that this method can only be used to indicate the presence of a field rather than to provide an accurate estimate of its strength.

The advantage of these methods is their success in detecting (or at least suggesting the presence of) complex magnetic-field structures in new types of objects (G, K and M dwarfs, T Tauri and Am stars), for which conventional polarimetric methods have failed (due to their essential limitation to large-scale field structures). Another advantage of these methods is that the instrument they require (a high-resolution spectrograph) is often available at most observatories. However, both methods only provide extremely limited spatial information on the details of the actual magnetic topology. In an attempt to solve this problem, Saar et al. (1992, 1994) proposed to couple these techniques with a stellar surface imaging package and get magnetic maps of faster rotators. This new method is however still subject to caution as no simulation demonstrating unambiguously that it can reliably recover such tiny magnetic broadening/amplification effects in Doppler broadened line profiles has been published to date. In any case, none

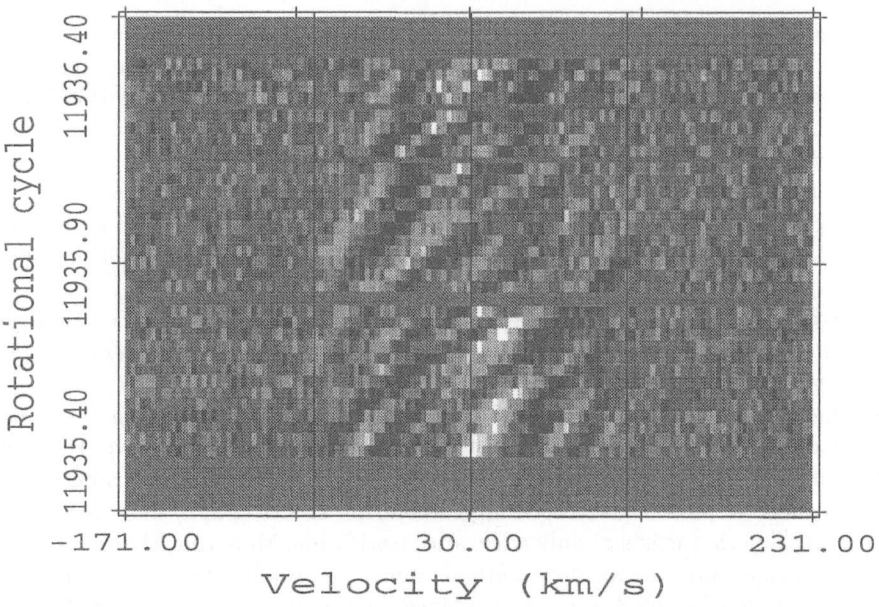

Fig. 1. Rotationally modulated Stokes V signature of AB Dor in 1996 December. Black and white code relative circular polarisation levels of –0.06 and 0.06% respectively. The central and side vertical lines depict the radial and rotational velocities of AB Dor

of these methods enables one to derive information on the *orientation* of field lines within magnetic regions.

2.3 Stokes Q, U & V Spectropolarimetry

The third type of method involves measuring circular and (whenever possible) linear *Zeeman signatures in line profiles*. Optimally, one would like to measure these signatures in individual line profiles, as one can do for solar magnetic regions for instance. However, this is only possible when such signatures are larger than about 0.5 to 1% peak to peak, i.e. in some chemically peculiar stars (Babcock 1947; Borra & Vaughan 1977; Glagolevskij et al. 1985; Mathys 1991, 1995). Most of the time, though (and in solar-type stars in particular where Stokes V signatures rarely exceed 0.3% peak-to-peak, Donati et al. 1992), one has to extract the desired polarisation information from thousands of spectral lines simultaneously, with the help of analogue or numerical cross-correlation tools (Brown & Landstreet 1981; Borra et al. 1984; Donati et al. 1997). As an illustration, Fig. 1 shows the rotationally modulated Stokes V signature (extracted from about 1,500 spectral lines with Least-Squares Deconvolution, Donati et al. 1997) for the rapidly rotating young K0 dwarf AB Dor (Donati et al. 1998a).

While the very first spectropolarimetric attempts were essentially restricted to measuring a mean longitudinal magnetic field (i.e. the first moment of the

Stokes V profile) and its rotational modulation (Babcock 1947; Brown & Land-street 1981; Borra et al. 1984; Mathys 1991), newer methods tend to analyse Stokes Q, U and V signatures themselves (e.g. Borra & Vaughan 1977) or at the very least higher moments (Mathys 1995) to make better use of the spatial information they contain. For fast rotators in particular, one can in principle even reconstruct from such data sets (with the help of a stellar surface ima-ging package) a full photospheric map of the vector magnetic field, i.e. both the photospheric distribution of magnetic regions and the *orientation* of field lines within them (Brown et al. 1991; Donati & Brown 1997).

Stokes Q, U and V spectra are certainly the most reliable source of inform-ation on stellar magnetic structures, *polarisation in line profiles* being the less ambiguous diagnosis for stellar magnetic fields. They also tend to provide the strongest constraints on magnetic topologies, and in particular on the small-scale details that are out of reach of all techniques mentioned above. However, recording such data is quite difficult as it requires a Cassegrain polarimeter with rotatable waveplates linked to a high-resolution spectrograph, an instrumental combination that exists at only a few sites worldwide. Moreover, the small size of Zeeman signatures makes such methods unusable on faint stars, with a limiting V magnitude of 11 on 4 m class telescopes for cool stars (Donati et al. 1997).

3   Results from Studies of Stellar Magnetic Topologies

Chemically peculiar (CP) and cool solar-type stars are the only two classes of magnetic stars that have been studied in some detail yet.

3.1   Chemically Peculiar Stars

This class (the first on which non solar magnetic fields were detected; Babcock 1947) includes Ap and He peculiar stars, for which extensive Hβ polarimetry re-vealed the presence of simple large-scale essentially dipolar magnetic field struc-tures, with polar fields ranging from as low as 200 G up to 34 kG (e.g. Borra & Landstreet 1980; Mathys 1991; Bohlender et al. 1987, 1993). In some rare cases, the variation of the longitudinal field with rotation phase departs significantly from a pure sinusoid, indicating that the associated field structure is somewhat more complex than a simple dipole (e.g. Mathys 1991).

More recently, broadband linear polarimetric studies demonstrated, through a detailed monitoring of rotational modulation in both Stokes Q and U com-ponents, that such field topologies more often depart from pure dipoles than previously thought (e.g. Leroy et al. 1994). Although these measurements sug-gest that such departures are largest at magnetic equator (Leroy et al. 1996; Wade et al. 1996), one should keep in mind that the limited spatial informa-tion content of broadband linear polarimetric data does not allow unambiguous modeling of small-scale magnetic features. Detailed Stokes Q, U and V spec-tropolarimetry of these objects is being undertaken now (Donati et al. 1998b; Wade & Donati 1998a), and should bring us one step further in describing the fine details of these magnetic distributions and quantifying their departures from dipolar structures.

Comparing magnetic topologies and abundance distributions at the surface of CP stars tells us that large-scale fields as weak as the equipartition value (as in ϵ UMa for instance, Donati et al. 1990) can freeze most large-scale transport processes (e.g. turbulence) in the atmosphere of such objects, and generate very high contrast abundance inhomogeneities at photospheric level. Although microscopic diffusion is probably a major clue to this problem, it is still not successful at explaining quantitatively detailed sets of observations (including in particular surface imaging results of numerous chemical species, e.g. Babel et al. 1995b).

For a few slowly rotating Am stars, Stokes I spectroscopy suggested (through both magnetic broadening and intensification effects) that these objects may host very complex magnetic structures with field strengths of a few kG (Mathys & Lanz 1990). However, one may be puzzled by the fact that positive detections are obtained for three stars only, calling into question the validity of the analysis itself. It nevertheless revives the old debate on the nature of magnetic fields in chemically peculiar stars, aimed at understanding whether these fields are pure fossil remnants or result from some exotic type of dynamo processes.

3.2   Solar-type Stars

The very first evidence for magnetic fields in cool stars other than the Sun was obtained from Stokes I spectroscopy (Robinson et al. 1980). The newest results from such methods confirm earlier claims that magnetic fields tend to be equal to the equipartition value, while the area covered by magnetic regions scales up with rotation rate (Saar 1996), in agreement with theoretical dynamo predictions.

Results from Zeeman-Doppler imaging (Donati et al. 1992b; Donati & Cameron 1997; Donati et al. 1998a; Donati 1998) reveals that magnetic images of rapidly rotating cool stars are indeed extremely complex (see for instance Fig. 2 for the K1 subgiant HR 1099) explaining a posteriori the null results obtained on such objects with conventional polarimetric methods. A particularly intriguing characteristic of such images is that they often include magnetic regions of mainly azimuthal field. Detecting such features at photospheric level is interpreted as strong evidence that such objects trigger dynamo processes which are not confined to an interface layer between the radiative and convective zones, but are distributed throughout the whole convective envelope. An independent confirmation of this conclusion is that rising flux tubes from a deep-seated magnetic structure located at the base of the convective zone are expected to emerge at latitudes higher than 50° or so in rapidly rotating late-type stars (Schüssler et al. 1996), in strong contradiction with the brightness, radial and azimuthal field maps we reconstruct for these objects that clearly show features at latitudes lower than 30° (Donati & Cameron 1997; Donati et al. 1998a; Donati 1998).

The polarity of these azimuthal field features is essentially longitude independent at a given latitude (e.g. clockwise and counterclockwise at high and low latitudes respectively on HR 1099, see Fig. 2). The latitude dependence is constant on a year-to-year basis and evolves on timescales of a few decades (Donati

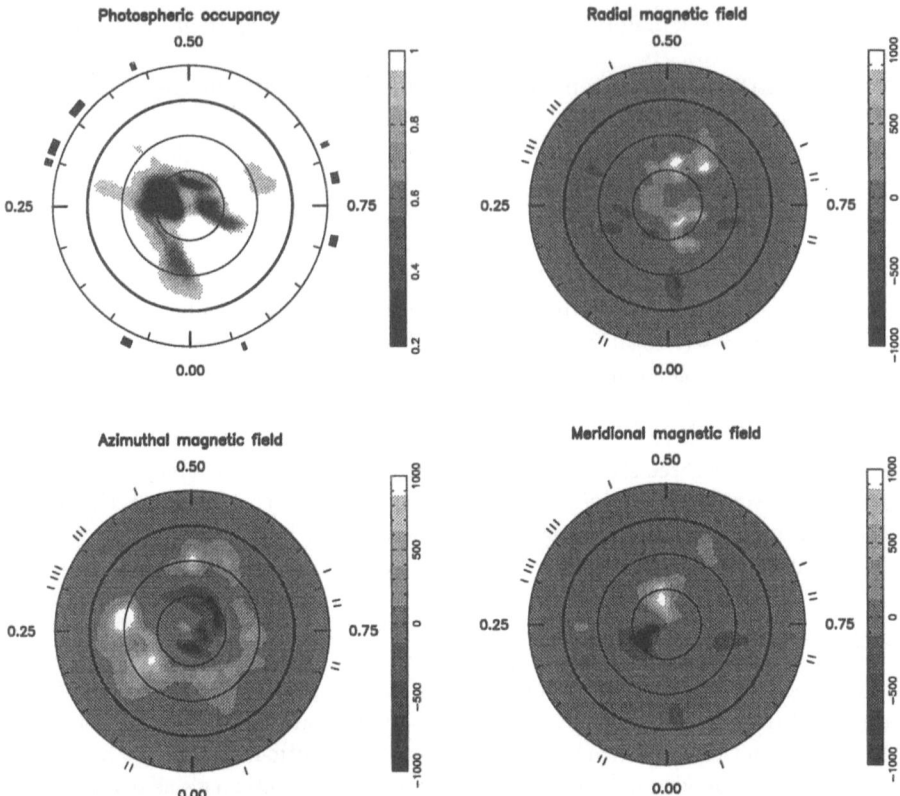

Fig. 2. Flattened polar view of the brightness (upper left), radial field (upper right), azimuthal field (lower left) and meridional field (lower right) distributions at the surface of the K1 subgiant of HR 1099 at epoch 1995.94. Parallels are shown every 30° down to a latitude of −30°. Magnetic fields are labelled in G

1998). This argues in favour that such features witness the *toroidal component* of the large-scale field itself and its modulation with the activity cycle. We also observe that the latitudinal polarity pattern of the large-scale toroidal field gets more complex with increasing rotation rates and convective depths.

4   Magnetic Topologies and Stellar Winds

Only very few papers focus on studying the effect of magnetic fields on stellar winds from an observational point of view, essentially because we lack magnetic measurements for most stars on which stellar winds are detected (and vice versa). I will nevertheless try to present, as a conclusion to my paper, the directions in which this relatively understudied field is likely to progress in the coming years.

4.1   Magnetically Confined Winds

He peculiar objects are the best example of stars showing evidence for stellar winds confined in large scale magnetic field structures (Shore et al. 1990; Shore & Brown 1990). Extensive Balmer polarimetric data sets indicate strong magnetic field organised in relatively simple large-scale structures tilted with respect to the stellar rotation axis (e.g. Bohlender et al. 1987, 1993). Rotational modulation from optical emission lines, UV wind lines and photospheric lines is usually very well defined and stable on fairly long timescales. Finally, these stars are also non-thermal radio (and probably X-ray) emitters (Linsky et al. 1992; Drake et al. 1994). The most recent model that tentatively reconciles all data together is that of Babel & Montmerle featuring a magnetically confined wind escaping the star from both magnetic poles and colliding with itself in the magnetic equatorial plane (Babel & Montmerle 1997a; 1997b).

One O star at least, θ_1 Ori C, is likely to host very similar processes as well, given the very similar observational phenomenology reported by Stahl et al. (1996). In this case as well, the model of Babel & Montmerle is able to reproduce quantitatively the observed rotational modulation in X-ray flux (Gagné et al. 1997), provided that the (assumed dipolar) photospheric field structure has a polar strength of about 300 G, very close to equipartition (Babel & Montmerle 1997b). No magnetic measurement is available to test this model yet, all attempts to date indeed reporting null longitudinal field detections at a 1σ level of about 200 G at best (e.g. Wade & Donati 1998b). As the detection of such a field is within reach of spectropolarimetric experiments on 4 m class telescopes (e.g. Donati et al. 1997), we should be able to progress on this topic fairly soon.

β Cep stars, whose position in the HR diagram overlaps with that of He strong stars, are yet another category of objects for which the observational phenomenology is similar to that of magnetic Bp stars as far as UV wind lines are concerned (Henrichs et al. 1993). This suggests that they also host a photospheric magnetic field strong enough to confine the wind, i.e. close to the equipartition value of a few hundred G. Once again, spectropolarimetric observations on 4m-class telescopes should be able to confirm this view.

Although the model of Babel & Montmerle was originally proposed to explain in a self-consistent way the X-ray emission of the hot magnetic Ap star IQ Aur (which also shows He peculiarities), it is getting increasingly obvious that it could very well be much more general and apply to the whole class of magnetic early-type stars. Additional observational material (and in particular more accurate magnetic field measurements) as well as detailed theoretical studies are obviously needed to validate this presumption.

4.2   Azimuthally Structured Winds

The best example of such objects are hot O and B supergiants. Although such stars also show evidence for rotationally modulated wind signatures (and in particular the presence of discrete absorption components migrating through UV lines from blue to red, e.g. Kaper et al. 1996; Howarth et al. 1995), the detailed

LSD profiles of HD 104237, 1993 Dec. 29 and 1995 Dec. 12   LSD profiles of SU Aur, CFHT, 1996 Nov. 19

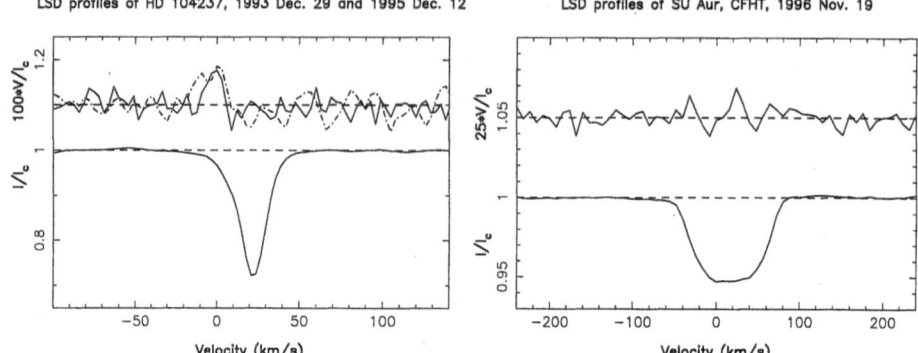

Fig. 3. Marginal detection of Stokes V signatures (with a false alarm probability of 0.001% and 0.2% respectively) in the spectrum of the Herbig Ae star HD 104237 (left) on 1993 Dec. (full line) and 1995 Dec. (dash-dot line), and in the spectrum of the classical T Tauri star SU Aur (right). Note the different Stokes V scaling factors

observational phenomenology is slightly different from that of stars with magnetically confined winds, and the long term behaviour much less stable (Prinja et al. 1995). There is now a considerable body of evidence that this variability is due to corotating wind structures (i.e. the stellar analogue of solar Corotating Interaction Regions or CIRs, Mullan 1986), rooted in stellar surface inhomogeneities (Owocki et al. 1995; Kaper et al. 1997). Several potential interpretations can be found in the literature on the physical nature of these surface inhomogeneities. One of them is the presence of a magnetic field weaker than equipartition (Henrichs et al. 1994; Howarth et al. 1995; Kaper et al. 1997), either with a simple decentred dipole structure tilted with respect to the rotation axis (as in Bp stars) or with a significantly more complex geometry. All attempts at detecting such fields have failed to date down to a 1σ level of about 70 to 100 G (Henrichs et al. 1995, 1998). Spectropolarimetric experiments on a 4m-class telescope such as those of Donati et al. (1997) should allow to decrease this threshold by typically an order of magnitude on the brightest O stars like ζ Pup or ξ Per.

Similar phenomenology is observed for several other classes of objects such as Be stars (e.g. Barker & Marlborough 1985; Smith 1989), Herbig Ae/Be stars (Praderie et al. 1986), A/B supergiants (Kaufer et al. 1996) and Wolf-Rayet stars (St. Louis et al. 1995). Although no magnetic field is detected yet for Be stars (Barker et al. 1985) nor for A/B supergiants and Wolf-Rayet stars, the marginal magnetic field detection obtained by Donati et al. (1997) on the Herbig Ae star HD 103237 (yielding a longitudinal field value of about 25 G, see Fig. 3) is very encouraging. Of course, more observations are required, in order to model the parent field structure of this star first, but also to attempt detecting fields on other stars from the above classes. In this context, β Ori, α Cyg, λ Eri and EZ CMa are obvious candidates for future spectropolarimetric experiments.

4.3   Magnetically Controlled Accretion/Ejection in TTS

T Tauri stars are also thought to be the place where magnetically controlled ejection (and accretion) phenomena take place (e.g. Ferreira & Pelletier 1993). Although magnetic fields are now reported in several weak-line T Tauri stars (Basri et al. 1992; Donati et al. 1997), only marginal detections exist for the subclass of classical T Tauri stars (Unruh et al. 1998, see Fig. 3) for which these magnetically controlled accretion/ejection processes are expected to play a significant role. The corresponding Zeeman signature has a fairly complex shape, indicating that the parent photospheric field structure is obviously not a large-scale fossil dipole, but rather a multipolar dynamo-type topology very similar to those detected in older active stars (see Sect. 3.2). Detailed monitoring of such Zeeman signatures simultaneously with profile variability in Balmer and He lines (informing us on accretion/ejection processes) should bring strong observational constraints for theoretical models on this subject.

Another important issue would be to detect fields from *accretion discs* in which all ingredients for a vigorous dynamo are present (Horne & Saar 1991). Magnetised discs are indeed expected to play an active role, not only through accretion/ejection phenomena, but also for rotational braking of the central star (Cameron & Campbell 1993). The best targets for such experiments are FU Oris, on which all attempts to date have yielded null results (Donati et al. 1997).

Acknowledgements I am very grateful for fruitful discussions with J. Landstreet, C. Catala, J. Babel and H. Henrichs on various points discussed in this paper.

References

Angel J.R.P, Landstreet J.D. (1970): ApJ 160, L147
Babcock H.W. (1947): ApJ 105, 105
Babcock H.W. (1949): ApJ 110, 126
Babcock H.W. (1958): ApJ 128, 228
Babel J., Donati J.-F., Gonzalez J.-F. (1995b): in: Strassmeier K.G. (ed.), IAU Symp. 176, "Stellar Surface Structure". Poster proceedings, p. 129
Babel J., Queloz D., North P., Mayor M. (1995a): in: Strassmeier K.G. (ed.), IAU Symp. 176, "Stellar Surface Structure". Poster proceedings, p. 15
Babel J., Montmerle T. (1997a): A&A 323, 121
Babel J., Montmerle T. (1997b): ApJ 485, L29
Barker P.K., Marlborough J.M. (1985): ApJ 288, 329
Barker P.K., Marlborough J.M., Landstreet J.D., Thompson I.B. (1985): ApJ 288, 741
Basri G., Marcy G.W., Valenti J.A. (1992): ApJ, 390, 622
Basri G., Valenti J.A., Marcy G.W. (1990): ApJ, 360, 650
Bedford D.K., Chaplin W.J., Davies A.R., et al. (1995): A&A 293, 377
Borra E.F., Edwards G., Mayor M. (1984): ApJ 284, 211
Borra E.F., Landstreet J.D. (1980): ApJS 42, 421
Borra E.F., Vaughan A.H. (1977): ApJ 216, 462
Bohlender D.A., Brown D.N., Landstreet J.D., Thompson I.B. (1987): ApJ 323, 325
Bohlender D.A., Landstreet J.D., Thompson I.B. (1993): A&A 269, 355

Brown D.N., Landstreet J.D. (1981): ApJ 246, 899
Brown S.F., Donati J.-F., Rees D.E., Semel M. (1991): A&A 250, 463
Cameron A.C., Campbell C.G. (1993): A&A 274, 309
Donati J.-F. (1998): MNRAS (submitted)
Donati J.-F., Brown S.F. (1997): A&A 326, 1135
Donati J.-F., Brown S.F., Semel M., et al. (1992b): A&A 265, 682
Donati J.-F., Cameron A.C. (1997): MNRAS 291, 1
Donati J.-F., Cameron A.C., Semel M., Hussain G.A.J. (1998a): MNRAS (submitted)
Donati J.-F., Catala C., Wade G.A., et al. (1998b): A&A (submitted)
Donati J.-F., Semel M., Carter B., et al. (1997): MNRAS 291, 658
Donati J.-F., Semel M., Rees D. (1992a): A&A 265, 669
Donati J.-F., Semel M., del Toro Iniesta J.C. (1990): A&A 233, L17
Drake S.A., Linsky J.L., Schmitt J.H.M.M., Rosso C. (1994): ApJ 420, 387
Faurobert-Scholl M. (1993): A&A 268, 765
Ferreira J., Pelletier G. (1993): A&A 276, 625
Gagné M., Caillault J.-P., Stauffer J.R., Linsky J.L. (1997): ApJ 478, L87
Glagolevskij Y.V., Piskunov N.E., Khokhlova V.L. (1985): Pis'ma AZh 11, 371
Horne K., Saar S.H. (1991): ApJ 374, L55
Henrichs H.F., Bauer F., Hill G.M., et al. (1993): in: IAU Coll. 139. CUP, p. 186
Henrichs H.F., Kaper L., Nichols J.S. (1994): A&A 285, 565
Henrichs H.F., Kaper L., Nichols J.S., et al. (1995): in: Strassmeier K.G. (ed.), IAU
    Symp. 176, "Stellar Surface Structure". Poster proceedings, p. 229
Henrichs H.F., and the MuSiCoS collaboration (1998): these proceedings
Howarth I.D., Prinja R.K., Massa D. (1995): ApJ 452, L65
Huovelin J., Linnaluoto S., Piirola V., et al. (1985): A&A 152, 375
Kaper L., Henrichs H.F., Fullerton A.W., et al. (1997): A&A 327, 281
Kaper L., Henrichs H.F., Nichols J.S., et al. (1996): A&AS 116, 257
Kaufer A., Stahl O., Wolf B., et al. (1996): A&A 305, 887
Kemp J.C., Wolstencroft R.D. (1974): MNRAS 166, 1
Landstreet J.D. (1980): AJ 85, 611
Landstreet J.D. (1992): A&ARv 4, 35
Leroy J.L. (1962): Ann. Ap 25, 127
Leroy J.L. (1995): A&AS 114, 79
Leroy J.L., Landolfi M., Landi Degl'Innocenti E. (1996): A&A 311, 513
Leroy J.L., Landstreet J.D., Bagnulo S. (1994): A&A 284, 491
Leroy J.L., Leborgne J.F. (1989): A&A 223, 336
Linsky J.L., Drake S.A., Bastian T.S. (1992): ApJ 393, 341
Mathys G. (1991): A&AS 89, 121
Mathys G. (1995): A&A 293, 733
Mathys G., Hubrig S., Landstreet J.D., et al. (1997): A&AS 123, 353
Mathys G., Lanz T. (1990): A&A 230, L21
Mathys G., Solanki S.K. (1989): A&A 208, 189
Mullan D.J. (1986): A&A 165, 157
Owocki S.P., Cranmer S.R., Fullerton A.W. (1995): ApJ 453, L37
Praderie F., Simon T., Catala C., Boesgaard A.M. (1986): ApJ 303, 311
Preston G.W. (1971): ApJ 164, 309
Prinja R.K., Massa D., Fullerton A.W. (1995): ApJ 452, L61
Robinson R.D. (1980): ApJ 239, 961
Robinson R.D., Worden S.P., Harvey J.W. (1980): ApJ 236, L155
Rüedi I., Solanki S.K., Mathys G., Saar S.H. (1997): A&A 318, 429

Saar S.H. (1988): ApJ 324, 441

Saar S.H. (1996): in: Strassmeier K.G., Linsky J.L. (eds.), IAU Symp. 176, "Stellar Surface Structure". Kluwer Academic Publishers, Dordrecht, p. 237

Saar S.H., Piskunov N.E., Tuominen I. (1992): in: Giampapa M.S., Bookbinder J. (eds.), 7th Cambridge Workshop. ASP Conf. Series 26, p. 255

Saar S.H., Piskunov N.E., Tuominen I. (1994): in: Caillault J.-P. (ed.), 8th Cambridge Workshop. ASP Conf. Series 64, p. 661

Saar S.H., Solanki S.K. (1992): in: Giampapa M.S., Bookbinder J.A. (eds.), 7th Cambridge Workshop. PASP Conf. Series 26, p. 259

Schüssler M., Caligari P., Ferriz-Mas A., et al. (1996): A&A 314, 503

Shore S.N., Brown D.N. (1990): ApJ 365, 665

Shore S.N., Brown D.N., Sonneborn G., et al. (1990): ApJ 348, 242

Smith M.A. (1989): ApJ 71, 357

Stahl O., Kaufer A., Rivinius T., et al. (1996): A&A 312, 539

St-Louis N., Dalton M.J., Marchenko S.V., et al. (1995): ApJ 452, L57

Tinbergen J., Zwaan C. (1980): A&A 101, 223

Unruh Y.C., Donati J.-F., and the MuSiCoS collaboration (1998): these proceedings

Valenti J.A., Marcy G.W., Basri G. (1995): ApJ 439, 939

Wade G.A., Donati J.-F. (1998a): in: Ziznovsky J., North P. (eds.), 26th European Working Group on CP stars (in press)

Wade G.A., Donati J.-F. (1998b): ApJ (submitted)

Wade G.A., Elkin V.G., Landstreet J.D., et al. (1996): A&A 313, 209

Ignace: Comment: A weak Zeeman effect, the Hanle effect, is another diagnostic of stellar magnetic fields using resonance line scattering polarisation with field sensitivities in the range 1–1000 Gauss. Question: Can you place a limit on the magnetic field strength from your marginal Stokes V detection in a Herbig star?

Donati: The magnetic detection we obtained on HD 104237 corresponds to a longitudinal field value of about 25 G, which demonstrates that Zeeman effect on line profiles can also be used to detect weak stellar magnetic fields.

Israelian: A. Collier-Cameron and R. Robinson have observed interesting variations of Hα in AB Dor. Absorptions found in the blue were gradually shifting to the red. Keeping these symmetric variations in mind, what kind of geometry would you suggest for the photospheric magnetic fields?

Donati: The Hα transients seen on AB Dor cross line profiles from blue to red as prominences trapped in the coronal field of AB Dor cross the stellar disc. Given the angle of 60° between the rotation axis of AB Dor and the line of sight as well as the distance of the prominences from the rotation axis ($\sim 2.5\ R_\star$), we obtain that these prominences must be anchored at high latitudes (or they would not be observed). This clearly indicates a complex field topology, in agreement with the magnetic maps we reconstruct. Actually, field extrapolation of our reconstructed surface magnetic topology is being undertaken at the moment. The first (preliminary) results indicate that the azimuth of the detected prominences are in good agreement with those of the main coronal loops we obtain from field extrapolation.

The Hanle Effect as a Probe of Stellar Magnetic Fields

Richard Ignace

U. Glasgow, Dept. Physics and Astronomy, Kelvin Bldg., Glasgow G12 8QQ, UK

Abstract. The Hanle effect is discussed as a means of diagnosing both the strength and geometry of stellar magnetic fields from measurements of resonance line polarization. The effect refers to the *modification* of resonance line scattering polarization in the presence of a magnetic field. Perhaps the most attractive feature of the Hanle effect is that magnetic sensitivities of typical resonance lines of astrophysical interest lie in the range 1–1000 G, a regime of weak field strengths that may be important for stellar variability, winds, and evolution, but which are generally difficult to detect using standard methods based on the more well-known Zeeman effect.

1 Introduction

Used by atomic physicists to measure the radiative decay rates of resonance lines and employed by solar physicists to measure magnetic fields in prominences and the solar atmosphere, the Hanle effect presents exciting new possibilities for probing the magnetic fields of stars.

The discovery and explanation of the Hanle effect has an intriguing history with repercussions on the early development of quantum theory. Rayleigh (1922) performed a series of experiments involving the scattering of light by mercury vapor. He observed that the 2537 Å line could be linearly polarized, but that the degree of polarization could change significantly between experiments. Subsequently, Wood & Ellett (1923) found that the observed polarization was sensitive to the orientation of the experimental setup to the Earth's magnetic field. In 1923, Hanle, who was a student of James Franck, was the first to correctly interpret the observations using a classical description of a damped atomic oscillator. In 1925 the results of the Hanle effect led Heisenberg to extend the correspondence principle to what is called "The Principle of Spectroscopic Stability." These historical developments are interesting because they involve so many prominent figures of the early quantum era, such as Planck, Sommerfeld, Bohr, Weisskopf to name a few. An excellent review on the history and applications of the Hanle effect can be found in Moruzzi & Strumia (1991).

Astrophysically, an important advance was made when for the first time Redman (1941) observed resonance line scattering polarization in the Sun. That same year, two papers by Zanstra described the theory of non-magnetic line polarization to explain the observations of Redman. House (1970) made another significant step toward producing an interpretative framework of astrophysical resonance line polarization with his considerations of linearly polarized light from scattering in weak magnetic fields. Since then, advances in the observations and theory of the Hanle effect in the solar context has proceeded rapidly, and far too many individuals have been involved than can be mentioned here. However, the book by Stenflo (1994) and the proceedings of a conference on solar polarization (Stenflo & Nagendra 1996) indicate its current status in that field.

2   The Hanle Effect

To describe the Hanle effect, Hanle's own approach is adopted, wherein atoms are regarded as damped oscillators. Figure 1 shows the scattering of light by a single atom located at the origin. The incident 100% polarized light is from the left and scatters to the observer through a right angle. The distribution of resonance scattered light is a superposition of a part E_1 that is dipolar and a part $1 - E_1$ that is isotropic, where $E_1 \in [0, 1]$ and is governed by the j_l and Δj of a particular transition (Chandrasekhar 1960). As an example, the case of $j_l = 1$ and $\Delta j = -1$ gives $E_1 = 0$ (isotropic scattering) because the light is emitted from a state $(j_u = 0)$ that is spherically symmetric. In contrast, the case of $j_l = 0$ and $\Delta j = +1$ gives $E_1 = 1$ (dipole scattering) because the light is emitted from a state $(j_u = +1)$ that is like a free electron.

For simplicity take the latter case and consider a magnetic field along the x-axis. Absorption by the atom induces an oscillation transverse to B_x, leading to a precession of the classical electron motion. Consequently, the atom emits at a sequence of position angles instead of one that is fixed in the z-direction as in the zero field case. The fundamental parameter governing the Hanle effect is the ratio of the Larmor frequency to the A-value of the transition:

$$\frac{\omega_L}{A_{ul}} \propto \frac{B}{A_{ul}},$$

This "Hanle ratio" is interpreted as that of the atomic precession rate to the radiative rate. Figure 2 shows how the Hanle effect alters the observed line polarization for three Hanle ratios. For a small Hanle ratio, there is little precession, and the scattered light is nearly 100% polarized in the z-direction. For a Hanle ratio of about unity, there is a significant component of the scattered intensity along the y-axis, hence a net depolarization. However, note that an observer located on the z-axis would now see scattered radiation that is linearly polarized along the y-direction. Thus, the Hanle effect results in depolarization for some viewing perspectives but enhanced polarization in others. In the limit of large Hanle ratios, light scattered in the x-direction is almost completely depolarized, for which the Hanle effect is "saturated" because there is no sensitivity to further increases of the field strength. The fact that different lines do not have the same Hanle ratio owing to differing values of A_{ul} is important for employing the Hanle effect as a magnetic diagnostic of stars.

3   Applications to Circumstellar Envelopes

An inspection of A-values for typical resonance lines of astrophysical interest reveals that the Hanle diagnostic can probe magnetic fields in the range 1–

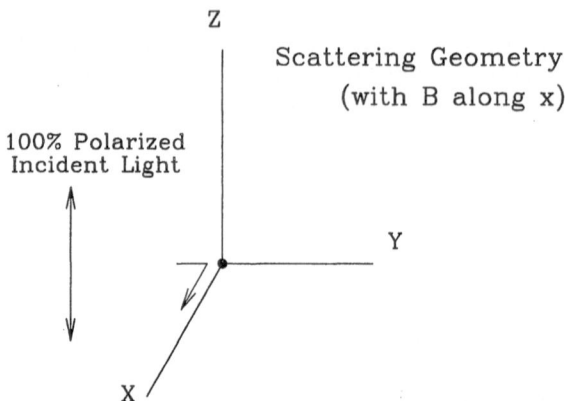

Fig. 1. A hypothetical scattering experiment with 100% polarized light incident on a scattering atom at the origin.

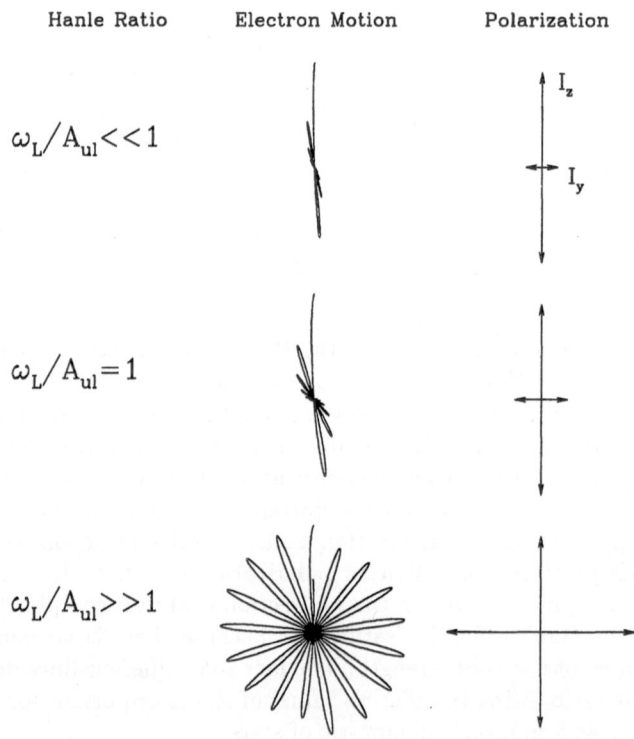

Fig. 2. The Larmor precession associated with three Hanle ratios for the geometry of Fig. 1. The left column gives the ratio ω_L/A_{ul}; the middle column shows the classical electron motion of the atom; and the right column indicates the intensities of the scattered light in two orthogonal directions.

1000 Gauss. Such fields may be sufficient to affect the conditions in outflows from local regions of a star, possibly inducing the co-rotating interaction regions (CIRs) that are thought to explain the cyclical occurrence of DACs observed in the UV lines of some hot stars (Cranmer & Owocki 1996). Other magnetic effects might also be expected, so that the direct detection of stellar magnetic fields at the 100 Gauss level and lower is important.

The discussion of the previous section focussed on the polarization of a single atom as modified by the Hanle effect. Ignace et al. (1997a) have considered the Hanle effect in extended stellar envelopes for line *integrated* polarizations, and Ignace et al. (1997b) have discussed the case of polarized profiles. Although simplified, their work suggests that the Hanle effect can be a viable magnetic diagnostic.

The key to deriving stellar magnetic properties from the Hanle diagnostic comes from a multiline approach. For a given stellar magnetic field, lines with different A-values will show different polarimetric signatures that can be used to infer both the surface field strength and the field topology. For example, consider two resonance lines formed in an equatorial flow with a significant toroidal field component. Suppose that the Hanle ratio of the first is small and the second is large (i.e., in the saturated limit). The position angle of the polarization for the second will be rotated by 90° relative to the first providing strong evidence of the Hanle effect in operation and placing both upper and lower limits to the surface field strength.

In closing it is hoped that 1998 will see the detection of magnetic fields using the Hanle effect in a star other than the Sun when Nordsieck (U. Wisc., USA) launches a rocket experiment with the Far Ultraviolet Spectro-Polarimeter (FUSP).

References

Chandrasekhar, S. (1960): *Radiative Transfer* (New York: Dover)

Cranmer, S.R., Owocki, S.P. (1996): ApJ **462**, 469

Hanle, W. (1923): Naturwissenschaften, 11, 690

House, L.L. (1970): J. Quant. Spectrosc. Radiat. Transfer **10**, 909

Ignace, R., Nordsieck, K.H., Cassinelli, J.P. (1997a): ApJ **486**, 550

Ignace, R., Cassinelli, J.P., Nordsieck, K.H. (1997b): in ASP Conf. Ser. 120, p.198

Moruzzi, G., Strumia, F. (1991): *The Hanle Effect and Level-Crossing Spectroscopy* (New York: Plenum Press)

Rayleigh (1922): Proc. R. Soc. London, Ser. A **102**, 190

Redman, R.O. (1941): MNRAS **101**, 266

Stenflo, J.O. (1994): *Solar Magnetic Fields* (Dordrecht: Kluwer)

Stenflo, J.O., Nagendra, K.N. (1996): Sol. Phys. **164**, nos. 1-2

Wood, R.A., Ellett, A. (1923): Proc. R. Soc. London, Ser. A **103**, 396

Zanstra, H. (1941): MNRAS **101**, 250

Zanstra, H. (1941): MNRAS **101**, 273

Henrichs: May there be existing data that contain the information to determine the field strength by means of the Hanle effect?

Ignace: Yes, I suspect that some data do already exist. For example in their paper on two quasars, Brotherton et al. (1997, ApJ, 487, L113) suggest a contribution to the C III] 1909 polarization (redshifted to visible wavelengths) due to resonance line scattering. This line is interesting since it is sensitive to micro-Gauss level fields. For the stellar objects that are dearer to our hearts, Clayton et al. (1997, ApJ, 476, 870) observe polarization position angle changes across the Na I D line. They interpret this change as arising from geometrical effects, but with a magnetic sensitivity of order 1 Gauss, perhaps the Hanle effect is playing a role too. I cite these papers as potential examples only. In some cases the S/N and/or the spectral resolution may not be sufficient to make any determination of the magnetic properties. I stress that from the perspective of the data analyst, it is recommended to make new or better observations with the intent of employing the Hanle effect.

Photometric Monitoring of Spotted T Tauri Stars

Th. Granzer and K.G. Strassmeier

Inst. für Astronomie, Universität Wien, Türkenschanzstr. 17, A-1180 Wien, Austria

Since 1996 November we have monitored four weak-lined, classical T Tauri stars – RY Tau, SU Aur, V410 Tau, and HDE 283572 – with the University of Vienna twin automatic photoelectric telescope (APT). Here we present the first long-term light curves of RY Tau and SU Aur for the 1996/1997 observing season. A description of the APT can be found in Strassmeier et al. (1997, PASP 109, 697); a survey with more photometric observations is described by Strassmeier et al. (1997, A&AS 125, 11). Further information is also available at http://www.ast.univie.ac.at/~kgs/APT.

1. **RY Tau:** RY Tau shows an average magnitude of $V = 10\overset{m}{.}5$ with erratic increases of brightness of around 1 magnitude. In the left panel of Fig. 1 we show V and $V - I$ colour versus MJD. During maximum light the colour index follows the V variations closely, while at minimum light a clear jump to very blue colours can be seen. This is probably due to scattering of starlight by circumstellar dust.

2. **SU Aur:** SU Aur is a rapidly rotating G2 star. Our measurements show an amplitude of almost $0\overset{m}{.}7$ in V, with a remarkable drop in brightness of $0\overset{m}{.}5$ around MJD 50400. In the right panel of Fig. 1 we plot the V magnitude and $V - I$ colour against MJD. The colour index follows the variations in V closely.

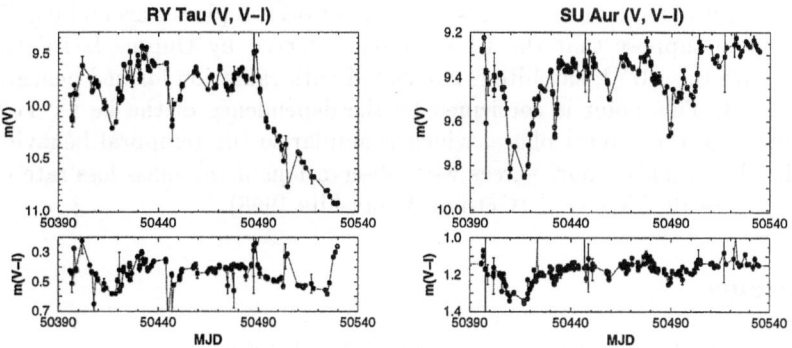

Fig. 1. V magnitude (upper) and $V - I$ colour (lower) of RY Tau (left) and SU Aur (right) against MJD. Note the blueness of RY Tau at minimum light.

The Activity Cycle and Interaction
of Stellar Winds in the Capella System

Maria M. Katsova[1] and Alexander G. Shcherbakov[2]

[1] Sternberg State Astronomical Institute, Moscow State University,
119899 Moscow, Russia
[2] Crimean Astrophysical Observatory, 334413 Nauchny, Crimea, Ukraine

We present observations of the He I $\lambda 10830$ line of Capella obtained between 1985 and 1994. An analysis of radial velocities shows that the main He I absorption is located in the outer atmosphere of the quiescent, cooler G6 giant. We confirm our earlier conclusion that the equivalent width (EW) of the He I line varies with the 104-day orbital period of the Capella system. The mean level of the EW changes from year to year, and is associated with a 5–6-year activity cycle. In addition, a longer 11-year cycle is traced in all our data sets since 1980 (Katsova & Shcherbakov 1998; Katsova 1997). The amplitude of the EW variations versus the phase of the 104-day orbit becomes lower when the mean EW is higher. During a season when the activity level of the Capella system was high, we were able to identify the fraction of the He I absorption formed in the chromosphere of the active, hotter F9 giant.

All these observations can be understood in the framework of the following model: the hot corona of the F9 giant and active regions therein are responsible for the activity cycle and are sources of enhanced stellar wind. A fraction of this wind penetrates the Lagrangian point L_1 to the circumstellar environment of the G6 giant. The interaction between this flow and the stellar wind or corona of the quiescent G6 giant leads to the formation of a shock wave. Most of the He I $\lambda 10830$ absorption arises in regions of the chromosphere of the cooler G6 giant illuminated by the source of EUV and soft X-ray radiation, which is located behind the shock front. This region is best seen at orbital phases around 0.5. There is reason to suppose that the EUV source observed by Dupree & Brickhouse (1995) with the EUVE satellite is associated with radiation formed beneath the shock front. This point is confirmed by the dependence of the Fe XX–Fe XXIII line fluxes on the orbital phase, which is similar to the temporal behaviour of the He I EWs. This model agrees with observations if the mass-loss rate of the F9 giant is $\sim 10^{-8} \, M_\odot \, \mathrm{yr}^{-1}$ (Getman & Livshits 1998).

References

Katsova, M.M., Shcherbakov, A.G. (1998): A&A 329, 1080
Katsova, M.M. (1997): Ap&SS 252, 427
Dupree, A.K., Brickhouse, N.S. (1995): in K. Strassmeier (ed.), Proc. IAU Symp. 176, Stellar Surface Structure (Posters), U. Wien, Vienna, p. 184
Getman, K.V., Livshits, I.M. (1998): Vestnik of Moscow State University, in press

Hα Line-Profile Variability
in Chromospherically Active Stars

M. Weber, A. Washuettl, and K.G. Strassmeier

Inst. für Astronomie, Universität Wien, Türkenschanzstr. 17, 1180 Wien, Austria

One of the major puzzles in the study of stellar surfaces has been the discovery of cool starspots at high latitudes, some of which even straddle the rotational poles of rapidly-rotating late-type stars. The puzzle comes because the Sun shows spots only in two narrow equatorial bands. The role of the Sun in understanding stellar magnetic activity is nevertheless important and intercomparisons should be made whenever possible.

Another important difference are the Hα profiles of active, RS CVn-type stars, which exhibit red-shifted absorption features that can be interpreted as an inward-pointing stellar wind at the Hα line-formation levels. We have compared photospheric Doppler maps of 10 active, late-type stars and their respective Hα spectra; one of them is shown in Fig. 1. For some of the targets we acquired time-series of Hα line profiles that exhibit red-shifted absorption features, which we interpret as being due to a chromospheric antiwind. We suggest that the origin of this wind is connected with the complex polar magnetic surface geometry indicated by our Doppler images.

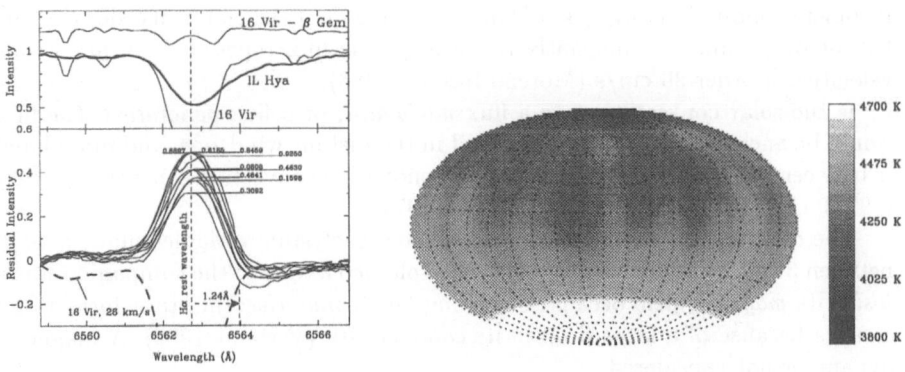

Fig. 1. Time-series Hα line profiles (left) and Doppler image (right) of the RS CVn star IL Hydrae ($P_{rot} = 12\overset{d}{.}9$. spectral classification K0 III–IV, $v \sin i = 26.5$ km s^{-1}).

Cyclical Variability in the Solar Wind

Wolfgang Kundt

Institut für Astrophysik der Universität, D-53121 Bonn, Germany

Abstract. Old and new arguments are presented in favour of the solar convection zone being a magnetic-flux modulator – rather than a magnetic-flux generator. Quite likely, the solar wind is centrifugally driven, forced by the torque of the dragged-out magnetic flux whose footpoints are frozen into the highly conductive core.

1   Stellar Magnetic Fields: Inherited or Acquired?

An extended literature considers stellar magnetic fields the result of a stochastic magnetic dynamo driven by *turbulence*: e.g. Parker (1971), Krause & Rädler (1980), Rädler (1990), but the alternative possibility of frozen-in magnetic fields since stellar formation has equally been proposed: Layzer et al. (1979), Roxburgh (1990), Kundt (1992, 1993). Diffusive flux escape from the solar radiative core takes $10^{10.9}$yr $1 \, {}^{2}_{11}T_{7}^{3/2}$ on length scales $1 - l_{11} := 1/10^{11}$cm – for an average interior electrical conductivity of $\sigma = 10^{15.5}s^{-1}T_{7}^{3/2}$, which largely exceeds the Sun's age for core temperatures $T \gtrsim 10^{7}$K. The solar core is thus likely to *anchor* a significant magnetic flux, whereas her convection zone is magnetically *transmittent* on the time scale of some 10 years: either on grounds of a (poor) turbulent conductivity $\sigma_{turb} \approx c^{2}/4hv_{turb}$ acting on scaleheights of order $h \approx 10^{2}$ km, or on grounds of buoyantly rising magnetic flux ropes drifting at viscous velocities of order 30 cm/s (Moreno-Insertis 1986).

Is the solar convection zone a flux *modulator*, or a flux *generator*? The flux would be anchored both in the core and in the distant windzone, and modulated in between, by the superrotating, poorly conducting convection zone; cf. Schatten (1973), cautioned (unduly?) by Gilman (1974).

The distribution of orbital periods of the cataclysmic variables shows a 'gap' between 3 and 2 hours which tends to be explained as due to the companion star's *losing* its *magnetic field* because of getting *fully convective*: the star's former flux escapes because of a large drop in its conductivity (Verbunt 1993). A magnetic dynamo is not considered.

A non-magnetized plasma acquires (small) magnetic fields both thermally – according to the *fluctuation-dissipation theorem*, important for high densities – and *convectively*, via conversion of turbulence into flux, important for low densities, whereby (Kippenhahn & Möllendorf 1975)

$$d/dt \int (\mathbf{B} + \mathbf{\Omega}) \, d^{2}\mathbf{x} = \int \mathbf{Q} \, d^{2}\mathbf{x} \tag{1}$$

with $\mathbf{\Omega} := (m_{i}c/Ze)\nabla \times \mathbf{v}$, $\mathbf{Q} := \nu_{M}\Delta\mathbf{B} + \nu\Delta\mathbf{\Omega} + (c/e(Z+1)) \, \nabla p \times \nabla n^{-1}$, $\nu_{M} := c^{2}/4\pi\sigma$ = magnetic viscosity, and with m_{i}, Z, n being the ionic mass, charge number, and number density of the stationary plasma respectively. I.e. the sum of magnetic flux and (suitably normalized turbulent) vorticity decays according to a diffusion equation, with both magnetic and kinematic viscosity at work, but can also grow for non-parallel gradients of pressure and density. According to Landau & Lifshitz VIII (1967) §55, the fluctuations of **B** behave stochastically like those of Ω for stationary turbulence. They take maximal values on the (small) *dissipation-length scale* l_{diss}, of $\lesssim \beta_{turb}^2 (\lambda/l)^{1/2}$ times thermal in energy (with $\beta_{turb} :=$ mean turbulent over thermal velocity, and $\lambda :=$ mean free path), beyond which they drop as $l^{-2/3}$ with increasing scale l, or as $k^{2/3}$ with decreasing wave number k. They tend to be ignorable on the (large) 'outer' scale on which the turbulence is excited. A generation of large-scale magnetic fields requires large-scale excitations.

2 Solar Regularities Unexpected for a Turbulent Dynamo

In present days, magnetic-dynamo theory has gained so much attention that earlier indications of a *frozen-in flux* in the solar core tend to be largely ignored, like Bracewell's (1953) 'flywheel', or Dicke's (1978) 'chronometer' deep inside the Sun, or like the more recent work quoted above. In this section, I want to summarize and update my earlier list of observed regularities:

(1) The solar surface is very *unevenly* covered with (discrete) protruding *flux loops*, the larger ones – known as 'sunspots' – clustering at lower latitudes, in the form of the well-known periodic 'butterfly' pattern, and with field-free regions in between. Why doesn't magnetic flux scale as vorticity, as expected for a stochastic dynamo? Flux ropes imply small-scale electric current loops which tend to be wiped out by the FOSA approximation of dynamo theory.

(2) There is even a *hemispheric asymmetry* of sunspots, the southern hemisphere showing fewer spots (by some 10%): Hathaway & Wilson (1990). At the same time, an inverse correlation is found between *sunspot area* and *rotation rate*, both spatially and temporally, amounting to variations by $\lesssim 0.5\%$ in rotation rate. Magnetized areas rotate (a little) more slowly.

(3) Sunspots vary in magnitude with a period of order 11 years, but a more accurate period is Hale's $P_{Hale} = (22.2 \pm 2)$ yr which takes care of the sign of the solar dipole, hence covers two successive 11-year cycles. At least 16 *Hale cycles* have been traced – Bracewell (1985) even identifies them in the $10^{7.5}$-cycles old Elatina formation – both in sunspots, bolometric flux, aurorae, droughts in high planes, D/H in bristal-cone pines, C^{14}, Be^{10}, neutral iron (Ulrich & Bertello 1995), 10.7 cm, X-rays, surface oscillations [item (5) below], vibrational p-modes, magnetic multipoles [item (6)], windspeed (Lotova 1988), and in 'cosmic rays' (Parker 1992). These well-defined oscillations of the solar appearance and output fluctuate strongly from cycle to cycle but have a remarkable *long-term stability*. Moreover, solar *activity* anticorrelates with *cycle length* (Calder, priv. comm.). A flywheel is indicated.

(4) The solar surface shows *equatorial superrotation* by $\lesssim 30\%$, w.r.t. its poles and perhaps even w.r.t. its deep interior: Elsworth et al. (1995). This remarkable

behaviour – which requires permanent excitation, and has been highlighted e.g. by Schröter (1985) – has the opposite sign to what results from any type of latitudinal mixing; it manifests a decelerating torque on the solar interior. Similar decelerating (magnetic) torques appear to be at work on the *outer planets* (Kundt & Lüttgens 1997) and on *Earth*, whose magnetic 'anomalies' drift westward. These torques couple the flux-anchoring core to the surrounding plasma.

(5) The solar surface shows not only (secular) equatorial superrotation but also *torsional oscillations* (on the Hale period, of amplitude some 3%), both rigid and twisting (Howard, 1984); they have been modelled by Volland (1992). Such oscillations require (magnetic) torques acting between the solar core and convection zone.

(6) The *magnetic-multipole* structure of the solar surface varies systematically with the Hale period, see Figure 1. The total magnetic energy varies much less than its dipole fraction, and both have a distinct one-year peak at solar minimum. Contrary to Earth, odd-order multipoles are much stronger than even-order ones, often by an order of magnitude.

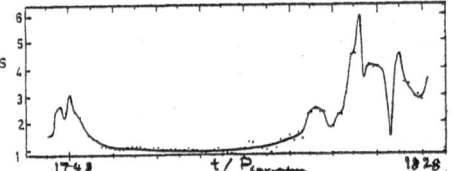 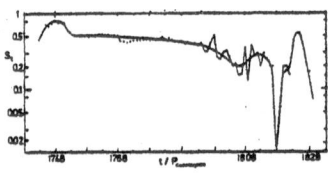

Fig. 1. Solar-surface magnetic energy $S = \sum S_k$ (in suitable units, $S_k :=$ k^{th} multipole fraction) and its dipole fraction S_1 as functions of time, expected to be quasi periodic; from Kundt (1993), after Hoeksema (priv. comm.). Note that the dipole energy is plotted logarithmically, and that it has a distinct one-year peak at solar minimum.

(7) The magnetic flux ropes piercing the solar surface show two different revolution periods, the long-term 'Carrington' period of $P_{Carr} = (27.3 \pm 0.5)d$, and a (longer) 'Snodgrass' period (characteristic of short-term motions) which is strongly latitude dependent, ranging from some 1% (longer) near the equator to almost 30% near the poles: Short-lived patterns ($\lesssim 10d$) are overtaken by long-lived patterns (Sawyer 1968; Bogart 1982; Stenflo 1989). Both periodicities can be traced through $\lesssim 16$ cycles; they reveal *flux-rope dragging* across the whole convection zone.

3   Cyclical Wind Variability

The preceding section summarized arguments in favour of solar variability being controlled by the magnetic flux frozen into the Sun's core. As argued by Indulekha et al. (1990), the dragged-out fraction of this flux may well be strong enough to force the solar wind into corotation out to Alfvén-radius distances. This conclusion has been corroborated by Lotova's (1988) scintillation data which locate the wind's transonic region at (20 ± 10) $R_\odot$, distinctly downstream from weak-magnetic predictions. Clearly, a variable emerging flux implies a variable solar-wind velocity structure. This behaviour would link the solar wind to the winds from Wolf-Rayet stars for which Indulekha et al. (1990) have concluded that they cannot be radiation driven: in particular I disagree with equ(1) in Gayley et al. (1995) which can violate the radial-momentum balance. Their cartoon in fig(1) – on multiple line-photon scattering – does not take proper account of overall momentum balance.

Acknowledgement

My thanks for a rewarding correspondence go to Joe Cassinelli, and for the manuscript to Hans Baumann, Harald Giersche, and Carsten van de Bruck.

References

Bogart R.S. (1982): Solar Physics 128, 299
Bracewell R.N. (1953): Nature 171, 649
Bracewell R.N. (1985): Australian J. Phys. 38, 1009
Dicke R.H. (1978): Nature 276, 676
Elsworth Y., Howe R., Isaak G.R., et al. (1995): Nature 376, 669
Gayley K.G., Owocki S.P., Cranmer S.R. (1995): ApJ 442, 296
Gilman P.A. (1974): Solar Phys. 36, 61
Hathaway D.H., Wilson R.M. (1990): ApJ 357, 271
Howard R. (1984): Ann. Rev. A&A 22, 131
Indulekha K., Kundt W., Shylaja B.S. (1990): Ap&SpSc 172, 1
Kippenhahn R., Möllenhoff C. (1975): Elementare Plasmaphysik, Bibliograph. Institut Mannheim
Krause F., Rädler K.-H. (1980): Mean-Field MHD and Dynamo Theory, Pergamon Press, London
Kundt W. (1992): Ap&SpSc 187, 75
Kundt W. (1993): in The Cosmic Dynamo, F. Krause et al. (eds.), IAU 157, p.77
Kundt W., Lüttgens G. (1997): Ap&SpSc, submitted
Layzer D., Rosner R., Doyle H.T. (1979): ApJ 229, 1126
Lotova N.A. (1988): Solar Phys. 117, 399
Moreno-Insertis F. (1986): A&A 166, 291
Parker E.N. (1971): ApJ 168, 239
Parker G.D. (1992): in Solar Wind Seven, E. Marsch, R. Schwenn (eds.), Pergamon, p. 233
Rädler K.-H. (1990): in Inside the Sun, G. Berthomieu and M. Ciribier (eds.), Kluwer, Dordrecht, p. 385
Roxburgh I.W. (1990): Phil. Trans. Roy. Soc. A 330, 641
Sawyer C. (1968): Ann. Rev. A&A 6, 115
Schatten K. (1973): Solar Phys. 32, 315
Schröter E.H. (1985): Solar Phys. 100, 141

Stenflo J.O. (1989): A&A 210, 403
Ulrich R.K., Bertello L. (1995): Nature 377, 214
Verbunt F. (1993): Ann. Rev. A&A 31, 93 (119)
Volland H. (1992): A&A 259, 663

Johns-Krull: I am confused by the first point that there are large regions at the solar surface free of any magnetic field. Alan Title's group at Lockheed have shown movies of magnetogram data from the MDI instrument on SOHO which appear to show a diffuse "magnetic carpet" everywhere on the Sun. They also find that this field regenerates every 40 hours, which they say means that the field is generated locally, just beneath the photosphere.

Kundt: Your phenomenon is new to me; thank you! It should not invalidate, however, the centuries-old wisdom that the solar-surface magnetic field emerges predominantly in the form of discrete flux ropes that cluster at low solar latitudes. Flux-tube "generation" can be easily confused with "migration".

Luminous Magnetic Rotator Theory

Joseph P. Cassinelli

University of Wisconsin, Astronomy Department,
475 N. Charter St., Madison, WI 53706, USA

Abstract. This paper describes the development of a wind theory that combines forces associated with magnetic rotators with those operating in line-driven winds. It has relevance to cyclic variability because if a rotating star has a magnetic field that changes from one longitude sector to the next, the wind speed can also change and this could lead to fast wind / slow wind co-rotating interaction regions. This paper focuses on effects that can increase the radial velocity of a line-driven wind within a sector.

1 Introduction

In this meeting on cyclic variability it is useful to consider ways by which the properties of a wind can vary over the surface of a rotating star. Here I consider a way in which wind speeds and mass fluxes might be enhanced in certain longitudinal sectors on a star with a radiatively driven wind. For winds driven by radiation alone, one might expect that the radiation field and the winds should be spherical or, at most, show a latitudinal variation owing to polar brightening. However, if a star also has magnetic field lines emerging from its surface, it is possible that circulation currents below the surface will cause the field strength to vary from one sector to the next. To illustrate what is required for significant variations in wind speeds and mass fluxes, I consider here the effects of magnetic rotator forces, and for simplicity, consider only the equatorial region of the star.

The basic physics for how an open magnetic field emerging from the star will affect the mass flux, and the radial and azimuthal components of velocity was first developed for the solar wind in the classic paper by Weber & Davis (1967). The equations describing how radiation forces can drive a flow from a luminous star was developed in another classic paper, by Castor, Abbott and Klein (1975, CAK). Luminous Magnetic Rotator (LMR) theory concerns the combination of the forces treated in these two papers. In section 2, some basic ideas regarding the effects of magnetic rotator forces on winds are described in the framework of the $\dot{M}$ versus v_∞ diagram. In section 3, several of the papers in which LMR theory was developed are described.

2 Basic Description of Hybrid Wind Models

Different wind mechanisms lead to different mass-loss rates and terminal velocities. So it is useful to have a way to isolate the effects of each of the driving forces. In the case of LMR winds we are dealing with the line-driven-wind mechanism which provides luminous stars such as O stars, OB supergiants and Wolf Rayet stars with large mass-loss rates in the range of about 10^{-7} to 10^{-5} $M_\odot$ yr^{-1}, and high speeds, of order 1000 km s^{-1}. The combination of line driven forces with magnetic-rotator forces was first suggested by Hartmann & Cassinelli (1981) as a possible explanation of the Wolf Rayet momentum problem (Cassinelli, 1982). An alternative explanation involving multiple scattering of photons has been developed by Abbott & Lucy (1985), and Gayley et al. (1995). Here we will consider the standard line-driven-wind theory and use it in conjunction with magnetic-rotator forces. Magnetic fields and rotation have also been considered for explaining the rotationally distorted winds of emission-line Be stars, Oe stars, and the supergiant B[e] stars.

A useful diagram for describing the effects of rotation and magnetic fields on a wind is an $\dot{M}$ versus v_∞ plot, as is shown in Figure 1. The box on the plot represents the values of $(\dot{M}, v_\infty)_P$ associated with the "primary wind mass-loss mechanism". In the case of the sun, this would be the coronal-wind mechanism, in the case of hot stars it would be the line-driven-wind mechanism.

There are three domains of magnetic-rotator theory that can be identified in regards to Figure 1. A star can be a *slow magnetic rotator* (SMR), a *fast magnetic rotator* (FMR), or a *centrifugal magnetic rotator* (CMR). An example of a SMR is the original magnetic-rotator model for the solar wind developed by Weber and Davis (1967). They found that the solution of a magnetic rotator wind radial velocity structure, $v_r(r)$, has two X-type singularities, instead of the one that occurs in Parker wind theory, and that there is also a critical point for the azimuthal speed $v_\phi(r)$, called the Alfvén point. The defining characteristic of an SMR wind is that the magnetic field and rotation have almost no effect on the values of either the mass-loss rate or wind terminal velocity. So, for an SMR, the values for $\dot{M}$ and v_∞ are completely set by the primary mechanism. However, even for an SMR wind there is a transfer of angular momentum from the star to the wind, which leads to a spin down of the star. For example, the Weber & Davis theory led to the realization that the sun could lose its angular momentum J on a characteristic time scale comparable to the age of the sun. This is a surprising result if you realize that the time scale for mass loss from the sun $(M_\odot / \dot{M}_\odot)$ is about 10^{14} years, or 4 orders of magnitude slower than the angular-momentum-loss rate. The magnetic field causes an angular momentum loss *as if* the wind were in solid body rotation with the star out to the Alfvén radius, which is at about 10-20 $R_\odot$ for the case of the sun. The wind does not actually co-rotate out to that distance. In fact, more than half of the angular momentum transferred from the star is carried by magnetic field stresses. However, our goal here is to investigate effects that increase v_∞ or $\dot{M}$, and the forces in an SMR model do not accomplish that.

Belcher & MacGregor (1976) developed fast-magnetic-rotator wind theory in their considerations of the early evolution of the solar wind, when the sun would have had a faster rotation speed and perhaps also a larger surface-magnetic-field strength. In the case of a FMR wind, there occurs a transfer of radial momentum from the star that can be described in terms of the Poynting energy flux. The

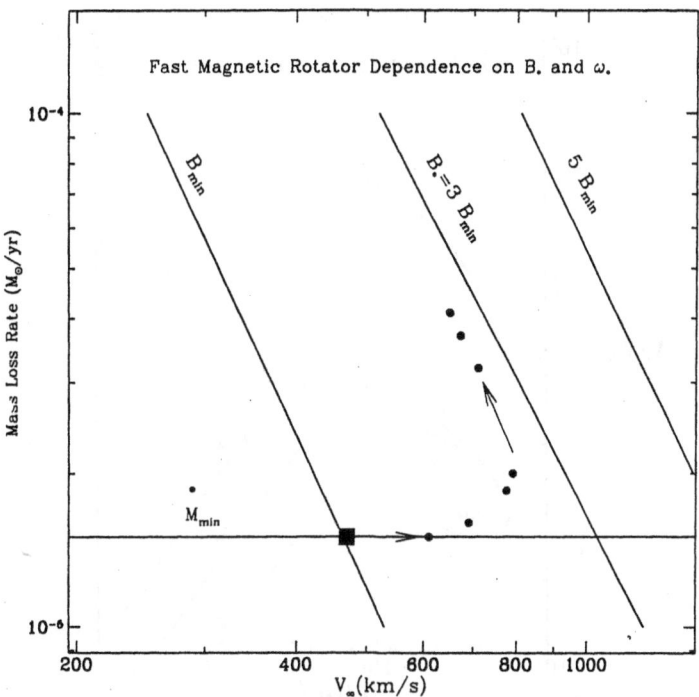

Fig. 1. A mass loss rate versus terminal velocity plot that is useful for discussing hybrid magnetic rotator models. The solid square represents the $(\dot{M}, v_\infty)$ values for the "primary mechanism", which for luminous magnetic rotators would be the modified CAK result. If the surface magnetic field exceeds the minimal field, B_{min}, then the LMR model can be in the fast magnetic rotator regime and can have a terminal velocity that is faster than the primary model, The v_∞ increases as Ω increases (as indicated by the arrows). If the rotation rate is rather large, the $\dot{M}$ will also increase and the star is in the centrifugal magnetic rotator regime. (From Lamers & Cassinelli, 1998)

deposition of the momentum occurs mostly in the region beyond the inner critical point, and as is well known in wind theory, this means that the terminal velocity is increased, but with little change in the mass-loss rate. Thus, in the models of Belcher & MacGregor the magnetic-rotator forces do not increase the mass-loss rate over that established by the primary wind mechanism, but the value of v_∞ is increased. This means that a point in Figure 1, representing a FMR model is located to the right of the point representing the primary mechanism. Some results from Belcher & MacGregor are shown in Figure 2. Notice that the slope, dv_r/dr of the velocity law is increased in the FMR regime. Later we will see that this change in the velocity gradient increases the radiative acceleration in luminous magnetic-rotator models.

Centrifugal magnetic-rotator theory was developed by Hartmann & Mac-

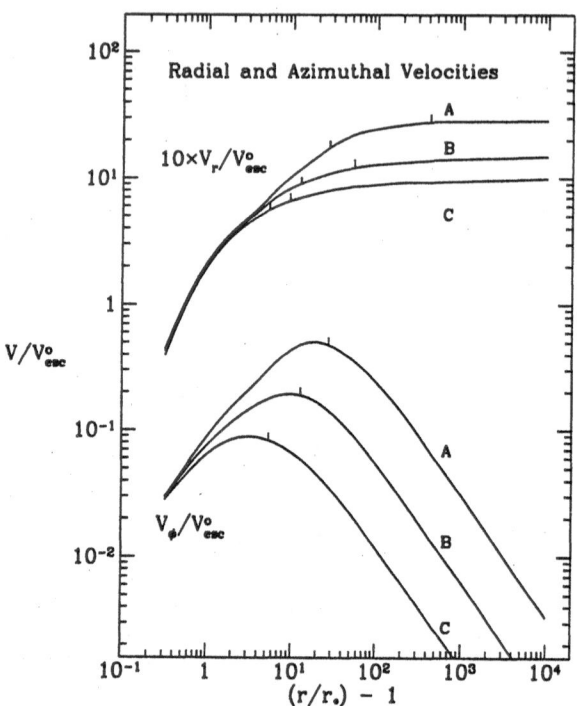

Fig. 2. Shows the radial and azimuthal velocity distributions as functions of radius for three different magnetic field strengths $B_{r,o}$(A=4.6, B=1.5 and C=0.46 Gauss) for a fast-magnetic-rotator model of the sun having a surface rotation speed $r_o\Omega = 0.01v_{esc}$. The tick marks on the v_ϕ curves indicate the location of the Alfvén radius, the tick marks on the radial velocity curves show both the Alfvén radius and the fast point radius where, $v(r) = v_M$. Note that as the magnetic field increases, both the terminal velocity and the slope d v_r/dr increase. In regards to the azimuthal speed, note that as the field increases, the region over which the rotation rate is nearly solid body becomes larger. (from Lamers & Cassinelli, 1998; adapted from Belcher & MacGregor 1976)

Gregor (1982), in a paper in which they considered the mass loss from rapidly rotating protostars. In these models the mass-loss rate associated with the primary mechanism (coronal wind) was small. So the CMR models produced major changes in both the terminal velocity *and* the mass-loss rate. Although they referred to their wind models as fast magnetic rotators, the name centrifugal magnetic rotator provides a better description. The CMR type of magnetic-rotator model is especially simple because the critical point is determined by the rotation rate of the star, which is the basic parameter for any rotating wind model. The magnetic field in a CMR model is assumed to cause solid-body rotation out to the inner critical point. Thus, there is a simple modification owing to centrifugal forces to the hydrostatic density distribution, $\rho_{HS}(r)$.

The inner critical point, which is also the sonic point in the Hartmann & Mac-Gregor CMR model, occurs at the radius where the solid body rotation speed $(V_{solid}(r) = r \times \Omega)$ equals the circular speed $(V_{circ}(r) = \sqrt{GM/r}$. This leads to the following expression for the radius of the sonic point.

$$(r_s/R_*)^3 = (\Omega_{max}/\Omega)^2 \qquad (1)$$

where Ω_{max} is the critical (or "breakup") rotation rate at which the equator of the star is rotating at circular Keplerian speed. The mass-loss rate (or $4\pi R_* \times$ the equatorial mass-flux rate) is determined by r_s (given by equation (1), the sonic speed (determined by the wind temperature), and the density at the sonic point. The latter is given, to a good approximation, by $\rho_s = 0.6 \times \rho_{HS}(r_s)$). Thus, to determine the mass-loss rate we only need to know the rotation speed. Note, in particular, that the magnetic field has no direct effect on the mass loss rate. The field must be sufficient to cause solid body rotation out to the inner critical point, but any further increase in the field has no effect on the mass-loss rate. However, the magnetic field does have a major effect on the speed of a CMR wind.

The terminal velocity of a centrifugal magnetic rotator is given to a very good approximation by the speed at the fast critical point, v_M, called the Michel velocity

$$v_M{}^3 = \left(R_*{}^4 \Omega^2 B_{r,0}{}^2 / \dot{M} \right) \approx v_\infty{}^3. \qquad (2)$$

Note here that the terminal velocity is fully determined by the basic parameters at the base of the wind, the rotation rate, Ω (which has already determined the mass loss rate $\dot{M}$), and the magnetic field at the surface of the star. So unlike $\dot{M}$, the value of v_∞ depends explicitly on the field strength; given a rotation rate, the higher the field the faster the wind.

Referring to Figure 1, if we follow the path corresponding to increasing the rotation rate of the star, we first find that the rotation and magnetic field have no effect on the $(\dot{M}, v_\infty)$ value, i.e. the star is a slow magnetic rotator. Then in the FMR regime the mass-loss rate is not affected, but v_∞ increases. Then the star enters the CMR domain and the mass-loss rate increases as the point moves along the line defined by the Michel velocity expression (equation 2). Note that different fields will lead to different iso-Michel velocity lines, but equal mass-loss rates correspond to equal rotation rates.

3 Luminous Magnetic Rotators

Now let us incorporate the effects of the radiation field. In regard to Figure 1, the first effect is to provide a starting point which is at a fairly high mass-loss rate and a fast speed. Given the $\dot{M}$ and v_∞ values expected from the primary mechanism, i.e. line-driven-wind theory, we can derive from a consideration of equation (2), a minimal value, B_{min}, for the surface magnetic field that will allow the star to be in the fast magnetic rotator regime. In the case of WR stars,

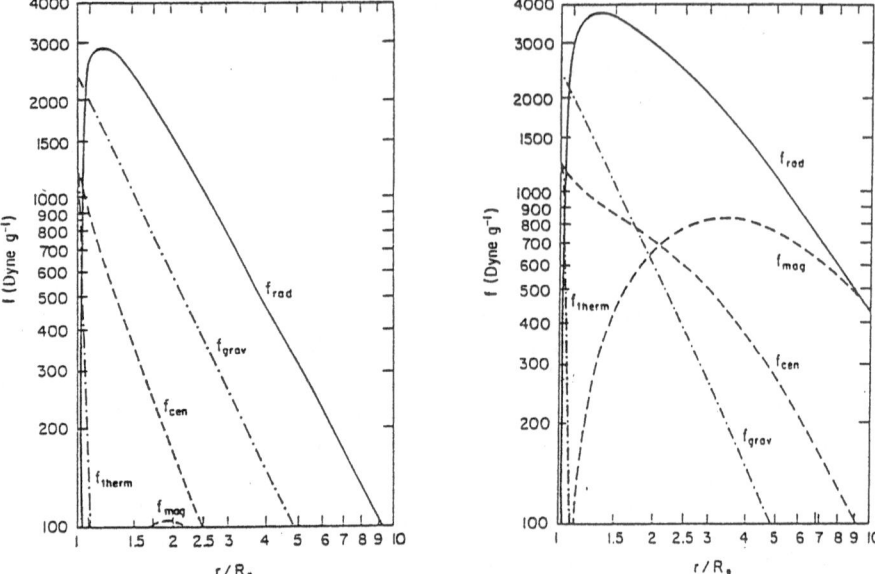

Fig. 3. The distribution, as a function of radius, of 5 forces per unit mass: thermal, magnetic, centrifugal, gravitational, and line radiation, for the Oe star λ Cep, which is assumed to rotate at 70 % critical speed. Panel (a), on the left, shows results for a model with $B_{r,o} = 200$ G, and we see there is a negligible magnetic force. The star has a terminal velocity of 1200 km s^{-1}, and $\dot{M} = 6.6 \times 10^{-6}\ M_\odot$ yr^{-1}. For Panel (b), on the right, $B_{r,o}$ has the much larger value, 1600 G, and we see the magnetic force is much larger, this leads to a larger velocity gradient, and thus the radiation force is also increased, as we see. This high $B_{r,o}$ model has a terminal velocity of 2900 km s^{-1}, much faster than the panel (a) model, but a mass-loss rate ($7.9 \times 10^{-6} M_\odot$ yr^{-1}) that is only slightly larger than the panel (a) model, as is characteristic of models in the FMR regime. (from Friend & MacGregor, 1984)

B_{min} is about 1000 Gauss, while for B[e] supergiants and Be stars B_{min} is about 100 Gauss and 10 Gauss, respectively.

Radiation also plays a role in increasing the velocity of the wind along the FMR portion of the track in the $\dot{M}$ versus v_∞ plot. The reason can be seen from results of the earliest luminous magnetic rotator calculations, by Friend & MacGregor (1984). These authors considered the O6.5e star λ Cep, and the results for the forces in the wind are shown in Figure 3. Note that because of the velocity-gradient effect, the radiative acceleration is enhanced by the effects of the strong magnetic field, as seen in Figure 3b. There is an increase in the radiation force because in Sobolev line-transfer theory the larger the velocity gradient the larger the amount of stellar radiation that is Doppler shifted into the line-absorbing region per unit length in the wind. Thus, the magnetic-rotator forces make the stellar radiation field more effective, per spectral line, in driving a wind.

In the Friend & MacGregor paper the original CAK formulation of line-driven-wind theory was used. The CAK theory assumes a point-source star, so the radiation field was everywhere radial, and thus is maximally efficient at driving mass. "Modified CAK", (or mCAK) theory, developed by Friend & Abbott (1986) and Pauldrach et al. (1986), accounts for the finite size of the star. The finite size affects the angular distribution of the radiation field most strongly in regions near the star. Mass-loss rates are also determined by conditions near the star, and it turns out that the mass-loss rate of mCAK theory is smaller (by about a factor of 2) than that from CAK theory. Since less mass is being driven out in the mCAK models, the radiative momentum flux from the star can drive a faster wind, (again by a factor of 2). Both of these modifications lead to much better agreement between line-driven-wind theory and observations. Poe & Friend (1986) used mCAK theory for their LMR models of the O star λ Cep, and for Be star, 59 Cyg. It led to important quantitative differences. Their paper also showed for the first time the transition from the FMR to the to CMR regime for luminous magnetic rotators. They found that the critical point of mCAK theory (at $r_{crit} \approx 1.03\ R_*$) is much closer to the star than in the CAK theory (with $r_{crit} \approx 1.75\ R_*$). Thus, their LMR models had conditions at the inner critical point that are similar to the coronal CMR models with strong winds.

Poe et al. (1989) applied LMR theory to explain the momentum problem of WR stars. The results are shown in Figure 4. The momentum problem can be explained as follows. The LMR model predicts a dense *equatorial* region. This produces most of the radio flux by which observers deduce that WR stars have large mass-loss rates. If the line of sight is along some intermediate latitude, the wind velocity derived from the P-Cygni profile absorption, will be that of gas in the high speed *polar* flow. As a result the observational derivations of both $\dot{M}$ and v_∞ will be overestimates of the star as a whole, and the combination leads to the Wolf Rayet wind-momentum problem.

4   Summary

In regards to the position of a star on the $\dot{M}$ vs. v_∞ diagram, these are the ways that high luminosity affects magnetic-rotator theory,

- The mCAK line forces provide a basal model with a high $\dot{M}$, and a fast v_∞.
- On the FMR segment, the increase in v_∞ mostly arises from the fact that radiative acceleration is proportional to the velocity gradient dv_r/dr.
- On the CMR segment of the plot, the inner critical point is at the mCAK theory critical point (at $\approx 1.03\ R_*$). The location of this point and the density conditions at the base of the wind determine the angular speed at which there is a rise in the $\dot{M}$ value.

We have not discussed explicitly the cyclical variability that might be caused by having a star with a field that varies across the face of the star. However, it

244   Joseph P. Cassinelli

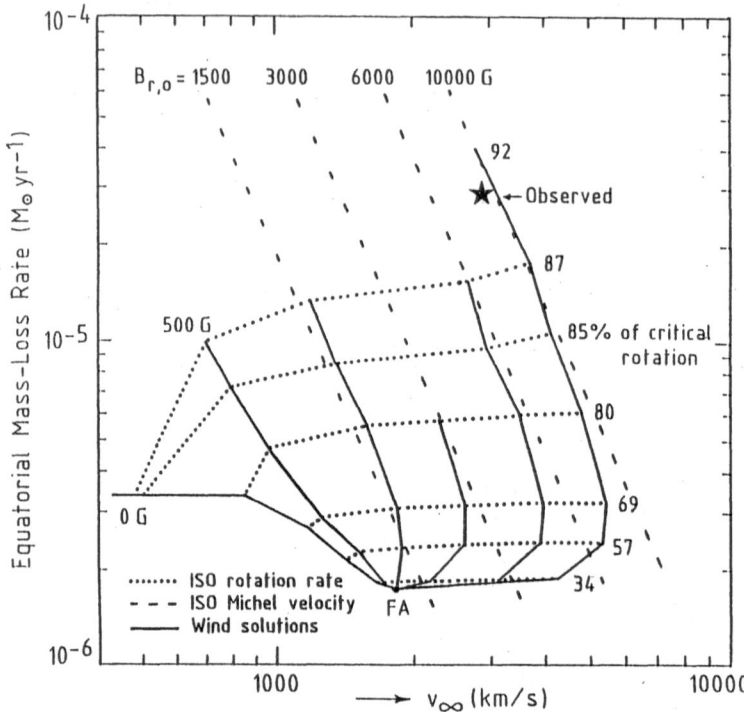

Fig. 4. The $\dot{M}$ versus v_∞ plot for a Wolf Rayet Star, that shows the transition from SMR to FMR to CMR regimes of magnetic rotators. The primary mechanism point in this case was found from the mCAK theory of Friend & Abbott (1986). The dashed lines show the iso-Michel velocity lines for the 4 values of the surface field as indicated, and the dots connect models that have the same rotation speed. For the WR star illustrated here, the winds reach the CMR regime if the rotation speed exceeds about 34 % critical speed. The zero field model shows that there can be a factor of two increase in mass loss rate as the rotation speed increases. The star symbol indicates the observational estimates of mass loss (obtained from radio observations) and terminal velocity (from UV line profiles) of the WR star. (From Poe et al. 1989)

is clear that the $\dot{M}$ versus v_∞ diagram forms a good framework for interpreting observational evidence for fast-wind / slow-wind interactions in co-rotating interaction models.

References

Abbott, D.C., Lucy, L.B. (1985): ApJ, 288, 679
Belcher, J.W., MacGregor, K.B. (1976): ApJ, 210, 498
Castor, J.I., Abbott, D.C., Klein, R.I. (1975): ApJ, 195, 157 (CAK)

Cassinelli, J.P. (1982): in *Wolf Rayet Stars: Observations, Physics and Evolution* Eds. C. de Loore and A. Willis (Dordrecht, Reidel) p. 173

Friend, D.B., Abbott, D.C. (1986): ApJ, 311, 701

Friend, D.B., MacGregor, K.B. (1984): ApJ, 282, 591

Gayley, K.G., Owocki, S.P., Cranmer, S.R. (1995): ApJ, 442, 296

Hartmann, L., Cassinelli, J.P. (1981): BAAS, 13, 795

Hartmann, L., MacGregor, K.B. (1982): ApJ, 259, 180

Lamers, H.J.G.L.M., Cassinelli, J.P. (1998): *Introduction to Stellar Winds*, (Cambridge, Cambridge), Chap. 9

Pauldrach, A., Puls, J., Kudritzki, R.P. (1986): A&A, 164, 86

Poe, C.H., Friend, D.B. (1986): ApJ, 311, 317

Poe, C.H., Friend, D.B., Cassinelli, J.P. 1989, ApJ, 337, 888

Weber, E.J., Davis, L. Jr. (1967): ApJ, 148, 217

Henrichs: 1) How should an observer identify your B_{min}, wouldn't it depend on the geometry?

2) The co-rotating stream model gives a different "terminal speed" than the ambient wind. Would this velocity difference, combined with the known rotation rate yield immediate information on the strength of the field at the base of the streamline?

Cassinelli: 1) For some stars seen equator-on, one can derive from the P-Cygni line profiles, both the speed of the equatorial wind and the speed of the non-equatorial or "polar wind" (e.g. HD 93521). In such cases one could use the equatorial v_∞ and the mass flux to derive B_{min} from the Michel velocity formula.

2) From the nature of the DAC's one should be able to derive the speed of the fast stream relative to the slower ambient flow. If so this would provide a value for V_{michel} in the fast stream. So if you can also estimate the mass flux in the fast stream you should be able to find B_{min}. For this purpose it might be sufficient to use the mass loss rate of the wind as a whole, and use just the V_{michel} value to obtain B_{min}. This is because, as we have seen, the mass loss rate tends to rise relative to that of the ambient flow only if the rotation rate is very large.

Owocki: These models are interesting in showing how magnetic fields and high rotation can influence hot-star mass loss. However, for cyclical variability, a perhaps easier way to modulate the wind speed is through the faster than radial expansion of flow tubes, as in solar coronal holes. Keith MacGregor showed a while ago that the lower density arising from the faster divergence would, within CAK line-driving theory, lead naturally to high speed flows in hot-star winds. Then even slow rotation will give CIR where such fast streams interact with slower ones.

Periodic Variability of θ^1 Ori C

Otmar Stahl

Landessternwarte, Königstuhl, D–69117 Heidelberg, Germany

Abstract. The young O star θ^1 Ori C shows periodic variability in the strength of optical emission lines (Hα, He II λ4686), UV stellar-wind absorption lines (e.g. C IV λ1548), photospheric absorption lines in the optical (e.g. O III, C IV, He I, He II), and the X-ray flux. The current best period is 15.426 $\pm$ 0.002 days. The emission lines, the X-ray flux and the photospheric absorption lines peak at phase 0. In contrast, the maximum strength of the stellar wind lines occurs around phase 0.5. We discuss these observations in the framework of an oblique magnetic rotator model. The radial velocity of θ^1 Ori C is also variable. Only small variations with the 15.4 day period are seen. These are probably due to line asymmetry changes. In addition, the radial velocity is variable on longer time scales.

1 Introduction

θ^1 Ori C is the brightest and hottest star in the Orion trapezium and also the main source of ionization of the Orion nebula. Thus it is important for models of H II regions. As a very young O7 main sequence star it is also important for the understanding of stellar evolution. For this meeting it is more relevant that it is also the only O-star known which shows strictly periodic stellar-wind variations.

The stellar variability was discovered by Conti (1972), who noticed variability with a timescale of days in the He II λ4686 emission. Conti found no significant radial-velocity variations. Particularly remarkable was the presence of an *inverse* P Cygni profile. This finding was confirmed by Walborn (1981), who also suspected a variable ratio of He II λ4541 / He II λ4471 with a similar timescale. Walborn et al. (1985) found the stellar wind to be weak and variable. Van Genderen et al. (1989) found no photometric variations.

Stahl et al. (1993) discovered strictly periodic Hα and He II λ4686 variations with a period of $P = 15.4$ days. They suggested an oblique magnetic rotator model. Later Walborn & Nichols (1994) found that the UV stellar wind lines follow the same period. Stahl et al. (1996) could improve the accuracy of the period sufficiently to allow to phase the optical and the IUE observations. They also found the photospheric lines to vary in strength with the same period.

Gagné et al. (1997) detected that the X-ray emission varies in phase with the optical emission lines. Babel & Montemerle (1997) could quantitatively explain the X-ray flux and its variations with a "magnetically confined wind shock" model. This gives strong support for a magnetic origin of the variability.

2   Evidence for Rotational Modulation

The strict periodicity of the variability of θ^1 Ori C requires a stable clock in the system. A binary scenario is unlikely for several reasons:

- Only small radial velocity variations (a few km sec^{-1}) with the 15.4 day period are observed. These variations do not require a binary motion, but are probably due to small line profile changes as a function of phase (Stahl et al. 1996).
- A binary scenario does not provide a reasonable modulation mechanism for the emission lines and the X-ray variations (Gagné et al. 1997).

Pulsation can also be excluded since the expected pulsation periods for a main-sequence O star are much too short.

Rotation, on the other hand, provides a natural mechanism to produce the smooth variability observed. A rotating surface feature also explains the observed line asymmetries (Stahl et al. 1996). Since we need a mechanism to lock the stellar wind to the stellar rotation, an oblique magnetic rotator seems the most plausible explanation. Absorption in a corotating wind, magnetospheric eclipses by the star and magnetospheric eclipses by the circumstellar disk are possible modulation mechanisms.

Stahl et al. (1996) have shown that the rotation of θ^1 Ori C is slow. However, if the measured line width is due to rotation only, the observed period of θ^1 Ori C is still too long by a factor of two. So, additional line broadening is required.

In this model the UV absorption lines arise from the polar region. The emission is produced by trapped gas near the magnetic equator. The enhanced photospheric absorption seems to originate from the same region.

3   New Observations

In order to check the stability of the variability pattern and to improve the accuracy of the period, we observed θ^1 Ori C in Sep./Oct. 1997 with the same instrument as described by Stahl et al. (1996).

From the equivalent width of Hα we now derive a period of 15.426 days. The equivalent widths have been measured by integrating from 6550 to 6575 Å, after removing the nebular emission lines. The zero point of the phase is $MJD = 48832.5$. Here MJD is defined as Julian date $-\, 2\,400\,000.5$. From Fig.1 it can be seen that the scatter in the phase plot is very small. This shows the remarkable stability of the behaviour of θ^1 Ori C.

Stahl et al. (1996) noted peculiar variability of the radial velocity of θ^1 Ori C. We can now study the long-term behaviour. Fig.2 shows the radial velocity as measured from He II λ5411. A long-term trend is clearly indicated. It shows that the "real" radial velocity of θ^1 Ori C with respect to the Orion nebula is difficult to determine.

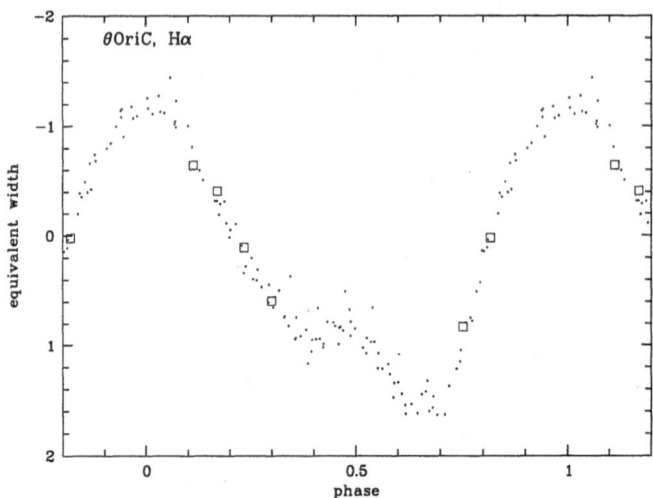

Fig. 1. Equivalent width in Å of the Hα line of θ^1 Ori C versus phase (for an adopted period of 15.426 days and a zero point of $MJD = 48832.5$). Data already published by Stahl et al. (1996) are shown as small symbols. The large symbols denote measurements obtained in Sep./Oct. 1997 from Heidelberg. The total time span is more than five years or about 120 cycles.

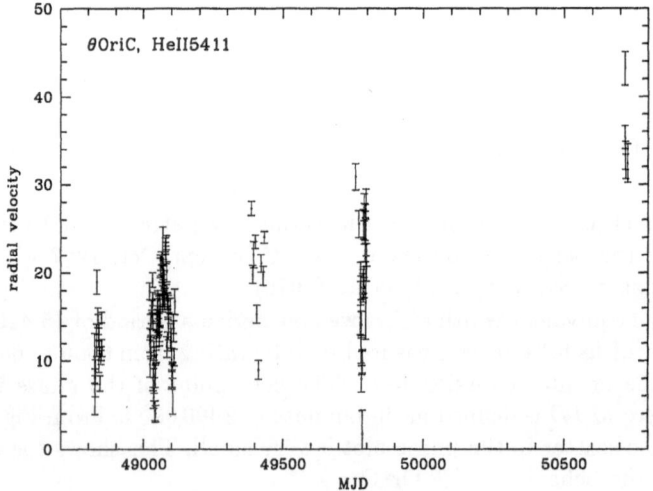

Fig. 2. Heliocentric radial velocity in km sec^{-1} of He II λ5411 versus MJD. Note the long-term trend in the radial velocity and the superimposed faster variability. The radial velocity of the Orion nebula is 28 km sec^{-1}.

4   Conclusions

θ^1 Ori C seems to be closely related to the He-variable magnetic stars. The relation to the cyclic variability common in other hot stars, which is interpreted as rotational modulation of stellar winds, is less clear. The long-term stability of the variability pattern in θ^1 Ori C is not observed in other O stars. The origin of the magnetic field and its influence on the evolution of early-type stars needs to be investigated. A measurement of the magnetic field would be very valuable, but is difficult to achieve.

An additional source of line broadening is required to explain the observed line widths. The long-term radial-velocity variations need further monitoring. It seems plausible that the additional line broadening and the radial-velocity changes both originate from large scale atmospheric motions.

References

Babel, J., Montmerle, T. (1997): ApJ 485, L29

Conti P., (1972): ApJ 174, L79

Gagné, M., Caillaut, J.-P., Stauffer, J.R., Linsky, J.L. (1997): ApJ 478, L87

Stahl O., Wolf B., Gäng Th., et al. (1993): A&A 274, L29

Stahl, O., Kaufer, A., Rivinius, Th., et al. (1996): A&A 312, 539

van Genderen A.M., Bovenschen H., Engelsman E.C., et al. (1989): A&AS 79, 263

Walborn N.R. (1981): ApJ 243, L37

Walborn N.R., Nichols-Bohlin J., Panek R.J. (1985): IUE Atlas of O-type spectra from 1200 to 1900 Å, NASA reference publication 1155

Walborn N.R., Nichols J.S. (1994): ApJ 425, L29

Discussion

Henrichs: Is the star chemically peculiar, especially He over- or underabundant?

Stahl: I am not aware of any abundance analysis of this star. Because of the variability and peculiar line profiles, it will be difficult to analyse it, but I agree that an attempt would be worthwhile.

Baade: At what level can you exclude the additional presence of nonradial oscillations, i.e. have you tried to pre-whiten the data for the rotational variability and then search for other periods?

Stahl: We did not detect other periods from the line-strength variations. It is difficult to put limits on the presence of non-radial oscillations since we have not searched specifically for these variations.

Rotating Loops in the Atmosphere of Rigel (β Orionis A)

Garik Israelian

Instituto de Astrofisica de Canarias,
Via Lactea S/N, E–38200 La Laguna, Tenerife, Canary Islands, Spain

Abstract. We discuss time series of Hα profiles of the late B-type supergiant Rigel (β Orionis A). Recent observations indicate the presence of variable and asymmetric outflows and infall of matter. We detected an extraordinary blue-shifted high-velocity absorption (HVA) component almost simultaneously with Kaufer et al. (1996). We found that the blue-shifted HVA disappeared completely in the course of one month, and a strong *red-shifted* absorption appeared at $+ 50$ km s^{-1}. Based on our observations, as well as those reported in the literature, we propose that extended rotating magnetic structures (closed loops) exist in Rigel. According to our model, the gradual decrease of the blue HVA simultaneous with the increase of the red HVA is not due to variable ejection/accretion rates, but to the variable effective thickness of the part of the loop which crosses our line of sight.

1 Introduction

The variability of the Hα profile of Rigel has been known for many years (Sanford 1947; Kikuchi 1968). None of the previous observations were systematic, and it was therefore impossible to follow the evolution of Hα, derive time-scales, or understand the physical processes responsible for the variations. High-velocity absorption (HVA) components on the *blue* side of line center have recently been reported for two stars, Rigel and HD 96919 (Kaufer et al. 1996). These features differ from other spectral patterns first of all because they appear suddenly over a wide velocity range (up to 200 km s^{-1}). Surface spots with enhanced mass loss can give rise to extended loops along magnetic lines of force. The rotation of these structures will bring about the sudden appearance and disappearance of HVAs over a large velocity range. Slow streams of matter emerging from the cool spot(s) will interact with fast wind giving rise to corotating interacting regions (CIRs) (Mullan 1984). From the observational point of view, CIRs may have a significant effect on Hα (formed in the inner wind), while UV resonance lines will be affected less since they form in the farthest part of the wind. Thus, Hα provides an opportunity to test the CIR idea, as well as models proposed by Underhill & Fahey (1984) and Cassinelli (1985).

Recently we reported observations of Rigel which suggest the existence of extended rotating closed loop(s) (Israelian et al. 1997).

2   Peculiar Variability of Hα

In Fig. 1 we show five Hα profiles to illustrate the development of a blue-shifted HVA in Rigel in late 1993. We succeeded in recording a final part of this extraordinary event (Fig. 2), a so called *red-shifted* HVA. The evolution of Hα between early November and early December is obvious: the strong blue-shifted absorption gradually diminished, and turned into emission by early December. In the meantime a *red-shifted* absorption feature started to increase and reached its maximum strength on 1993 December 3.

Comparison of the first (November 1993) and second (February 1994, Kaufer et al. 1996) HVA events shows that (a) the residual intensity of the red-shifted HVA in December 1993 reached 0.65, i.e. much stronger than during the second HVA event in 1994, and (b) the duration of both events (counted from the date of maximum depth of the blue absorption until the blue emission wing appears) in 1993 November and 1994 February were almost equal. In addition, the separation time of the first and second blue-shifted HVAs is very close to the estimated maximum rotation period (suggesting that the same structure appeared after one rotational cycle).

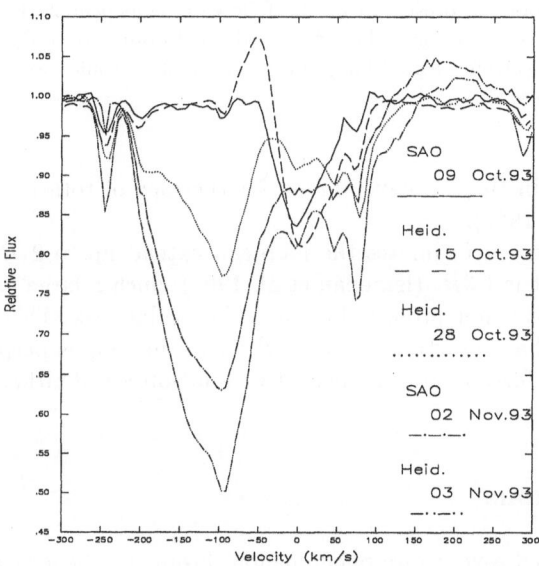

Fig. 1. Hα profiles observed in Rigel during of the first blue-shifted HVA event. Observations were done in Heidelberg (Kaufer et al. 1996) and at the Special Astrophysical Observatory (Israelian et al. 1997). Positions of the profiles were not corrected for the system velocity (20 km s^{-1}).

We have shown that the red-shifted HVA of Hα cannot be produced by the free fall of matter. Instead, based on common kinematic properties of cool solar

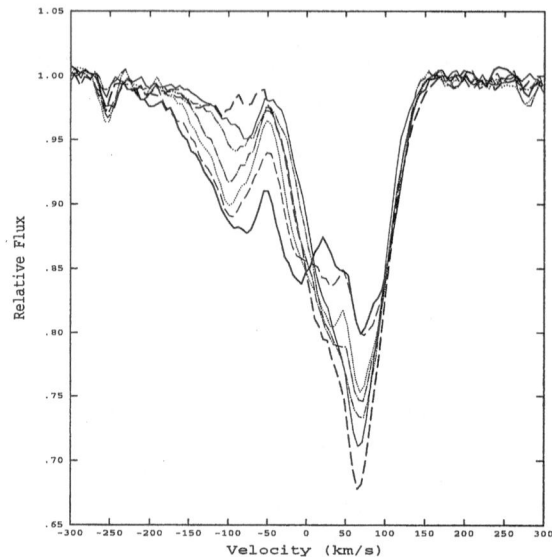

Fig. 2. Variability of Hα during the period from 1993 November 24 to 1993 December 3 (SAO). Plots are presented in the following sequence: 24 (heavy full line), 25 (dashed), 26 (dotted), 27 (dash-dotted), 29 (dashed double dotted), 30 (light full line) and December 3 (heavy dashed line). Positions of the profiles were not corrected for the system velocity.

loops observed in Hα, we have argued for the idea of rotating loop(s) in Rigel (Israelian et al. 1997).

According to our estimates the loop can extend up to 2 R_* if the distance between the feet is $1.3R_*$ (Israelian et al. 1997). Such a large loop may produce emission when it is not projected against the stellar disk. This emission will be unshifted (and have maximum intensity) when the loop is perpendicular to the line of sight. Unfortunately this period was not observed either by us or by the Heidelberg group.

3   Conclusions

We have observed a very rare phenomenon. From the theoretical point of view, the origin of magnetic loops in hot stars is not clear, although the existence of loops in the atmospheres of hot stars was proposed by Cassinelli (1985) and Underhill & Fahey (1984). Perhaps differential rotation in a non-convective star helps to create magnetic loops. However, in spite of the theoretical difficulties, it seems that a rotating loop model is able to explain at least qualitatively the observations of HVAs in Rigel. Many interesting observations of Rigel have been reported previously. Severny et al. (1974) found strong (260 Gauss) and variable magnetic fields in Rigel, Takeda et al. (1995) derived a large value of the macroturbulent velocity $\zeta = 43$ km s^{-1}, the existence of a chromosphere has been suggested by Ruban (1994), and X-rays from Rigel have been recorded by ROSAT (Berghoefer et al. 1996). It is possible that we have found the first clear evidence of loops from Hα spectroscopy of Rigel.

References

Berghoefer, T., Schmitt, T., Cassinelli, J. (1996): A&AS, **118**, 481

Cassinelli, J. (1985): in *Non-radiative Activity in Hot stars*, eds. A. Underhill & A. Michalitsianos, SP-NASA 2358, Washington DC, p. 1

Israelian, G., Chentsov, E., Musaev, F. (1997): MNRAS, **290**, 521

Kaufer, A., Stahl, O., Wolf, B., et al. (1996): A&A, **314**, 599

Kikuchi, S. (1968): PASJ, **20**, 190

Mullan, D. (1984): ApJ, **283**, 303

Ruban, E. (1994): Astron. Astrophys. Transec., **6**, 59

Sanford, R. (1947): ApJ, **105**, 222

Severny, A., Kuvshinov, V., Nikulin, N. (1974): Bull. of Crimean Astr. Obs., **50**, 3

Takeda, Y., Sadakane, K., Takada-Hidai, M. (1995): PASJ, **47**, 307

Underhill, A., Fahey, R. (1984): ApJ, **280**, 712

Kundt: According to your outflow/infall interpretation, the transition from blue- to red-shifted absorption should pass through a phase of large unshifted absorption. Prediction: it won't exist.

Israelian: Of course there is some material with zero Doppler shift and the phase you have mentioned does exist. It is also clear (see Israelian et al. MNRAS 290, 521) that the absorption at 0 km/s cannot be as strong as the blue/red-shifted HVAs, which is additional support for the loop model.

Unruh: From the depth of absorption components that you observe, it appears to me that the amount of material that you need to produce the absorption is huge. Is this commensurate with a magnetically confined loop model? And would you not expect to pick up some of the scattering from this material when it is off disk? Aren't you worried by the lack of data as the transient supposedly crosses the disk? The absorption signatures you observe are NOT very similar to what is observed on AB Dor.

Israelian: Yes, we have a massive, dense and very extended $(2R_*)$ loop. However, the cross-sectional area of the legs of the loop is not large and when out of the line of sight they won't produce any observable emission. The large amount of cool (10^4 K) and dense gas is concentrated along but not across the legs. In addition, the presence of a velocity gradient (which is observed) will spread this emission out so that it might be hard for us to detect it.

Constraints on Wind Structure from the Infrared and Radio Continuum of Hot Stars

Mark C. Runacres and Ronny Blomme

Koninklijke Sterrenwacht van België, Ringlaan 3, B–1180 Brussel, Belgium

Abstract. For 18 galactic hot stars (spectral type O-B3), we compare continuum observations gathered from the literature with the predictions of a NLTE model of a smooth wind. We find that for stars which have unambiguous observations at wavelengths beyond $\sim 20\ \mu m$, a smooth wind model systematically underestimates the observed emission. We take this to be a confirmation of the presence of considerable structure in the stellar wind.

1 Introduction

Although the most distinctive fingerprints of structure observed in hot-star winds are doubtlessly the Discrete Absorption Components in ultraviolet spectral lines, considerable insight can be gained from a careful study of the infrared and radio continuum. We have gathered all the infrared, millimetre and radio observations from the literature for a sample of 18 galactic O-B3 stars, as well as broadband visual photometry.

2 Smooth Wind Model

We compare the continuum observations of our sample stars with a NLTE model of a smooth wind (i.e. density and velocity given by the force multiplier formalism). The B to L filters are used to determine the shape of the interstellar extinction curve. If radio observations are available, we adjust the mass-loss rate of the model until the theoretical and observed radio fluxes agree. Otherwise, mass-loss rates derived from $H\alpha$ are used.

The infrared and millimetre observations are unconstrained. It is their agreement or disagreement with the observations that will tell us whether the model is correct. We judge the significance of possible discrepancies according to three criteria:

- there must be more than one discrepancy, as a single difference can always be due to a mismeasurement. Preferably, these discrepancies should show a trend with wavelength
- the discrepancies should be large compared to the scatter on the observations
- the observed infrared and millimetre fluxes should be due to the star and stellar wind and not from some extended object close to it.

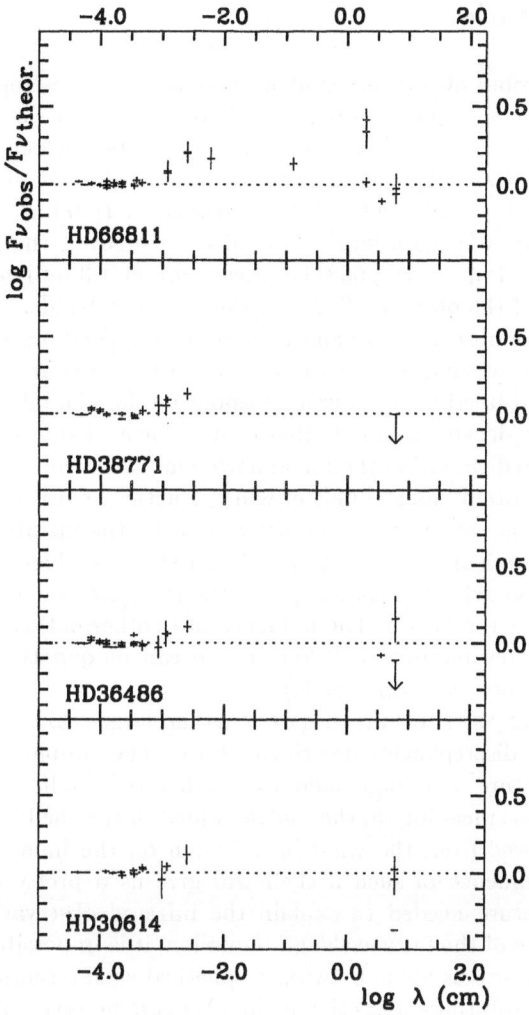

Fig. 1. The difference in log $\dot{F}_{\nu}$ between the observations and the smooth wind model, as a function of wavelength, for the four stars from our sample showing unambiguous discrepancies.

When we apply these criteria to the sample stars we find unambiguous discrepancies for 4 stars. These discrepancies are shown on Fig. 1. It is clear that the model systematically underestimates the observed infrared fluxes, pointing to extra emission in excess of that expected from a smooth wind. For the other stars, the presence of additional emission could not be confirmed, due to the lack of observations beyond 10 μm (where the wind contribution starts to become detectable) or due to the presence of a nearby source. It is important to realise that, for these stars, additional emission is *not* excluded.

3   Interpretation

There are a number of possible explanations for these discrepancies. One possible explanation could be a more gradual acceleration of the stellar wind. The present radiatively driven stellar-wind theory predicts a velocity law corresponding closely to a β-law with $\beta \approx 0.8$. To explain the present discrepancies at 12, 25 and 60 μm we would need a $\beta \approx 2$ (Runacres 1997). It is not clear what the physical cause of such a large β would be. Moreover, this large β does still not explain the ζ Pup (= HD 66811) observation at 1.3 mm and causes a slight overestimation of the observed fluxes in the L and M bands.

In view of the observational and theoretical evidence supporting it, we consider the presence of wind structure to be the most likely source of extra emission. The free-free and free-bound emission responsible for the infrared fluxes is proportional to the density squared. Hence, it is clear that a clumped wind will emit more infrared radiation than a smooth wind with the same mass-loss rate. There are two quite different kinds of wind structure. On the one hand there is the relatively small, stochastic structure caused by the instability of the radiative driving mechanism (see Owocki 1994). On the other hand there is the large scale structure, somehow co-rotating with the star, that is currently believed to be responsible for the DACs. The influence of (isothermal) stochastic structure on the infrared continuum is well known and can be quantified by the volume filling factor (Lamers & Waters 1984).

In the present work we investigate whether large-scale, localised structure can explain the discrepancies described above. The simplest example of such a localised structure is a single spherical shell in the stellar wind. In order to make comparisons meaningful, the matter added in the shell is compensated for by matter removed from the wind in a trough on the inner side of the shell. Studying the influence of such a shell will give us a pretty clear idea of the amount of structure needed to explain the infrared observations. Due to the integrated nature of the far wavelength emission, it is impossible to differentiate from these observations alone between a spherical shell structure and the spiral structure of a co-rotating interaction region. We can, however, obtain an estimate of the density contrast and dimension of the structure involved. Using the shell strength formalism developed in Blomme & Runacres (1997), we find that a shell at 2 R_* with a strength of 7.5 (e.g. a density contrast of 25 and a shell width of 0.1 R_*) can explain the observed fluxes at 12, 25 and 60 μm for ζ Pup. A second shell is needed to explain the observation at 1.3 mm.

4   Conclusions

We have found that at least some (possibly all) of the infrared and millimetre observations in our sample require a structured wind. We have derived a position and a strength for the shell that can reproduce the observations. This shell strength can be interpreted literally (i.e. as the combination of the width and density contrast of a single shell). The shell strength we derive could also refer to the effect of stochastic structure in the infrared formation region or to the combined effect of a number of partial shells (i.e. that fill a solid angle $< 4\pi$). The possibility of a single shell can be checked by searching for variability. The time coverage of our present far infrared data is insufficient to decide on this matter.

Acknowledgements. We thank Stan Owocki, Joachim Puls and Henny Lamers for stimulating discussions and useful suggestions.

References

Blomme R., Runacres M.C. (1997): A&A **323**, 886
Harries T.J., Howarth I.D. (1996): A&A **310**, 533
Lamers H.J.G.L.M, Waters L.B.F.M. (1984): A&A **136**, 37
Owocki S.P. (1994): ApSS **221**, 3
Runacres M.C. (1997): PhD Thesis, Vrije Universiteit Brussel

Emerson: Can you rule out the possibility that the far-infrared (IRAS) and 1.3 mm excess is due to dust grains?

Runacres: Dust would have to be cool (~ 100 K) to show its maximum at ~ 30 μm. At such temperatures it would have to be at large distances from the star, where the density is insufficient for dust to form.

Kaufer: For me a discrete shell is something time-dependent. The question is if your data from all wavelength regimes were taken simultaneously?

Runacres: Our conclusions are mainly based on how the IRAS data differ from the "extrapolation" (by means of a model) of the visual data. The IRAS data were taken simultaneously, and almost no variability is apparent in various determinations of the visual data.

Ignace: Does your shell width parameter extend over just the density enhancement or also over the trough?

Runacres: By the width of a shell we mean the width of the density enhancement only (from which the width of the trough can be derived).

Prinja: What effects would departures from spherical symmetry in the ambient, underlying wind - i.e. equatorial compression - have on your derived results?

Runacres: The presence of a wind-compressed disk or zone could in principle explain the discrepancies. However, spectropolarimetry of ζ Pup by Harries & Howarth (1996) shows that the density is only slightly enhanced (a factor ~ 1.3). Such a density contrast is insufficient to produce the additional emission needed.

Clumping-Corrected Mass-Loss Rates of WR Stars: Dependence on Stellar Parameters

Tiit Nugis

Tartu Astrophysical Observatory, EE2444 Tõravere, Estonia

Nugis (1989) derived the universal formula for determining stellar mass-loss rates. The parameters of WR stars (which formed the basis for the derivation of that formula) are now much better known and we tried to find a new and more accurate approximation formula. Clumping-corrected mass-loss rates for WR stars of different subtypes were determined by Nugis et al. (1998). Luminosities of WR stars were found either by using the masses predicted by evolutionary theory (Schaerer & Maeder 1992) or from the subclass-dependent mean bolometric corrections. First, we tried to find the dependence of $\dot{M}$ on luminosity, effective temperature, and the mass fractions of helium and metals (i.e., elements heavier than helium). We adopted $T_{\text{eff}} = T_{\text{eff}}(\tau = 2/3)$ and found the first formula, which has the form:

$$\dot{M} = \dot{M}_\odot (L/L_\odot)^{1.5} \, (Y/Y_\odot)^{2.5} \, (Z/Z_\odot)^{0.6} \, (T_{\text{eff}}/T_{\text{eff}}^\odot)^{-1} \; . \tag{1}$$

The parameter $T_{\text{eff}}(\tau = 2/3)$ is not reliably known for WR stars. It is therefore desirable to find a formula that does not depend on T_{eff}. We found a second formula of the form:

$$\dot{M} = c_1 (L/L_\odot)^{1.5} \, Y^{4.5} \, (1-Y)^{0.7} \, Z^{0.6} \; , \tag{2}$$

where $c_1 \approx 1.1 \times 10^{-11}$ if $\dot{M}$ is in $M_\odot$ yr^{-1}. These formulae also approximate the mass-loss rates of O stars and LBVs quite well.

References

Nugis, T. (1989): Sov. Astron. Lett. 15, 19
Nugis, T., Crowther, P.A., Willis, A.J. (1997): A&A 333, 956
Schaerer, D., Maeder, A. (1992): A&A 263, 129

Part 4

MODELLING
CYCLICAL VARIABILITY
IN STELLAR OUTFLOWS

Formation of Clumps in the Wind of Wolf-Rayet Stars by Thermal Instability and Alfvén Waves

Denise R. Gonçalves, Vera Jatenco-Pereira, and Reuven Opher

Instituto Astronômico e Geofísico, Universidade de São Paulo,
Av. Miguel Stéfano, 4200, 04301-904, São Paulo, Brazil

The observed infrared emission (at the base of the wind) of early-type stars is often larger than expected. Thus, there might be clumps in the wind, which increase the mean density and hence the emission measure. The presence of these clumps explains both the observed infrared excess and the enhancement of the blue wing of the Hα line.

We have investigated the possible importance of Alfvén waves in the creation of inhomogeneities in the winds of Wolf-Rayet stars via thermal instability. This study is based on the wind acceleration model for Wolf-Rayet stars developed by dos Santos et al. (1993), in which Alfvén waves act jointly with the radiation pressure.

The clumps we are considering are formed via thermal instability. The heat-loss function, $H(T, n)$, includes: thermal bremsstrahlung, H_B; radiative losses via resonant transitions, H_{RT}; heating from recombination and photoionization, H_R; Compton heating-cooling, H_C; and heating by Alfvén waves, H_a. We study nonlinear ($H_a = H_{NL}$) and turbulent ($H_a = H_T$) Alfvénic heating.

We looked for an equilibrium set of physical parameters that satisfy:

$$H(T, n) = H_B + H_{RT} + H_R + H_C + H_a = 0 \ . \tag{1}$$

We found solutions that show three equilibrium regions: (i) one stable region representing the diffuse medium; (ii) one unstable region; and (iii) another stable region representing the condensations.

From the isobaric instability criterion, we require that the clouds and the hot medium coexist at the same pressure. For values typical of Wolf-Rayet stars, the pressure in the hot atmosphere is $P = 2\,n_H\,k\,T_H \sim 5.5 \times 10^4$ dyn cm^{-2}.

We show that the two stable equilibrium regions can exist over the range of pressures that describe the diffuse medium and the clumps.

References

dos Santos, L.C., Jatenco-Pereira, V., Opher, R. (1993): ApJ 410, 732

Understanding the Solar Wind

Egil Leer[1], Viggo H. Hansteen[1], and Thomas E. Holzer[2]

[1] Institute for Theoretical Astrophysics, University of Oslo, P O Box 1029,
N–0316 Oslo, Norway
[2] High Altitude Observatory, National Center for Atmospheric Research, P O Box 300,
Boulder, CO 80307, USA

Abstract. Heating of the extended solar corona leads to high proton temperatures and relatively low electron temperatures. This is due to the low heat conductivity in the proton gas as compared to the electrons. To a fairly good approximation we can say that the energy flux added to the electrons is conducted back into the transition region and lost as radiation, whereas the energy flux deposited in the protons is lost as kinetic and gravitational energy flux in the solar wind flow. How this energy flux is divided between gravitational and kinetic energy flux (i.e. how large are the particle fluxes and flow speeds) depend upon details of the heating process. However, we find that if energy is dissipated over more than one coronal scale height most of the energy goes into accelerating the solar wind. The low electron temperature that results in this sort of model is consistent with a small polarization electric field and a small electron heat flux in the solar wind acceleration region. Thus, the electron gas may play only a minor role for the force and energy balance of the solar wind.

1 Introduction

The solar corona shows considerable variation over the eleven-year sunspot cycle. Near solar maximum the structure of the solar magnetic field is quite complex, and there are many coronal mass ejections per month. Later in the cycle the magnetic field assumes a dipolar form, with a streamer belt near the solar equator. Although the coronal mass ejections are more or less of the magnitude as the mass ejections near solar maximum, in the range $10^{15} - 10^{16}$ g, there are much fewer mass ejections near minimum. The flux of ultraviolet radiation from the corona is also much higher around maximum than near minimum. In spite of the significant variations of the solar corona between maximum and minimum, the solar wind mass and energy flux are surprisingly constant. Near minimum the Ulysses spacecraft found that the solar wind has a high, constant, asymptotic flow speed, $750 - 800$ km/s, at latitudes beyond 20 degree from the ecliptic. In the ecliptic the solar wind is slower, around 400 km/s, and it is more variable (Phillips et al. 1995). Near maximum, the flow speed, at high latitudes, is lower than near minimum (Kojima & Kakinuma 1987, Rickett & Coles 1991).

In the present paper we discuss the basic physics of thermally driven solar wind in light of these facts. Our emphasis will be to show that the energy and mass flux contained in the solar wind result in a simple way from the energy inserted into the corona and the form of the conductive flux that channels energy back towards the chromosphere. The force balance of the solar wind was originally discussed by Parker (1958) almost 40 years ago, but an understanding of the energy balance requires that we understand the energy deposition in the corona, and how this energy flux is distributed between kinetic and potential solar wind energy flux.

2   Energy Balance in a Static Corona

In order to understand the energy balance of the corona-solar wind system it may be of some help to first consider a simpler system, namely the static corona. This system is relevant for closed coronal regions, where the energy flux that is heating the coronal gas is lost mainly as heat conductive flux to the lower layers. Radiative losses are another potential energy-loss mechanism, but for the (relatively) low-density loops we are considering here they are insignificant.

Let us consider a spherically symmetric corona, heated by an energy flux from the sun, $F_0 = 4\pi R_s{}^2 f_0$, where f_0 is the energy flux density at the solar surface. In order to make the model as simple as possible we will assume that all this energy flux is dissipated at the location $r = r_d$, where we will have a temperature maximum, $T = T_d$. A "lid", which prevents energy or mass flow, is put on the corona at this location. Since we are ignoring (the insignificant) coronal radiative losses, all the energy flux deposited in the corona is conducted back towards the transition region and lost as radiation as the density increases. This occurs at some $r = r_0 \approx R_S$, where $T = T_0 \ll T_d$.

In a time-independent system the divergence of the energy flux is zero and we can make use of the conservation of the energy flux, F_E, to find how the coronal temperature, T_d, varies with the input energy flux heating the corona,

$$F_E = F_0 + F_{q0} = 0, \quad \text{in} \quad r_0 < r < r_d. \tag{1}$$

The heat conductive flux in a spherically symmetric corona is

$$F_q = 4\pi r^2 (-\kappa \frac{dT}{dr}) \tag{2}$$

where $\kappa = \kappa_0 T^{5/2}$ for an ionized gas. The heat conductive flux, is constant in $r_0 < r < r_d$, and $T(r_0) = T_0 << T(r_d) = T_d$.

This inward heat flux is the only energy loss from the corona, so an energy flux density from the sun, f_0, deposited at $r = r_d$, gives a temperature maximum, T_d:

$$T_d \approx \left(\frac{7}{2} \frac{f_0}{\kappa_0} \frac{(r_d - R_s)}{(\frac{r_d}{R_s})} \right)^{\frac{2}{7}} \tag{3}$$

For a dense corona, where the electrons and protons collide frequently, it is the electron heat conductivity that determines the coronal temperature, and we may set $\kappa_0 = \kappa_{0e}$ where κ_{0e} is the heat conduction coefficient for electrons. This expression shows that the coronal temperature is *not* a sensitive function of either the location of the heat input, r_d, nor of the magnitude of the heat input f_0. Let us fix the location of the heat input $r_d = 2R_s$ and vary the energy flux density, f_0 heating the corona. We then find that $f_0 = 0.001\text{W m}^{-2}$ gives $T_d \approx 0.1 \times 10^6$ K, $f_0 = 100\text{W m}^{-2}$ gives $T_d \approx 2.0 \times 10^6$ K, while $f_0 = 6 \times 10^7$ W m^{-2} (all the energy produced in the sun) gives $T_d \approx 100 \times 10^6$ K,

The relatively small variation of T_d given the large variation of f_0 illustrates that a "cold" ionized gas is a very good insulator, whereas a "hot" ionized gas is a very good conductor of heat and also that the coronal temperature is almost invariably of the order 10^6 K given a "reasonable" heat input.

For a low density corona, where heat added to the protons (electrons) is conducted away by the protons (electrons), we find that

$$T_{de} \approx \left(\frac{7}{2} \frac{f_{0e}}{\kappa_{0e}} \frac{(r_{1e} - R_s)}{\left(\frac{r_d}{R_s}\right)} \right)^{\frac{2}{7}} \tag{4}$$

$$T_{dp} \approx \left(\frac{7}{2} \frac{f_{0p}}{\kappa_{0p}} \frac{(r_{1p} - R_s)}{\left(\frac{r_d}{R_s}\right)} \right)^{\frac{2}{7}} \tag{5}$$

Here, $f_{0p(e)}$ is the energy flux density from the sun that is deposited in the proton(electron) gas, and $\kappa_{0e} = 7.8 \times 10^{-12}$, and $\kappa_{0p} = 3.1 \times 10^{-13}$. As $(\kappa_{0e} / \kappa_{0p})^{2/7}$ = 2.5, the proton temperature, T_{dp}, is a factor 2.5 higher than the electron temperature, T_{de}, for $f_{0p} = f_{0e}$.

3   Energy Balance in a Corona – Solar Wind System

The energy balance in an expanding corona can be written as

$$F_E = F + F_q + F_{SW} = \text{constant} \tag{6}$$

where F is the deposited coronal energy flux, F_q is the heat conductive flux, and F_{SW} is the solar wind energy flux. We introduce $(-\dot{M})$ for the solar wind mass flux and write the energy flux, close to the Sun, as:

$$F_E = F_0 + F_{q0} + (-\dot{M})(-\frac{GM_s}{r_0}) \tag{7}$$

Far from the Sun the deposited coronal energy flux is dissipated and the heat conductive flux has been converted to flow energy so that the energy flux is

$$F_E = 0 + 0 + (-\dot{M})(\frac{1}{2}u_\infty^2) \tag{8}$$

Thus, the energy flux, F_0, that is deposited in the corona is lost as inward heat conductive flux, $-F_{q0}$, and as solar wind energy flux:

$$F_0 = -F_{q0} + (-\dot{M})(\frac{GM_s}{r_0} + \frac{1}{2}u_\infty^2). \tag{9}$$

3.1   One-fluid Model

Let us now make an estimate of the energy losses from a corona with temperature T_C: Based on the inward heat flux from a static corona, heated to a temperature T_d at $r = r_d$, we estimate the heat conductive flux from a quasi-static corona, with a "mean" temperature T_C, to be (3)

$$-F_{q0} \approx \frac{8\pi\kappa_0 R_s}{7} T_C^{\frac{7}{2}}. \tag{10}$$

The heat flux density from the (spherically symmetric) corona,

$$-f_{q0} = -F_{q0}/(4\pi R_s{}^2) \approx \frac{2\kappa_0}{7R_s} T_C^{\frac{7}{2}} \tag{11}$$

determines the pressure in the chromosphere-corona transition region (e.g. Landini & Monsignori-Fossi 1973);

$$p_0 \approx -C f_{q0} \tag{12}$$

where $C \approx 7 \times 10^{-5}$ s m^{-1}, and the pressure in the transition region determines the electron density at the "base" of the corona.

$$n_{C0} \approx \frac{p_0}{2kT_C} \tag{13}$$

As the electron density in the inner corona is determined by the coronal temperature, T_C, we can find the solar wind proton flux as a function of coronal temperature. The flux from an electron-proton corona with a coronal base electron density n_{C0} and a "mean" temperature $T_{Ce} = T_{Cp} = T_C$ is

$$-\dot{M} = 4\pi n_{C0} m_p \left(\frac{GM_s m_p}{4kT_C}\right)^2 \left(\frac{2kT_C}{m_p}\right)^{\frac{1}{2}} \times \exp\left(-\frac{GM_s m_p}{2kT_C R_s} + \frac{3}{2}\right) \tag{14}$$

The solar wind energy flux is given by

$$F_{SW} = (-\dot{M})(\frac{GM_s}{R_s} + \frac{1}{2}u_\infty{}^2) \tag{15}$$

In order to make an estimate of the solar wind energy flux we set $\frac{1}{2}u_\infty{}^2 \approx GM_s/R_s$, and we find that the equation

$$f_0 = (-F_{q0} + F_{SW})/(4\pi R_s{}^2) = -f_{q0} + f_{SW} \tag{16}$$

where

$$-f_{q0} \approx \frac{2\kappa_{0e}}{7R_s} T_C^{\frac{7}{2}} \tag{17}$$

and

$$f_{SW} \approx \frac{(-\dot{M})}{4\pi R_s{}^2} \left(\frac{2GM_s}{R_s}\right) \tag{18}$$

determines the coronal temperature, T_C. For a given value of f_0 we can also find the distribution of the energy flux between inward heat flux and solar wind energy flux, and we can determine the transition-region pressure.

In Table 1 we make use of T_C as a parameter, and find the inward heat-flux density, $-f_{q0}$, the electron density, n_{C0}, at the coronal base, the solar wind energy flux, f_{SW}, mapped back to the sun, and the energy flux density, $f_0 = -f_{q0} + f_{SW}$, that is consistent with this temperature. For a small energy flux density, f_0, a low temperature corona is created, and inward heat conduction is the only energy loss. For a larger energy flux density the coronal temperature increases, and a larger fraction of the energy flux is lost in the solar wind. In the one-fluid model, discussed here, the heat conductive loss and the solar wind energy loss are roughly equal for the energy fluxes that give a solar wind energy flux that is comparable to the observations, $f_{SW} \approx 100 \text{W m}^{-2}$. The corresponding transition-region pressure is $p_0 \approx 7 \times 10^{-3} \text{ Nm}^{-2}$, and the electron density at the $T = 10^6$ K level is around $n_0 = 10^{14} \text{ m}^{-3}$.

$T_C (10^6 K)$	$n_{C0}(\frac{1}{m^3})$	$-f_{q0}(\frac{W}{m^2})$	$f_{SW}(\frac{W}{m^2})$	$f_0(\frac{W}{m^2})$
0.5	1.6(12)	0.29	2.3(-5)	0.29
1.0	9.0(12)	3.3	1.1	4.4
2.0	5.1(13)	37	28	65
3.0	1.4(14)	150	220	370
4.0	2.9(14)	420	2020	2440
5.0	5.0(14)	910	16700	17600

Table 1. Variation of solar wind parameters for a one-fluid model with coronal temperature T_C.

3.2   Two-fluid Model

The coronal electron density in the one-fluid models is not high enough to give a collision rate that is larger than the expansion rate, so the electrons and protons are not thermally coupled. Heating of the extended corona will lead to a proton temperature that is higher than the electron temperature, and if most of the energy flux is deposited in the proton gas, the temperature difference may be significant. Let us therefore consider a model where *only* the protons are heated, and where the coupling to the electrons is so weak that the coronal proton temperature is much higher than the electron temperature, $T_{Cp} >> T_{Ce}$. In such a model the energy loss from the corona is in the form of proton heat conduction and solar wind energy flux. The inward heat flux density is

$$-f_{q0} \approx \frac{2\kappa_{op}}{7R_s} T_{Cp}^{\frac{7}{2}} \tag{19}$$

268   Egil Leer, Viggo H. Hansteen, and Thomas E. Holzer

This heat flux density determines the transition region pressure, $p_0 \propto -f_{q0}$, and the coronal "base" electron (proton) density, $n_{C0} = p_0/(kT_{Cp})$. The solar wind mass flux from such a corona, with base density n_{C0}, and a much higher mean proton temperature, T_{Cp}, than electron temperature, is

$$-\dot{M} \approx 4\pi n_{C0} m_p \left(\frac{GM_s m_p}{2kT_{Cp}}\right)^2 \left(\frac{kT_{Cp}}{m_p}\right)^{\frac{1}{2}} \times \exp\left(-\frac{GM_s m_p}{kT_{Cp}R_s} + \frac{3}{2}\right) \quad (20)$$

The solar wind energy flux, mapped back to the sun, is

$$f_{SW} \approx \frac{(-\dot{M})}{4\pi R_s^2}\left(\frac{2GM_s}{R_s}\right) \quad (21)$$

We make use of the energy equation,

$$f_0 = -f_{q0} + f_{SW} \quad (22)$$

and the expressions we have found for the inward heat flux from the corona and the solar wind energy flux, to make a new table, with T_{Cp} as a parameter, to see how the energy flux from the sun, deposited in the proton gas, is distributed between inward heat flux and solar wind energy flux. The energy flux density observed in the solar wind, $F_{SW}/(4\pi R_s^2) \approx 100$ Wm^{-2}s^{-1}, is consistent with a high coronal proton temperature, 3–4 $\times 10^6$ K, and a low electron density in the inner corona. Only 10% of the energy loss from the corona is in the form of inward heat conduction; around 90% of the energy flux is lost in the solar wind.

$T_{Cp}(10^6 K)$	$n_{C0}(\frac{1}{m^3})$	$-f_{q0}(\frac{W}{m^2})$	$f_{SW}(\frac{W}{m^2})$	$f_0(\frac{W}{m^2})$
1.0	7.0(11)	0.13	2.0(-6)	0.13
2.0	3.8(12)	1.4	.46	1.9
3.0	1.1(13)	6	34	40
4.0	2.2(13)	16	300	320
5.0	3.8(13)	35	1160	1200
6.0	6.0(13)	69	3090	3200

Table 2. Variation of solar wind parameters for a two-fluid solar wind model with coronal proton temperature T_{Cp}. In these models the coronal electron temperature is much lower than the proton temperature.

4   Discussion

The goal of this paper is to illustrate the basic physics of thermally driven solar wind models. In order to understand the energy balance of the corona-solar wind system, coronal heating and solar wind acceleration must be treated as *one* problem. Unfortunately, we do not know the mechanisms for energy transport from the sun and how and where this energy flux is transferred to the coronal plasma. Here, we focus on heating of the corona. There may, of course, also be energy transfer in the form of direct acceleration, but a significant energy transfer as "work" on the gas in the quasi-static region of the flow will lead to a large density scale height and a very large solar wind proton flux. Direct acceleration of the supersonic flow will lead to an enhanced asymptotic flow speed (Leer & Holzer 1980).

If some type of low frequency Poynting flux, with frequencies well below the electron gyro frequency, is responsible for heating of the corona it is likely that ions receive more of the energy flux than the electrons. When the energy flux is transferred to the very inner corona, most of the energy is lost as heat conduction into the transition region (Hammer 1982, Withbroe 1988, Hansteen & Leer 1995), the pressure in the transition region is high, and the electrons and protons are thermally coupled in the inner corona. The solar wind emanating from the corona has a low asymptotic flow speed (Hansteen & Leer 1995). When the energy flux from the sun is dissipated over a scale length that is large compared to a density scale height, most of the energy flux is lost in the solar wind. Significant proton heating leads to a high proton temperature, much higher than the electron temperature, and a low coronal electron density. A high speed solar wind outflow is found when a significant fraction of the energy flux from the sun is deposited in the region of the corona where the protons are collision-less (Evje & Leer 1997). Then, the inward heat flux is small, the transition region pressure is small, the coronal density is low, the solar wind proton flux is low, and the energy per unit mass is large enough to drive high speed streams.

The models presented in this paper are simplified as much as possible, but they are "consistent" is the sense that the electron density in the inner corona is consistent with the heat flux into the transition region, and they illustrate, quite well, how an energy flux from the sun, dissipated in the extended corona in magnetically open regions, is distributed between inward heat flux and solar wind energy flux. In models with proton heating most of the energy flux is lost in the solar wind. We have not made an effort to find the distribution of the solar wind energy flux between gravitational and kinetic energy flux. This is the goal of most solar wind models.

In many model studies of the solar wind the electron density at an inner boundary, the "coronal base", is set, rather arbitrarily, to some "reasonable" value, and it is not consistent with the energy deposition in the extended atmosphere (e.g. Holzer & Leer 1980, Sandbæk et al. 1994, Esser & Habbal 1996, Habbal et al. 1996). Such models may shed some light on the energy balance in the corona and the solar wind, but the solar wind proton flux depends on the coronal base electron density, so we cannot expect to derive reliable information about the distribution of the solar wind energy flux, between kinetic and potential energy flux, in such models. In other models the proton flux is specified (e.g. McKenzie et al. 1995). If the solar wind energy flux also is specified in these models, and the energy loss into the transition region is neglected, the asymptotic flow speed is also specified. Such models may shed light on the relation between the coronal (proton) temperature and the asymptotic flow speed. However, it is not satisfactory to specify the solar wind proton flux, independent of the coronal heating function.

Heating of the corona and acceleration of the solar wind can be treated as one problem in models where the inner boundary is placed in the chromosphere and the ionization and recombination balance of hydrogen and helium are included (e.g. Hansteen et al. 1997). It is found that the essential characteristics of the wind are determined by the energetics of the corona. As an alternative, one can therefore place the inner boundary in the upper transition region and adjust the density at the boundary such that the inward heat flux balances the radiative loss from the transition region and the solar wind enthalpy flux at the inner boundary (e.g. Wang 1993).

5   Conclusion

In order to improve our understanding of the solar wind we have to consider the formation of the corona and acceleration of the solar wind as one problem. This requires that we understand the energy transfer to the extended solar atmosphere, and how this energy flux is lost. There is abundant supply of plasma to the corona and the solar wind, from the lower solar atmosphere, so we should not focus on the ionization and recombination processes in the lower layers of the atmosphere: It is the energy flux from the sun, and how and where this energy flux is deposited, that determines the structure of the corona and the solar wind mass flux and asymptotic flow speed.

Acknowledgements

This work was supported by the Norwegian Research Council (NFR) under contracts 115828/431 and 115831/431 and by NASA under contract W-18786.

References

Esser, R., Habbal, S.R. (1996): in *Proc. of the 8th International Solar Wind Conference*, American Institute of Physics, Conf. Proc. 382, p. 133
Evje, H.O., Leer, E. (1997): *A&A*, in press
Habbal, S.R., Esser, R., Guhathakurta, M., Fisher, R. (1996): in *Proc. of the 8th International Solar Wind Conference*, American Institute of Physics, Conf. Proc. 382, p. 129
Hammer, R. (1982): *ApJ* **259**, 767
Hansteen, V.H., Leer, E. (1995): *J. Geophys. Res.* **100(A11)**, 21577
Hansteen, V.H., Leer, E., Holzer, T.E. (1997): *ApJ* **482**, 498
Holzer, T., Leer, E. (1980): *J. Geophys. Res.* **85**, 4665
Kojima, M., Kakinuma, T. (1987): *J. Geophys. Res.* **92**, 7269

Landini, M., Monsignori-Fossi, B. (1973): *A&A* **29**, 9

Leer, E., Holzer, T.E. (1980): *J. Geophys. Res.* **85(A9)**, 4681

McKenzie, J.F., Banaszkiewicz, M., Axford, W.I. (1995): *A&A* **303**, L45

Parker, E.N. (1958): *ApJ* **128**, 664

Phillips, J., et al. (1995): *Sci* **268**, 1030

Rickett, B.J., Coles, W.A. (1991): *J. Geophys. Res.* **96**, 1717

Sandbæk, Ø., Leer, E., Hansteen, V.H. (1994): *ApJ* **436**, 390

Wang, Y.-M. (1993): *ApJ* **410**, L123

Withbroe, G.L. (1988): *ApJ* **325**, 442

Linsky: You have not mentioned the role of magnetic fields in your models of the solar corona and wind. As I understand it, the hottest regions of the corona have closed magnetic fields and little mass loss. This is contrary to your models. Please explain.

Leer: The closed regions have the highest electron density and (probably) the highest electron temperature, but the gas pressure is not large enough to "break open" the magnetic field. Therefore, the mass loss from closed regions is small.

Owocki: First a quick comment. Linsky noted that you never mentioned magnetic fields. I would like to call attention to the fact that you never mentioned critical points. To me, this suggests that for understanding the physics of the solar wind the mathematical subtleties of critical points are not essential.

My question regards your remarks that heavy ions will naturally become very hot. Did I understand you that this is a natural consequence of their low thermal conductivity, so that one no longer needs to consider heating models in which the heating rate itself is proportional to mass ?

Leer: Energy deposition in the quasi-static corona (inside the critical point) and in the supersonic region of the flow have different effects on the solar wind energy balance. So it may be useful to introduce the "critical point" to understand the solar wind energy balance, but it is not essential.

In the models I have discussed the ion temperature is higher than the electron temperature, the polarization electric field is small, and the electron heat flux into the solar wind is small compared to the solar wind energy flux. The ions are heated, and their temperature increases until the energy loss, in the form of heat conduction, collisional loss, and solar wind energy loss balance the heating. When heating heavy ions in the extended corona, the losses in the form of heat conduction and collisions are small, and the ions reside in the corona until their "temperature" is high enough to allow them to escape in the solar wind.

Cassinelli: I thought that high speed streams were caused by Alfvén wave momentum deformation, but now you are saying the high speeds are produced by the high temperature protons. Do Alfvén waves have any role in your model in producing the high speed wind?

Leer: I have tried to illustrate how proton heating of the extended corona can produce quite large proton temperatures, and high speed wind, without invoking

Alfvén waves. If Alfvén waves are present we can drive the high speed wind with lower coronal temperatures. I am not saying that Alfvén waves do not play a role in accelerating high speed wind, but that they are not necessary.

Kundt: I have a difficulty with your gedanken-experiment of a "static" corona, which is argued to reach temperatures above photospheric. Doesn't that conflict with the second law? You reply by invoking mechanical heating. But a "static" sun avails only of nuclear energy whose conversion yields $L_\odot$ in the form of radiation (at $10^{3.8}$ K), all other flows being minor contributions.

Leer: The solar photon flux does not interact with the corona, but a low frequency Poynting flux may be damped by electrons and ions in the coronal plasma. It is the transfer of such an energy flux that gives rise to the high coronal temperature.

Foing: You described a tool permitting to calculate how the energy deposition (where and how much) translates into temperature or solar wind. How do you specify realistically where this energy is deposited?

Leer: The goal of this study is to place constraints on the coronal heating. For significant electron heating a large fraction of the energy flux is lost as heat conductive flux into the transition region, the coronal electron density is quite high, the solar wind mass flux is high, and the energy per unit mass in the flow is not large enough to drive high speed streams. Heating of the coronal protons may lead to high proton temperatures, a small conductive loss into the transition region, low coronal electron density, a relatively small solar wind proton flux, and an energy per unit mass that is large enough to drive high speed wind. In order to drive streams with speeds of 800 km/s or so, most of the energy flux must be deposited in the proton gas, in the outer corona, where the protons are collision-less, and the proton heat flux is very small.

MacGregor: You noted in your talk that model calculations indicated that flows with properties similar to those of high-speed solar wind could be produced by depositing energy in the protons over a length scale $\gtrsim R_\odot$. Have you given any thought to specific mechanisms for heating the proton component of the gas over such a length scale?

Leer: Sorry, I have no ideas that are any better than the suggestions made by Parker, Axford and others. But I feel that it is very important that we think, carefully, about experiments/observing programs that can discriminate between different possible coronal heating mechanisms. If we understand the energy transfer to the coronal gas, I think that we will also understand the acceleration of the solar wind.

The Solar Wind Modelled
by Exact MHD Solutions *

João J.G. Lima[1] and Kanaris Tsinganos[2]

[1] Centro de Astrof. and Dep. de Matem. Apl., Universidade do Porto,
Rua do Campo Alegre, **823**, 4150 Porto, Portugal
[2] Dep. of Phys., Univ. and Res. Cent. of Crete,
P.O. Box 2208, 710 03 Heraklion, Crete, Greece

Abstract. Recent *Ulysses* data have shown that the solar wind's macroscopic physical quantities deviate substantially from the 1-D spherical symmetry. We propose a 2-D (axisymmetric) MHD model for the solar wind constructed by neglecting the meridional components of the magnetic field and flow. The solution, found by using a nonlinear separation of the variables, yields three anisotropy parameters that control the dependences of the physical quantities with latitude and which can be fixed by observations. The wind speed is topologically controlled by an X-type critical point filtering a unique wind-type solution, and located where the radial component of the flow velocity equals the fast MHD wave speed in that direction. Choosing appropriately the three anisotropy parameters, we are able to reproduce the variations of the wind speed, density, radial component of the magnetic field and mass flux with latitude, as observed by *Ulysses*.

1 A 2D MHD Model

Our aim is to find solutions of the steady MHD equations describing the dynamical interaction of an inviscid, compressible and highly conducting plasma with an axially symmetric magnetic field created by a central rotating object. We shall solve only the dynamical equations, considering the self-consistent energy balance *a posteriori*. We assume a helicoidal geometry ($V_\theta = B_\theta = 0$). The technique used follows that in Lima & Priest (1993) in the context of an hydrodynamic model. It is based on a nonlinear separation of the variables and yields the following solution (Lima et al. 1997):

$$V_r = V_0 Y(R)\sqrt{\frac{1 + \mu \sin^{2\epsilon}\theta}{1 + \delta \sin^{2\epsilon}\theta}}, \qquad V_\phi = \lambda V_0 \frac{R \sin^\epsilon \theta}{\sqrt{1 + \delta \sin^{2\epsilon}\theta}}\left(\frac{Y_* - Y}{1 - M_A^2}\right) \quad (1)$$

$$B_r = \frac{B_0}{R^2}\sqrt{1 + \mu \sin^{2\epsilon}\theta}, \qquad B_\phi = \lambda B_0 \frac{\sin^\epsilon \theta}{R}\left(\frac{R^2/R_*^2 - 1}{1 - M_A^2}\right) \quad (2)$$

$$\rho = \frac{\rho_0}{Y R^2}\left(1 + \delta \sin^{2\epsilon}\theta\right), \quad (3)$$

* J. Lima wishes to acknowledge support by JNICT Project PESO/C/PRO/1032/94 and PRAXIS XXI. K. Tsinganos wishes to acknowledge support by the Greek GGET

where $R = r/r_0$ is the radial distance normalised to the base r_0 of the outflow and $M_A = (V_r/V_A)$ is the radial Alfvén Mach number. This solution involves three anisotropy parameters: δ, ϵ, μ. For R and θ to separate, we must have $Q(R, \theta) = Q_0(R) + Q_1(R) \sin^{2\epsilon} \theta$, where $Q(R, \theta) = 2p(R, \theta)/\rho_0 V_0^2$ is the dimensionless pressure. Substitution into the momentum equation yields a differential equation for $Y(R)$ which can be integrated numerically. Topologically, the radial speed shows a singular point at the Alfvénic radius. There is also a second singular point (Figure 1), wherein the r-component of the flow speed equals the fast MHD mode wave speed in that direction (Tsinganos et al. 1996). Only the r-direction plays a role here since it is perpendicular to both the directions of axisymmetry (ϕ-direction) and self-similarity (θ-direction in this model).

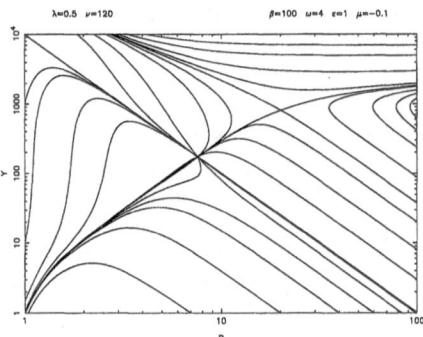

Fig. 1. Topology of the radial dependence of the radial speed $Y(R)$ for a solar-type highly magnetized star: $\lambda = 0.5$, $\nu = 120$, $\beta = 100$, $\delta = 4$, $\epsilon = 1$ and $\mu = -0.1$.

In order to make our particular dynamical solution also a solution of the energy equation, we must include the following equation for the conservation of energy and then check if the obtained energy distribution is physically reasonable: $(\mathbf{V} \cdot \nabla)[p/(\Gamma - 1)\rho] + (p\mathbf{V} \cdot \nabla)[1/\rho] = \sigma$, where σ is the heating/cooling per unit mass and Γ the ratio of specific heats. Figure 2 shows the deduced heating rate.

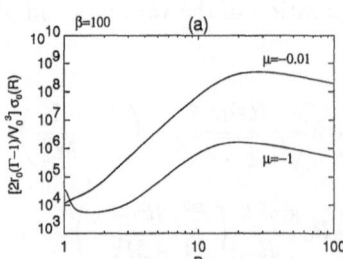 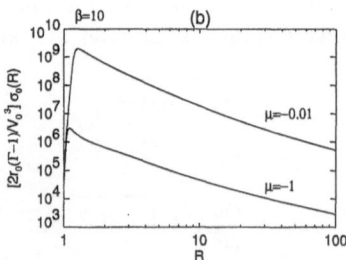

Fig. 2. Radial dependence of the dimensionless heating rate along the polar axis for: $\lambda = 0.5$, $\nu = 120$, $\delta = 4$, $\epsilon = 1$ and $\beta = 100$ in **(a)**, or, $\beta = 10$ in **(b)**.

2   Application to the Solar Wind

In situ observations of the solar wind by *Ulysses* as well as observations from *SOHO* have confirmed beyond doubt that the solar wind's macroscopic physical quantities deviate substantially from the 1-D spherical symmetry.

From the *Ulysses* data in Phillips et al. (1995) averaged over the solar rotation, we fitted (using their median curves) the scaled ion density and wind speed with our model (Figure 3). Also, it is interesting to note that *Ulysses* observations of the radial component of the heliospheric magnetic field show very little evidence of a latitudinal gradient (Smith et al. 1996). In fact, Figure 4a shows that for the values of μ and ϵ that best fit the profiles in Figure 3, B_r is constant from around 40° to 90° latitude in both hemispheres.

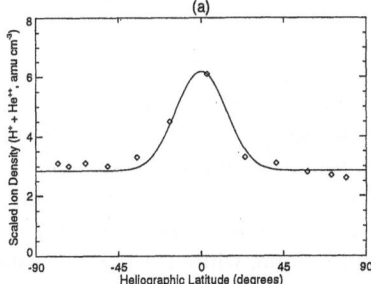 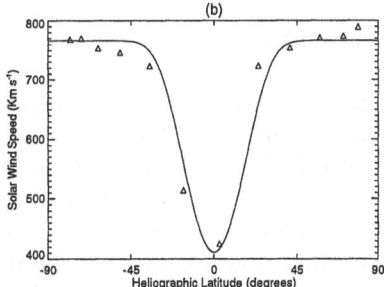

Fig. 3. Solar rotation averaged *Ulysses* data (Phillips et al, 1995) on top of our model. In **(a)** the curve corresponds to $n_0 = 2.85$, $\delta = 1.17$ and $\epsilon = 8.60$; in **(b)** the model curve is obtained by fixing $\delta = 1.17$ and $\epsilon = 8.60$ and estimating $V_0 = 766$ km s^{-1} and $\mu = -0.38$.

Another important quantity is the mass efflux $(\rho V_r r^2)$ or, equivalently, the mass-loss rate per infinitesimal solid angle $d\Sigma$ at angle θ. This is a function of θ alone, $\dot{m}(\theta) = \rho_0 V_0 r_0^2 \sqrt{(1 + \mu \sin^{2\epsilon} \theta)(1 + \delta \sin^{2\epsilon} \theta)}$. Early observations by the Soviet Prognoz satellite in 1976-1977 together with the Mariner 10, 1974 observations have shown that there is more Lyα emission near the ecliptic poles than predicted by an isotropic solar wind. These Lyα photons are photospheric UV photons scattered by neutral H atoms of interstellar origin. Where the solar-wind mass flux is largest, the neutral H atoms are destroyed and the H I concentration the lowest. Thus, maximum Lyα emission in some direction is correlated with a minimum solar-wind mass efflux in this direction. These Lyα observations imply that the solar wind mass efflux should be minimum at the poles and maximum at the equator, a trend fully confirmed by *SWAN/SOHO* observations (Bertaux et al. 1997) and by *Ulysses* in situ observations out of the ecliptic plane (Phillips at al. 1995). With the parameters from Figure 3b, we are able to reproduce qualitatively such a trend in the mass efflux profile (Figure 4b).

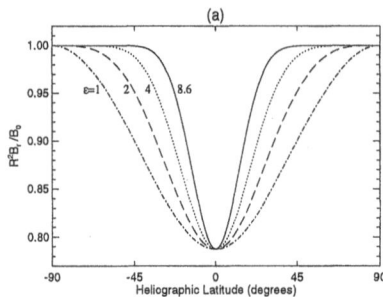
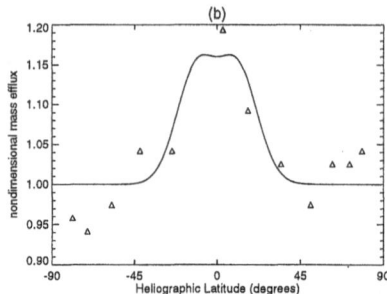

Fig. 4. In **(a)** the latitudinal profile of the radial magnetic field is plotted for $\mu = -0.38$ and different values of ϵ. In **(b)**, *Ulysses* data on the mass efflux are plotted together with the corresponding curve of our model for $\delta = 1.17$, $\epsilon = 8.60$ and $\mu = -0.38$.

3   Summary

We have constructed an exact and simple MHD solution for the magnetized/rotating solar wind which is a generalisation of the classical Weber & Davis (1967) model. Similarly to this model, a unique physical solution is selected by critical points while asymptotically the Alfvén number is proportional to the radial distance. A well known difficulty of the Weber & Davis (1967) 1-D model is that it is not valid outside the equatorial plane, while the present solution is inherently fully 2-D. We took advantage of the availability of observations of the 2-D dependence (radial+latitudinal) of the solar wind quantities, such as the radial speed and magnetic field, density, mass efflux, and angular velocity. The latitudinal dependence of these quantities was found to fit rather well the corresponding observations if the three integration parameters have the values $\delta = 1.17$, $\epsilon = 8.60$ and $\mu - 0.38$.

References

Bertaux, J.-L., Quemerais, E., Lallement, R., et al. (1997): in *Proc. Fifth SOHO Workshop*, ESA SP-404, 29

Lima, J.J.G., Priest, E.R. (1993): A&A **268**, 641

Lima, J.J.G., Priest, E.R., Tsinganos, K. (1997): in *Proc. Fifth SOHO Workshop* ESA SP-404, 521

Phillips, J.L., Bame, S.J., Barnes, A., et al. (1995): Geophysical Research Letters, **22**, 3301

Smith, E.J., Neugebauer, M., Balogh, A., et al. (1996): Space Sci. Rev. **72**, 165

Tsinganos, K., Sauty, C., Surlantzis, G., et al. (1996): in *Solar and Astrophysical Magnetohydrodynamic Flows*, K. Tsinganos (ed.), (Kluwer Academic Publishers), 427

Weber, E.J., Davis, L.J. (1967): ApJ, **148**, 217

Kubát: In the title of your talk you claim that you are using "exact" MHD solutions. Nevertheless, the pressure in your MHD equations is considered as a scalar quantity, not a tensor one. Can you comment on this?

Lima: I've used the term "exact solutions" meaning not numerical. In fact, we are able to deduce an analytical solution up to a system of first-order differential equations that can be solved using a standard integrator. The pressure is in fact assumed as a scalar quantity.

MacGregor: You noted that the energy addition to the flow (i.e., the quantity σ) implied by your treatment of the wind energy balance could be checked *a posteriori*. What did this turn out to be for the model results you discussed?

Lima: For a solar-type star, the energy addition to the flow is highly concentrated at the base of the wind, reaching a maximum within 1 solar radius, as in Fig. 2b. This yields a rapid acceleration of the wind. Further downstream, σ decays rapidly with distance and is unable to sustain the rapid acceleration. For a more magnetized star, σ increases slowly near the base reaching a maximum at larger distances from the star, Fig. 2a.

Mullan: A previous 2-D MHD model of solar wind expansion was presented by Pneuman & Kopp: this included a helmet streamer structure. How does your model compare with the Pneuman & Kopp results?

Lima: The Pneuman & Kopp (1971) model is a numerical one based on an iterative procedure to integrate the system of partial differential equations. They considered a helmet streamer configuration, thus allowing for the presence of closed field regions, not possible in the present analytical solution. Their sample solution shows that, close to the base of the wind, the radial velocity near the equator can exceed the polar one, whereas after a certain distance the inverse is true. In our solution, we always have a higher radial speed at the pole than at the equator.

Time-Dependent Behavior of Cool-Star Winds

Peter Woitke

Institut für Astronomie und Astrophysik, TU Berlin, Sekr. PN 8–1,
Hardenbergstraße 36, 10623 Berlin, Germany

Abstract. The present status of time-dependent modeling of the circumstellar envelopes of cool pulsating stars on the AGB is reviewed. Such model calculations nowadays investigate the complex interplay between hydrodynamics, thermodynamics, chemistry, dust formation and radiative transfer. We report on the scientific insight provided by such models and show the physical mechanisms which control the dynamical behavior of these envelopes and drive the massive winds of these stars. The common results as well as the contrary aspects obtained by different groups are summarized and discussed. Observational and theoretical constraints on the temperature structure in these winds are outlined which allow for some important conclusions concerning the wind driving mechanism.

1 Introduction

Long-period variables (LPVs) and Miras on the Asymptotic Giant Branch (AGB) belong to the class of stellar objects which exhibit the largest mass loss rates known ($10^{-7}\ldots 10^{-4}\,M_\odot/\mathrm{yr}$). Such massive outflows are certainly capable of returning a considerable fraction of the original stellar material back to the interstellar medium. Since the overwhelming part of the newly born stars[1] evolve along the AGB at the end of their lifetime, these winds provide the most important source of new fuel for star formation. Understanding the final stages of stellar evolution and the circuit of matter therefore requires detailed knowledge about the mass-loss mechanisms of AGB stars.

Considerable progress has been achieved in the last decade concerning the dynamic modeling of the atmospheres and circumstellar envelopes (CSEs) of LPVs and Miras. This review intends to summarize the physical insight provided by time-dependent model calculations performed in *Iowa* (Bowen 1988, Willson et al. 1997), in *Berlin* (Fleischer et al. 1992, Winters et al. 1994, Fleischer et al. 1995) and in *Vienna* (Feuchtinger et al. 1993, Höfner et al. 1995, Höfner & Dorfi 1997). Despite this progress, the basic wind driving mechanism is still a matter of debate. The most promising processes which can generate the wind are (i) stellar pulsations, (ii) radiation pressure on newly formed dust grains and — most probably (iii) a combination of both. Atmospheric levitation due to radiation pressure on molecules has been discussed (e.g. Jørgensen & Johnson 1992), but adequate dynamical model computations are still lacking. Magnetic fields and stellar rotation are usually considered to play a minor role for the wind generation from these late-type giant stars (Dorfi & Höfner 1996).

---

[1] More than 97% based on the assumptions $M_i \leq 8\,M_\odot$ (Reimers & Köster 1979), initial mass function $dN/d\log M_i \propto M_i^{-1.7}$ (Miller & Scalo 1979) and $M_i \geq 1\,M_\odot$.

2   The Models: common Aspects and Distinct Approaches

First we discuss the *common aspects of the model computations* performed in *Iowa, Berlin,* and *Vienna.* A typical model covers the region from about 0.9 to 30 R$_*$, containing the stellar atmosphere and the inner part of the CSE where the wind is generated. The interior driving zone of the stellar pulsation is *not* included in any of the models[2]. The set of hydrodynamic equations

$$\frac{d\rho}{dt} = -\rho \operatorname{div} u \tag{1}$$

$$\frac{du}{dt} = -g + a_{\text{dust}} - \frac{1}{\rho} \operatorname{grad} p \tag{2}$$

$$\frac{de}{dt} = -p \frac{dV}{dt} + \frac{1}{\rho} Q_{\text{rad}} \quad , \tag{3}$$

comprising the equation of continuity, the equation of motion and the energy equation, are simultaneously solved together with the radiative-transfer problem. As acceleration terms in the equation of motion (2), all groups account for gravitation g, gas pressure gradient and radiation pressure on dust grains a_{dust}. Only adiabatic and radiative (Q_{rad}) heating/cooling rates are considered in the energy equation (3). Due to the large time consumption of the computer models, all groups so far rely on spherical symmetry and on the grey approximation of the radiative-transfer problem. Its solution yields the local radiative-equilibrium temperature T_{RE} which enters into the determination of Q_{rad}. For the numerical treatment of the discontinuities caused by shock waves, an artificial viscosity is included. In order to simulate the interior pulsation of the star, the so-called piston approximation $u_1(t) = \Delta u \cdot \cos(2\pi t/P)$ is applied. It adopts a sinusoidal motion of the inner boundary with Δu being the amplitude and P the period of the stellar pulsation. All calculations start from an initially hydrostatic model given by the static solution of Eqs. (1–3) without piston motion and excluding the presence of dust.

The distinct model assumptions are summarized in Table 1. The main difference between the *Iowa* approach on one hand and the *Berlin* and *Vienna* approaches on the other hand concerns the treatment of the dust complex. In the pioneering work of Bowen (1988), a parameterized step-function for the dust opacity is used which solely depends on T_{RE}, whereas in the *Berlin* and *Vienna* models, the dust formation process is modeled relying on chemical equilibrium, classical nucleation theory and a time-dependent moment method from which the dust properties (e.g. a_{dust}) are consistently calculated. Up to now, the latter detailed description of the dust complex has been applied to carbon stars only, because the details of the nucleation and growth of inorganic grains in an oxygen-rich environment are still not fully understood.

---

[2] Therefore, an artificial approximation for the inner boundary is applied (the "piston" approximation), which introduces two additional parameters to the models.

Table 1. Summary of the different model approaches

	Iowa	Berlin	Vienna
chemistry	—	equilibrium	
nucleation	—	modified classical nucleation theory (Gail et al. 1984)	
dust	$f(T_{RE})$ (Wood 1979)	time-dependent moment method (Gail & Sedlmayr 1988)	
Q_{rad}	$Q_{rad} \propto \rho^2$ (non-LTE)	— (1)	$Q_{rad} \propto \rho$ (LTE)
radiative transfer	grey Eddington (Chandrasekhar 1934)	grey Eddington (Lucy 1976) (Unno & Kondo 1976)	time-dependent grey variable Eddington (Yorke 1980)
num. scheme	explicit Lagrange (Richtmyer & Morton 1967)		implicit adaptive grid (Dorfi & Feuchtinger 1991)

[1] : Iowa's or the LTE approach may be considered optionally

Another difference concerns the treatment of the radiative heating/cooling function $Q_{rad}[\mathrm{erg\,s^{-1}\,cm^{-3}}]$: The *Iowa* group uses an approximate formula adjusted to the non–LTE behavior of permitted line transitions which scales as $Q_{rad} \propto \rho^2$. The *Vienna* group relies on the assumption of LTE resulting in a heating/cooling rate which roughly scales as $Q_{rad} \propto \rho$. Most of the published *Berlin* models use the "isothermal" approximation based on the assumption that the gas relaxes instantaneously towards radiative equilibrium $(T = T_{RE})$. In this case, the energy equation (3) becomes obsolete and the temperature structure is an immediate result of the solution of the radiative-transfer equation. In all cases where efficient dust formation takes place, the different approaches for Q_{rad} adopted do not result in substantial deviations from the published Berlin models (Gauger et al. 1993, Fleischer & Gauger 1997).

3   Basic Physical Results

Shock waves: generation and propagation. A simple but very important implication from all the models is that propagating shock waves are almost inevitably present in the atmospheres of pulsating stars. According to the large density gradient present in the photosphere, even small-scale (subsonic) waves steepen into strong shocks, the propagation of which is demonstrated in Fig. 1. Inside the photosphere these shocks cause large time variations of the atmospheric structure which, nevertheless, can still be considered as a "disturbance" of the static stratification governed by gravitation and pressure gradients. However, in the outer atmosphere, the shock waves create a completely different dynamical behavior of the gas which no longer can be understood on the basis of static or even stationary models, but is only revealed by time-dependent models.

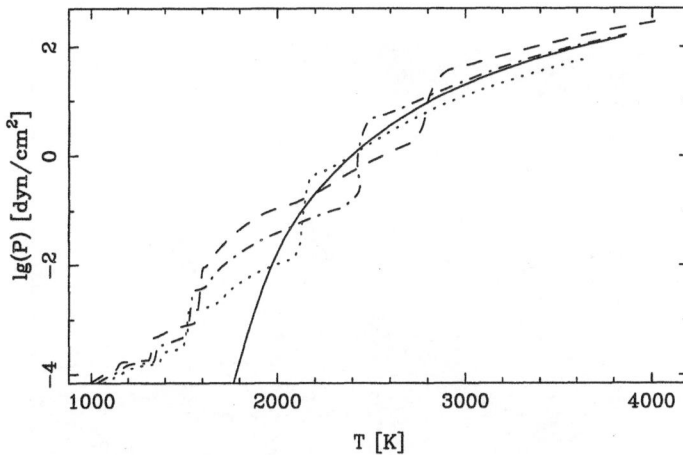

Fig. 1. The initial model (full line), the propagation of shock waves through the atmosphere (dashed: phase 0.5, dash-dotted: phase 0.75 and dotted: phase 1.0), and the levitation of the outer atmosphere (Höfner et al. 1997). Parameters: $T_* = 2880\,\mathrm{K}$, $L_* = 7000\,\mathrm{L_\odot}$, $M_* = 1\,\mathrm{M_\odot}$, $C/O = 1.4$, $P = 330\,\mathrm{d}$ and $\Delta u = 4\,\mathrm{km\,s^{-1}}$.

Levitation and dust formation: The shock waves introduce an outward directed, periodically acting, additional force on the gas which causes a levitation of the outer atmosphere resulting in densities being larger by orders of magnitude as compared to a hydrostatic model (cf. Fig. 1). Levitation alone, however, is not the primary cause of the mass loss. It only provides larger densities at distances where efficient mass-loss mechanisms may set in and, therefore, helps to drive more *massive* outflows. For example, since effective dust formation from the gas phase requires large densities, the driving of stellar winds by radiation pressure on dust grains is strongly supported by matter levitation. The two-step process of levitation and dust (+ wind) formation is demonstrated by the emergence of two distinct radial zones as shown in Fig. 2 for a typical *Berlin* model.

1) In the inner levitated regions ($r \lesssim 2\,R_*$), the gas elements are periodically hit by shock waves, which causes instantaneous outward accelerations. In the meantime, the fluid elements fall back towards the star following roughly ballistic trajectories.

2) On top of these levitated layers, effective dust formation takes place, radiation pressure on dust accelerates the gas and causes the transition into the wind region ($r \gtrsim 2\,R_*$).

Further details of this model are shown in Fig. 3. An unexpected but typical feature being common to all models which include a time-dependent dust formation is that the emerging dust is not distributed uniformly. The step-like behavior of the degree of condensation f_c (i.e. the amount of condensible material condensed into grains) indicates that the dust is present in discrete, onion-like layers. When new dust forms close to the star, an optically thick shell appears which blocks the

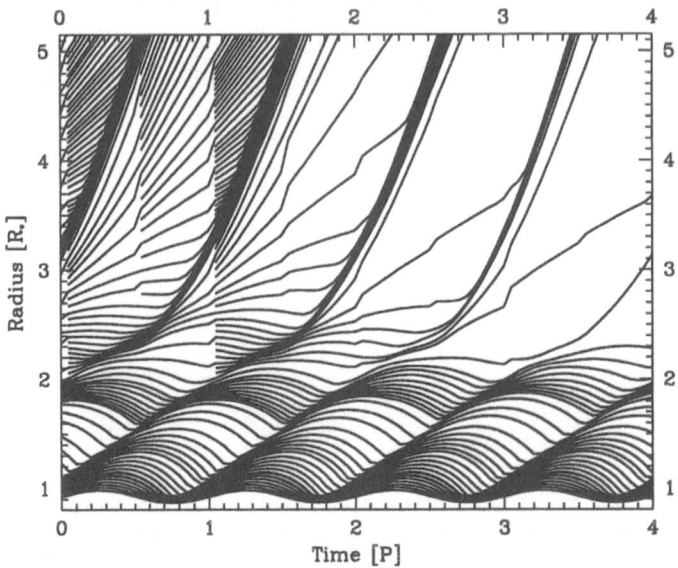

Fig. 2. A "Lagrange-plot" of a typical Berlin model indicating the radial positions of fixed fluid elements (in units of the initial stellar radius R_*) as function of time (in units of the pulsational period P). The sinusoidal variation at the bottom reflects the prescribed motion of the "piston". The CSE is clearly divided into two distinct zones: the levitated layers at the bottom and the wind region on top. Parameters: $T_* = 2600$ K, $L_* = 10000$ L$_\odot$, $M_* = 1$ M$_\odot$, C/O $= 1.8$, $P = 650$ d and $\Delta u = 2$ km s^{-1}.

radiation and re-radiates it towards the star. Consequently, the region inside of this layer heats up ("backwarming", see T-plateau at about 2000 K in Fig. 3) and further nucleation is temporarily inhibited. The optically thick layer is driven out by radiation pressure, radially dilutes and finally becomes transparent again. Accordingly, the inner region cools down and a new cycle of dust formation may start. This "exterior κ-mechanism", which acts in combination with pulsations in the depicted model, is thoroughly described in Fleischer et al. (1995) and in Höfner et al. (1995). In fact, the model described here belongs to the most simple class of *single periodic* models. The dust complex introduces an additional characteristic time scale to the system which is given by the time between the formation of two subsequent dust layers. Coupling with the pulsational period may cause a *multiperiodic* or even chaotic behavior of the CSE.

Wind driving mechanisms: pulsation and/or dust formation? Besides the dust-driven winds supported by pulsation (*Berlin, Vienna*), a quite different mass-loss mechanism has been proposed by the *Iowa* group (Bowen 1988): According to the assumption $Q_{\rm rad} \propto \rho^2$, the efficiency of radiative cooling of the gas ($\propto Q_{\rm rad}/\rho$) decreases linearly with decreasing density in the CSE. Consequently, there exists some point in the envelope beyond which the gas is too diluted to radiate away the excess internal energy dissipated by a shock wave and to re-

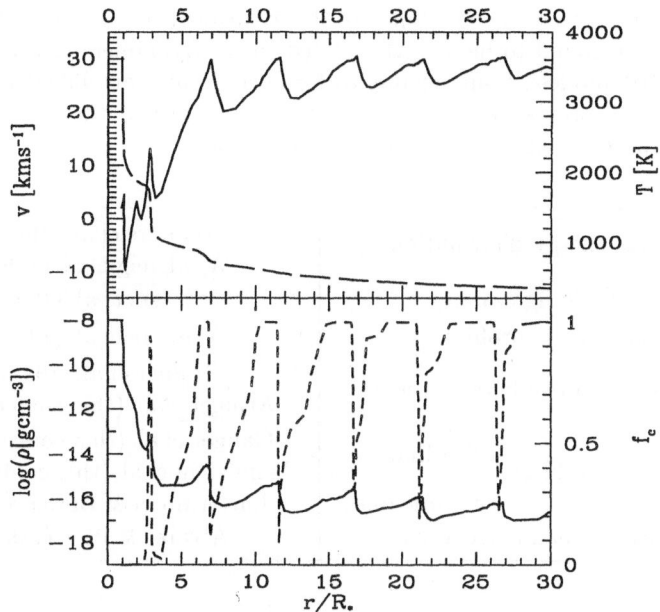

Fig. 3. Velocity v (upper diagram, full line), temperature T (upper diagram, dashed line), mass density ρ (lower diagram, full line) and degree of condensation f_c (lower diagram, dashed line) as function of radius at a fixed time. The remarkable steps in the f_c-curve reflect the onion-like, layered dust distribution. Parameters as in Fig. 2.

establish radiative equilibrium (RE) within one period. Hence, the gas behaves almost adiabatically. In this case, the propagating shocks produce a so-called "calorisphere" (cf. Fig. 5b) having temperatures always much higher than in RE, and the large pressure gradients alone may cause substantial mass-loss rates, *even without dust*. The mass-loss rates of such *pulsation induced, thermally driven* winds critically depend upon the details of the evaluation of Q_{rad} (Willson 1998, this conference). If dust formation is included (in the approximate way described above), the dust may take over the driving of the wind and may increase the mass-loss rate. In this case, the calorisphere disappears because of the additional adiabatic expansion in the enhanced dust-driven outflow and the *Iowa* results look similar to the corresponding *Berlin* and *Vienna* results.

4   Synthetic Observations

A recent development concerning the research on AGB stars is to use the time-dependent models as a basis for the calculation of various synthetic observations. The common procedure is to perform a frequency-dependent radiative-transfer calculation[3] on top of the model structures (density, temperature, velocity, dust).

---

[3] In contrast to the grey approximation used in the time-dependent model computations.

In spite of the inherent inconsistency of this treatment, important observational facts have been found to be directly related to the dynamical characteristics of the calculated models. Some examples for such results are listed below. This success in turn proves the quality and the capability of the time-dependent models compared to static or stationary approaches.

spectral energy distribution	Winters et al. (1997)
	Windsteig et al. (1997)
synthetic lightcurves	Winters et al. (1994)
brightness profiles	Winters et al. (1995)
synthetic photospheric spectra	Höfner et al. (1997)
	Aringer et al. (this conference)
synthetic line profiles: • CO lines	Gauger et al. (this conference)
	Windsteig et al. (this conference)
• atomic lines	Luttermoser & Bowen (1990)
molecular band strengths	Alvarez & Plez (1997)

5   Mass-loss Rate Formula for C–rich AGB Stars

Besides the studies of the physical processes in single models and their links to observations, the behavior of a sample of models yields important informations, for example the functional dependence of the resulting quantities on the input parameters. In particular, the value of the mass-loss rate is a result of each consistent model calculation, which implies that the influence of the stellar parameters on $\dot M$ can be determined. Based on a statistical analysis of 48 *Berlin* models, Arndt et al. (1997, see also this conference) have recently extracted such a "model" mass-loss rate formula for C-rich AGB stars, which could serve e.g. as input for stellar-evolution calculations, where the effect of mass loss has to be taken into account quantitatively. The mean error of this formula for $\log \dot M$ relative to the individual *Berlin* model values is less than 3%. As long as the same input physics is used, the *Vienna* results agree within about 6%.

A comparison between calculated and observed mass-loss rates for C–stars is shown in Fig. 4. A clear correlation exists between $\dot M$ and the $K-L$ color index, which is an approximate measure of the dust optical depth of the CSE. The *Berlin* models tend to predict mass-loss rates which are, on average, too large by a factor of two. One should keep in mind, however, that these models have been calculated *self-consistently*, i.e. from first principles. Accordingly, there are no additional parameters which could be tuned. The existence of such a correlation suggests that the winds of C–stars, in fact, are dust-driven or at least strongly determined by the emerging dust component. The consistency of the slope of the correlation between observations and models moreover indicates that the wind driving mechanism is modeled in an appropriate way.

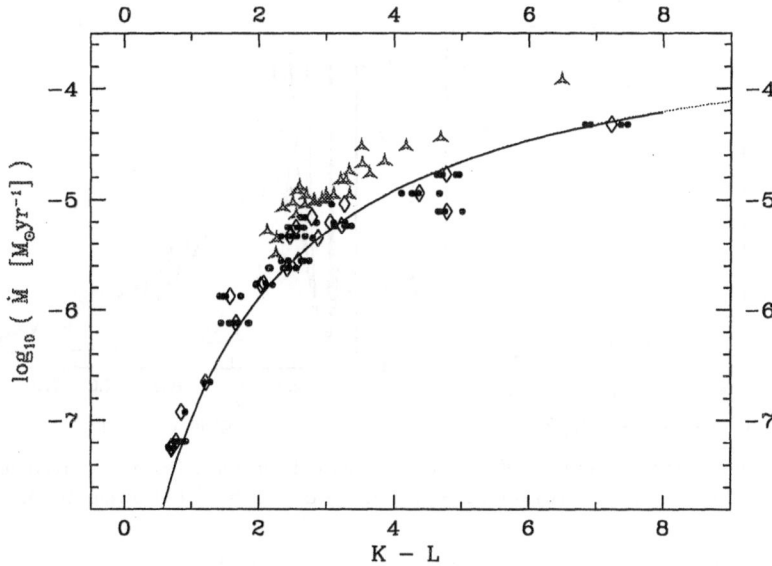

Fig. 4. The observed mass-loss rates of C–stars (open diamonds and dots) as a function of the $K-L$ color index compared with the results of the Berlin models (open triangles). The full line indicates the correlation obtained from the observations (Le Bertre 1997).

6   Discussion: Hot or Cool Envelopes?

The basic question whether the winds of LPV and Mira stars are driven by pulsation or by dust is closely related to the slope of the temperature profile in the envelope. The thermal driving by pulsation is causally connected with the formation of a "calorisphere" (cf. Fig. 5b) with temperatures several thousand degrees above $T_{\rm RE}$ in the wind driving zone. This requires a *low efficiency of radiative cooling*, because the energy dissipated by the shocks must not be radiated away, but is to be converted into work done on the gas via adiabatic expansion, thereby driving the wind. Thus, the debate on the driving mechanism may be decided by means of reliable information about $T(r)$ either obtained by observations or by theory.

Radio observations: Recently, Reid & Menten (1997) detected and partially resolved centimeter-wavelength emissions from a sample of nearby long-period (Mira and semi-regular) variables. They found an optically thick "radio photosphere" (caused by H^- and H_2^- free-free transitions) located at about $2\,R_*$ and derived brightness temperatures of $1500\pm570\,\mathrm{K}$. Furthermore, they argue for temperatures $\lesssim 1100\,\mathrm{K}$ at radial distances $r \gtrsim 10^{14}\mathrm{cm}$. Crosas & Menten (1997), analyzing CO rotational lines, come to similar conclusions concerning the LPV carbon star IRC+10216. These observations are compatible with the dust-driven models, but seem to contradict the purely pulsation-driven models.

Shock models: The efficiency of radiative cooling under conditions typical for CSEs has scarcely been investigated. Some models of radiative, plane-parallel,

286   Peter Woitke

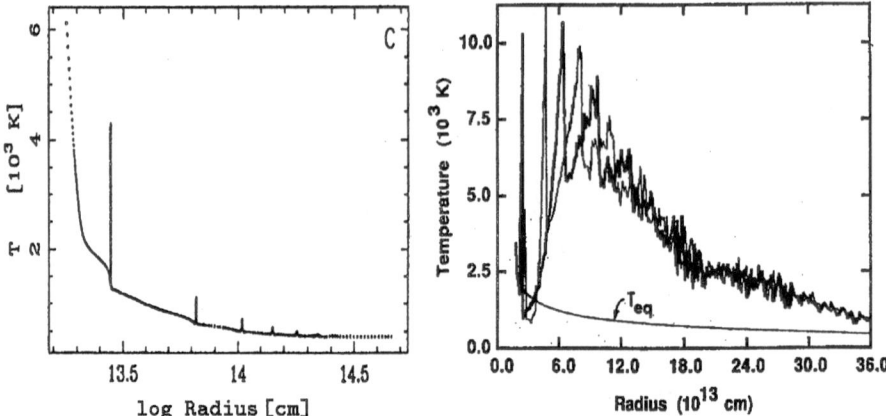

Fig. 5. Temperature structures of different model calculations: (a) dust-driven model (Feuchtinger et al. 1993) — (b) dust-free, pulsation-driven model (Bowen 1988).

stationary shock waves are available nevertheless, which contain important information about the cooling time scale $\tau_{\rm cool}$, after which a shocked gas element returns to temperatures of the order of a few 1000 K. The corresponding hydrogen column density behind the shock $N_{\rm H} = \int n_{\rm H} v\, dt = n_{\rm H,1} v_1 \tau_{\rm cool}$ proves to be fairly independent of the pre-shock density $n_{\rm H,1}$ and the shock velocity v_1.

reference	$n_{\rm H,1}[{\rm cm}^{-3}]$	$v_1[{\rm km\,s}^{-1}]$	$N_{\rm H}[{\rm cm}^{-2}]$
Hollenbach & McKee (1989)	10^{5}	80	$10^{20.2}$
Neufeld & Hollenbach (1994)	10^{9}	50	10^{21}
Fadeyev & Gillet (1997)	$10^{13.6}$	25	$\approx 10^{19.6}$

Assuming $N_{\rm H} \approx 10^{21} {\rm cm}^{-2}$ (worst case) and considering $v_1 = 20\,{\rm km\,s}^{-1}$ and $P = 1\,{\rm yr}$ as typical values for Miras and LPVs, the critical pre-shock density $n_{\rm cr,1}$ where $\tau_{\rm cool}$ equals P is given by

$$n_{\rm cr,1} = \frac{N_{\rm H}}{v_1 P} \approx \frac{10^{21} {\rm cm}^{-2}}{20\,{\rm km\,s}^{-1} \cdot 1\,{\rm yr}} = 10^{7.2} {\rm cm}^{-3} \quad . \tag{4}$$

Evaluating various radiative heating/cooling rates and relying on time-scale arguments, Woitke et al. (1996) arrived at similar conclusions (critical post-shock density $10^{6...8} {\rm cm}^{-3}$). The formation of a calorisphere requires $\tau_{\rm cool} > P$ or, equivalently, $n_{\rm H} < n_{\rm cr,1}$. Such small densities, however, are only present at radial distances as large as $\gtrsim 3 \cdot 10^{14} {\rm cm}$ (calculated for $\dot{M} = 4\pi r^2 \rho v_\infty = 10^{-6} {\rm M}_\odot/{\rm yr}$ and $v_\infty = 20\,{\rm km/s}$), i.e. at distances which are unlikely to be relevant for the wind generation and where the formation of dust is expected to be almost completed. Thus, theoretical constraints on the efficiency of radiative cooling do not favor high mass-loss rates caused by pulsations alone.

Acknowledgments The author would like to thank E. Sedlmayr, J. M. Winters and P. Cottrell for valuable comments on the manuscript. This work has been supported by the Deutsche Forschungsgemeinschaft (grants Se 420/8–1, Se 420/15–1) and by the BMBF (grant 05 3BT13A 6).

References

Alvarez R., Plez B. (1997): A&A, accepted
Arndt T.U., Fleischer A.J., Sedlmayr E. (1997): A&A, 327, 614
Bowen G.H. (1988): ApJ, 329, 299
Chandrasekhar S. (1934): MNRAS, 94, 444
Crosas M., Menten K.M. (1997): ApJ, 483, 913
Dorfi E.A., Feuchtinger M.U. (1991): A&A, 249, 417
Dorfi E., Höfner S. (1996): A&A, 313, 605
Fadeyev Y.A., Gillet D. (1997): A&A, submitted
Feuchtinger M.U., Dorfi E.A., Höfner S. (1993): A&A, 273, 513
Fleischer A.J., Gauger A., Sedlmayr E. (1992): A&A, 266, 321
Fleischer A.J., Gauger A., Sedlmayr E. (1995): A&A, 297, 543
Fleischer A.J., Gauger A. (1997): private communications
Gail H.-P., Sedlmayr E. (1988): A&A, 206, 153
Gail H.-P., Keller R., Sedlmayr E. (1984): A&A, 133, 320
Gauger A., Fleischer A.J., Winters J.M., Sedlmayr E. (1993): AG Abstr. Ser., 9, 127
Höfner S., Feuchtinger M.U., Dorfi E.A. (1995): A&A, 297, 815
Höfner S., Dorfi E.A. (1997): A&A, 319, 648
Höfner S., Jørgensen U.G., Loidl R. (1997): in *Proc. ISO's view on Stellar Evolution*, Waters R., Waelkens C., van der Hucht K. (eds.), Kluwer Academic Publishers, in press
Hollenbach D., McKee F. (1989): ApJ, 342, 306
Jørgensen U.G., Johnson H.R. (1992): A&A, 265, 168
Le Bertre T. (1997): A&A, 324, 1059
Lucy L.B. (1976): ApJ, 205, 482
Luttermoser D.G., Bowen G.H. (1990): in *Proc. Sixth Cambridge Workshop on Cool Stars, Stellar Systems, and the Sun*, Wallerstein G. (ed.), 491
Mihalas D. (1978): *Stellar Atmospheres*, W.H. Freeman, San Francisco, 2nd edition
Miller G.E., Scalo J.M. (1979): ApJS, 41, 513
Neufeld D.A., Hollenbach D.J. (1994): ApJ, 428, 170
Reid M.J., Menten K.M. (1997): ApJ, 476, 327
Reimers D., Köster D. (1979): A&A, 202, 77
Richtmyer R.D., Morton K.W. (1967): *Difference methods for initial-value problems*, John Wiley & Sons, New York, 2nd edition
Unno W., Kondo M. (1976): PASJ, 28, 347
Willson L.A., Struck–Marcell C., Bowen G.H. (1997): in *Proc. Cosmic Winds and the Heliosphere*, Jokipii J.R. et al. (eds.), University of Arizona Press, in press
Windsteig W., Dorfi E., Höfner S., et al. (1997): A&A, 324, 617
Winters J.M., Fleischer A.J., Gauger A., Sedlmayr E. (1994): A&A, 290, 623
Winters J.M., Fleischer A.J., Gauger A., Sedlmayr E. (1995): A&A, 302, 483
Winters J.M., Fleischer A.J., Le Bertre T., Sedlmayr E. (1997): A&A, 326, 305
Woitke P., Krüger D., Sedlmayr E. (1996): A&A, 311, 927
Wood P.R. (1979): ApJ, 227, 220
Yorke H.W. (1980): A&A, 86, 286

Puls: One of the major ingredients in the hydro-description is the temporal behavior of the radiation pressure on the dust. What is the basic "trick" of calculating this quantity accounting for the NLTE-dust formation investigated by your group?

Woitke: The time-dependent treatment of the dust complex is based on the moment method developed by Gail & Sedlmayr (1988) and the way it is included into the hydro-description is shown in Fleischer et al. (1992). Four additional differential equations for the dust complex are solved together with the usual hydrodynamic equations. Applying the small particle limit of Mie theory, the radiation pressure on dust grains can be expressed in algebraic terms of the third moment of the dust distribution function.

Owocki: I was intrigued by your scaling formula, particularly by the −8 power dependence on temperature, and the +1.5 power on the luminosity, which is reminiscent of scaling in hot stars. But for stars of a given evolutionary status, must there not also be some relation among these parameters, and that this will be relevant when comparing predicted scaling to observations?

Arndt: Regarding the extreme −8 power for T_*, I think it is (mostly) due to the extreme temperature dependence of dust formation and growth. Dependences between stellar parameters, arising from evolutionary considerations, are part of planned studies – but right now the shown $\dot{M}$-scaling formula is restricted to C-rich AGB stars. The observed stellar parameters of C-stars are somewhat a matter of debate, because their CSEs tend to be enshrouded by circumstellar dust, often causing them to be pure IR-objects. Therefore, I feel personally that "observed" relations may be misleading in the parameter range of "our" $\dot{M}$-formula.

Linsky: Tom Ayres, David Muchmore and others have called attention to the thermal instability that results from the strong cooling by CO and SiO that can lead to rapid formation of these molecules. Their model atmospheres are bistable in the sense that two solutions can result. In your models of carbon-rich Mira variables do you see evidence for this bistability?

Woitke: The work of T. Ayres and D. Muchmore is mainly devoted to solar-type stars, where in addition to the warm atomic, almost molecule-free gas phases, cool molecule-rich phases may coexist which are thermally stable as well. According to my personal studies, C-star envelopes do not show much of these thermal bifurcations, because the gas is molecule-rich anyway, in particular CO-rich. Nevertheless, some minor instabilities may be present, e.g. due to SiS formation. This effect has to be further investigated and may be very important, if another heating mechanism is active, for example the heating by propagating shock waves.

Atmospheric Dynamics and Stellar Winds of Long-Period Variables

Susanne Höfner and Uffe Gråe Jørgensen

Niels Bohr Institute, Astronomical Observatory,
Juliane Maries Vej 30, DK-2100 Copenhagen, Denmark

Abstract. The atmospheres and circumstellar envelopes of AGB stars are characterized by complex physical phenomena like shock waves caused by stellar pulsation or
formation of molecules and dust which often lead to heavy mass loss by slow stellar
winds. To allow a physical interpretation of various observations we have constructed
improved dynamic models of atmospheres and circumstellar dust shells of long-period
variables. We investigate the dependence of the wind characteristics on stellar pulsation, molecular opacities and dust formation and discuss the time-dependent behaviour
of the circumstellar envelopes and its influence on observational properties.

1 Introduction

Large-amplitude pulsations with periods of about 100 to 1000 days are a characteristic feature of stars on the asymptotic giant branch (AGB). The stellar
pulsation induces shock waves in the atmosphere of such long-period variables
(LPVs) which play a crucial role for the atmospheric structure and the mass
loss by a stellar wind. The strong non-linear interaction of dynamics, formation
of molecules and dust grains and the stellar radiation field presents a challenge
for the self-consistent modelling of such objects and makes the interpretation of
observational results difficult. However, in recent years some progress has been
achieved in modelling various time-dependent phenomena and understanding
the wind mechanism of LPVs (see related reviews in this volume).

We present improved dynamic models for atmospheres and stellar winds of
pulsating AGB stars which are based on a more realistic treatment of the gas
opacity than most earlier time-dependent models in the literature. As discussed
by Höfner et al. (1998) this has a dramatic influence on the atmospheric structure
and the resulting observable properties. Our models show a good qualitative
agreement with various observational results. In this contribution we investigate
the wind characteristics of our new models and compare them to the results of
Höfner & Dorfi (1997).

2 Atmospheric Structure and Dynamics

We obtain the variable spatial structure of the atmosphere and circumstellar envelope (density, temperature, degree of condensation, etc.) by solving the equations of grey radiation hydrodynamics together with a time-dependent descrip-

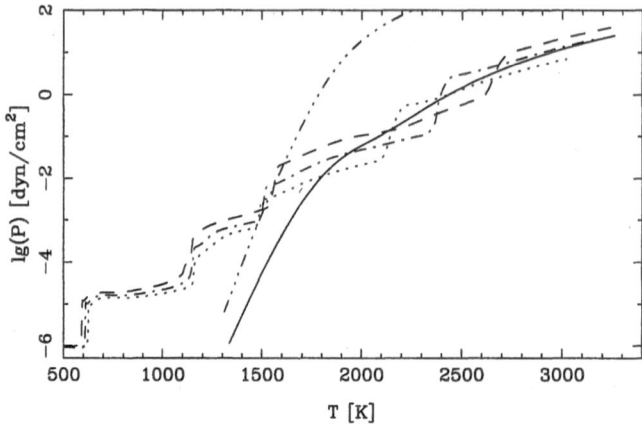

Fig. 1. Atmospheric structure (gas pressure vs. temperature) of the model with $L_\star = 13000\ L_\odot$ and $\Delta u_p = 4$ km/s at different phases (0.50: dashed; 0.75: dash-dotted; 1.00: dotted) and of the corresponding hydrostatic initial model (full line); phase 0.5 corresponds to minimum light, 1.0 to maximum light. The "steps" in the pressure indicate shock waves. Their propagation through the atmosphere can be seen by tracking the progress of individual features from higher pressures and temperatures to lower ones (i.e. from right to left) with increasing time (phase 0.5 to 1.0). To demonstrate the influence of the gas opacity on the atmospheric structure we plot a hydrostatic initial model calculated with the same stellar parameters but a constant gas opacity ($-\cdots-$; see text) for comparison with the new initial model (full line).

tion of the dust formation (cf. Höfner & Dorfi 1997 and references therein for details of the modelling method).

The dynamic calculations start with an initial model which represents the full hydrostatic limit case of the grey radiation hydrodynamics equations (including a variable Eddington factor) and which is determined by the following parameters: luminosity $L_\star$, mass $M_\star$, effective temperature $T_\star$ and the carbon-to-oxygen ratio $\varepsilon_C/\varepsilon_O$ (all element abundances except carbon are assumed to be solar). The structure of this hydrostatic initial model can be directly compared to standard model atmospheres. We find that using Planck mean absorption coefficients based on detailed molecular line data (SCAN, Jørgensen 1997) yields reasonably realistic atmospheric structures which are largely comparable to standard MARCS model atmospheres based on the corresponding frequency-dependent opacities (cf. Höfner et al. 1998). On the other hand, models calculated with a constant gas absorption coefficient of $2 \cdot 10^{-4}\ \mathrm{cm^2/g}$ (as introduced by Bowen 1988 and used by many time-dependent calculations in the literature, e.g. Fleischer et al. 1992, Höfner & Dorfi 1997) have a much higher overall density since they tend to underestimate the gas opacity in most zones (cf. Fig. 1). The resulting synthetic spectra exhibit far too strong molecular features.

The stellar pulsation of the LPV (which is simulated by a variable boundary

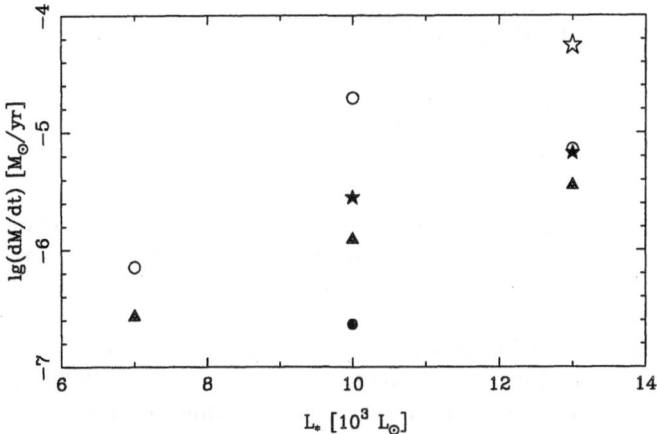

Fig. 2. Comparison of mass-loss rates: selected models of Höfner & Dorfi (1997; open symbols) and this work (full symbols). The shape of the symbol indicates the piston velocity amplitude of the model (Δu_p: $\circ$ = 2 km/s, $\triangle$ = 4 km/s, $\star$ = 6 km/s), the other stellar parameters (which are correlated to $L_\star$) are given in the text.

located beneath the stellar photosphere, moving with a velocity amplitude Δu_p and a period P) creates shock waves in the atmosphere which basically influence the structure in two ways (see Fig. 1): In the inner part of the model the passing shocks cause a periodic modulation of the pressure-temperature structure which varies around that of the corresponding hydrostatic initial model. Further out (below $\approx$ 2000 K in Fig. 1) the shocks induce a levitation of the atmosphere, providing at all phases and temperatures much higher densities than in the hydrostatic model, hereby favouring the formation of molecules and dust grains, and eventually leading to the onset of a dust-driven wind.

These dynamic phenomena have a strong impact on the resulting synthetic spectra (Loidl et al. 1998). The variations of optical and near-infrared molecular features show a qualitative agreement with observations (Hron et al. 1997, Aringer et al., this volume). In particular, our dynamic models are able to reproduce observed out-of-phase variations of different features (causing e.g. loops in colour-colour diagrams) which reflect the effects of shock waves running through the atmospheres and cannot be simulated with hydrostatic model atmospheres (e.g. Aringer et al. 1997, Alvarez & Plez 1997). We have also started to investigate line profile variations of synthetic CO lines which probe the velocity structure of the atmosphere and circumstellar envelope (cf. Windsteig et al., this volume).

3   Wind Properties

We have calculated new dynamic models based on mean molecular opacities using the same combinations of stellar parameters as in Höfner & Dorfi (1997):

$L_\star$ [$L_\odot$]	$M_\star$ [$M_\odot$]	$T_\star$ [K]	$\varepsilon_C/\varepsilon_O$	P [d]
7000	1.0	2880	1.80	390
10000	1.0	2790	1.80	525
13000	1.0	2700	1.40	650

Figure 2 shows a comparison of the resulting mass-loss rates. As expected from the lower overall density of the new models (cf. Sect. 2) the mass-loss rates are considerably lower for similar model parameters. This effect can partly be compensated by increasing the piston velocity amplitude Δu_p (cf. Fig. 2), which affects the density in the wind acceleration zone (levitation). All new models show significantly lower final degrees of condensation (< 0.3) than those of Höfner & Dorfi (1997), reflecting reduced grain growth rates due to the lower gas density. The wind velocities, however, are approximately the same as before (or at least not reduced by the same amount as the degree of condensation). Since, in a dust-driven wind, the outflow velocity should be closely correlated to the degree of condensation (cf. Höfner & Dorfi 1997), this seems to indicate that in our present models radiation pressure on molecules contributes to driving the stellar winds.

Acknowledgments

This work is supported by the Austrian *Fonds zur Förderung der wissenschaftlichen Forschung* (grant J1487-PHY).

References

Alvarez, R., Plez, B. (1998): A&A 330, 1109
Aringer, B., Jørgensen, U.G., Langhoff, S.R. (1997): A&A 323, 202
Bowen G.H. (1988): ApJ 329, 299
Fleischer, A.J., Gauger, A., Sedlmayr, E. (1992): A&A 266, 321
Höfner, S., Dorfi, E.A. (1997): A&A 319, 648
Höfner, S., Jørgensen, U.G., Loidl, R. (1998): Proc. of *ISO's View on Stellar Evolution*, eds. R. Waters, C. Waelkens, K.A. van der Hucht, Kluwer, in press
Hron, J., Loidl, R., Kerschbaum, F., Jørgensen, U.G. (1997): Proc. of *ISO's View on Stellar Evolution*, eds. R. Waters, C. Waelkens, K.A. van der Hucht, Kluwer, in press
Jørgensen, U.G. (1997): in *Molecules in Astrophysics: Probes and Processes*, E.F. van Dishoeck (ed.), Kluwer, p. 441
Loidl, R., Hron, J., Aringer, B., et al. (1998): Proc. of *ISO's View on Stellar Evolution*, eds. R. Waters, C. Waelkens, K.A. van der Hucht, Kluwer, in press

Linsky: In your models, does the mass-loss rate measured far from the star vary with pulsation phase?

Höfner: Close to the star (on spatial scales of several stellar radii) the variations with phase are more or less pronounced, depending on the stellar parameters. However, as the material moves away from the star the shocks are damped and the variations become smaller and smaller.

Mullan: You have examined different models with various amplitudes of piston motion. Do you calculate the velocity widths of spectral lines which are formed in these pulsating atmospheres?

Höfner: We have recently started to do detailed radiative transfer calculations (including velocity fields) for our dynamic models. One example is shown in the poster of Windsteig et al. where we discuss the variability of CO line profiles. The first overtone lines presented there show dramatic variations due to the dynamics of the stellar atmosphere.

Pulsation and Stellar Winds:
Some Lessons Learned from Dynamical Models
of the Atmospheres of Cool Stars

L.A. Willson[1] and G.H. Bowen[1]

Department of Physics and Astronomy, Iowa State University, Ames IA 50011

Abstract. Stellar pulsation is associated with some of the most dramatic epochs of mass loss from stars. Detailed numerical modeling calculations that reproduce the observed characteristics of such mass losing stars necessarily include a number of key physical processes. The reader is referred to detailed calculations of dynamical atmospheres for Mira variables (Bowen 1988 (=B88) and 1998 (=B98)). A review of physical processes with examples from a variety of computations for Miras, Cepheids, and RR Lyrae stars is given by Willson, Struck & Bowen (1997 = WSB), and evolutionary consequences are outlined by Willson, Bowen & Struck (1996 = WBS). This review focusses on (1) the importance of departures from LTE, radiative equilibrium, and chemical equilibrium and (2) the treatment of "piston" boundary conditions. We then compare the results of different modeling calculations with observed Mira period-luminosity relations. Finally, we summarize the implications for "cyclical variations" in stellar winds.

1 Departures from Equilibrium
in Pulsating Star Atmospheres

Great changes in density over the cycle accompany the passage of pulsation-induced shock waves in the atmosphere. Thus, PdV work and interaction of the gas with the radiation field both are important in determining the temperature structure, and the radiative relaxation time determines during how much of the cycle the gas can remain near the radiative equilibrium temperature, T_{RE}. The radiative relaxation time exceeds the time between shocks at low densities (Section 3). Another slow process that is important for LPVs is the recombination of molecular hydrogen, which is a significant factor in the energy equation. The abundances of H and H_2 also affect the grain formation process; even if the other molecular species approach their equilibrium values rapidly, they are doing so in a mix that contains more H and less H_2 than in strict molecular equilibrium. (For a detailed discussion and results of coupled dynamical and H $\leftrightarrow$ H_2 computations see B98.)

For dusty winds, optical properties of the grains are important, not only because they determine the radiation force on the grains, but because they affect the temperature, hence the formation and survival of grains. At low gas densities, where collisional heat exchange between gas and grains is very slow, grain temperatures are determined mostly by somewhat complicated radiation effects. Silicate grains, whose absorption/emission coefficient is larger around $10\mu m$ than for most of the stellar spectrum, exhibit an "inverse greenhouse effect" (Gilman 1974); their temperature can run several hundred Kelvins below the local black-body value. Carbon grain temperatures tend to be above blackbody values[1]. In both cases the grain size plays a role.

2   Pulsation Driving and Atmospheric Dynamics

Pulsation alters the atmospheric structure and thus the boundary conditions for the internal pulsation. Periods and growth rates from internal calculations available to date ignore the effects of pulsation on the boundary. Analysis of the dependence of atmospheric structure on pulsation period and amplitude by Bowen (1990) indicate the importance of the minimum in $P_{\text{acoustic}}(= 4\pi H/c_{\text{sound}})$ in the atmosphere: For periods shorter than $P_{\text{acoustic}}^{\min}$, traveling waves are formed below the photosphere. This leads to very heavy "leakage" of pulsation energy into the atmosphere. This applies for overtone modes in most low-mass AGB stars. For periods long compared with $P_{\text{acoustic}}^{\min}$, there is strong reflection, and standing waves are formed below the photosphere. This allows the pulsation amplitude to be large at the photosphere. This applies for fundamental mode in most AGB stars. Thus short-period pulsations lead to much greater atmospheric dissipation, possibly enough to play a key role in mode selection for the LPVs. For a detailed discussion, see Bowen (1990).

There has been concern that "piston-driven" models may give misleading results, especially if inappropriate piston amplitudes are used. In Bowen's models (B98, Bowen & Willson 1991) the driving amplitude used for each model in comparative or evolutionary studies is set by a uniform physical criterion based on the energy available for driving pulsation. The immediate source of driving energy is necessarily the internal energy residing in the driving region, which is used during the half cycle of outward motion (expansion) to do work on the atmosphere outside it. That energy is mostly returned by the atmosphere during inward motion (compression) in the next half cycle, but small losses into the atmosphere are readily made up from the outward energy flow from the deep stellar interior. Following Cox (1980, equation 10.2; or Cox & Giuli 1968, equation 27.186) we assume that the maximum pulsation amplitude can be no greater than about the value at which internal energy available in the driving region is exhausted during each half cycle of expansion. It corresponds to an average mechanical power equal to the stellar luminosity L, for the half cycle of expansion; for sinusoidal pulsation the corresponding peak power is $L(\pi/2)$. Indeed, models driven at that level are typically beginning to exhibit slightly erratic behavior, and that becomes rapidly more extreme if the driving amplitude is made larger. Models driven a little less strongly are very well behaved, however, with stable and reproducible behavior over wide ranges of the stellar parameters. All of Bowen's models in recent years have been adjusted (automatically and precisely) to have a specified value of the mean driving power during

---

[1] This is one of several reasons why mass loss models for M stars cannot be assumed to apply to C stars and vice versa.

the expansion half cycle, usually chosen to be $L(2/\pi) = 0.637L$, corresponding to a peak value of L for sinusoidal motion. The results appear dependable.

At that driving amplitude the power loss into the atmosphere is generally less than 1% of L for fundamental-mode Mira models (which typically have pulsation periods of at least twice the acoustic cutoff period, hence strong reflection of traveling waves in the region just outside the photosphere). That is enough to produce strong shocks in the atmosphere, but not enough to damp the pulsation significantly. The same models driven at the shorter first overtone period, for which reflection is much weaker, have much greater power loss into the atmosphere, typically at least 10% of L, which is enough to damp the pulsation heavily, possibly preventing sustained OT-mode pulsation entirely in complete stellar models. It should be emphasized that these values are in no way prescribed for the model; they are inevitable consequences of the physics present in the dynamical equations, as the numerical integration of those equations develops a detailed solution consistent with the given boundary conditions.

Careful measurement of the mass-loss rate (and other modeling results) as a function of the piston amplitude has shown that these are not especially sensitive to the amplitude, as long as the "overdriven" region described above (average driving power during expansion $> L$) is avoided. The principal conclusions this work leads to regarding evolution with mass loss would not be significantly changed, for example, if all piston amplitudes were changed by a factor of two.

This coupling of the atmosphere and the interior suggests that inclusion of the driving zones in the interior with the nonlinear modeling of atmospheric dynamics is desirable. This would make it possible to determine more precisely the power being transferred into the atmosphere as PdV work for each kind of star. It is required for a firm theoretical understanding of mode selection and limiting amplitudes, and also to "predict" light curves. It may *require* time-dependent convection treatment. However, the fundamental conclusions about the mechanisms and magnitude of mass loss are unlikely to be changed (B98).

3   Radiative Relaxation and Non-LTE

Relaxation time scales (from known rates for the change in number density n of an atom, molecule, or ion: $\tau \equiv n/(dn/dt)$) for different processes behind the shock are used to estimate the exchange of energy with the radiation field. Using finer grids near the shock to trace the rates of microprocesses in a global hydrodynamical model will enable a closer coupling of radiative transfer and dynamical models, particularly for modeling emission lines and other phenomena developing close to the shock. For higher-gravity stars with larger amplitude shocks better shock treatment may be important, and it is also required before detailed predictions can be made of the postshock emission spectrum. However, such refinements in the shock treatment are not expected to make a large difference in the dynamical structure and mass-loss rates for AGB stars; this statement is based on considerable testing of the sensitivity of models to the parameters used. *While we do not follow every process in detail, but rather use reliable first-order*

(= dominant term) approximations, we also do not assume equilibrium for any process until calculation establishes that this is a good assumption.

Is the appropriate description LTE or non-LTE? When collisional de-excitation rates are higher than radiative de-excitation rates, LTE is established. In a two-level atom, the condition $C_{ul} = R_{ul}$ defines the critical value[2] ρ_x; for $\rho > \rho_x$ collisions dominate and the level populations are in LTE. For complex atoms, the characteristic ρ_x depends on conditions. In stellar atmospheres permitted transitions establish the population levels. For these, the critical density $\rho_x \approx 10^{-10}$ gm cm^{-3} (using the Saha equation to estimate n_e/n_H and hence collision rates).[3]

Underestimating the critical density (and thus assuming LTE when it isn't appropriate) can give a large error in the net cooling and heating rates. A first-order approximation for the source function (valid for a two-level atom) is

$$S/S_{LTE} \approx \frac{1}{(1 + \rho_x/\rho)} \tag{1}$$

Thus $S/S_{LTE} \to \rho/\rho_x$ when $\rho/\rho_x \ll 1$.

Dynamical models computed with non-LTE cooling have thicker regions of high $T_{kinetic}$. However, these regions have lower $T_{excitation}$; necessarily, the same amount of energy is emitted by these thicker regions (to first order[4]) as in the LTE-cooling models.

It has been argued that models computed with LTE cooling rates are closer to reality because, *when combined with LTE radiative transfer*, the results of non-LTE modeling produce too much emission from warm material. However, as noted above, applying LTE radiative transfer to dynamical models computed with non-LTE is *inconsistent;* more energy is found to be radiated from the extended $T_{kinetic}$ region than was put there in the dynamical model. LTE radiative transfer applied to dynamical model atmospheres computed with finite thermal relaxation violate conservation of energy! Using LTE radiative transfer with dynamical models computed with LTE cooling functions is *consistent.* (It is analogous to "jumping over" detailed relaxation in a shock using continuity

---

[2] To avoid confusion, we note that in these proceedings, Woitke discusses a different "critical density": The density at which the thermal relaxation time is equal to the period. These are closely related, since the thermal relaxation time only becomes long for densities $\ll \rho_x$.

[3] Woitke (these proceedings and references therein) argues for much lower values of the critical density, but his argument is based on calculations with model atoms that include ONLY forbidden transitions, except for H and He which lack low-lying energy levels. Such artificial model atoms necessarily give low ρ_x, but they are not appropriate for the conditions in a stellar atmosphere, where the background radiation field is less dilute and the density higher than in the interstellar case for which these model atoms were originally constructed.

[4] Only to first order since the cooling affects the dynamical structure; however, the radiation we see is dominated by photons emitted near the photosphere, where the density is high and the structure most similar between LTE and non-LTE models.

relations for "isothermal shocks".) It tells you nothing about which cooling rate is correct, but produces a first-order fit to the observed spectrum.

Two otherwise identical models, one with $\rho_x = 10^{-10}$ gm cm^{-3} and the other with $\rho_x = 10^{-16}$ gm cm^{-3} are displayed in Fig. 1 – 3. The high-critical-density model is typical of results obtained by Bowen; the low-critical-density model is very similar to results published by the Vienna and Berlin groups (see Woitke's paper for references). Clearly this single factor is critical in determining the appearance of the models. Note that for the purposes of this comparison all other model parameters were kept fixed – including the prescription for dust formation around two stellar radii. In the large-ρ_x case there is a very cool region close to the star; in that region, SiO vapor is highly supersaturated and grains are expected to nucleate. In contrast, in the small-ρ_x case, dust formation has been artificially introduced in spite of the fact that nowhere in that model was SiO supersaturated. While the carbon chemistry studied by the Berlin and Vienna groups is significantly different, it is also true that in order to get large amounts of mass loss with dust they need to go to models with very low T_{eff} and/or very large L.[5] In both models dust is driving the final outflow, so *it is not true that the treatment of the thermal physics is unimportant when dust is important.* Note also that some of the phenomena found by Fleischer et al. (1995) – particularly the "onion-skin" structure for dusty winds – depend on having $T \approx T_{RE}$ through most of the envelope, and are therefore not expected to be present in models with correct thermal physics.

4   Models vs. Observations for Pulsating AGB Stars

Perhaps the best observational constraint on the Miras is the period-luminosity relation for Miras in the LMC. The $P - L$ relation results from a combination of the evolutionary constraints on the stars ($L - T - M - Z$ relations), the dependence of pulsation periods on stellar parameters ($P - M - R$ relations), and the dependence of mass loss on stellar parameters. The Bowen (1995) grid consisted of models constrained to lie on a consistent set of evolutionary tracks and with periods specified by standard (fundamental mode) $P - M - R$ relations. Thus the mass-loss rate can be traced as a function of evolutionary time (or L or R or P) for a given initial mass M and metallicity Z. The resulting mass-loss rate is found to increase *steeply* with increasing evolutionary time, L, R or P. For example, for a given mass star evolving up the AGB we find $\dot{M} \propto L^a$ with $a \sim 12$ to 16. The evolution terminates with loss of the entire envelope as the star evolves through the region of the "cliff", defined to be where $\dot{M}/M = \dot{L}/L$. (See WBS and B98 for details.) To compare this with observed properties, we argue that stars along the "cliff" are most likely to be selected as (a) mass losing red giants and/or (b) Miras.

The steep dependence of $\dot{M}$ on L (including co-evolving R) and M means that errors in $\dot{M}$ of the order of a factor of 2 (such as may result from uncertain-

---

[5] Most of their models also assume much higher carbon abundances than are generally observed; higher C/O gives more mass loss if all other parameters are kept the same.

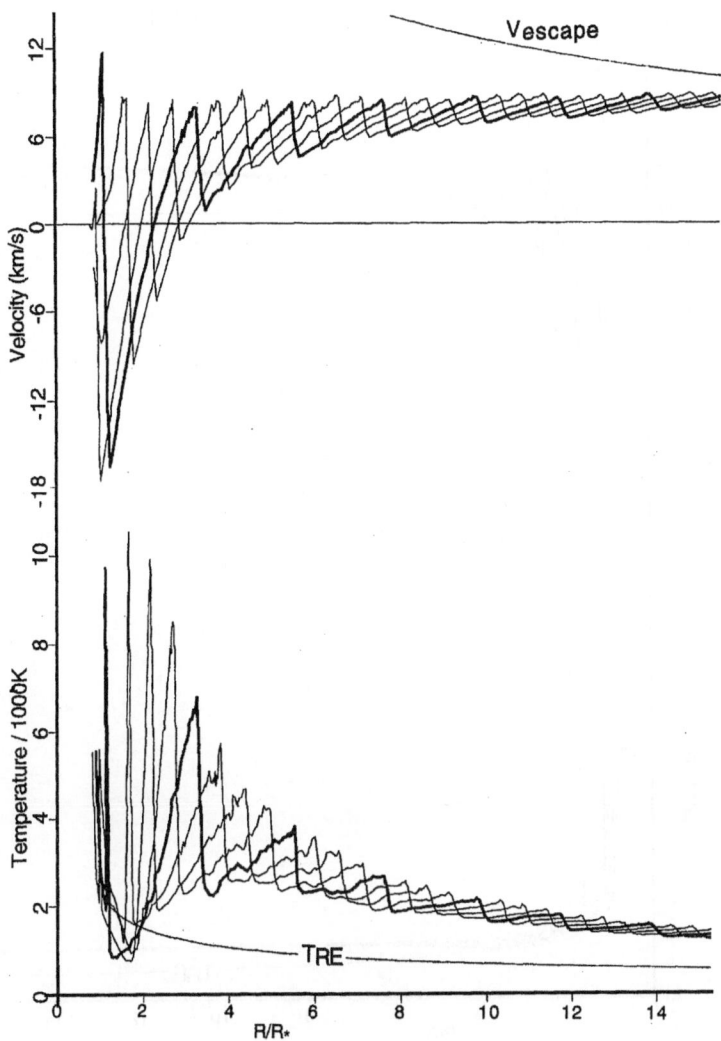

Fig. 1. Velocity (top) and temperature (bottom) vs. radius for a model with $\rho_x = 10^{-10}$ gm cm^{-3}. The structure of this model is typical of the "non-LTE" case: There is a deep minimum in the temperature of postshock gas at about two stellar radii, corresponding to the location where dust is expected to form. Farther out, slow relaxation to radiative equilibrium leads to temperatures that remain above the radiative equilibrium temperature during much of the cycle; this effect would be greater were it not for the expansion of the atmosphere produced by the radiative forces acting on the dust.

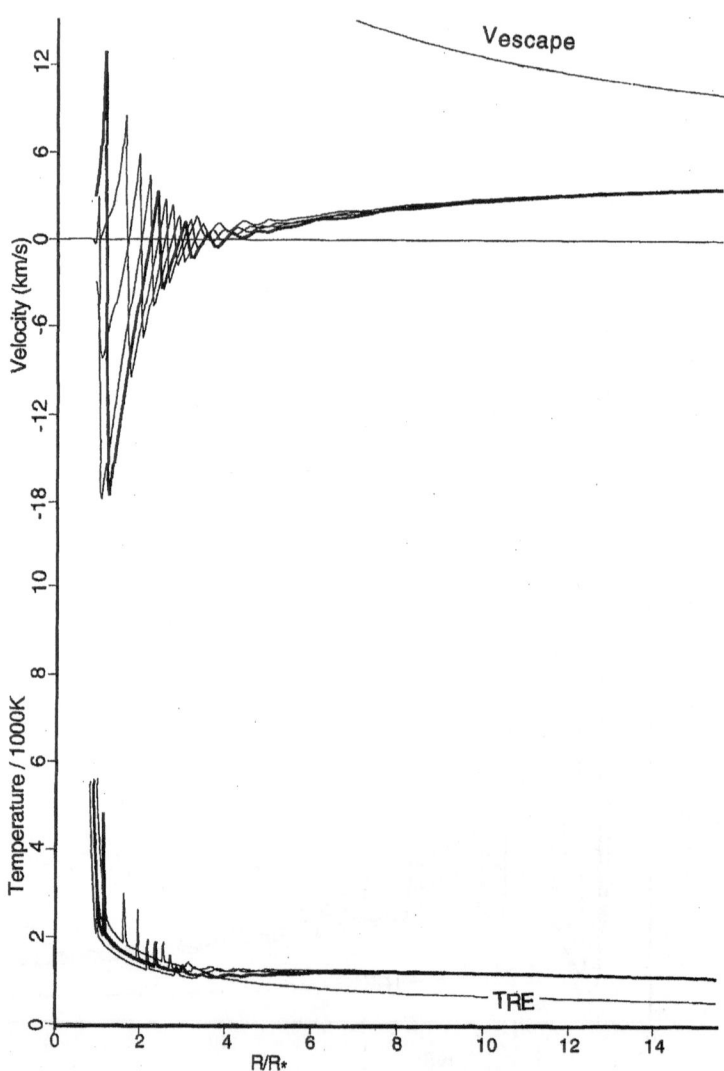

Fig. 2. Velocity (top) and temperature (bottom) vs. radius for a model that is identical to the one in Figure 1 in every way except that $\rho_x = 10^{-16}$ gm cm^{-3}, corresponding to assuming non-LTE sets in only when radiative transitions are faster than collisional transitions for forbidden lines. The resulting temperature structure is typical of "LTE" models: Rapid recovery from shock heating, no super-cooled regions, and $T \approx T_{RE}$ nearly everywhere. The variation of T_{RE} with phase in the inner atmosphere is in response to an assumed phase dependence of the photospheric radiation field.

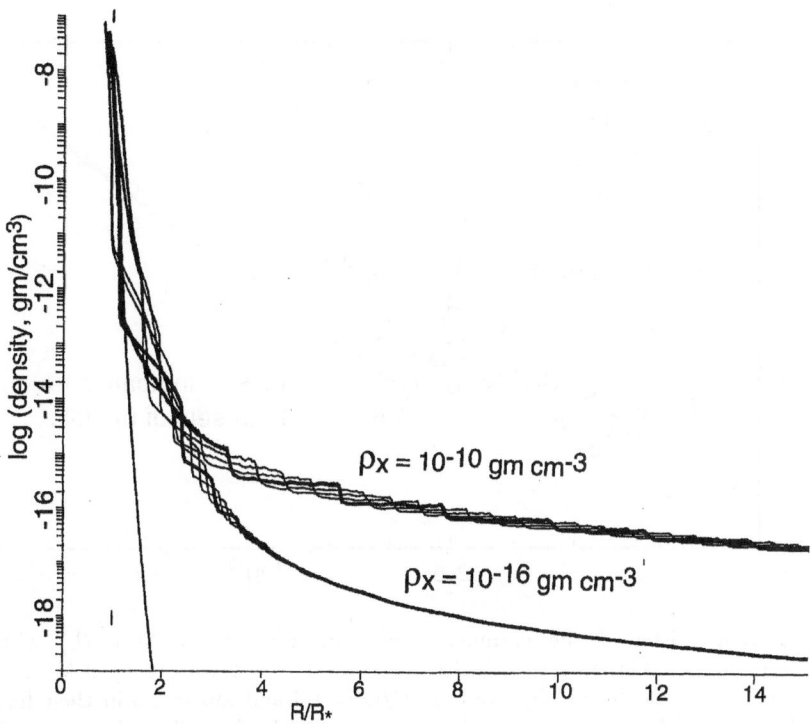

Fig. 3. Density vs. R for the two models shown in Fig. 1 & 2. For a given set of stellar parameters, much lower mass-loss rates result from LTE treatment (with mechanical energy being converted int radiation much more efficiently) than in the non-LTE case.

ties in model parameters discussed above) only result in errors in $\log L_{cliff} \lesssim 0.02$. Detailed studies of the effects of each model parameter studied separately show that within reasonable physical bounds, errors in $\dot{M}$ should be no greater than a factor of 2 in each case. Parameters investigated include: driving period; driving amplitude; evolutionary model parameter $\alpha = \ell/H$; critical density for the cooling law; and dust parameters.

The 1995 Bowen models provide an effortless fit[6] to the period-luminosity relation for Magellanic Cloud Miras (Fig. 4). To demonstrate that the agreement with the observed $P - L$ relation is not inevitable for any "reasonable" mass-loss law, we have included the result of applying the same analysis to the mass-loss law presented by Arndt et al. in these proceedings. The reason that their relation lies where it does is that dust-driven mass loss in models with $T \approx T_{RE}$ in most of the atmosphere only becomes large with much cooler, larger stars. To force

---

[6] No parameters were adjusted to obtain this fit. After the model grid was computed, the empirical data were located and compared to the model results and Figure 4 was the result.

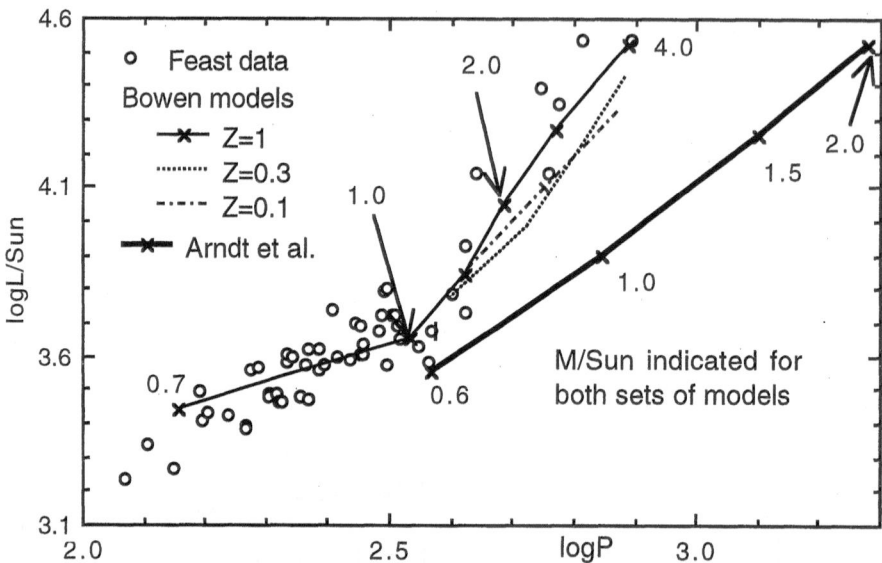

Fig. 4. Comparison of the parameters of stars with $\dot{M} = M\,(\dot{L}/L)$ with the period-luminosity relation established by observations by Feast (1989) of Miras in the LMC. For the Arndt et al. line we took C/O = 1.5 and $\Delta u = 2.0$ in their formula. The Bowen models reproduce the observations remarkably well, with no adjustment of modeling parameters required, while the alternative mass-loss law predicts critical mass loss at much longer periods and higher luminosities for a given stellar mass.

the Arndt et al. relation to match observations would require at least a different choice for the pulsation periods, with the resulting pulsation period considerably less than the photospheric P_{acoustic}.

Mass-loss rates vs. LM/R at the "cliff" are in almost perfect agreement with the empirical relation derived by Reimers (1975; see WBS for the comparison plot and more discussion). We conclude that Reimers' relation is telling us something about the properties of stars near critical rates of mass loss rather than (as has been nearly universally assumed) telling us how mass loss varies with stellar parameters. This is analogous to the two interpretations of the significance of the main-sequence correlation of M and L: Before the late 1930s, as an evolutionary track; after 1940, as the locus of stars of different masses while they convert H to He in their cores.

5   Implications for "Cyclical Variations"

Phenomena occurring at periods other than P in the models (see B88 and WSB for details and examples) include: (1) For $P \gg P_{\text{acoustic}}^{\min}$, multiple shocks are seen each cycle. (2) In the region of strong shocks, the time between shock passages is not precisely constant, and this "jitter" increases with r/R_*. (3) The structure of models with dust-driven winds is different from those with thermally driven winds. If the dust takes several pulsation periods to build up to a size capable of driving the flow, this may lead to semi-periodic dust "episodes". The above may lead to cyclic variation in the outflow at periods other than the pulsation period, but typically still with a factor of a few of the pulsation cycle. Only (3) is likely to produce observable structure in the resulting wind, however.

Some fundamental lessons from pulsating star models with implications for "cyclical variations" in the winds:

1. There is a big difference between steady (periodic) pulsation and the transient waves induced by sudden changes in R or L.
2. It is critically important to treat the thermal relaxation with care.
3. Small changes in stellar parameters can lead to large changes in the appearance of the atmosphere by influencing $\rho(r)$ and thus $\tau(r)$.
4. Mechanisms for mass loss are not always simple ("either dust or pulsation") but may be synergistic.
5. Pulsation alone does *not* produce "cyclically variable winds" in a quasi-steady state (= periodically varying). Rather, density and velocity vs. r are nearly smooth and monotonic with steady (periodic) pulsation.
6. Pulsation does lead to mass-loss rates that are very sensitive to stellar parameters (L, M, R, Z and the resulting pulsation period and amplitude). Thus circumstellar "shell" structures may be traced to modest changes in stellar parameters amplified into high density contrast in the wind.

References

Bowen, G.H. (1988): ApJ 329, 299
Bowen, G.H. (1990): in *Numerical Modeling of Nonlinear Stellar Pulsations: Problems and Prospects*, Buchler, J.R., ed., Kluwer Academic Publishers, 155
Bowen, G.H. (1998): (in preparation for submission to ApJ)
Bowen, G.H., Willson, L.A. (1991): ApJ 375, L53
Cox, J.P. (1980): *Theory of Stellar Pulsation*, Princeton University Press, 1980
Cox, J.P., Giuli, R.T. (1968): *Principles of Stellar Structure. Vol. 2. Applications to Stars*, Gordon & Breach
Feast, M. (1989): MNRAS 241, 375
Fleischer, A.J., Gauger, A., Sedlmayr, E. (1995): A&A 297, 543
Gilman (1974): ApJS 28, 397
Reimers, D. (1975): in *Problems in Stellar Atmospheres and Envelopes*, Baschek, B., Kegel, W.H., Traving, G., eds., 229
Willson, L.A., Bowen, G.H., Struck, C. (1996): in *From Stars to Galaxies*, ASP Conf. Ser. 98, C. Leitherer, U. Fritze-v. Alvensleben, J. Huchra, eds., 197
Willson, L.A., Struck, C., Bowen, G.H. (1997): in *Cosmic Winds and the Heliosphere*, Jokipii, J.R., Sonnett, C.P., Giampapa, M.S. eds., U. of Arizona Press, 155

Gauger: How were the evolutionary tracks with mass loss obtained? Do they stem e.g. from the calculations of AGB evolution including mass loss of Blöcker?

Willson: Because the mass loss is so precipitous, we prefer to use evolutionary tracks without an applied mass loss law. In effect, we assume that as M decreases the star continues to obey the same L, T, M relation. We have been using a simple formula from Iben 1984 (ApJ 277, 333).

Linsky: Centimeter wave continuum radio emission has been observed in 6 long period variable stars by Reid & Menton (ApJ 476, 327 (1996)). This emission is sensitive to the ionized component of the mass flux and thus the temperature of the wind far from the stellar photosphere. Reid & Menton find no radio flux variations from Miras at the 15% level. I suggest that you compute the radio flux for your models for comparison with observations.

Willson: Yes, the existence of an extended "calorisphere" is quite characteristic of the models with non-LTE cooling. I expect that those observations will provide a very useful constraint on the cooling treatment. We should compute the expected signal from Bowen's models and compare them with observations; we'll do that soon. Thanks for the suggestion.

Grinin: Did you consider the non-LTE chemistry in the quasi-state approximation? Is it a good approximation at all distances from the star?

Willson: I can show you one example of non-equilibrium chemistry: H_2. Computing the rate of recombination and the equilibrium abundance as functions of ρ and T, we find, for densities below about 10^{-11} gm/cm^3, that H_2 that is dissociated in a shock does not have time to recombine. So only in a limited range of T and ρ will there be equilibrium and high abundance of H_2. Where H_2 is present, it plays a significant role in the energy balance. Hydrogen is special because recombination requires a three-body interaction; other species may well be able to reach equilibrium, but where H_2 is underabundant, the other species' equilibrium abundances must be calculated for an environment with hydrogen out of equilibrium. This includes most of the "dust factory" region.

Little-Marenin: Can you give us a feeling of why you use non-LTE models whereas the Berlin group uses LTE models?

Willson: I think initially they used LTE because it was convenient. More recently there has been an effort to calculate cooling rates, including all kinds of effects – dust, molecules, atoms, ions ... a massive effort. For the role of lines in atoms and ions, they essentially extrapolated the interstellar case, considering only forbidden transitions in their model atoms (i.e. they used model atoms with NO permitted levels, directly from Hollenbach & McKee, except for H, He, and He$^+$). From previous studies I have done, I am quite sure that at the relevant densities (10^{-18} to 10^{-12} gm/cm^{-3}) permitted transitions are important. We have been having rather intense email discussion over the last couple of weeks on this point, but they haven't really had enough time to deal with it, so I'd prefer not to comment further publicly at this time.

Mullan: To help in addressing the differences between the results obtained by various modellers as regards cooling behind shocks, is it possible to use data on chromospheric emission? For example, Mg II emission in H and K lines is seen in essentially all pulsating stars, and it requires temperatures of at least 6000 K. The intensity of the observed emission is a measure of the amount of material in shocked gas at $T \gtrsim 6000$ K.

Willson: Looking at chromospheric emission indicators such as Mg II is a very good idea. In the Miras, the observed Mg II emission is phase dependent, with maximum intensity about 0.2 after the peak Balmer emission. So that gives some indication of where in the atmosphere the shock is when you see the Mg II emission. Luttermoser has been doing some calculations to get line emission and profiles from Bowen models with cooling appropriate for the high critical density; he could easily look at models with very low N_{critical} (near LTE), and/or at some of the Berlin/Vienna models. I'll suggest it to him.

The Mass-Loss Rate of C-Rich Mira Stars

T.U. Arndt, A.J. Fleischer, and E. Sedlmayr

Technische Universität Berlin, Institut für Astronomie und Astrophysik,
PN 8-1, Hardenbergstr. 36, D-10623 Berlin, Germany

We present an equation that relates the parameters of dynamical wind models to the resulting mass-loss rate. These parameters are the stellar temperature, the stellar luminosity, the stellar mass, the abundance ratio of carbon to oxygen, the pulsational period, and the velocity amplitude of the pulsation.

The equations are constructed by applying the maximum–likelihood method to a grid of 48 dynamical models. These models of circumstellar dust shells around carbon-rich, long-period variables are based on the results of time-dependent hydrodynamical simulations that include a detailed treatment of dust formation, growth, and evaporation.

Since the resulting equations are based on physical models, they are free of additional arbitrary parameters and could therefore serve as a more reliable description of, e.g., the mass-loss rate in stellar evolution calculations.

However, we caution that the following approximation for the mass-loss rate is only valid within the parameter range given by Arndt et al. (1997).

$$\log \dot{M}_{\text{fit}} = -4.95 - 9.45 \cdot \log \left(\frac{T[\text{K}]}{2600} \right) + 1.65 \cdot \log \left(\frac{L[L_\odot]}{1 \cdot 10^4} \right) \tag{1}$$

$$-2.86 \cdot \log \left(M[M_\odot] \right) + 0.470 \cdot \log \left(\frac{\varepsilon_C/\varepsilon_O}{1.8} \right)$$

$$-0.146 \cdot \log \left(\frac{P[\text{d}]}{650} \right) + 0.449 \cdot \log \left(\frac{\Delta u [\text{km s}^{-1}]}{2.0} \right) \quad .$$

The main influence in (1) is due to the stellar temperature, followed by the stellar mass, and luminosity. The influence of the carbon abundance, the pulsational period, and the amplitude of the piston is almost negligible.

By applying the same method, we also constructed two equations that relate the parameters of dynamical wind models to the averaged outflow velocity and to the averaged dust-to-gas ratio, respectively. These equations and a comparison of the approximations with models obtained by a different modelling method (Höfner & Dorfi 1997) are presented by Arndt et al. (1997).

References

Arndt, T.U., Fleischer, A.J., Sedlmayr, E. (1997): A&A 327, 614
Höfner, S., Dorfi, E.A. (1997): A&A 319, 648

Molecular and Dust Features in Mass-Losing AGB Stars: Models and ISO Observations

Bernhard Aringer[1], Franz Kerschbaum[1], Susanne Höfner[2],
Uffe G. Jørgensen[2], Rita Loidl[1], and Walter Windsteig[1]

[1] Institut für Astronomie der Universität Wien, Vienna, Austria
[2] Niels Bohr Institute, University of Copenhagen, Copenhagen, Denmark

Until recently, synthetic spectra for AGB stars have usually been calculated by using hydrostatic model atmospheres like the MARCS code. This is problematical, since AGB stars are dominated by dynamical phenomena like pulsation, shock waves, dust formation, and heavy mass loss. As a consequence, classical hydrostatic models fail to reproduce the observed intensities and variations of molecular features. To address these problems, we have computed synthetic spectra based on dynamical atmospheres from Höfner et al. (1998; these proceedings) by calculating chemical abundances of molecular species, continuum absorption, and molecular opacities for a given temperature-pressure structure (COMA code). For carbon stars we can also include dust opacities. The results are then taken as input for a radiative transfer code.

In this paper we want to focus on the molecular features of oxygen-rich stars calculated from line lists with high spectral resolution. At present, data for SiO, CO, TiO, and H_2O are included. The results can be directly compared to our ISO observations of AGB stars, which cover the spectral range from 2.5 to 45 μm and show a large number of different molecular (SiO, CO, OH, CO_2, H_2O) and dust (silicate and corundum) features. Since some of these stars have been observed more than once, it is also possible to study their time-dependent behaviour. As it turns out, the variability of the molecular features can only be explained by the dynamical model atmospheres. In addition, the MARCS atmospheres produce a very strong SiO absorption, which does not agree with the observations. In contrast, we can easily get much weaker features with dynamical models because of their larger extension and spherical radiative transfer effects, and their different atmospheric structure (due to pulsation and shocks). Thus, the new method offers a possible explanation for the variable and relatively weak SiO bands of AGB stars.

Another important point is the effect of water lines. It turns out that for very cool stars (T<3000 K) the intensity of water absorption depends very strongly on temperature. Atmospheric extension also plays a role. Consequently, water opacity dominates the infrared spectrum, especially for very cool Miras. It should be mentioned that it is really necessary to include a very large number of water lines (more than 10 million) in order to reproduce the effects on the continuum correctly.

More information on this poster can be found at:
http://www.ast.univie.ac.at/ fzi/AGB/aringer/isospec.html

Variability of CO Line Profiles in AGB Stars

W. Windsteig, S. Höfner, B. Aringer, and E.A. Dorfi

Institut für Astronomie, Türkenschanzstr. 17, A-1180 Wien, Austria

The structure of the atmospheres of cool, pulsating stars differs considerably from hydrostatic or stationary states due to the presence of shock waves. Dynamic model calculations that include a detailed description of dust formation predict a more or less complex spatial structure of the circumstellar dust shell on relatively small scales, which is due to the complicated, time-dependent interactions between gas dynamics, dust formation, and radiation. The structure of the atmosphere obtained from dynamic models results in various observable features like spectral energy distributions (Windsteig et al. 1997), spatial brightness profiles, line profiles, and molecular spectra (Loidl et al. 1997). In this contribution we present synthetic CO line profiles based on the models of Höfner & Dorfi (1997).

This work is supported by the Austrian *Fonds zur Förderung der wissenschaftlichen Forschung* under project numbers S7305-AST, S7308-AST, and J1487-PHY.

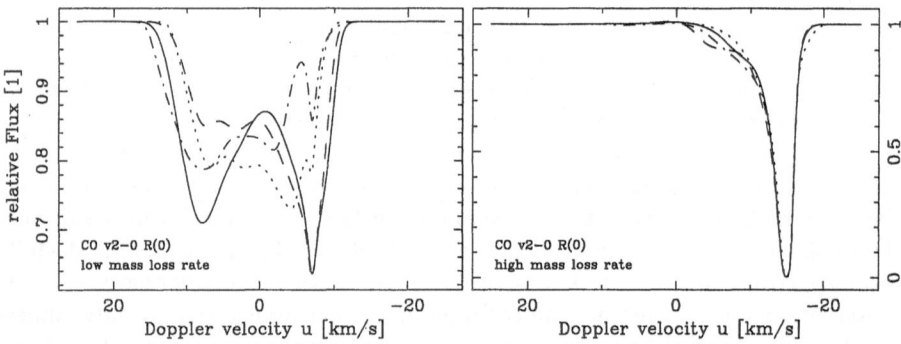

Fig. 1. Variablity of synthetic CO line profiles of the first overtone v2-0 R(0) at $2.34531\,\mu$m for models with a low ($5.8 \times 10^{-7}\,M_\odot\,\mathrm{yr}^{-1}$, left panel) and a high ($8.3 \times 10^{-5}\,M_\odot\,\mathrm{yr}^{-1}$, right panel) mass-loss rate: minimum (full line), ascent (dashed line), maximum (dotted line) and descent (dash-dotted line) of the light curve (envelope extension $\approx 1700\,R_\star$). Note the different scales of the two plots.

References

Höfner, S., Dorfi, E.A. (1997): A&A 319, 648
Loidl R., Hron J., Aringer B., et al. (1997): Ap&SS, in press
Windsteig, W., Dorfi, E.A., Höfner, et al. (1997): A&A 324, 616

Temporal Variations of CO Infrared Lines in Cool-Star Winds

Andreas Gauger[1], Jan Martin Winters[2], Axel Fleischer[2], and John J. Keady[3]

[1] MPI für Radioastronomie, Auf dem Hügel 69, 53121 Bonn, Germany
[2] Institut für Astronomie und Astrophysik, Technische Universität Berlin,
   PN8-1, Hardenbergstr. 36, 10623 Berlin, Germany
[3] Los Alamos National Laboratory, Theoretical Division, Group T–4,
   MS B212, Los Alamos, NM 87545, USA

High-resolution infrared spectroscopy of molecular lines provides a powerful diagnostic probe to investigate the spatial structure and temporal evolution of the cool, dusty winds of Miras and long–period variables, such as the carbon rich IR-Mira IRC+10216 (Keady et al. 1988). For this object, high-resolution spectra of the CO fundamental and first overtone transitions at 5μm and 2μm have been obtained repeatedly over an interval spanning more than ten years. The unsaturated overtone lines reflect the actual conditions and their temporal changes in the acceleration region of the wind. The line profiles show a multi-component absorption structure, and spectra from different epochs reveal changes of the line profiles (e.g., the emergence of a new absorption component) on time scales that are much longer than the period of the star (Sada 1993).

In order to model these observations, we have calculated the synthetic CO fundamental and first overtone line spectra for a dynamical model of the circumstellar dust shell around IRC+10216, which was obtained from the consistent treatment of time-dependent hydro- and thermodynamics, radiative transfer, chemistry, and carbon grain formation (Fleischer et al. 1992). Due to the interaction between interior stellar pulsation and grain condensation in the circumstellar shell, such wind models are characterized by an inhomogeneous, onion-like grain distribution, strong shocks accelerated by radiation pressure on dust, and cycle-to-cycle variations of the spatial structure (either periodic or non-periodic).

Our synthetic CO first overtone line profiles resemble the multi-component absorption structure and the time variations of the corresponding observed line spectra, which thus can be interpreted as a result of the time-dependent dynamics in the inner parts of the wind. In particular, the emergence and subsequent evolution of the low-velocity absorption feature is reproduced by the model and can be explained by the formation of a new dust layer and the subsequent acceleration of the matter by radiation pressure on dust.

References

Fleischer, A.J., Gauger, A., Sedlmayr, E. (1992): A&A 266, 321
Keady, J.J., Hall, D.N.B., Ridgway, S.T. (1988): ApJ 326, 832
Sada, P.A.V. (1993): PhD thesis, New Mexico State University

Cyclical Variability in MHD Disk Winds

Rachid Ouyed[1,2] and Ralph E. Pudritz[3]

[1] Dept. of Astronomy and Physics, St-Marys University,
Halifax, Nova Scotia, B3H 3C3 Canada
[2] CITA National Fellow, University of Toronto, Toronto, Ontario, M5S 1A1 Canada
[3] Dept. of Physics and Astronomy, McMaster University,
Hamilton, Ontario, L8S 4M1 Canada

Abstract. Jets in star forming regions are observed to be intrinsically time-dependent phenomena whose episodic eruptions lead to moving knots in the flow. We demonstrate using time-dependent magnetohydrodynamical simulations that magnetic fields anchored in a gaseous accretion disks orbiting a star can accelerate and collimate a bipolar jet which originates from the surface of the disk. Moreover, for certain magnetic field geometries, we find that episodic eruptions originate within the jet close to the central object. The link between the accretion disk and the wind variability is analysed.

1 Introduction

Episodic jets are observed in regions of star formation (O'Dell & Wen 1994). These jets have knots (bright regions that move along with the jet) that travel at speeds of $200-300$ km/s. For most jets, a pre-main sequence star or an embedded young stellar object (YSO) has been identified as the candidate source (e.g., Edwards et al. 1993). Furthermore, some of these infrared sources have been found to be X-ray sources as well, indicating significant magnetic activity (Feigelson et al. 1993). Whenever one observes a jet, there is good evidence that an accretion disk is also present; a fact that is probably not fortuitous. Current models for outflows invoke magnetic fields that thread Keplerian disks. They are of two types; (i) *hydromagnetic disk winds* wherein the engine consists of a Keplerian disk threaded by a magnetic field that is either generated in situ, or advected in from larger scales (e.g., Blandford & Payne 1982; Camenzid 1987; Lovelace et al. 1987; Heyvaerts & Norman 1989; Pelletier & Pudritz 1992; Li 1995; Appl & Camenzid 1993; Königl & Ruden 1993); or (ii) *X winds*, which are magnetized stellar winds where the interaction of a protostar's magnetosphere with a surrounding disk results in the opening of some of the magnetospheric field lines (Shu et al. 1987).

Perhaps the most important difference between these two classes of models lies in the role of the central object. For disk winds, only the depth of the gravitational well created by the central object is of any importance. The energy source for the flow is the gravitational energy release of material in the Kepler disk as the wind torque extracts its angular momentum. For X wind models on the other hand, the magnetization and structure of the central object is critical. Its magnetic field strength must be sufficient to carve a magnetosphere inside the disk and outflow requires that the magnetopause and co-rotation radii of the star are virtually identical; $R_m \simeq R_{co}$. Here, we consider winds emanating directly from the surface of Keplerian accretion disks surrounding young stellar objects.

2   Set Up and Simulations

Numerical simulations of disk winds by Ouyed et al. (1997) and Ouyed & Pudritz (1997a,b&c, hereafter OPI, OPII and OPIII) were run in order to see whether or not time-dependent calculations would yield jets that are truly episodic. The simulations have an initial state consisting of a central point mass, the surface of a surrounding Keplerian accretion disk (inner radius r_i), and a disk corona that is in exact (analytical and numerical) hydrostatic balance in the gravitational field of the central object and in pressure balance with the accretion disk below. The disk and corona is threaded by a magnetic field configuration chosen to have initial current $\mathbf{J} = 0$ so that no magnetic force is exerted initially upon the corona. Cylindrical coordinates (r,ϕ,z) are used.

One of the simplest magnetic configurations we have investigated consists of a constant uniform magnetic field that is parallel to the z-axis and perpendicular to the disk. This configuration was chosen because no outflow is expected in steady state theory. The models depend on 5 parameters; three prescribe the initial corona (ratio of gas to magnetic pressure (β_i), rotational to thermal energy density (δ_i), and the ratio of the disk density to the density of the base of the corona (η_i); all these measured at r_i), one gives the ratio of the toroidal to poloidal field strength in the disk (μ_i), and a final parameter measures the speed at which mass is injected from the disk into the base of the corona $(v_{inj.})$. This latter speed is taken to be a tiny fraction of the local Kepler speed. All lengths in our simulations are in units of r_i, and all times (τ) are in units of the Kepler time $t_i = r_i/v_{K,i}$ at the inner edge of the disk (for a $0.5M_\odot$ object, $r_i = 0.04$ AU, $v_{K,i} = 120$ km/s and $t_i = 0.6$ days).

We performed a series of 6 simulations (Table 1) with parameter values $(\delta_i, \beta_i, \eta_i, \mu_i) = (100.0, 1.0, 100.0, 1.0)$, except the injection speed is varied from a value of 0, to a maximum of $10^{-2}v_{K,i}$. The domain of simulation is $(z,r) = (20, 10)r_i$ with a resolution of $(500, 200)$ cells, run up to 50 time units. Table 1 demonstrates that for injection velocities $v_{inj.} \geq 5 \times 10^{-3}$, *stationary* outflows did indeed develop whose terminal speeds were $\simeq v_{K,i}$ while for lower injection speeds, the outflows were episodic and had terminal speeds and knot properties similar to those reported in OPII. In the simulation where we had *no* mass injection, a transient outflow composed of coronal gas developed which gradually died out as the initial coronal material in the jet zone was depleted.

Figure 1 (for $v_{inj.} = 10^{-3}$) shows that outflow occurs even for our initially uniform magnetic field configuration. The highly collimated, jet-like outflow is in this case dominated by a series of dense knots. The left panels show the poloidal magnetic field structure of the flow at four times during which a new knot is formed. The right panels show the opening angle of the magnetic field lines as a

Table 1. Effect of Varying the Injection Velocity v_{inj}^a (adapted from OPIII)

$v_{inj.}$	W^b	N_K^c	$z_{gen.}/r_i$	$\delta r_{wind}/r_i$	$\Delta z_{knot}/r_i$	$\Delta t_{knot}/t_i$	$\bar{v}_\infty/v_{K,i}$
0.0	3	1	4	—	—	—	0.3
10^{-4}	2	2	5	1	7	15	0.3
5×10^{-4}	2	3	7 – 8	3 – 4	5 – 7	11 – 13	0.5
10^{-3}	2	3	7 – 8	3 – 4	5 – 7	11 – 13	0.5
5×10^{-3}	1	0	—	10	—	—	0.8
10^{-2}	1	0	—	10	—	—	0.9

[a]The rest of the parameters are the standard ones. Namely, $\beta_i = 1.0$, $\delta_i = 100.0$, $\eta_i = 100.0$ and $\mu_i = 1.0$ (see OPI & OPII).
[b]Wind properties: $1 \equiv$ steady; $2 \equiv$ episodic; $3 \equiv$ transient.
[c]Number of knots: Total time of simulation $t_S = 50\ t_i$. The simulated region is $(r \times z) = (10 \times 20)r_i$ with a resolution of (200×500) cells.

function of their footpoint radius r_o on the surface of the accretion disk. One sees from these graphs that the field lines have been pushed open in a small region of the disk, making an angle of $50° - 60°$ with respect to the disk surface. These field lines have been opened up by the toroidal magnetic field pressure arising from the Keplerian rotation of each field line. Since Kepler rotation is faster at smaller radii, one expects that torsional waves introduce stronger toroidal field into the corona at smaller radii. This creates the radial gradient in toroidal field pressure that opens the field lines to less than the critical angle of $60°$ from the axis (Blandford & Payne 1982).

We found that knots are produced whenever the toroidal field in this inner region is sufficient to recollimate the newly accelerated gas back towards the outflow axis (see Ouyed et al. 1997). The gas necessarily speeds up. Because the gas is rotating, however, it encounters a centrifugal barrier at $r \geq r_o$ (r_o is the footpoint of a field line on the disk surface). As it reflects off of this barrier, it collides with the slower gas around it and shocks. The shocked gas regions move away from this generator region, and are kept coherent by strong enhancements of the toroidal field both ahead of it, and behind it. The knots, which are the overdense regions, have low toroidal field strengths, and conversely, the space between the knots is dominated by high toroidal field strength. The time scale for the passage of an Alfvén wave (in the toroidal field) from the jet radius towards the axis and back again, turns out to be precisely the knot generation time.

In general, the knots are produced periodically on a time scale of $t_{knot} \simeq 11 t_i$ (6.6 days for a typical $0.5 M_\odot$ object). Knots are produced in a generating region close to the central source; at a distance $z_{knot} \simeq 7 - 8 r_i$ (0.25 AU for a typical $0.5 M_\odot$ object). Knots continue to be produced for as long as we have run our simulations, up to 1000 time units, and so they are truly generic and are not transients.

Figure 2, for $v_{inj.} = 10^{-2}$, shows that a steady solution is also achieved. The

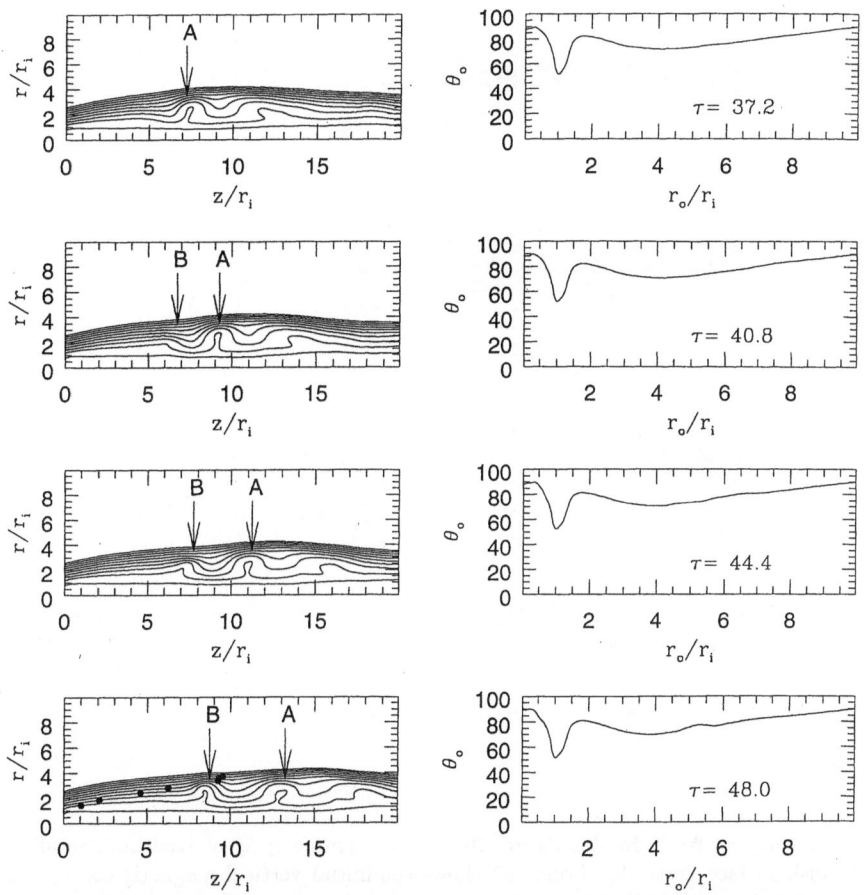

Fig. 1. $v_{inj.} = 10^{-3}$: In this figure, the axis of symmetry is plotted horizontally and the disk surface vertically. The left panels show the magnetic field structure of the knot generating region, at the four times; 37.2, 42.6, 44.0 and 48.0 inner time units. The right panels show the angle θ_o of field lines at the disk surface, at these times. Note the narrow band of field lines which is sufficiently opened ($\theta_o \leq 60^o$) so as to drive the outflow. Only field lines involved in the knot generation process are shown; field lines at larger disk radius stay reasonably vertical as seen in the right panels (from OPII).

position of the Alfvén and fast magnetosonic (FM) surfaces, where the outflow speed achieves the propagation speeds of two of the three important wave speeds in magnetized gas, are shown. The data are compared with the position of the Alfvén point on each field line in the simulation as predicted by steady-state theory (e.g., Blandford & Payne 1982). The agreement is very good. This and many other diagnostics (see OPI & OPIII) show that there is good agreement with steady-state disk-wind theory and our simulations.

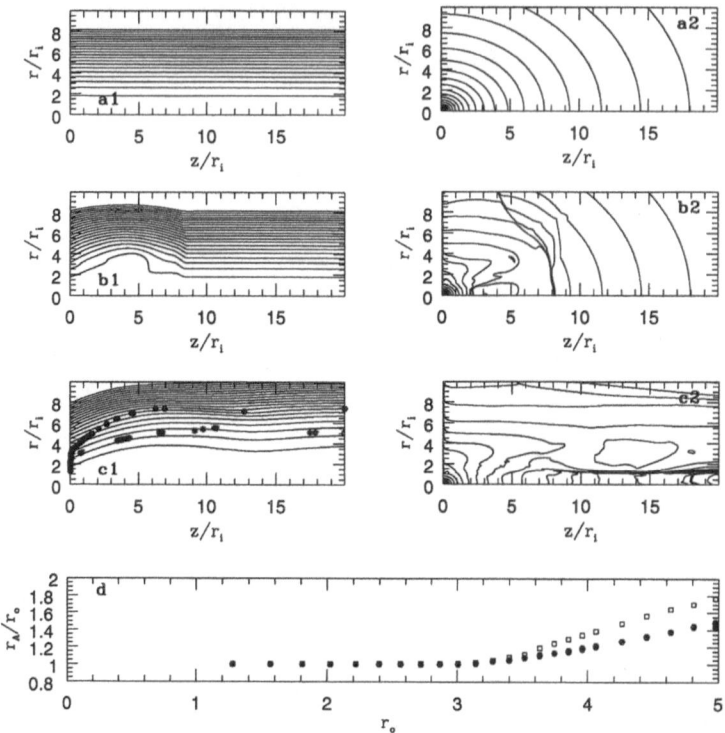

Fig. 2. $v_{inj.} = 10^{-2}$: In this figure, the axis of symmetry is plotted horizontally and the disk surface vertically. Frame a1 shows the initial vertical magnetic configuration; Frame a2 shows the initial isodensity contours of the corona. Frames b1&b2 and c1&c2 show the evolution of the initial magnetic and density structure at 75.0 and 150.0 inner time units. Frame c1 shows the location of the Alfvén critical surface (filled hexagons) and of the FM surface (stars). Frame d compares the Alfvén lever arm r_A/r_o (the footpoint of each field line on the disk is r_o) for each field line in the simulation (filled hexagons), with predictions of steady-state theory (squares) (from OPIII).

3   Discussion: Variability-Disk Connection

It is difficult to understand the complete physics that is responsible for the nature of disk winds. However, we find the nature of the flow (steady or episodic) to be controlled by the interplay between the collimating effects of the toroidal magnetic field and the flow's thrust. A general result can be stated as follows; for any given magnetic structure (or β distribution), one can find a threshold in the flow's injected thrust above which the twisting of the magnetic field is strong enough for the confining effects to control the dynamics (that is, and this is important, after the acceleration regime): the flow is then non-steady. For high injection thrust, we find the wind is accelerated by both the centrifugal and the magnetic force (∇B_ϕ^2) allowing for steady state to set in.

Steady-state theory indicates that recollimation of the flow should occur slightly beyond the FM surface at FM Mach numbers between 1.5 and 2 (e.g. Pelletier & Pudritz 1992). We are currently working to see whether this prediction can account for our discovery that episodic behaviour (which as we said is a consequence of magnetic recollimation) is strongly linked to the rate at which the disk moves gas into the corona. Clearly, one must ultimately remove the constraint of keeping the disk as a fixed boundary condition in the problem if one hopes to explore this idea by numerical simulation. Such a task requires computational resources beyond those available for this study.

For the moment, because our simulations have much smaller scales than can be directly imaged by HST and because the knots velocities are too small to match even the slowest structures seen in observed YSO jets, it is a logical next step to simulate larger scales; making direct contact with observations and studying the evolution of the knots far from the source (Ouyed & Pudritz 1998, in preparation).

A dynamic view of the simulations presented in the conference can be found at "http://www.physics.mcmaster.ca/Grad_Ouyed/ROuyed.html".

R.O. thanks the organizers of this most stimulating conference for the invitation to give this talk. R. O. also acknowledges the financial support and stimulating environment of McMaster University where this work was done. The research of R.O. is supported by St-Marys University and the Canadian Institute for Theoretical Astrophysics (CITA).

References

Appl, S., Camenzid, M. (1992): A&A, 256, 354

Blandford, R.D., Payne, D.R. (1982): MNRAS, 199, 883

Camenzind, M. (1987): A&A, 184, 341

Edwards, S., Ray, T.P., Mundt, R. (1993): in *Protostars and Planets III*, eds. E.H. Levy and J.I Lunine (Tucson: Univ. Arizona Press), 567

Feigelson, E.D., Casanova, S., Montmerle, T., Guibert, J. (1993): ApJ, 416, 623

Heyvaerts, J., Norman, C. (1989): ApJ, 347, 1055

Königl, A., Ruden, S.P. (1993): in *Protostars and Planets III*, eds. E.H. Levy and J.I. Lunine (Tucson: Univ. Arizona Press), 641

Li, Z-H. (1995): ApJ, 444, 848

Lovelace, R.V.E., Wang, J.C.L., Sulkanen, M.E. (1987): ApJ, 315, 504

O'Dell, C.R., Wen, Z. (1994): ApJ, 436, 194

Ouyed, R., Pudritz, R.E. (1997a): ApJ, 482, 712 (OPI)

Ouyed, R., Pudritz, R.E. (1997b): ApJ, 484, 794 (OPII)

Ouyed, R., Pudritz, R.E. (1997c): ApJ, submitted (OPIII)

Ouyed, R., Pudritz, R.E., Stone, J.M. (1997): Nature, 385, 409

Pelletier, G., Pudritz, R.E. (1992): ApJ, 394, 117

Shu, F.H., Adams, F.C., Lizano, S. (1987): ARA&A, 25, 23

Accretion Mass-Loss Relation and Variability in T Tauri Stars

Erik Gullbring

Harvard-Smithsonian Center for Astrophysics,
60 Garden St., Cambridge, MA 02138, USA

Abstract. We discuss the different line-emission regions in T Tauri stars and summarize their properties. Results from modeling of emission line profiles and fluxes from the accreting gas and from shock-excited regions close to the stellar surface show that the emission lines originate in the accretion flow. The wind reveals itself, in the permitted lines, as a blue-shifted absorption component. We discuss, in particular, aspects in separating the line contribution from the infall and from the wind. In regard to this, we emphasize the importance in using line-variability to disentangle the wind and infall regions and to gain further understanding in the complex boundary between the in- and outflows around T Tauri stars. This is illustrated by presenting observations of variability in both the infall-emission and wind-absorption line regions.

1   Introduction

Due to the general predominance of normal P Cygni profiles among T Tauri stars (TTS), wind models were generally adopted to explain the origin of the line-emission. Although the wind calculations put forward during the last 10 − 15 years succeeded in reproducing the line luminosities, the line profiles were not successfully predicted. In particular, self-consistent wind models failed to reproduce 1) the observed sharply peaked profiles, 2) the low-velocity blue-shifted absorption components, in the case of accelerating wind models, 3) the decreasing depths of the blue-shifted absorption with higher Balmer lines. In the last few years it was realized that infall through a stellar magnetosphere provides the velocity field needed to account for the observed emission line profiles (Calvet & Hartmann 1992; Hartmann et al. 1994; Muzerolle et al. 1997; Calvet 1997 and references therein).

2   General Properties of Classical T Tauri Stars

In the magnetospheric picture, the disk material is elevated from the disk plane by the magnetic field lines at a few stellar radii. The accreting material then flows along the magnetospheric field lines and eventually goes through a strong shock with peak temperatures of several 10^6 K, before settling down on the stellar surface. A strong support for the infall model was provided by the observations of Edwards et al. (1994) showed that ~87% of the accreting TTS show evidence of inverse P-Cygni profiles. Thus, for TTS, the inflowing gas in the magnetosphere,

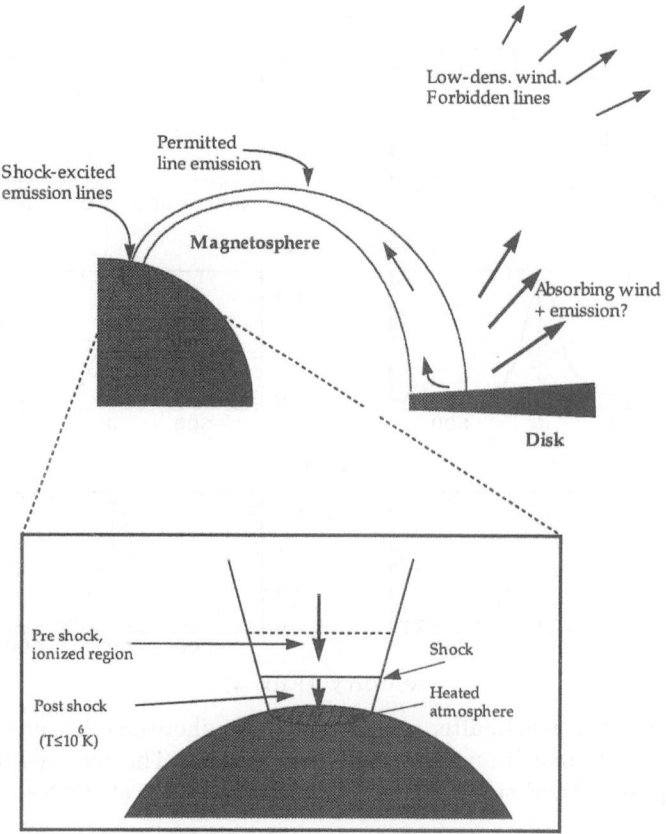

Fig. 1. Illustration of an accreting T Tauri star

and the shock-heated region close to the stellar surface provide the line-emission, while a dense wind close to the star produces blue-shifted absorption. A low-density wind is normally traced by forbidden-line emission (Fig. 1).

3   Predictions of Emission-line Properties

3.1   Calculated Emission Line Profiles from the Magnetosphere

Calvet & Hartmann (1992) and Hartmann et al. (1994) calculated Balmer line profiles from infall using two-level model atoms, and showed that the general characteristics of TTS Balmer line profiles were recovered. Also, given that the bulk of the line-emission comes from the magnetosphere, it is straight forward to explain the decrease in the depth of the blue-shifted absorption components with higher Balmer lines, since it just reveals the relative changes between the source functions in the magnetosphere and in the wind.

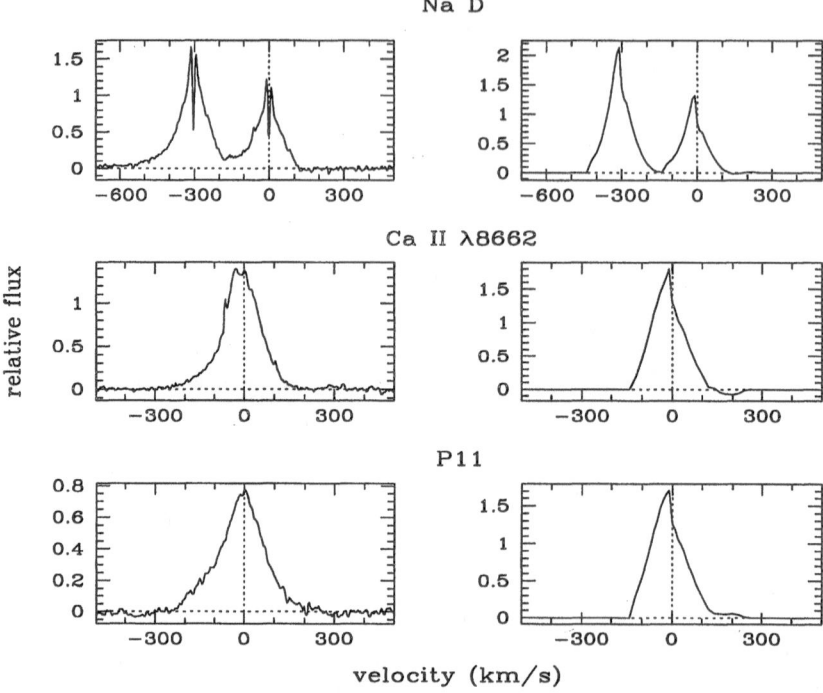

Fig. 2. Observed (left panel), after subtraction of the photospheric contribution, and calculated (right panel) magnetospheric line profiles. The observed lines are from the classical T Tauri star BP Tau. (From Muzerolle et al. 1998.)

Recently, a refined magnetospheric model, using multi-level atoms, has been calculated by Muzerolle et al. (1997) and was used to predict profiles and line luminosities for a wide range of hydrogen lines. The model has also been applied to other elements, and successfully reproduces the observed lines of, for instance, Na I, Ca II and He II. In Fig. 2, the results of these calculations are presented. Comparison with the predicted line profiles with observed profiles, show a convincing similarity between the calculated and observed profiles over a range of elements and ionization stages.

3.2 Shock Excited Line Emission

Many emission lines from T Tauri stars show a narrow feature, centered around the line central wavelength, superimposed on a broad component (see Fig. 3). In addition to the broad emission lines, as discussed in Sect. 3.1, the magnetospheric model predicts the existence of narrow components: before the gas hits the stellar surface it will go through a standing shock after which the gas will have a much lower infall velocity than before the shock. This region should provide a temperature-density environment in which a manifold of emission lines can

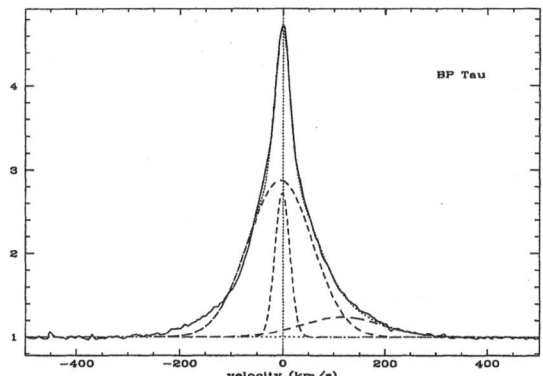

Fig. 3. A Ca II (8542Å) line decomposed into a broad and a narrow component. The red-shifted broad component (long-dashed line) shows the contribution of a Paschen line.

form. To investigate this in detail, we developed a shock-model of this region (Calvet & Gullbring 1998; Gullbring & Calvet 1998 in preparation). In this model we numerically solved for the shock structure and calculated the total emission from three regions: the post-shock, in which the gas settles down on the stellar surface; the underlying stellar atmosphere, which is heated by the shock radiation; and the pre-shock region, heated by the upwardly directed shock-emission (see Fig. 1). This model shows that, indeed, both the line luminosity as well as the low radial velocity (at most a few km/s) of the narrow emission lines were recovered. In addition, the model predicts the spectral energy distribution of the UV-continuum emission of TTS.

4   Line Variability

As discussed in Sect. 2, the wind reveals itself in the permitted lines as blue-shifted absorption features, and in the forbidden lines as asymmetric profiles with most of the emission at blue velocities. The forbidden line profiles were early on interpreted in terms of a circumstellar disk obscuring the receding part of the wind so that only the part approaching the observer is seen (Appenzeller et al. 1984; Edwards et al. 1987). As discussed by Hartmann (this volume), there is evidence that the wind, traced by the forbidden lines, is powered by the accretion.

Both the gas flow in the magnetosphere and in the wind are likely to suffer from a range of instabilities, producing inhomogeneous flow patterns. These instabilities should be observable as time-dependent changes in the emission lines. To illustrate this, the left panel in Fig. 4 shows a large set of Hα line profiles of the CTTS BP Tauri. These lines were sampled at a time-resolution ranging from

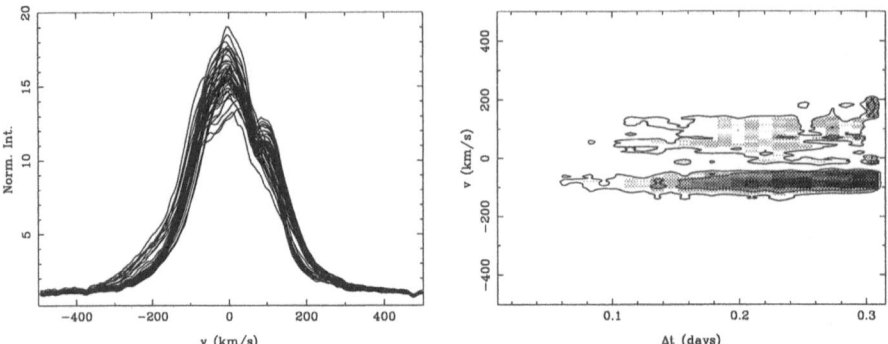

Fig. 4. Left panel: Hα line-profiles of BP Tau collected at time-resolutions from 20 minutes to several days. Right panel: the amount of changes in line-profile as a function of line velocity and duration. (From Gullbring et al. 1996)

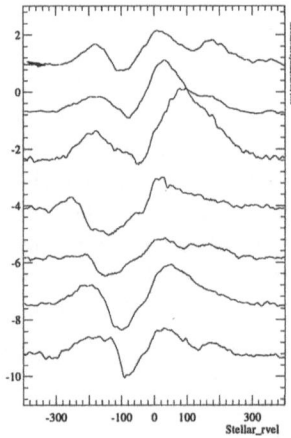

Fig. 5. A series of Hα line profiles of SU Aur, collected at one day intervals (with increasing time downwards). Note, in the last four spectra, how a blue-shifted absorption component appears at $v \sim -200$ km/s and then gradually moves towards lower velocities. The same behavior are seen at other epochs. (From Petrov et al. 1996)

~ 20 minutes to a few days, and it is clearly seen that the line shows variability, to some degree, both in the shape of the line-profiles and in the equivalent width (if the change in the equivalent width is caused by a variable line strength or the level of the underlying continuum emission is not known, however). In general, these changes are not large enough as to completely alter the appearance of the line, but acts mostly as a small-scale variability superimposed on an otherwise stable line-profile. This suggests that the magnetospheric configuration is essentially stable, but contains gas flows with a certain degree of inhomogeneity.

What this means in terms of temperature and/or density fluctuations needs to be modeled in detail. It is interesting to note (right panel of Fig. 4), that significant variations are present on time-scales less than the infall time scale through a magnetosphere ($\sim 2 - 3$ hours). Therefore, in principle, it should be possible to monitor the change of a given (or several) emission lines as the gas flows through the magnetosphere down to the central star.

The wind components in the line profiles (i.e. the blue-shifted absorption features) also show variability, although in general on a different time-scale. As an illustration, we show in Fig. 5 a set of Balmer line profiles of the CTTS SU Aur (for a thorough discussion on this star see the chapters by Johns-Krull and by Unruh). The Hα spectra in Fig. 5, collected at one day intervals, indicate that parts of the wind-absorption show a drift in radial velocity with time: during a few days the feature moves from ~ 150 km/s to a few tens km/s, after which a new higher-velocity component seems appear. This behavior can be interpreted in two ways: either as decelerating gas shells (Petrov et al. 1996) or as large rotating structures coupled to the stellar rotation.

I would like to thank the organizing committee for financial support, Lee Hartmann and Nuria Calvet for valuable discussions, and James Muzerolle for providing the figs. 2 and 3. This work was partly financed by the Swedish Natural Science Research Council.

References

Appenzeller I., Östreicher R., Jankovics I. (1984): A&A 141, 108

Calvet, N. (1997): in Proc. IAU Symp. 182, eds. B. Reipurth & C. Bertout (Dordrecht: Kluwer), p. 417

Calvet N., Hartmann L., Hewett R. (1992): ApJ 386, 229

Edwards S., Cabrit S., Strom S.E., et al. (1987): ApJ 321, 473

Edwards, S., Hartigan, P., Ghandour, L., Andralis, C. (1994): AJ, 108, 1056

Gullbring, E., Hartmann, L., Briceno, C., Calvet, N. (1997): ApJ, in press

Gullbring, E., Petrov, P.P., Ilyin, I., et al. (1996): A&A, 314, 835

Hartigan, P., Edwards, S., Ghandour, L. (1995): ApJ, 452, 736

Hartmann L., Hewett R., Calvet N. (1994): ApJ 426, 669

Muzerolle, J., Calvet, N., & Hartmann, L. (1997): ApJ, in press

Muzerolle, J., Calvet, N., & Hartmann, L. (1998): in preparation

Petrov, P.P., Gullbring, E., Ilyin I., et al. (1996): A&A, 314, 821

Ignace: Your paradigm for moving material from the disk to the star seemed to be a magnetic flux tube. Why not an axisymmetric sheet? Such a sheet would yield a ring of shocked material at the star instead of a spot.

Gullbring: This is really what it is. It's just my poor cartoon that makes it appear as a flux-tube. The real picture is a magnetosphere that is azimuthally symmetric, thus resembling a sheet.

Johns-Krull: Regarding the magnetospheric accretion model: because we observe such dynamic variability on nightly time scales and the loss of phase coherence in the stars that show periodicity, these stellar magnetospheres must be very dynamic, evolving structures. The dipole models are a nice theoretical starting point, but they should not be taken too literally.

Gullbring: This was the point I made by showing the short-term profile variability of the Hα line in BP Tauri: it shows considerable small-scale variations on many time scales, without completely changing the appearance of the line. This suggests that the emitting magnetosphere is overall unstable in some sense but, at the same time, preserves its basic configuration.

Linsky: The spectrum that you showed of the Balmer emission lines and continuum from a T Tauri star with accretion looks very similar to that of an M dwarf star during a flare. Thus a bright Balmer emission line spectrum may not be a unique indicator of an accretion disk.

Gullbring: It may appear similar but the excess continuum is for sure related to accretion: it is only for T Tauri stars where you know they have accretion disks that you see the excess emission. T Tauri stars without disks show neither continuum nor line excess emission.

The Wind of FU Orionis:
Modelling the Atmosphere

G. D'Angelo[1], L. Errico[2], M.T. Gomez[2], L.A. Smaldone[1], M. Teodorani[2]
and A.A. Vittone[2]

[1] Dipartimento di Scienze Fisiche, Mostra d'Oltremare Pad. 19, I-80125 Napoli, Italy
[2] Osservatorio Astronomico di Capodimonte, Via Moiariello 16, I-80129 Napoli, Italy

The FU Orionis objects are a class of accreting pre-main sequence stars that undergo dramatic outbursts in their brightness. They also show signatures of very strong winds, as indicated by the P Cygni profiles of Balmer lines and the complex blue-shifted absorptions of the Na I resonance lines. These winds carry matter at rates several times larger than those estimated for T Tauri stars.

We computed synthetic line profiles for $H\alpha$ and the Na I D lines of FU Orionis by using the non-LTE MULTI code (Carlsson, Rep. No. 33, Uppsala Astr. Observatory, 1986). We assume that the optical spectrum of FU Ori comes from an accretion disk atmosphere. The atmospheric structure we chose for this object is similar to the one proposed by Croswell et al. (1987, ApJ. 312, 227) except that we use photospheric models by Kurucz. The most prominent part of our atmospheric models is a plane-parallel, quasi-isothermal, low temperature wind. Between these two regions we suppose the existence of a thin chromospheric layer in which the temperature grows.

We found that the emission component of the $H\alpha$ profile is very sensitive to the temperature gradient and the maximum temperature reached in this layer. In contrast, the blue absorption component of the profile depends only on the velocity field of the wind. We also noticed that the transition between absorption and emission is produced where the chromosphere ends and the wind begins.

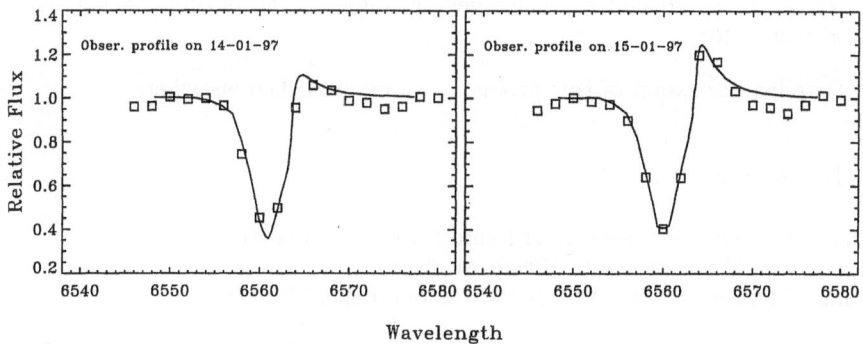

Fig. 1. Observed $H\alpha$ profiles for FU Ori (squares) on two nights. Synthetic profiles (solid lines) refer to chromospheric temperatures of 8000 K with $v_\infty = 400\,\mathrm{km\,s^{-1}}$ (left panel) and 11000 K with $v_\infty = 500\,\mathrm{km\,s^{-1}}$ (right panel).

Numerical Modeling of Anisotropic Stellar Winds in Herbig Ae/Be Stars

Larisa Tambovtseva[1] and Vladimir Grinin[2]

[1] Central Astronomical Observatory Pulkovo, St. Petersburg, Russia
[2] Astronomical Institute of SPb University, St. Petersburg, Russia

According to the results of statistical analyses by Grinin & Rostopchina (1996), the variety of Hα line profiles observed in the spectra of HAEBE stars can be described in the framework of a model that includes an accretion disk and an axisymmetric outflow viewed at different inclination angles to the line of sight. Here we test a simplified model of this kind by using non-LTE calculations for hydrogen and helium lines based on the Ferland (1993) and Grinin & Mitskevich (1990) numerical codes.

We assume for simplicity that the accretion disk is flat. The model parameters are: the geometrical thickness h and temperature T_d of the disk, the mass accretion rate $\dot{M}_a$, the wind temperature T_w, the mass-loss rate $\dot{M}_w$, and the wind opening angle, 2ϕ. The velocity law for the accretion disk was taken to be $u(r) \sim v(r) \sim r^{-1/2}$, while the velocity law for the wind was $u(r) \sim r^{-1}$ and $v(r) \sim r^{-1/2}$. A few tens of different combinations of parameters were considered and the preliminary results can be summarized as follows:

1. The He I $\lambda5876$ line arises in the hot ($T_e = 17,000 - 20,000$ K), compact region of the accretion disk near the stellar surface and has a small (0.1–0.2) covering factor. The rotation velocity in this region is much smaller than the corresponding Keplerian velocity.

2. The two-component and P Cygni line profiles of Hα, Hβ, and Hγ (and their intensities) can be explained by different inclinations of the gaseous envelope with respect to the line of sight for $\dot{M}_a = 10^{-7} - 10^{-8}$ M$_\odot$, $\dot{M}_w \leq \dot{M}_a$ and $T_w \approx T_a \approx 10,000$ K. The opening angle of the outflow varied within: $\phi = 20 - 40°$.

Detailed discussion of this problem will be published elsewhere.

References

Ferland, G.J. (1993): University of Kentucky, Internal Report
Grinin, V.P., Mitskevich, A.S. (1990): Afz 35, 61
Grinin, V.P., Rostopchina A.N. (1996): Astron. Report 40, 171

Modelling Variability in Hot-Star Winds

Stanley P. Owocki

Bartol Research Institute, University of Delaware, Newark, DE 19716, USA

Abstract. I review 2-D hydrodynamical simulations of rotating hot-star winds with azimuthal structure induced by modulation of the radiative driving force near the wind base. As a first step toward examining more realistic perturbation mechanisms (e.g., nonradial pulsations, or magnetic fields), the driving modulation here is taken to arise from bright and dark spots in the stellar photosphere. These spots induce decreases or increases in wind flow speed, and as the star rotates, spiral "Co-Rotating Interaction Regions" (CIRs) form, much as in the solar wind, from interaction between fast and slow flow streams. A new feature unique to line-driven flow is a velocity-gradient kink that propagates inward from interaction fronts at a fast radiative-acoustic mode speed. The slowly evolving velocity plateaus that form behind such kinks give rise to absorption features with a slow apparent acceleration, much like the Discrete Absorption Components (DACs) often observed in UV wind lines from hot stars. In simulation models with base driving sinusoidally modulated between increase and decrease, there arise alternating spiral streams of enhanced or decreased density, associated respectively with decreased or increased flow speeds. These speed variations have substantial impact on the line profile, and so these dynamical simulation are not as successful as analogous kinematic models of corotating density streams in reproducing the "phase-bowing" of periodic absorption modulations observed in the recent IUE 'Mega' project.

1 Introduction

Several papers at this meeting (e.g. by Prinja, Kaper, Kaufer, Massa, Henrichs, and Wolf) have summarized the extensive observational evidence for explicit variability, cyclical or otherwise, in the winds from hot, luminous, early-type (OB) stars. A general challenge for theory is to understand the nature of the physical changes and/or structures associated with this variability. To be visible as direct variations in line-profiles formed from globally integrated radiative flux, the associated flow structure must be on a relatively large scale, on order the stellar radius. This makes it unlikely that such variations could stem from processes entirely *intrinsic* to the wind outflow, such as the inherent instability in the line-driving of the wind to small-scale perturbations (Rybicki 1987; Owocki et al. 1988; Owocki 1994b). The dynamical evolution of such large-scale structure can be simulated using the local, computationally efficient, CAK/Sobolev expressions for the line force (Castor et al. 1975; Sobolev 1960), making it feasible to carry out multidimensional simulations of the wind structure resulting from large-scale perturbations from the underlying, rotating star.

In this review, I will focus on recent efforts to develop initial dynamical models for two distinct classes of such variability, namely the 'classical' *Discrete*

Absorption Components (DACs) and the more recently identified *Periodic Absorption Modulations* (PAMs) discovered in the IUE 'Mega' project (Massa et al. 1995). Unlike the quasi-episodic, slowly evolving, net absorption enhancements that characterize most DACs, the PAMs recur regularly at periods a low order fraction of the rotation period, include both reductions and enhancements of absorption, and evolve relatively quickly over the line profile. Indeed, in contrast to the slow blueward evolution of DACs, the PAMs in one case (B0 I star HD 64760) show a "phase-bowing" that reflects apparent redward as well as blueward propagation (Owocki et al. 1995; Fullerton et al. 1997). These distinct observational characteristics likely reflect differences in the underlying perturbation mechanisms, perhaps, for example, with the more stochastic DACs being induced by magnetic activity, and the regular PAMs being initiated by Non-Radial Pulsations (NRPs). But given the present uncertainty, the initial simulations here simply induce wind variations rather artificially, through direct modification of the radiative driving in the inner wind, much as might occur from "spots" on the underlying star (Cranmer & Owocki 1996, hereafter CO).

I first (§2) describe the effect of *isolated* spots, both brighter and darker than the ambient photosphere. As the star rotates, *Co-rotating Interaction Regions* (CIRs) form along spiral patterns by collision between faster and slower wind streams originating from different longitudes relative to the spot. These simulations thus represent the first dynamical test of the original proposal by Mullan (1984a,b; 1986) that the wind density enhancements in such CIRs could cause the DACs. A central goal here is to determine whether key characteristics of observed DACs, particularly their apparent slow acceleration, can be reproduced in synthetic line-profiles generated from dynamical models with CIRs.

I also examine (§3) the effect of a sinusoidal *modulation* of the radiative driving near the wind base, assuming a fixed number ($m = 4$) of nodes around the star. This is intended as a dynamical version of the simple kinematic picture proposed to explain the "phase bowing" of the PAMs in HD 64760 (Owocki et al. 1995). In this picture the PAMs arise from corotating streams of alternating increased or decreased density, within the key simplification that the velocity is fixed to a specific law, unaffected by the density perturbations. The dynamical simulations here self-consistently include such velocity variations, and so allow us to examine their effect in the line-formation.

2   CIRs Induced by Isolated Spots

Let us first review the CO simulations of wind structure induced by isolated spots on a rotating hot star. The aim is to mimic physical processes – e.g. magnetic field, NRPs – that might increase or decrease the mass flux emerging from some localized region of the star, and then study how such variations are propagated through the radiatively driven stellar wind. To reduce the complexity and computational costs, the simulations are confined to 2-D variations in radius and azimuth (r, ϕ) within the equatorial plane. The line-driving force is computed using the standard CAK/Sobolev formalism (Castor et al. 1975; Sobolev 1960),

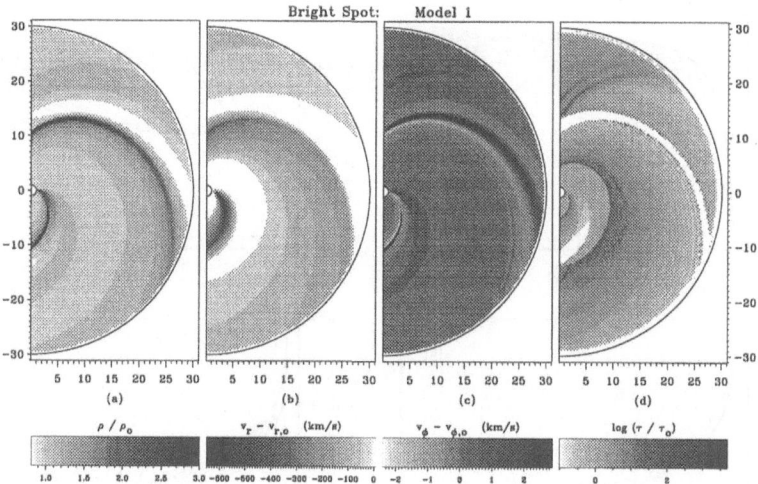

Fig. 1. Grayscale polar plots showing the spatial dependence of the deviations from a steady model for a. density, b. radial velocity, c. azimuthal velocity, and d. Sobolev optical depth.

modified by the finite-disk correction factor for a spherical outflow (Friend & Abbott 1986; Pauldrach et al. 1986). This ignores the azimuthal force that should arise from sideview perspectives of the spot, and simply modifies the radial force by a fixed enhancement/reduction factor determined by the relative proximity to the spot.

The spot parameterization allows specification of both a longitudinal width Φ and an amplitude A, with $-1 < A < 0$ for dark spots and $A > 0$ for bright spots. Of the models listed in Table 1 of CO, I confine attention here to the first two, for which $\Phi = 20^o$ and $A = \pm 0.5$, representing a standard bright and dark spot. The stellar parameters represent those for the canonical O-type supergiant ζ Puppis, with a rotation speed $V_{rot} = 230$ km/s, corresponding to a rotation period of $P \approx 4.2$ days. For convenience, let us assume two identical spots positioned on opposite hemispheres, allowing restriction of the computation to an azimuth range of just 180° with periodic boundary conditions.

For the bright spot case, Figure 1 shows grayscale plots of the resulting density, radial velocity, azimuthal velocity, and radial Sobolev optical depth, all relative to values in the corresponding steady, spherically symmetric, CAK wind model without any spot. The spot here is centered on the x-axis, at the origin of the spiral pattern of enhanced density that characterizes the resulting CIR. Note that regions of enhanced density generally correspond to regions of lower radial velocity. Because of the enhanced brightness over the spot, the wind driven from there initially has a higher mass flux, and thus higher density. As the enhanced brightness fades with increasing height above the spot, the higher density of this

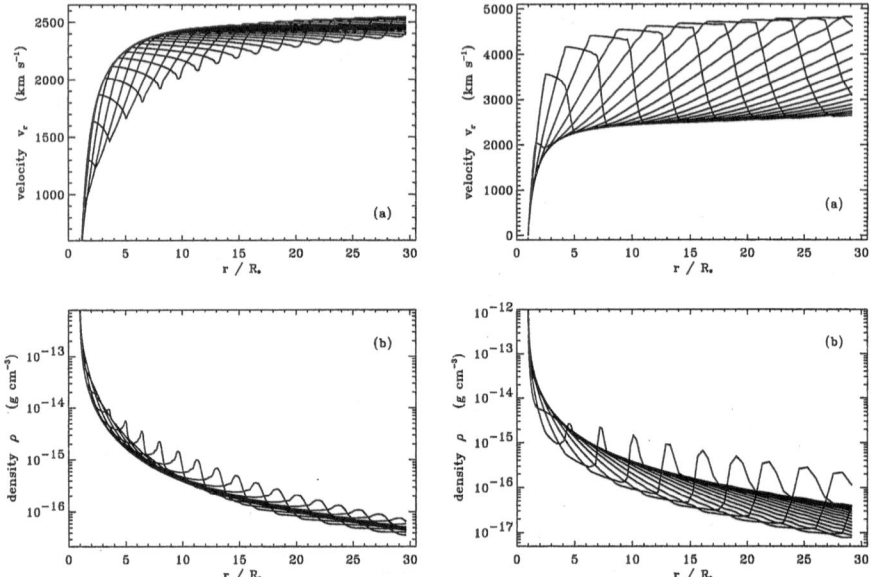

Fig. 2. Left panels: Radial variation of velocity (upper panel) and density (lower panel) along selected, fixed azimuthal angles in the bright spot model.

Fig. 3. Right panels: Same as Fig. 2, but for the dark spot model.

material makes it harder to accelerate, reflecting the characteristic scaling of the line acceleration with the inverse of the density ρ,

$$g_{lines} \sim \left[\frac{1}{\rho} \frac{\partial v_r}{\partial r} \right]^{\alpha} \tag{1}$$

where α is the usual CAK exponent. The lower velocity gradient $\partial v_r / \partial r$ associated with the lower acceleration also contributes, through Eq. (1), to a further reduction in the acceleration. As the stellar rotation brings this higher-density, lower-speed material into interaction with faster ambient wind originating away from the spot, there results a shock compression of the gas into the dense spiral stream that characterizes the CIR.

Though the general CIR formation here is analogous to that occuring in the solar wind (Hundhausen 1972; Zirker 1977), there are important differences. The much higher density of hot-star winds makes radiative cooling very efficient, and so the shocks are effectively isothermal. Unlike the nearly adiabatic shocks in the solar wind, which have a maximum compression ratio of four in the strong shock limit, the density compression in such isothermal shocks scales with the square of the Mach number, and so can be arbitrarily large. Moreover, whereas the nearly adiabatic shocks of the solar wind propagate backward into the faster wind at roughly a quarter of the shock speed, these isothermal shocks remain in close proximity to the compressed layer.

Another significant difference arises when we consider the nature of the *pre-shock* flow. In ordinary gas-dynamics (or even MHD) shocks, the preshock flow is assumed to be completely unaffected by the impending shock, simply because the speed of this material exceeds the sound speed (or in MHD, the fast-mode speed), preventing any upwind information propagation that a disturbance lies ahead. However, as first shown by Abbott (1980), the characteristic speeds $c_\pm$ in a *line-driven* flow are substantially modified by the operation of the line force, given by

$$c_\pm = -U/2 \pm \sqrt{(U/2)^2 + a^2}, \tag{2}$$

where a is the ordinary (isothermal) sound speed, and $U \equiv \partial g_{lines}/\partial(\partial v_r/\partial r)$ is a new characteristic speed, nowadays often called the "Abbott speed". From Eq. (1) we find $U \sim g_{lines}/(\partial v_r/\partial r)$, but from the radial equation of motion, we expect $v_r \partial v_r/\partial r \sim g_r$, which suggests $U \sim v_r$, i.e. this Abbott wave speed is typically of order the radial flow speed. Since most of the wind is supersonic, we find $v_r \approx U \gg a$, implying that *inward* wave propagation can now occur at a speed $c_- \approx -U \approx -v_r$ that is much *faster* than the propagation of an ordinary sound wave (Owocki & Rybicki 1985, 1986; Rybicki et al. 1990).

Figure 2 shows line plots of the radial variation of (a) velocity and (b) density along selected, fixed azimuthal angles. The location of the compressed CIR is apparent from the velocity minima and density maxima. But note that the velocity has a *decreasing* outward gradient over a broad region ahead of this CIR, extending back to a velocity gradient discontinuity, or "kink", that marks the connection to the unperturbed, outward-accelerating wind. This weak, kink discontinuity propagates inward (relative to the wind outflow) at roughly the characteristic speed $c_- \approx -U$, and since this is nearly as fast as the local outflow speed, it has only a slow net outward propagation in the fixed stellar frame.

Such fast inward-propagation of a velocity-gradient kink is a novel and quite unique feature of a line-driven flow, and it has potentially important consequences for interpreting observed line-profile features, particularly DACs. To identify which wind structures should yield the most prominent profile variations, let us examine the radial Sobolev optical depth,

$$\tau_r(r, \phi) = \frac{\kappa v_{th} \rho(r, \phi)}{|\partial v_r/\partial r|}. \tag{3}$$

The line absorption coefficient κ and the ion thermal speed v_{th} are assumed to have the same, constant value in the perturbed and unperturbed wind. Thus the variations in optical-depth ratio plotted in Figure 1d reflect changes in the ratio of density over velocity-gradient. Surprisingly, we see from comparison with Figure 1a that the regions of strongest optical-depth enhancements do not occur within the dense CIR compression, but rather within the shallow velocity-gradient region immediately after the kink.

The emergent line profiles from this wind structure can be readily synthesized using the standard SEI (Sobolev source function with Exact Integration) method of Lamers et al. (1987), generalized for 3-D integration as described in CO. For this purpose, the 3-D latitudinal structure is derived by interpolation from the

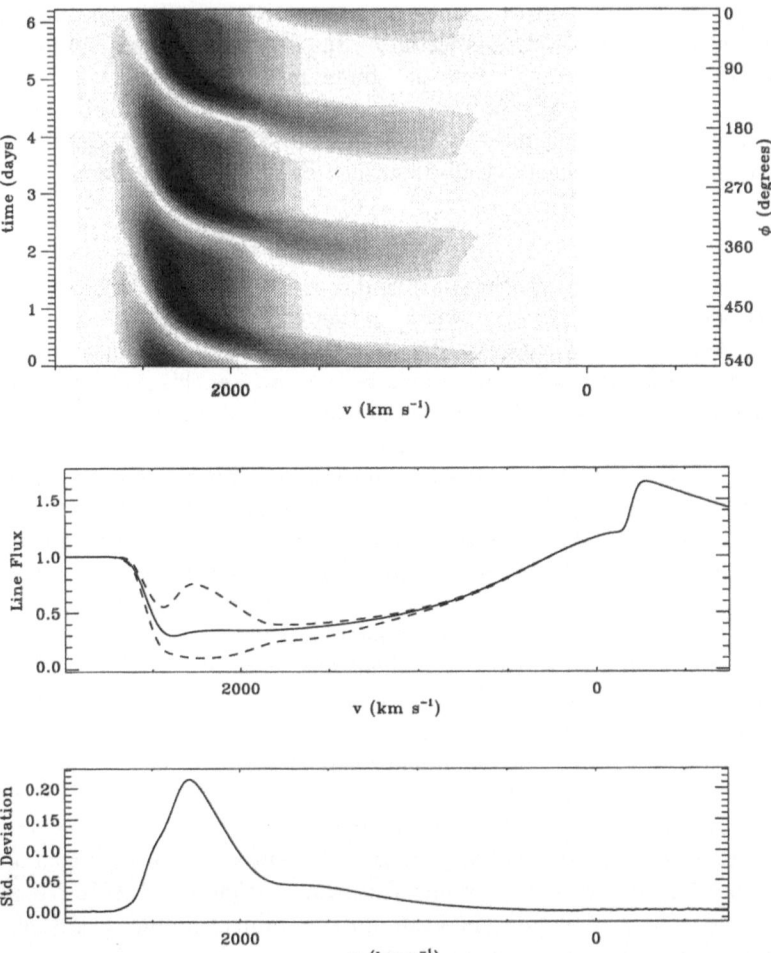

Fig. 4. Line profile variation for marginally thick line as a function velocity and time. In the middle box, the solid line shows the unperturbed wind profile, and the dashed lines show the extrema for each velocity. The bottom box shows the profile standard deviation.

equator to a polar model characterized by a 1-D, unperturbed wind solution. For a moderate strength line that is marginally optically thick in the wind, Figure 4 shows a grayscale plot of the profile variation as a function of frequency (in velocity units) and time. The latter is derived from simple mapping of the azimuthal coordinate ϕ using the star's rotation frequency Ω, i.e. $t = (\phi - \phi_o)/\Omega$, where ϕ_o is defined such that the observer is directly over the spot at the initial time $t = 0$. Following observational convention, the gray scale here is scaled to the range from minimum to maximum flux at each frequency, with the maximum flux

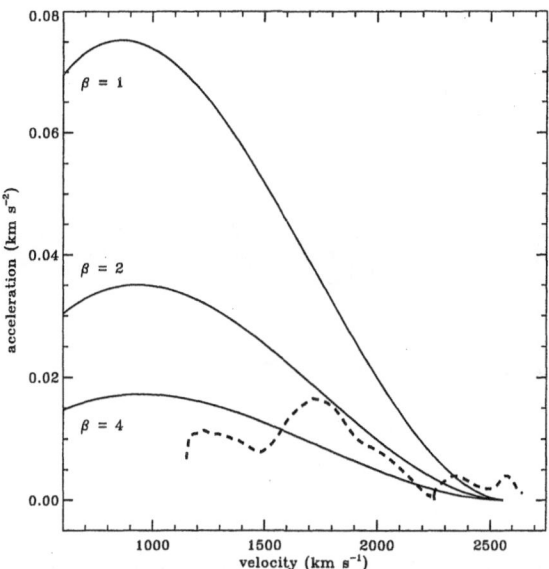

Fig. 5. Apparent acceleration vs. velocity for bright-spot model DAC, compared to velocity laws with various exponents β.

(i.e., minimum absorption) corresponding to the lightest shade. This convention is suitable for describing the observed DACs that represent mainly enhanced absorption relative to the background wind. But the synthetic profile variations here also show intervals of *reduced* absorption, reflecting the relative reduction of the density in the region just beyond the CIR.

As noted above, the absorption enhancements here arise primarily from regions of shallow velocity-gradient near the reverse mode kinks. Because of the slow net outward propagation of these kinks, the associated synthetic DACs migrate quite *slowly* blueward across the profile over roughly ~ 3.9 day, in general agreement with the slow apparent acceleration of observed DACs. Often this slow acceleration is characterized in terms of large values of the exponent β in the canonical velocity law given by $v(r) = v_{\infty}(1 - R_*/r)^{\beta}$. Figure 5 plots the apparent acceleration vs. velocity for the synthetic DAC compared with results for velocity laws with various β. Whereas an unperturbed wind is expected to have a velocity law with $\beta \approx 1$, Figure 5 shows that the DAC in this model corresponds more closely to higher exponents $\beta \approx 2 - 4$, much as inferred for observed DACs.

Let us next briefly consider results for the case of a "dark" spot, wherein the radiative driving is reduced at the base of the wind outflow. In this case, the mass flux emerging from the spot region is now reduced, and as the radiative driving recovers at greater heights above the spot, the reduced density causes an enhanced acceleration, which then results in a very fast stream. Figure 3 shows the radial variation of velocity and density at selected fixed angles. Note the

Density   Velocity   Optical
                                       Depth

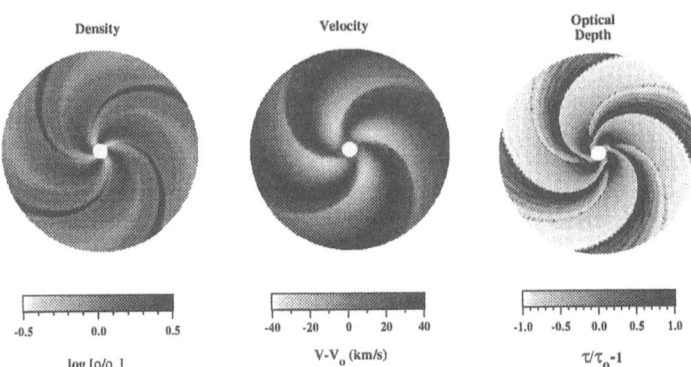

-0.5   0.0   0.5   -40 -20 0 20 40   -1.0 -0.5 0.0 0.5 1.0

log [ρ/ρ₀]   V-V₀ (km/s)   τ/τ₀-1

Fig. 6. Spatial variation of density, velocity and Sobolev optical depth, normalized relative to the unperturbed model, for the case with $m = 4$ sinusoidal modulation in azimuth.

velocities ranging up to 5000 km/s, more than double the maximum speed in the unperturbed wind, and much higher than is ever observed in line profiles of such an O supergiant. This implies that such reduced mass flux regions cannot be very prominent in hot-star winds. In particular, it argues against the importance of hot-star analogs of solar coronal holes (Zirker 1977), with very high-speed wind streams arising from regions of rapidly diverging open magnetic field (cf. MacGregor 1988).

3   Co-rotating Streams from Sinusoidal Modulations

An important result of the IUE 'Mega' campaign was the identification of the clearly cyclical PAMs in wind line profiles of several stars. In the case of the B0 I star HD 64760, these PAMs show a "phase-bowing" that suggests apparent redward as well as blueward propagation. Owocki et al. (1996) have shown that such bowing can be reproduced from a simple kinematic model with co-rotating spiral streams of alternating regions of increased and reduced density. This kinematic model ignores any dynamical effect these density variations might have in inducing corresponding velocity changes. As an initial attempt to extend this kinematic picture into a dynamically self-consistent model, let us next briefly consider the response of a line-driven wind to a perturbation that varies *sinusoidally* in azimuth between enhanced and reduced radiative driving.

Figure 6 shows grayscale plots of the changes in density, velocity, and Sobolev optical depth for such a dynamical model with an azimuthal period of $90°$ and

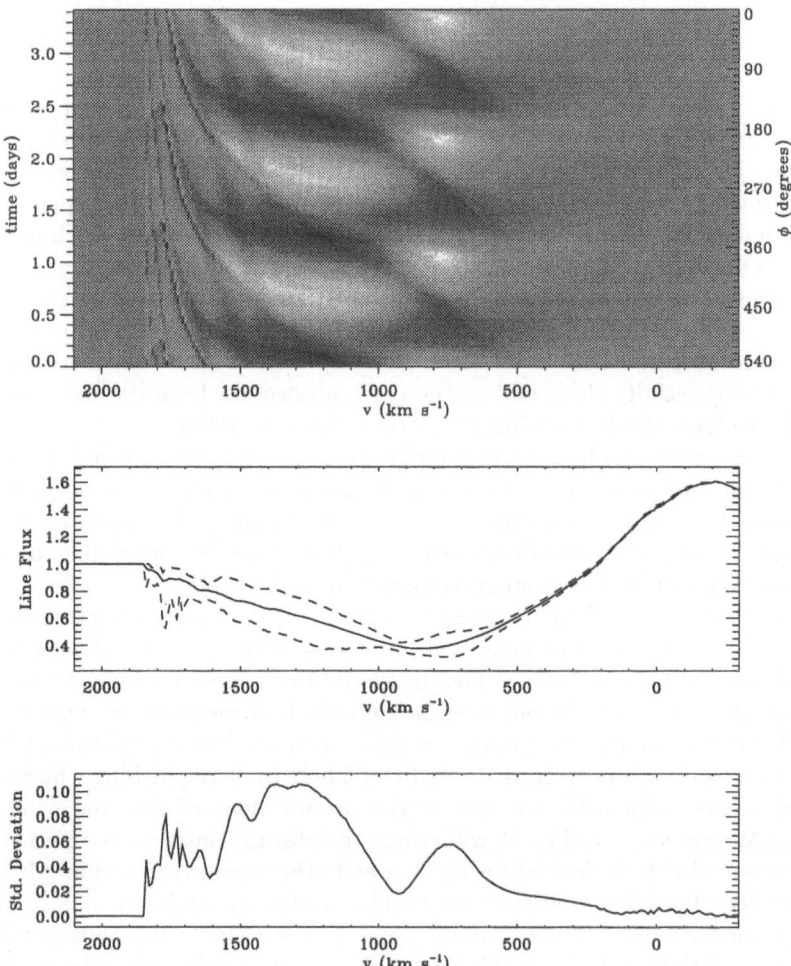

Fig. 7. As in Figure 4, but for the sinusoidally modulated model.

perturbations amplitude of $A = 0.5$. Note that there are substantial effects on the velocity as well as density, and that the optical-depth variations no longer just follow the density, as was assumed in the heuristic corotating stream model for the observed phase bowing. Figure 7 shows the quite complex profile variability for the line profile of a moderately strong line in this sinusoidal modulation model. The complexity reflects the intricate interdependence of flow variations in density, velocity, and velocity gradient. The comparatively simple phase-bowing pattern of the kinematic model is not apparent here. Indeed, it is not clear what sort of dynamically consistent structure could give the observed phase bowing.

4   Concluding Remarks

The simulations here demonstrate that the physical dependencies of line driving play an important role in the dynamical evolution of wind structure, and lead to phenomena, e.g., the reverse-mode kink, that have no analog in ordinary gas dynamics. The inward propagation of this kink yield the slow net outward propagation of a velocity plateau flow pattern that causes the enhanced absorption, providing one possible explanation for the relatively slow apparent acceleration in observed DACs. However, such models typically also predict line-profile features that are not commonly seen, most notably *reduced* absorption components. Thus, an alternative mechanism for producing slow DACs is a substantial increase, i.e. a factor of two or more, in the mass being ejected into the wind, without a compensating increase in the driving. The acceleration of such material can be especially slow, and so the extra absorption from its higher density could also form slowly evolving DACs (Owocki et al. 1994).

As for the PAMs observed in the IUE Mega project, we find that dynamical models with periodic modulation typically show substantial velocity variations that complicate the line-profile signature. In particular, the simple, phenomenological, kinematic picture developed to explain phase bowing cannot be easily adapted to a fully self-consistent dynamical model.

The work described here represents only the first steps in trying to develop dynamically self-consistent models of the flow structure associated with the observed variability of hot-star winds. In the future, models should address more directly the two most obvious possible perturbation mechanisms, namely non-radial pulsations and surface magnetic fields, with particular emphasis on identifying observational characteristics that could help in distinguishing which mechanism is most responsible for each of the various kinds of observed wind variability. Within such studies, it will remain helpful to contrast the inferred flow structures with those derived for the sun and other cool stars, and in particular to identify which characteristics are peculiar to the dynamics of radiative driving by line-scattering. The cyclical variability of hot-star winds thus provides a unique potential for studying this important aspect of radiation hydrodynamics.

Acknowledgements

This work was supported in part by NASA grant NAGW-2624. Supporting computations were carried out using an allocation of supercomputer time from the San Diego Supercomputer Center. Much of the research reviewed here was carried out in collaboration with S. Cranmer, as part of his Ph.D. thesis. I also acknowledge numerous helpful discussions with A. Fullerton, K. Gayley, H. Henrichs, L. Kaper, and J. Puls.

References

Abbott, D.C. (1980): ApJ, 242, 1183
Castor, J.I., Abbott, D.C., Klein, R.I. (1975): ApJ, 195, 157 (CAK)
Cranmer, S.R., Owocki, S.P. (1996): ApJ, 469, 469
Friend, D.B., Abbott, D.C. (1986): ApJ, 311, 701
Fullerton, A.W., Owocki, S.P. 1992, in
Fullerton, A.W., Massa, D.L., Prinja, R.K., et al. (1997): A&A, 327, 699
Hundhausen, A.J. (1972): Coronal Expansion and Solar Wind (Berlin: Springer-Verlag)
Lamers, H.J.G.L.M., Cerruti-Sola, M., Perinotto, M. (1987): ApJ, 314, 726
MacGregor, K.B. (1988): ApJ, 327, 794
Massa, D., et al. (1995): ApJ.L., 452, L53
Mullan, D.J. (1984a): ApJ, 283, 303
Mullan, D.J. (1984b): ApJ, 284, 769
Mullan, D.J. (1986): A&A, 165, 157
Owocki, S.P. (1994): in Proc. IAU Symp. 162, eds. L.A. Balona, H.F. Henrichs, & J.M.
    Le Contel (Dordrecht: Kluwer), 475
Owocki, S.P. (1994): Ap. Sp. Sci., 221, 1
Owocki, S.P., Castor, J.I., Rybicki, G.B. (1988): ApJ, 335, 914
Owocki, S.P., Cranmer, S.R., Fullerton, A.W. (1995): ApJ.L., 453, L37
Owocki, S.P., Fullerton, A.W., Puls, J. (1994): Ap. Sp. Sci., 221, 437
Owocki, S.P., Rybicki, G.B. (1985): ApJ, 199, 365
Owocki, S.P., Rybicki, G.B. (1986): ApJ, 209, 127
Pauldrach, A.W.A., Puls, J., Kudritzki, R.-P. (1986): A&A, 164, 86
Pizzo, V.J. (1982): J. Geophys. Res., 87, 4374
Rybicki, G.B. (1987): in Proc. Instabilities in Luminous Early Type Stars, ed.
    H.J.G.L.M. Lamers & C.W.H. de Loore (Dordrecht: Reidel), 175
Rybicki, G.B., Owocki, S.P., Castor, J.I. (1990): ApJ, 349, 274
Sobolev, V.V. (1960): Moving Envelopes of Stars (Cambridge: Harvard University
    Press)
Zirker, J.B., ed. (1977): Coronal Holes and High Speed Wind Streams (Boulder: Col-
    orado Associated University Press)

Henrichs: We, as observers, would like to thank you very much for the so many years of fundamental calculations you did, that helped enormously to interpret the data on stellar-wind variability. You showed even an example that is far ahead of any observations I have seen so far. For obvious reasons you put your perturbing spot on the stellar equator, but in practice such a spot is most likely at higher latitude. Would that not change the predicted behavior drastically? Could you comment?

Owocki: Perturbations away from the equator obviously require a 3-D simulation, which with modern computers is readily doable, though not yet attempted. In the context of CIRs in the solar wind, V. Pizzo in the 1970's already developed 3-D models, showing some interesting new effects not possible in 2-D, e.g. flow around the CIR at other latitudes. In rapidly rotating hot stars, an additional effect may be deflection toward or away from the equator. This remains a problem to be investigated.

Semi-empirical Modelling of Cyclical Hα Line Profile Variability of O Stars

Jeroen A. de Jong

Astronomical Institute, University of Amsterdam, The Netherlands

We present a semi-empirical, 2D model for the cyclical Hα line profile variability of O stars. We assume that the variability is caused by features on the stellar surface that locally perturb the base of the stellar wind. The origin of these features is not specified, but might be magnetic patches and/or non-radial pulsations. The line profiles were synthesized by computing the flux from the surface of constant line-of-sight velocity that contributes to each velocity bin. The density was computed from the continuity equation and a phase-dependent perturbation. We adjusted the parameters of the model to match qualitatively observations of the O7.5 III(n)((f)) star ξ Per that were obtained during a multisite campaign in 1994 October (see Henrichs et al. 1998, these proceedings). Although our 2D model does not permit quantitative analysis, the models do show redward and blueward moving features that are similar to what is observed (see Fig. 1). Our future aim is to derive physical parameters from fits to the observed profiles.

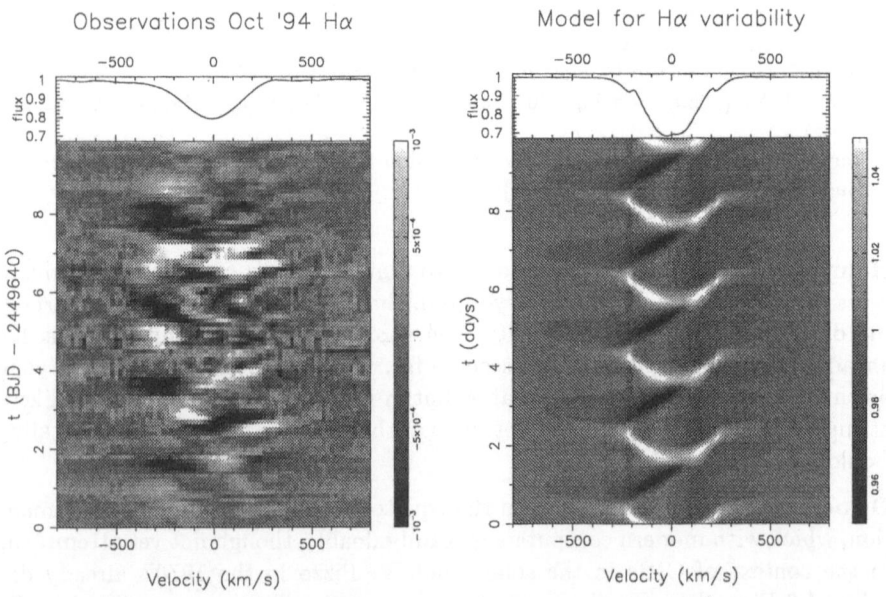

Fig. 1. Comparison between the observed spectra (inverse of the CLEANed Fourier transform) of ξ Per (left) and the model computations (right).

Long-Term Cyclic V/R Variations in Be Stars

Gerrit Jan Savonije

Astronomical Institute "Anton Pannekoek", University of Amsterdam, Kruislaan 403, 1098 SJ Amsterdam

Abstract. This paper discusses the model in which the cyclic long-term V/R-variations of the Balmer emission lines observed in many Be-stars are explained in terms of a slowly precessing one-armed density wave in a semi-Keplerian circumstellar disc. By including the gravitational effect of the rotational distortion of the rapidly spinning Be star, one can naturally explain the basic features of the observed long-term V/R variations (Papaloizou et al. 1992). We will discuss some new calculations for circumstellar discs with a radial density and temperature gradient and consider differential rotation of the Be star to boost the value of the apsidal motion constant.

1 Introduction

Many of the well observed Be stars exhibit (quasi-)cyclic variations in the relative intensity of the doubly peaked Balmer emission line profiles, the so-called V/R ratio. The main properties of these variations are (e.g. Dachs 1987):

1. The (quasi)-cycles show characteristic timescales of $\tau \simeq 2 - 13$ years.
2. The V/R-variations occur in a region confined to roughly 10 stellar radii from the star.
3. The whole V/R-profile shifts in direction of the weakest peak, i.e. the profile shifts blueward when the red-component is stronger and vice versa.
4. Sometimes a switch to a different period is observed (e.g. PP Car from 10 to 2 years).

Okazaki (1991) was the first to study $m = 1$ density waves in circumstellar discs in relation to the V/R variations. Because there can be different formation sites for the various Balmer-lines along the $m = 1$ mode, phase differences in the V/R-ratio are readily explained in this model. In Okazaki's calculation pressure effects induce a retrograde precession of the $m = 1$ modes, i.e. in the direction opposite to the orbital motion of the gas elements. In order to confine these retrograde modes to the region close to the star, Okazaki had to truncate the disc at the required distance from the star. Papaloizou et al. (1992) noted the importance of taking into account the deformation of the rapidly rotating Be star. Inclusion of the quadrupole potential due to the stellar distortion tends to make the precession of $m = 1$ modes prograde and this naturally confines the modes to the region close to the central star. For Be star/disc systems which happen to have their orbital plane close to the line of sight, we can observationally discriminate between prograde- and retrograde modes. When a gas element, in its fast orbital motion, speeds through the high-density region associated with the slowly precessing one-armed mode, it gets compressed, heated and emits more energy. This extra energy is emitted with a Doppler-shift corresponding to its orbital motion and gives rise to either a stronger blue- or red-shifted emission peak. When the gas element is speeding towards us when it passes through the high-density region in the disc we observe a phase with V>R, while when moving away from us a phase with R>V. One can readily see that for a prograde mode the R>V phase should be followed by a (partial) eclipse of the high-density region when the slow mode precession has brought this region behind the star. Because the enhanced emission is then obscured by the central star the eclipse phase is characterized by V≃R. For retrograde modes the mode eclipse should be preceded by a phase with V>R. For Be star/disc systems observed with smaller inclination angles the situation is more complicated. Observations seem to indicate that the order of V/R-phases is consistent with prograde modes: Telting et al. (1994), Reig et al. (1997), Mennickent et al. (1997).

2 Character of $m = 1$ Modes in the Potential of a Rotationally Distorted Central Star

The external potential of the rapidly rotating Be star can be written as a multipole expansion which we truncate after the quadrupole term:

$$\Psi = -\frac{GM_S}{r} - \frac{k_2 \Omega_S^2 R_S^5}{3r^3}.$$ (1)

where r is the radial coordinate centred at the centre of the Be star, Ω_S is the rotational angular velocity of the (uniformly rotating) Be star, M_S its mass, R_S its radius and k_2 its apsidal-motion constant. For a free particle in a nearly circular orbit about the Be star the apsidal line of its orbit precesses when the stellar potential deviates from that of a point mass and the epicyclic frequency κ and orbital frequency Ω are no longer equal. The particle precession rate $\Omega_p = \Omega - \kappa$ shifts from zero to a positive value when the quadrupole term is introduced in the potential. By inserting the expression for κ corresponding to (1) we find for the free particle precession frequency (Papaloizou et al. 1992):

$$\Omega_p \simeq \frac{k_2 \Omega_S^2 R_S^5}{\Omega r^5} = \frac{k_2 f^2 \Omega_c^2 R_S^5}{\Omega r^5}$$ (2)

Since $\Omega^2 \simeq GM/r^3$ we find that the particle precession frequency decreases steeply ($\propto r^{-\frac{7}{2}}$) with distance from the Be star.

Let us now consider the effect of the gravitational quadrupole term on the low frequency $m = 1$ density waves in a circumstellar disc. The difference with free particles is that there is an extra pressure force on gas elements in the disc. A coherently precessing wave can set itself up by adapting the pressure perturbations so that all gas elements taking part in the wave motion precess with the same angular frequency σ. For very low modal pattern speeds ($\sigma \ll \Omega$), consistent with the observed long-term V/R-variations, the local dispersion relation for inviscid linear one-armed density waves which vary with azimuth and time as $\exp[i(\sigma t - \varphi)]$ can be expressed as (Papaloizou et al. 1992):

$$\sigma - \Omega_p \simeq -\frac{c^2 k_r^2}{2\Omega} \tag{3}$$

k_r being the radial wavenumber, c the soundspeed and σ the frequency of the mode in the inertial frame. Equation (3) indicates that there can be prograde $m = 1$ modes ($\sigma > 0$) if $\sigma < \Omega_p$. Such prograde modes have a turning point ($k_r^2 \to 0$) where $\sigma = \Omega_p$, i.e. at the position where the free particle precession frequency equals that of the mode. Beyond the turning point the prograde modes become evanescent and rapidly decay outward. It is precisely this feature that leads to a confinement of these modes and of the associated long-term V/R-variations. Furthermore, we observe that in hotter (large c^2) discs and/or for higher harmonics (with larger k_r) the modes will become retrograde, because of the large negative right hand side in (3). Such retrograde modes have no turning point and can propagate throughout the disc.

3   Numerical Results

In view of the uncertain structure of circumstellar discs about Be stars, and to keep the analysis as simple as possible, we shall adopt a simple power law for the radial variation of the surface density $\Sigma = \Sigma_0 r^{-\alpha}$ and a simple polytropic equation of state for the gas in the disc $P = K\Sigma^\gamma$, where P is a z-averaged pressure. The standard linearized equations for non-axisymmetric modes in a thin non-self gravitating disc are then numerically integrated in the way described in Savonije & Heemskerk (1993).

Table 1. Periods (years) of the fundamental $m = 1$ modes in a circumstellar disc around a $9M_\odot$ evolved ($X_c = 0.43$) MS star.

| $k_2 f^2$ | $\alpha = 1.0$ | | | $\alpha = 1.5$ | | | $\alpha = 2.0$ | | |
	$\gamma = 1.5$	$\gamma = 1.75$	$\gamma = 2.0$	$\gamma = 1.5$	$\gamma = 1.75$	$\gamma = 2.0$	$\gamma = 1.5$	$\gamma = 1.75$	$\gamma = 2.0$
0.010	−14.2	18.3	7.4	26.2	5.8	3.4	9.7	3.5	2.3
0.008	−9.0	174.4	21.1	−23.5	14.2	6.4	41.1	6.6	3.7
0.006	−7.3	−17.6	137.1	−15.0	64.9	15.7	−36.9	16.8	7.4
0.004	−6.5	−13.9	3028.5	−12.4	896.5	64.5	−25.0	75.0	20.3

The results of the numerical calculations are summarized in Table 1 where the precession periods of the fundamental $m = 1$ modes are given as a function of the strength of the quadrupole term parametrized by $k_2 f^2$, where $f = \Omega_s/\Omega_c$ is the fraction of the stellar rotation rate in terms of the break-up speed. The disc structure is varied by considering different values for α and γ, whereby the temperature scale is set by adopting the stellar effective temperature as the temperature at the disc's inner radius $R_1 = R_S$. The corresponding Mach-number at the inner disc edge is about 28. The outer disc radius is adopted to be $R_2 = 30 R_S$.

The negative periods in Table 1 correspond to retrograde modes. For these modes the associated V/R-variations can occur all over the disc, see Fig. 1. Because the discs around Be stars appear in general larger than 10 stellar radii, this is in conflict with the confinement feature mentioned in the introduction.

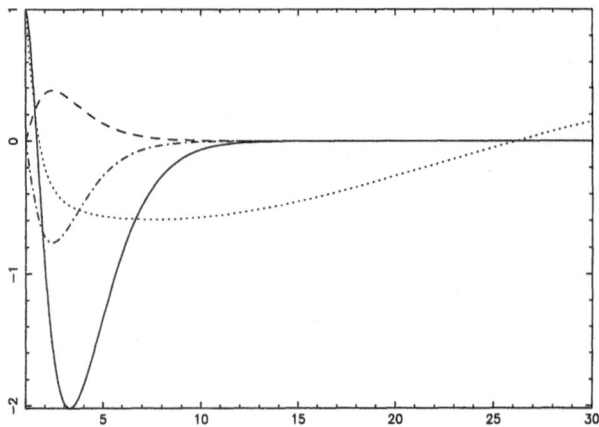

Fig. 1. Surface density perturbation Σ'/Σ versus r/R_S for a confined prograde mode (P=7.4 yrs: solid line) and a retrograde mode (P=−7.3 yrs: dotted line) for $k_2 f^2 = 0.006$. For the prograde mode we also show the velocity perturbations v'_r/v_φ (dash-dot) and v'_φ/v_φ (dashed line) with $v_\varphi = r\Omega(r)$.

Strongly confined (prograde) modes occur for the larger values of the quadru-pole term $k_2 f^2$ and/or larger values of γ, as predicted by (3). Figure 1 also shows the eigenfunctions for a prograde mode with a period of 7.4 yrs. In contrast with the retrograde modes, this mode is strongly confined to the region $r \lesssim 10 R_S$. For decreasing values of $k_2 f^2$ the frequency of the prograde mode gets smaller and smaller, passes through zero, and becomes negative: eventually it adopts a retrograde character. The prograde modes with the lowest frequencies ($P > 50$ yrs), although peaked towards the inner disc, extend through the whole disc, like the retrograde modes.

The perturbations Σ' and v'_φ differ in phase by π, so that in a high density region ($\Sigma' > 0$) the azimuthal velocity perturbation $v'_\varphi < 0$. This implies that in the phase with $V > R$ the whole double peak profile is shifted redward, while for $V < R$ it is shifted blueward, consistent with the observations. Finally, the sometimes observed abrupt period changes of the V/R cycles may be explained by a restructuring of the inner disc. Table 1 shows that relatively minor changes in the disc's radial (temperature) gradient can cause significant period changes.

4   Evolution and Differential Rotation of Be Star

The numerical results indicate that relatively large values of $k_2 f^2$ are required to obtain confined (prograde) $m = 1$ modes in the circumstellar disc. Differential rotation of the Be star, with larger rotation rates in the contracting helium core, gives rise to a larger value for the apsidal motion constant k_2. To this end a 9 $M_\odot$ star was evolved from the ZAMS (assuming uniform rotation at the surface break-up speed for $t = 0$) whereby spherical shells in the radiative envelope were assumed to conserve their angular momentum. The convective core, on the other hand, was assumed to remain in uniform rotation. The effects of rotation on the stellar structure and evolution were neglected. As a result of the subsequent evolution with contracting core and expanding envelope a state of differential rotation is defined for which we can determine the rotational deformation and the external gravitational potential (to first order in the distortion). This must be described elsewhere, by lack of space. Table 2 shows the result of these calculations at four stages during core hydrogen burning.

Table 2. Apsidal motion constant for a differentially rotating $9M_\odot$ star ($Z = 0.02$) evolving off the Main Sequence. X_c denotes the hydrogen abundance in the convective core, k_2^* is the apsidal motion constant for the differentially rotating star, while k_2 applies to the same (unperturbed) stellar structure, but is the standard value for uniform rotation (with the same Ω_s).

X_c	$R_S/R_\odot$	$\log_{10}[T_{eff}]$	$\log_{10}[L/L_\odot]$	$f = \Omega_s/\Omega_c$	k_2	k_2^*
0.70	3.91	4.363	3.591	1	0.0106	0.0106
0.50	4.71	4.349	3.696	0.91	0.0076	0.0091
0.36	5.41	4.334	3.756	0.85	0.0061	0.0083
0.21	6.41	4.311	3.811	0.78	0.0048	0.0075
0.01	7.57	4.293	3.885	0.72	0.0037	0.0066

Dividing the distortion part of the external gravitational potential by $\Omega_s R_S^5/(3r^3)$, Ω_s being the rotation speed at the stellar surface, we obtain an equivalent of k_2 for differential rotation, which is denoted by k_2^* in Table 2. Note that the assumption of initial (ZAMS) uniform rotation with $f = 1$ seems not far off to explain the estimated surface rotation speeds of somewhat evolved Be stars. Note also that k_2 and k_2^* are independent of the adopted scaling for f. The results show that $k_2^* f^2$ attains sufficiently large values to invoke prograde $m = 1$ modes, at least for the more favourable cases (large γ) shown in Table 1.

References

Dachs J. (1987) Proc. IAU Colloq. No. 92, 'Physics of Be Stars', Slettebak A. and Snow T.P. (Eds.), p. 149

Mennickent, R.E, Sterken, C. and Vogt, N. (1997) A&A **326**, 1167.

Okazaki A.T. (1991) PASJ **43**, 75

Papaloizou, J.C., Savonije, G.J. and Henrichs, H.F. (1992) A & A **265**,L45.

Telting, J.H., Heemskerk, M.H.M., Henrichs, H.H and Savonije, G.J. (1994) A&A **288**, 558.

Reig, P., Fabregat, J., Coe, M.J., Chakrabarty, D., Neguerela, I. and Steele, I. (1997) A&A 322, 193.

Savonije, G.J., Heemskerk M.H.M., (1993) A&A **276**, 409.

Hummel: How does the differential stellar rotation influence the value of the turning point radius?

Savonije: The differential rotation gives rise to a larger value for the apsidal motion constant which induces a faster precession speed and a smaller turning point radius.

Peters: The observations of V/R behavior in Be stars are considerably more complex than the predictions from your model, although I commend you for presenting some fresh new ideas on a phenomenon that has been known for a long time. The period for a V/R reversal can change in an individual Be star and the transition from V/R> 1 to V/R< 1 is not uniform - frequently a star will show a particular V/R ratio for a long time and then undergo an abrupt change. I think the activity is necessitated by changes in the wind (see e.g. papers by Doazan).

Savonije: Our model is fairly simple and I am sure it cannot explain all the observations without putting more physics in it. Regarding the phenomenon mentioned by you, the period of a one-armed mode can be much longer than the canonical few years. When the period is, say, 60 years we will observe a particular V/R ratio for a long time span. However, during such long periods the disc may change its structure resulting in a shorter modal period and thus a shorter V/R cycle time. The ensuing phase with reversed V/R ratio may then last (substantially) shorter.

Ignace: Is your model for explaining V/R variations consistent with continuum polarization measurements?

Savonije: I have no idea, but it seems interesting to check.

Large, Oscillating Be Star Circumstellar Disks

W. Hummel[1], R.W. Hanuschik[2], and M. Vrancken[3]

[1] Universitätssternwarte München, Scheinerstr. 1, D-81679 München, Germany
[2] Astronomisches Institut, Ruhr-Universität Bochum, D-44780 Bochum, Germany
[3] Royal Observatory of Belgium, Ringlaan 3, B-1180 Brussels, Belgium

We present numerical model calculations (3D, NLTE radiative line transfer) for the cyclic, long-term variations of Be-star emission lines. These variations are believed be formed in quasi-Keplerian disks with single-armed global disk oscillations (Okazaki 1991, PASJ 43, 75). In our recent study (Hummel & Hanuschik 1997, A&A 320, 852), we showed that global disk oscillations in disks with $R_{\mathrm{d}} = 5\,R_*$ can indeed account for both the observed line-profile asymmetries and their cyclic evolution.

In this study we present theoretical emission lines from oscillating disks with larger radii of $R = 10$ and $R = 15\,R_*$, but confine the disk wave within $R < 5\,R_*$. The equivalent width and shape of the computed Hα emission line profiles of the larger circumstellar disks are in much better agreement with the observed emission lines (Fig. 1).

However, the calculated profiles are, on the average, less asymmetric than the observed emission lines. We conclude that density waves are not confined to the inner disk regions alone. This was recently confirmed by Okazaki (1997, A&A 318, 548) and Savonije (1998; these proceedings), who showed that density waves can extend up to $R \leq 18R_*$.

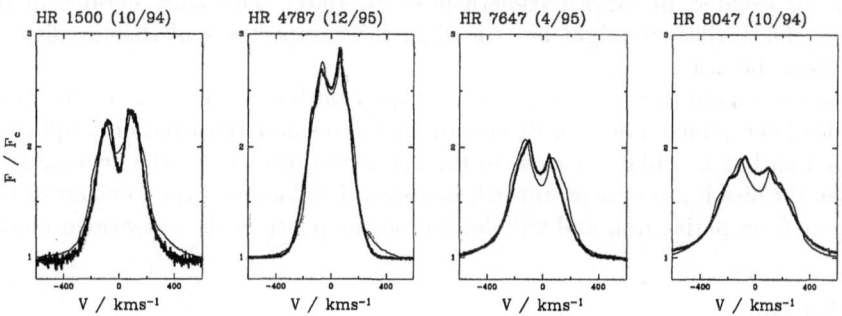

Fig. 1. High-resolution (R=70,000), high S/N long-term variable asymmetric Hα emission lines of Be stars (thick lines), fitted with theoretical emission lines of a non-axisymmetric circumstellar model disk (thin lines).

Variabilities in Be-Star Disks in Be/X-Ray Binaries

Atsuo T. Okazaki

College of General Education, Hokkai-Gakuen University,
Toyohira-ku, Sapporo 062, Japan

We examine long-term, cyclic variabilities of the outflow in the disks of Be stars in Be/X-ray binaries in the context of the viscous decretion disk scenario. In this scenario, the matter ejected from the equatorial surface of the star with the Keplerian rotation velocity drifts outward because of viscosity and forms a disk (Lee et al. 1991).

We consider an isothermal disk truncated at a radius where the tidally-induced eccentric instability occurs. We adopt the Shakura-Sunyaev α-viscosity prescription.

We find that the outflow in disks formed by viscous decretion is highly subsonic. Roughly speaking, the outflow velocity increases as r and the surface density decreases as r^{-2}. We also find that the decretion disks are in general overstable for one-armed ($m = 1$) spiral modes. The growth rate is on the order of $\alpha (H/r)^2 \Omega$, which is about 1–10 yr for $\alpha \sim 0.1$, where H is the scale-height of the disk and Ω is the angular frequency of disk rotation. Note that disks perturbed by these modes become eccentric. The characteristics of these modes agree with the long-term, cyclic variations of the Balmer line profiles observed for some Be/X-ray binaries, which are similar to those of isolated Be stars except that the periods of the variations are much shorter for Be/X-ray binaries than for isolated Be stars (Negueruela et al. 1997). The short periods of the line-profile variations might be related to the compactness of Be-star disks in Be/X-ray binaries.

In our model, the orbital phase of Type I outbursts depends on the amplitude of one-armed modes in Be-star disks. For a small perturbation amplitude, Type I outbursts will occur close to the periastron passages of the neutron star. When the disk is strongly perturbed, however, the phase of Type I outbursts will deviate from periastron, and will depend on the phase of the one-armed mode.

References

Lee, U., Saio, H., Osaki, Y. (1991): MNRAS 250, 432
Negueruela, I., Reig, P., Coe, M.J., Fabregat, J. (1998): A&A, in press

V/R Variability in Be/X-Ray Binaries

Ignacio Negueruela

Astrophysics Research Institute, Liverpool John Moores University, Byrom St., Liverpool, L3 3AF, U.K.

Long-term spectroscopic monitoring of Hα and other emission lines in the spectra of the optical components of Be/X-ray binaries during 1990–1997 has revealed the frequent presence of quasi-cyclic V/R ratio variations, which are similar to those seen in single Be stars (Negueruela et al. 1998; Stevens et al. 1997; Reig et al. 1997).

The observed quasi-periods are as follows: approximately 1 year for BQ Cam (V 0332+53); ~18 months (during 1990–1994) and ~1 year (after 1994) for HD 245770 (A 0535+262); $\lesssim$ 3 years for V801 Cen (4U 1145−619); and ~3 years for LS I +61°235 (RX J0146.9+6121). There is no hint of any sort of modulation in any parameters of the emission lines with the orbital motion of the neutron star companion, contrary to the predictions of the model by Apparao & Tarafdar (1986). All the observed quasi-periods concentrate near the short end of the period distribution for isolated Be stars (from ~2 years to decades). This could be related to the claim by Okazaki (1997) that radiation pressure gives rise to shorter periods in early-type Be stars, since all the objects are earlier than B2.

Fast variability in the shape of emission lines has been observed during giant X-ray outbursts of A 0535+262 and 4U 0115+634, which shows that large changes in the physical conditions of the circumstellar disc occur at the time of the outbursts. All our results are consistent with the idea that asymmetric lines are due to non-axisymmetric density distributions in the disc (Hanuschik et al. 1995) and V/R variability is caused by the presence of one-armed global oscillations (Okazaki 1991; Savonije & Heemskerk 1993).

References

Apparao, K.M.V., Tarafdar, S.P. (1986): A&A 161, 271
Hanuschik, R.W., Hummel, W., Dietle, O., Sutorius, E. (1995): A&A 300, 163
Negueruela, I., Reig, P., Coe, M.J., Fabregat, J. (1998): A&A, in press
Okazaki, A.T. (1991): PASJ 43, 750
Okazaki, A.T. (1997): A&A 318, 548
Reig, P., Fabregat, J., Coe, M.J., et al. (1997): A&A 322, 183
Savonije, G.J., Heemskerk, M.H.M. (1993): A&A 276, 409
Stevens, J.B., Reig, P., Coe, M.J., et al. (1997): MNRAS 288, 988

Part 5

THE MUSICOS 1996 CAMPAIGN

MUlti-SIte COntinuous Spectroscopy: Previous MUSICOS Campaigns

Claude Catala

Observatoire Midi-Pyrénées, 14 Avenue Edouard Belin, F–31400 Toulouse, France

Abstract. MUSICOS is a project for promoting and facilitating multi-site continuous spectroscopy. The scientific goals and achievements of the first 3 MUSICOS campaigns are briefly outlined, and the instrumental developments linked to the project are described.

1 Introduction

The goal of the MUSICOS project (**MU**lti **SI**te **CO**ntinuous **S**pectroscopy) is to facilitate multi-site, multi-wavelength observations in stellar spectroscopy. Several workshops dedicated to this project (Catala & Foing 1988; Catala & Foing 1990) showed that this type of observations is definitely needed for many scientific programs, in particular those related to stellar magnetic activity, stellar winds, and stellar oscillations. Most of these programs have similar instrumental needs, namely a world network of spectroscopic facilities providing a resolving power above 30,000, on telescopes of the 2m class.

A rapid survey of the existing telescopes of the 2m class and of their available instrumentation leads to the conclusion that: 1) they can provide a good longitude coverage, and 2) many 2m telescopes lack adequate spectroscopic facilities.

The MUSICOS project was born from this conclusion, and its general strategy was defined in three steps: (1) Organize multi-site, multi-wavelength campaigns with the existing instruments, and use transportable fiber-fed spectrographs on 2m telescopes without adequate spectrographs; (2) Define, design and build the prototype of a cheap spectrograph meeting the requirements of the scientific programs needing multi-site observations; (3) Duplicate this spectrograph and install the copies on telescopes of the 2m class around the world.

2 The MUSICOS Campaigns

2.1 General Organization

The general organization of MUSICOS campaigns is described in details in Catala et al. (1993).

The basic strategy is to gather the efforts of several groups of people interested in different scientific programs, all of them needing multi-site spectroscopic observations. Each scientific program is coordinated by a Principal Investigator, responsible for the scientific choices concerning the targets, the observing modes, the observing strategy. After the campaign, the PI receives raw data from all participating observers, he (she) is responsible for data reduction in a homogeneous way, he (she) organizes data analysis with interested participants, and is finally responsible for publication of the results.

The observing strategy is very clearly defined before the campaign, so that a good compromise can be found for each instrument between the requirements of the scientific programs and the instrumental limitations. For each program, the PI defines three priorities among the desired characteristics of the spectra, for instance (1) S/N ratio > 100; (2) resolving power $R > 30,000$; (3) exposure time $t < 30$ min. The rule is to try to fulfill the three desiderata. If not possible, the third priority is released within allowed limits. If the goal is still not reached, the second priority has to be released, and so on.

Also, each PI defines an optimal observing sequence, indicating how often and how many calibration spectra must be obtained, defining comparison stars, and how often they must be observed, as well as any particular precaution needed for the program. This standardization of the calibration and observing sequences is a first step toward an homogenization of the observations obtained at different sites.

In the few weeks before the campaign, a general coordinator is responsible for circulating all necessary information between the PIs and the observers. During the campaign, this coordinator gathers information on the observation status for each participating site, and circulates a daily report via email to all participants.

We are trying to organize a MUSICOS campaign every other year or so. During the intervening year between 2 campaigns, a MUSICOS workshop gathers the participants of previous campaigns, as well as potential participants to future campaigns. The data reduction and analysis, as well as preliminary or definitive results from previous campaigns are presented by the PIs, and discussed. Potential scientific programs for future MUSICOS campaigns are presented and discussed. A steering committee makes a proposal for the scientific programs to be selected for the upcoming campaign. The practical aspects of the campaign organization are then discussed among the participants.

2.2   Scientific Programs

Three scientific programs were carried out during each of the three previous MUSICOS campaigns. Each time, these 3 programs were related to the 3 main topics in which MUSICOS is supposed to bring valuable input: stellar magnetic activity, nonradial pulsations, and structures of stellar winds and circumstellar envelopes. This section presents a very brief summary of these 9 scientific programs, as well as some highlights of the scientific return of the MUSICOS campaigns for each of them.

MUSICOS 1989:

Flares and photospheric Doppler imaging of HR1099 (PI: B.H. Foing): HR1099 is a well-known RS CVn system, with a rotation period of 2.8 days. The goal of the campaign was twofold: (1) monitor the Hα line, as well as the UBV photometric bands and the Ca II K line, for searching for flares; (2) monitor the Ca I and Fe I photospheric lines near 6430 Å for deriving a Doppler image of the photosphere of the primary. This program was very successful, since 2 major flares were detected during the campaign, and their energy budget, time evolution and dynamical properties could be determined from the photometric and spectroscopic observations (Foing et al. 1994). Good photospheric data were also obtained, offering an excellent phase coverage of the star's rotation, leading to a Doppler image of the primary (Jankov & Donati 1994).

Search for nonradial pulsations in the Be star 48 Per (PI: A.M. Hubert): Rapid variability in Be stars has been interpreted in various ways, including nonradial pulsations and rotational modulation. The goal of these observations was to monitor the He I 6678 Å line of this star, to analyze the variations of its profile, in view of testing these 2 conflicting hypotheses. This program was a great success, and achieved a duty cycle of the order of 80%. Line profile variations as weak as 0.5% of the continuum were detected and analyzed. Multiple periods were discovered in the data, which rules out the interpretation in terms of rotational modulation, and confirms the nonradial-pulsation model. The modes responsible for the observed variations were identified as tesseral modes (Hubert-Delplace et al. 1997).

Search for corotating interaction regions in the wind of AB Aur (PI: C. Catala): AB Aur is the prototype of the pre-main sequence Herbig Ae/Be stars. The wind of AB Aur is structured in fast and slow streams, as the solar wind. In the solar wind, when fast streams overtake slow streams, a pair of shocks is formed, between which one finds a corotating interaction region (CIR), where the velocity is intermediate between that of fast and slow streams, and where the density and temperature are way above those of the unperturbed wind. The goal of this program was to search for signatures of these CIRs in the Ca II K line, as for example discrete absorption components within the line profile. Unfortunately, this program was unsuccessful due to bad weather on almost all the sites involved in the campaign, resulting in a very bad duty cycle. Some partial results of these observations are presented in Catala et al. (1993).

MUSICOS 1992:

Doppler imaging of HR 1099 (PI: T. Simon and J. Neff): The RS CVn system HR 1099 was revisited 3 years after the first campaign, in order to derive another photospheric Doppler image. In spite of several nights lost to bad weather on some of the sites, the resulting phase coverage was acceptable, and the derived Doppler image meaningful. A comparison of this image with previous ones, including that of the MUSICOS 89 campaign, indicated a long-term evolution of the active region pattern at the surface of this star (Jankov & Donati 1994). In addition to the MUSICOS data, HR 1099 was also observed at the 4m AAT with a Zeeman-Doppler setup, leading to a magnetic image of the star, which can be compared to the temperature image (Donati, this volume).

Oscillation modes of θ^2 Tau (PI: T. Kennelly, G. Walker): θ^2 Tau is a δ Scuti star, with multiperiodic photometric variations, interpreted as due to nonradial pulsations. The goal of the campaign was to monitor radial velocity and line-profile variations, in order to detect oscillation modes and to measure their frequency accurately. Besides, the analysis of line-profile variations provides a tool to identify the modes, since the details of the line profiles contain information on the spatial distribution of the oscillation pattern. This campaign was a great success, with an overall duty cycle approaching 80%. Seven different modes were detected and identified, bringing our knowledge of the oscillations of this star to an unprecedented level of details (Kennelly et al. 1996).

Rotational modulation of spectral lines in AB Aur (PI: T. Böhm, C. Catala): The Herbig Ae star AB Aur was also revisited during the MUSICOS 92 campaign. The goal was to search for rotational modulation in the chromospheric He I D3 lines, formed close to the stellar surface, in order to test whether the structures observed further out in the wind are linked to phenomena occurring at the stellar surface. The success of this part of the campaign was partial, with a duty cycle of only about 50%. However, the strong variations observed in the He I D3 lines were shown to be consistent with rotational modulation (Böhm et al. 1996). Moreover, variations of photospheric lines were also discovered, and shown to be consistent with rotational modulation (Catala et al. 1996). These results indicate that all the layers of AB Aur's atmosphere (photosphere, chromosphere, wind) show the same kind of structure, involving streams.

MUSICOS 1994:

Doppler imaging and prominence mapping of AB Dor (PI: A. Collier Cameron): AB Dor is an ultra-fast rotating active K0 dwarf. The goal of the campaign was to monitor photospheric lines in order to derive a Doppler image of starspot distribution at the star's surface, and to trace the evolution of circumstellar prominences for several consecutive rotations (P_{rot}=12.4 hrs) by monitoring the Hα line. In spite of some bad weather conditions on some of the sites, this program was successful: a Doppler image was reconstructed from the photospheric data, while several circumstellar prominences were observed and studied (Collier Cameron et al. 1997). Several flaring events were also detected during the campaign.

Line profile variations of γ Dor (PI: L. Balona): γ Dor is a variable star with a period which is too long to be considered as a δ Scuti star. The goal of the campaign was to test the two hypotheses for the observed variability: rotational modulation and nonradial pulsations. The star was monitored spectroscopically and photometrically. Several periods were found in the data, which confirms that the variations are due to NRP. The spectroscopic data allowed us to identify the observed modes (Balona et al. 1996).

Structure of the circumstellar environment of β Pic (PI: A.M. Lagrange): β Pic is an A-type star with a thin circumstellar disk, seen almost edge-on. Infalling

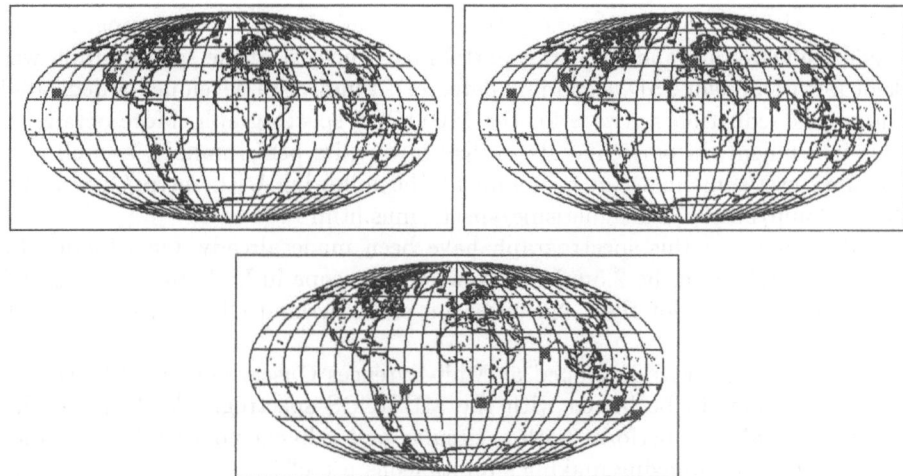

Fig. 1. Main sites involved in the MUSICOS campaigns. Upper left: 1989 campaign; Upper right: 1992 campaign; Lower middle: 1994 campaign

comets have been detected in its circumstellar environment. The goal of the campaign was to monitor the star continuously for several days, in several spectral lines, to detect cometary events and to follow them with no gap, in order to determine their dynamical properties. The data for this part of the campaign are still being analyzed.

2.3   Sites and Instruments

Figure 1 shows the location on the Earth of the main sites involved in the three MUSICOS campaigns. Additional details on the instrumentation used at these sites for each one of the campaign can be found at the following URL: http://www.obs-mip.fr/omp/umr5572/magnetisme/musicos.html

3   The MUSICOS Spectrograph

The MUSICOS spectrograph was developed in the framework of the MUSICOS project. A prototype was built, and has been in regular operation at the "Bernard Lyot" 2m telescope at Pic du Midi since 1991. This prototype is also transported to remote sites for MUSICOS campaigns.

This spectrograph (Baudrand & Böhm 1992) was designed to meet the requirements of scientific programs that need multi-site observations. It is a cross-dispersed spectrograph, covering most of the visible domain in 2 exposures (380-540 nm and 540-870 nm), with a resolving power of 38,000. The spectrograph is fed by 2 fibers, giving the possibility to record simultaneously 2 spectra with interleaved orders. Other characteristics of this spectrograph include high stability and easy transportability. The cost of the spectrograph was kept low (< 50,000 USD, excluding detector). Finally, we developed a dedicated automatic reduction software for the MUSICOS spectrograph. More details on this instrument, in particular its optical layout, its performances, as well as a detailed user's manual can be found at the following URL: http://www.obs-mip.fr/omp/umr5572/magnetisme/spectromus.html

Two copies of this spectrograph have been made already. One, funded by ESA, is installed on the 2.5m Isaac Newton Telescope in La Palma. The second one was developed for the 1.9m telescope of Sutherland Observatory in South Africa.

A polarimeter was developed at Midi-Pyrénées Observatory by J.F. Donati and the author, to be linked with the MUSICOS spectrograph. This new instrument is able to perform both a circular and a linear polarization analysis. Zeeman-Doppler imaging making simultaneous use of the very numerous spectral lines in the visible spectrum of late-type stars is becoming possible with this instrument, and pushes up the performances of this technique by several magnitudes. All details of this polarimeter can be found at the following URL: http://www.obs-mip.fr/omp/umr5572/magnetisme/polarmus.html.

References

Balona L.A., Böhm T., Foing B.H., et al. (1996): MNRAS, 281, 1315
Baudrand J., Böhm T. (1992): A&A, 259, 711
Böhm T., Catala C., Donati J.-F., et al. (1996): A&AS, 120, 431
Catala C., et al. (1993): A&A, 275, 245
Catala C., Böhm T., Donati J.-F., et al. (1997): A&A, 319, 176
Collier Cameron A., Walter F.M., Vilhu O., et al. (1997): AJ, in press
Foing B.H., et al. (1994): A&A, 292, 543
Hubert-Delplace A.M., et al. (1997): A&A, 324, 929
Jankov S., Donati J.F. (1994): in: "Proc. fourth workshop on Multi-Site Continuous Spectroscopy", L. Huang, D.S. Zhai, C. Catala & B.H. Foing (eds.), p. 143
Kennelly E.J., Walker G.A.H., Catala C., et al. (1996): A&A, 313, 571

MUSICOS Observations of SU Aurigae

Yvonne C. Unruh[1], J.-F. Donati[2], and the MUSICOS[3] collaboration

[1] Institut für Astronomie, Universität Wien, Türkenschanzstr. 17, 1180 Wien, Austria
[2] Observatoire Midi-Pyrénées, F–31400 Toulouse, France

Abstract. We present results from the MUSICOS (MUlti-SIte COntinuous Spectroscopy) campaign 1996 on the classical T Tauri star SU Aurigae. Our observations support the suggestion that the wind and infall signatures are out of phase on SU Aur. We find that the wind is structured: there are two redward moving features in the blue wings of Hα and the Na D doublet.

1 SU Aurigae

SU Aur is a relatively bright and hot classical T Tauri star (CTTS) that has been the subject of several studies (Johns & Basri 1995, Petrov et al. 1996). It is thought that SU Aur has a magnetic dipole that is slightly inclined with respect to the rotation axis of the star. Material from the disk accretes along magnetic field lines, but because of the inclination of the dipole, accretion is only seen for one half of the stellar rotation, whereas a wind is observed during the other half (Johns & Basri 1995, Johns-Krull 1998).

2 MUSICOS Observations

SU Aur was observed for approximately 10 days from 5 different observatories (Bejing Astrophysical Observatory in Xinglong (Xin), Observatoire de Haute Provence, France (OHP), Isaac Newton Telescope in La Palma (INT), MacDonald Observatory, Texas (MDO) and at the Canada-France-Hawaii Telescope in Hawaii (CFH)). The time coverage was very good during the first part of the campaign when one complete rotation of SU Aur was covered, but deteriorated during the second part. All telescopes were equipped with high-resolution spectrographs ($\geq$ 30,000). At the CFHT, a polarimeter was used to obtain Stokes V data. For ease of comparison, the data presented here have all been binned to a uniform resolution of 30,000.

---

[3] L. Balona (SAAO), H. Cao (Bejing Observatory), C. Catala (Observatoire Midi-Pyrénées), A. Collier Cameron (University of St Andrews), P. Ehrenfreud (University of Leiden), B. Foing (ESTEC/ESA), J. Hao (Bejing Observatory), A. Hatzes (University of Texas), H. Henrichs (University of Amsterdam) C. Johns-Krull (Berkeley), J. de Jong (University of Amsterdam), T. Kennelly (HAO), J. Landstreet (University of Western Ontario), C. Mullis (University of Toledo), J. Neff (Stony Brook) J. Oliveira (ESTEC/ESA), C. Schrijvers (University of Amsterdam), T. Simon (University of Hawaii), E. Stempels (University of Uppsala), J. Telting (ING, La Palma), and N. Walton (ING, La Palma)

3   The Non-photospheric Lines

SU Aur only shows Hα in emission; the higher Balmer lines, Na D, and He I (587.6 nm) are in absorption with varying degrees of infilling. Our spectral format covers Hα, Na D, and He I (587.6 nm) at all observatories. In addition, Hβ was observed simultaneously at the CFHT, OHP, and at the INT. Fig. 1 shows the stacked profiles of the sodium doublet, Hα, and Hβ. Spectra within 4-hour bins have been grouped together. The offset between the profiles is proportional to the time separation between the exposures.

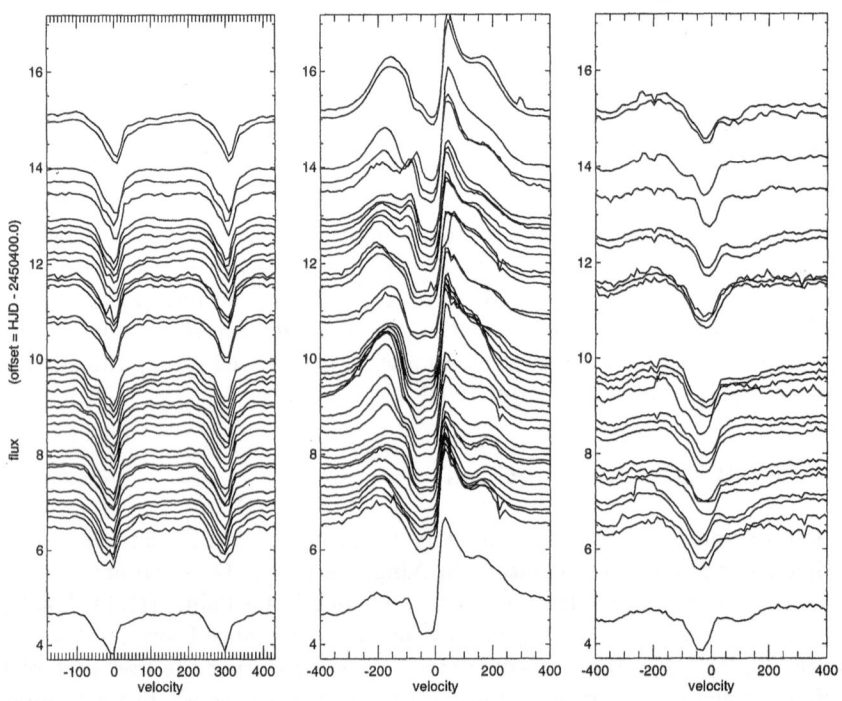

Fig. 1. The variations in the Na D lines (left), Hα (middle), and Hβ (right).

In the blue wings of the profiles we can trace two redward-drifting profile deformations, starting approximately at HJD 2450408 and 2450411.5. The deformations are seen at higher blueshifts in Hα and Hβ than in Na D. In Hα, there is a flux enhancement at HJD 2450409 that is approximately contemporaneous with a drop in brightness of SU Aur (Granzer & Strassmeier 1998), but there is no evidence for a simultaneous equivalent width enhancement in any line other than Hα. As first noted by Johns & Basri (1995), the Hβ lines show redshifted absorption components that are out of phase with the blueshifted absorption (see also Fig. 3).

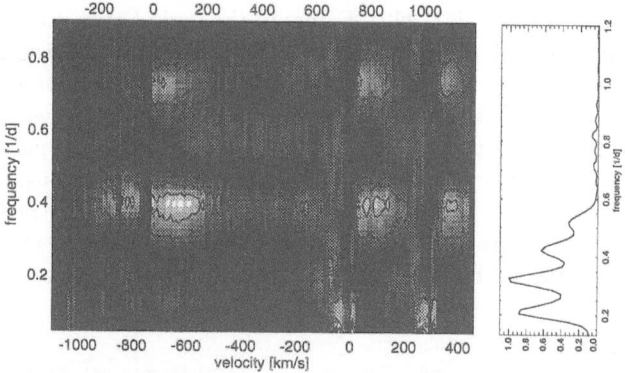

Fig. 2. The image on the left-hand shows the Lomb-Scargle periodogram of the intensity variations in the He I (587.6 nm) line (upper scale) and the sodium D lines (lower scale for Na D$_1$). The power scale ranges from 0 (black) to 47 (white). The graph on the right shows the window function of our observations.

3.1   Period Determinations

Photometric period determinations are difficult as SU Aur's brightness tends to vary rather erratically (see e.g. Granzer & Strassmeier 1998). Photometric period measurements range from 1.55 d and 2.73 d (Herbst et al. 1987) to an upper limit of 3.4 d (Bouvier et al. 1993). Several additional period measurements have been obtained from analyses of the Balmer emission lines, all of them indicating periods around 3 d (Giampapa et al. 1993, Johns & Basri 1995, Petrov et al. 1996). These measurements are hampered by the fact that the recurrence timescale of the rotationally modulated features seems to be rather short, perhaps due to unsteady accretion.

Line	Velocity [km s^{-1}]	Period [d]	Range [d]
Hα	−250...−200	2.8	2.5...3.4
Hα	−180...−140	5.0	3.9...6.6
Hβ	50...150	2.6	2.3...3.0
He D$_3$	0...200	2.5	2.3...3.1
Na D	50...150	2.5	2.2...3.1

Table 1. Table listing the periods found in normalised Lomb-Scargle periodograms. The false-alarm probabilities are all below 10^{-5}.

The periodicity that is most easily seen in our data is the one of the He I (587.6 nm) line, where it manifests itself as an equivalent width change. As the equivalent width increases, the line also becomes more redshifted. The periodogram for the helium line and the sodium doublet is shown in Fig. 2. Table 1 shows a summary of our period determinations; they suggest a period of less than 3 days. The periodograms were calculated using the Lomb-Scargle algorithm as

358   Yvonne C. Unruh, J.-F. Donati, and the MUSICOS collaboration

implemented by Press et al. (1992). Errors in the period were estimated from the
width of the power peak at half maximum. Na D and He I show secondary peaks
at a frequency of about 0.75 d^{-1}, corresponding to a period of 1.35 d, roughly
half the main period. The secondary peaks may suggest the presence of two infall
columns, though it has to be pointed out that the secondary periods were not
found in Hα or Hβ. The 5 d period in Hα may well be spurious, seen that our
observations only cover 10 d. There is an indication that the period increases
from He I and Na D through to Hβ and Hα, though the errors are large enough
to allow a single period for all lines.

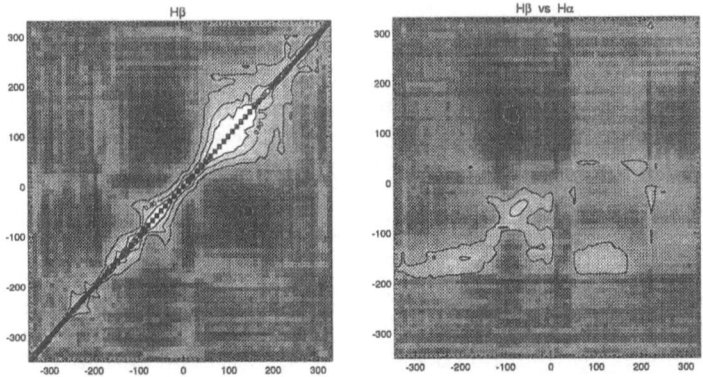

Fig. 3. Correlation matrices for Hβ (left) and Hβ versus Hα (right). For both matrices
the Hβ velocities are plotted along the y-axis. Black is at a correlation coefficient of
-0.7 and white for a correlation coefficient of 1.0.

3.2 Correlations

Fig. 3 shows two example correlation matrices. On the left is the correlation mat-
rix for Hβ. The squarish appearance of the contours between 0 and 200 km s^{-1}
shows that the red wing is correlated over a relatively large velocity range.
The blue-wing velocities are less well correlated with each other and there is a
good indication of a significant anticorrelation (false alarm probability $\approx$ 0.03 %)
between the Hβ blue and red wing, confirming the model of Johns & Basri (1995).

The plot on the right shows the correlation between Hβ and Hα. The red wing
of the Hα line is uncorrelated with the Hβ line (corresponding plots show that
it is also not correlated with the He I 587.6 nm or the sodium doublet). The blue
wings of Hα and Hβ are well correlated and there is an anticorrelation between
the red wing of Hβ and the blue wing of Hα. In summary, we find that there is a
strong correlation between the red wings of Hβ, Na D, and the total profile of the
He I line over a large velocity range (of the order of 100 km s^{-1}). Anticorrelated
with these are the blue wings of Hα, Hβ and Na D. The blue wings are well

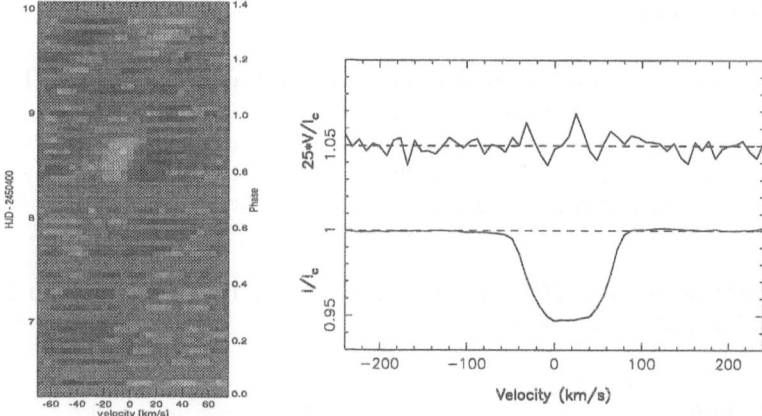

Fig. 4. LHS: Residual profiles of the least-squared deconvolved profiles (HJD 2450406 to 2450410). Black and white are at the $\mp 1.5\%$ level of the continuum. The right-hand scale indicates phase assuming $P = 2.7$ d. The drift rate of the features is in agreement with this period, but there is additional structure (e.g. between 20 and 40 km s^{-1}, HJD 2450408) that cannot easily be explained by surface features. RHS: Plot of the deconvolved Stokes V and I profiles on Nov. 19 (HJD 2450407). The Stokes V profile has been enhanced by a factor of 25; note that the profile is not indicative of a Dipole field, but suggests the presence of complex magnetic field structures.

correlated with each other, albeit over much smaller velocity ranges than the red wings.

4 Photospheric Lines and the Magnetic Field

From photometric observations, it has been inferred that TTS show sizeable surface spots. Large spots produce line-profile deformations which can be used to reconstruct an image of the stellar surface. The size of the profile deformations depends on the spot size and contrast, but is generally of the order of one percent. For a relatively faint target such as SU Aur, the S/N required to detect profile deformations cannot be achieved for individual spectral lines, unless one is using the largest telescopes. One way out is to combine the information contained in all spectral lines by using least-squares deconvolution (Donati et al. 1997). Fig. 4a shows the residual spectra after least-squares convolution.

The observations at the CFHT were taken with the MUSICOS spectropolarimeter (Donati et al. 1998), so that circularly polarised data spanning the whole 410 to 810 nm range were obtained. Due to bad weather, only 9 polarised spectra were recorded with S/N ratios of the order of 120 per 3.8 km s^{-1} bin. Even though more than 3,000 spectral features could be used for least-squares deconvolution, the moderate spectrum quality only allowed us to obtain one marginal detection (false alarm probability of 0.2%) on Nov. 19 (see Fig. 4b), but nothing on the other four clear nights.

5   Conclusions

We detect periodic variations of 2.5 to 2.8 d in all chromospheric lines. The photospheric lines show profile deformations indicative of surface structure, though the low S/N still poses problems for Doppler imaging. In agreement with the model proposed by Johns & Basri (1995) we see an anticorrelation between the positive velocities of $H\beta$, Na D, and He, and the blue wings of $H\alpha$, $H\beta$, and Na D.

Acknowledgements: YCU acknowledges support from the Austrian Science Foundation under grant S7302.

References

Bouvier, J., et al. (1993): A&AS **101**, 485
Donati, J.-F., et al. (1997): MNRAS **291**, 658
Donati, J.-F., et al. (1998): A&A submitted
Giampapa, M.S. et al. (1993): ApJS **89**, 321
Granzer, T., Strassmeier, K.G. (1998): *these proceedings*
Gullbring, E. (1998): *these proceedings*
Herbst, W. et al. (1987): AJ **94**, 137
Johns, C.M. and Basri, G. (1995): ApJ **449**, 341
Johns-Krull, C.M. (1998): *these proceedings*
Petrov, P. et al. (1996): A&A **314**, 821
Press, W. H. et al. (1992): *Numerical Recipes*, CUP

Hartmann: Is SU Aur varying so rapidly that you can't do Doppler imaging?

Unruh: This is indeed a fear that I have as I have so far not been able to unambiguously identify features that return after one rotation. It is our plan to produce Doppler images at different phases and times to check the recurrence. With respect to the very rapid darkening during the campaign, I do not believe that this will make Doppler imaging impossible, the darkening is too large to be produced by a surface spot. There is no change in the equivalent widths of the photospheric lines and it appears that material away from the star blocks off some of the light without distorting the line profiles

Johns-Krull: Comment: The time series of data I have taken should show periods $3\overset{d}{.}0$ and longer quite easily, if they maintain phase coherence over long time scales. Since I don't see these, if they are present they don't maintain phase coherence. Question: Do you have TiO in all the spectral formats that could be used to track spot behavior?

Unruh: Yes, we have TiO data for all the observatories, and I agree that it is a very good idea to try and trace the spots in TiO as this should give us constraints on the temperature.

Short-Term Spectroscopic Variability in the Pre-Main-Sequence Herbig Ae Star AB Aur During the MUSICOS 96 Campaign

Claude Catala[1], Jean-François Donati[1], Torsten Böhm[1], and the MUSICOS collaboration

Observatoire Midi-Pyrénées, 14, avenue Edouard Belin, F–31400 Toulouse, France

Abstract. We present preliminary results of the spectroscopic monitoring of AB Aur obtained during the MUSICOS 96 campaign. The analysis is mainly focussed on the He I D3 line, on the Hα line, and on a set of photospheric lines. The star was monitored irregularly for more than 200 hours.

We confirm the high level of variability of spectral lines in AB Aur. We find that the photospheric lines have a profile differing significantly from a classical rotational profile. The dominant features are a blue component with a velocity modulated with a 34hr period, and a red component of variable intensity, but with no clear periodicity. The He I D3 line exhibits two well-defined components: a blue component with a velocity modulated with a 43hr period, and a red component of variable intensity, occuring at a fixed velocity. The Hα line is also very much variable, but its variability does not appear to be correlated with that of the other lines.

These results cannot be understood in terms of a simple model of azimuthal structures in the wind of AB Aur. The different behaviors of the different lines suggest a more complex geometry involving an equatorial wind or a disk.

1 Introduction

The Herbig Ae/Be stars are pre-main sequence objects with masses ranging from 2 to 5 $M_\odot$. A significant fraction of them show conspicuous signs of strong stellar winds and of chromospheric activity (Finkenzeller & Mundt 1984, Catala et al. 1986a).

AB Aurigae is often considered as the prototype of the whole class. A detailed analysis of its line profiles and continua led to a model of its outer layers, including a wind with a mass-loss rate of 10^{-8} $M_\odot yr^{-1}$, and an extended chromosphere with a maximum temperature of 17,000 K overlying a photosphere at 10,000 K (Catala & Kunasz 1987).

This wind is certainly not spherically symmetric, nor stable, as witnessed by the periodic modulation observed in wind lines (Praderie et al. 1986, Catala et al. 1986b), or by longer-term changes in the shape of some lines (Beskrovnaya et al. 1991). The picture which emerges from these previous analysis is that of a complex wind, with both a latitudinal and azimuthal structure, probably controlled by a magnetic field. However, this model is far from being well established and needs observational confirmation. In particular, rotational modulation of lines formed in the photosphere or at the base of the wind, where the structured magnetic field is supposed to control the topology of the outflow, still needs to be investigated. In addition, no magnetic detection of AB Aur has been reported so far, in spite of several attempts.

AB Aur was monitored in the He I 587.6 nm line during the MUSICOS 92 campaign. The data show a spectacular variability of this line (Böhm et al. 1996). Whether the variations are periodic or not could not be firmly concluded on the basis of these data, although they present some indication of a periodicity near 32 hours. A high level of intrinsic periodicity is also present in addition to the possible periodic modulation. Finally, some low-level variability was also discovered in the photospheric lines of AB Aur during the MUSICOS 92 campaign (Catala et al. 1997). Again, the data were not sufficient to conclude on the periodicity of these variations.

The MUSICOS 96 campaign on AB Aur constituted a major effort to better understand the photosphere and wind of this star, and a further attempt at detecting directly a surface magnetic field. The main goal of the AB Aur observations during the MUSICOS 96 campaign was the monitoring of photospheric lines, and a selection of chromospheric and wind lines, like the He I D3 lines, and the Hα line.

2   Observations and Data Reduction

The participating telescopes were: the 3.6m CFHT in Hawaii, equipped with the transportable MUSICOS spectrograph, the 2.1m telescope at McDonald, with the Sandiford spectrograph, the 2.5m INT in La Palma, equipped with the ESA-MUSICOS spectrograph, the 1.93m OHP in France, with the Elodie spectrograph, and finally the 2.16m telescope in Xinglong (China), with the echelle spectrograph. Additional data were obtained at Ritter Observatory with the 1m telescope.

All the instruments used during this campaign were cross-dispersed echelle spectrographs. The data obtained at OHP, Hawaii, and La Palma cover a very wide wavelength domain, giving access to many photospheric lines, mainly in the blue, and to several lines formed in the wind and chromosphere, like He I 587.6 nm, Hα, and Fe II 501.8 nm. The spectrographs in use at Xinglong and McDonald cover a narrower spectral range, but sufficiently wide to contain all wind and chromosphere lines of interest.

The MUSICOS spectropolarimeter (Donati et al. 1998) was transported to Hawaii to be used on the 3.6m CFHT. This particular instrumental setup is designed for the study of stellar magnetic fields through the measurement of linear (Stokes Q and U) and circular polarisation (Stokes V) Zeeman signatures in line profiles.

The data obtained from Xinglong, McDonald, and INT were reduced with the "Esprit" reduction software developed by one of us (Donati et al. 1997). The data obtained at OHP were reduced on-site, using the automatic INTER-TACOS procedure (Baranne et al. 1996).

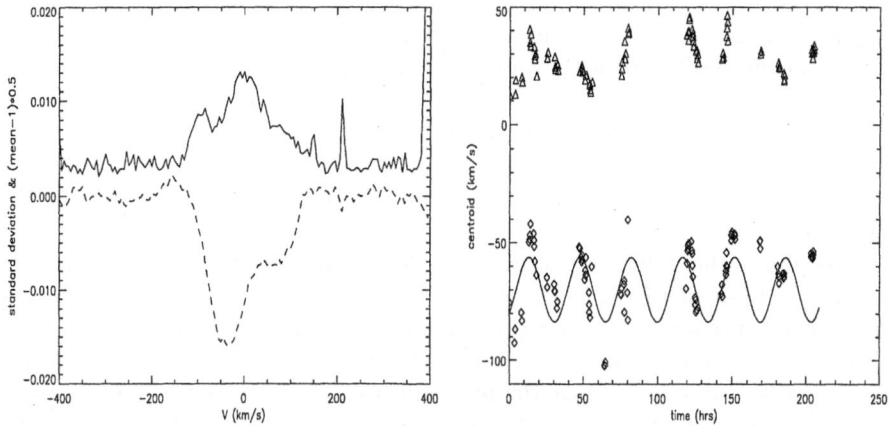

Fig. 1. *Left:* the standard deviation across the photospheric line profile, for the whole series (full line); the mean profile, averaged over all photospheric lines and over all spectra obtained during the campaign, is given for reference (dashed line). *Right:* the time variation of the centroids of the blue (diamonds) and red (triangles) photospheric components; a best-fit sine wave to the blue component centroid is also indicated (full line).

Finally, the spectra obtained at CFHT with the MUSICOS polarimetric setup were reduced following a dedicated procedure for extracting Stokes V & I parameters (Donati et al. 1997).

The wavelength scales of the stellar spectra were subsequently transformed to a frame linked to the interstellar Na I D lines, which are assumed to have a constant velocity in the star's rest frame.

3   The Photospheric Lines

Three of the five instruments used during this campaign have a wide enough spectral coverage to give access to a large number of photospheric lines: these are the 2 MUSICOS spectrographs (Hawaii and Canaries), and the OHP Elodie spectrograph. The time coverage provided by these 3 sites is only partial, due to their longitude distribution, and to the bad weather conditions at CFHT and OHP during the campaign. However, some very important conclusions can be drawn from these data.

We used Least-Square Deconvolution (Donati et al. 1997) to analyze the variations of a "mean" photospheric line. A total of 75 lines were used in this analysis.

All the photospheric profiles obtained during the campaign are highly asymmetric, with a blue component much deeper than the red component, as shown in the mean profile displayed in Fig. 1 (left), averaged over all photospheric lines and over all spectra obtained during the campaign. The red edge of the line is roughly constant over the whole series, while the centroid of the blue component is moving back and forth, and the depth of the red component is highly variable. Figure 1 also displays the the standard deviation across the line profile for the whole series. It can be readily checked that the line variations are real, with a main peak near zero velocity, corresponding to the red component discussed above, and a secondary peak near -80 km s^{-1}, corresponding to the blue component.

The photospheric line is always wider than a rotational profile of 85 km s^{-1} (the upper limit of Böhm & Catala 1993). Assuming a larger value for $v \sin i$ would be inconsistent with these previous observations, and also with the profile of the Fe II line at 501.8nm observed once as a symmetrical absorption by Catala et al. (1993). One way to make the photospheric lines observed here compatible with this value of $v \sin i$ is to assume a broader local profile for the line through turbulent motions, or to invoke strong velocity fields in the photosphere. Since the red edge of the line is constant throughout the campaign, we may assume that this part of the line is not affected by strong and variable velocity fields, and reproduce it assuming only turbulent broadening, as well as a radial motion between the photosphere and the IS medium producing the Na I D lines that have been used to set up the wavelength scale.

Turbulent velocities as high as 45 km s^{-1} were evidenced in the wind of AB Aur (Catala & Kunasz 1987). Besides, Catala et al. (1997) also concluded that radial velocity fields of the order of 100 km s^{-1} are likely to be present in the photosphere. We may therefore expect the photospheric regions to be submitted to high turbulent motions. This high level of turbulence in the photosphere is predicted by the model of mixing layer of Lignières et al. (1996). We note that strong turbulence is also present in the photospheres of O-type stars (Howarth et al. 1997), which share a lot of properties with the Herbig Ae/Be stars. On the other hand, the A4 Herbig star HD 104237 has very narrow photospheric lines (Donati et al. 1997), indicating at the most very small turbulent velocities, which would argue against the idea of highly supersonic turbulence in Herbig stars. Conversely, HD 104237 may be a peculiar Herbig Ae star. We also note that it is probably seen pole-on, so that its photospheric lines may not be affected by turbulence located in the near-equatorial regions.

We find that the red edge of the photospheric line is reproduced satisfactorily if we assume turbulent velocities of 40 km s^{-1} and a radial velocity of 13 km s^{-1} between the star and the interstellar medium responsible for the Na I D lines. This solution is not unique though, and a larger value for the turbulent velocity would lead to a smaller value for the radial velocity. On the other hand, the results presented below concerning the time variations of the photospheric lines are not affected by the choice of turbulent velocity and radial velocity.

The total equivalent width of the average photospheric line is also highly variable, The peak-to-peak amplitude of this variation is about 75%.

We subtracted the profile computed assuming rotational + turbulent broadening from each profile in the series. The residuals show almost systematically a blue absorption component, at a velocity which seems to be modulated with a period near 30 hrs, and a red emission component, with no clear modulated displacement but with strong intensity variations. These conclusions about the variability of the profile, as well as most of the time analysis which is presented further below, are not affected by the assumptions concerning the turbulent velocity within the photosphere and the radial velocity of the star with respect to the Na I D line forming region.

We studied separately the blue and red components of the average photospheric line. In each spectrum of the series, we fitted each one of these components by a gaussian, and determined its centroid, intensity and width. The most interesting result of this analysis is displayed in Fig. 1 (right), which shows the centroid of both components as a function of time. This plot shows that the blue component has a radial velocity modulated between -100 km s^{-1} and -40 km s^{-1}, while the variations of the velocity of the red component seem less important. We note also a very strong feature near t=65 hrs, leading to a deep absorption in place of the red component. The centroid of this red absorption is around 75 km s^{-1}, and is not represented in Fig. 1. This dramatic event is related to the same type of behavior seen in the He I D3 lines, described later in this paper.

The blue component seems periodically modulated. We searched these data for periods, using a simple periodogram method: we fitted the data with a sine wave of a given period, with the amplitude and the phase as free parameters, and let the period vary; for each fit, we computed the residual $R = \sum [y_i - f_i]^2 / \sum y_i^2$, where the y_i are the data points and the f_i the values of the fitting sine wave. Periodicities in the data reveal themselves as minima of R. The best result is obtained when the last few data points (after t=160 hrs) are given a zero weight in the fitting, as there appears to be a phase shift around that time, or a doubling of the modulation. We finally find a period of 35±2 hours. This period is close to, and in any case within the error bar of, that found by Catala et al. (1986b) for the Ca II K line of AB Aur (32 ± 4 hrs). The resulting sine wave fit (period=34.6 hrs) is plotted in Fig. 1. Note that the few data points near t=65 hrs, $v = -100$ km s^{-1} are not real, as the blue component disappears at that time, during the event giving rise to the strong red absorption mentioned in the previous paragraph. This fit is not completely satisfactory, in particular because the variation near the beginning of the series seems to be much more complex than a simple sine wave. However, in spite of this difficulty, we believe that the periodicity of the variation is well established, although a certain level of "intrinsic" variability must be introduced in addition to a strictly periodic modulation.

Finally, the MUSICOS spectropolarimeter was used on the 3.6m CFHT for this campaign, in the hope of detecting directly a surface magnetic field in AB Aur, through the measurement of circular polarisation Zeeman signatures in the line profiles. However, the seeing and transparency conditions experienced at Mauna Kea during the campaign were particularly bad, and the signal-to-noise ratios obtained in our spectra did not meet our expectations, by a large factor.

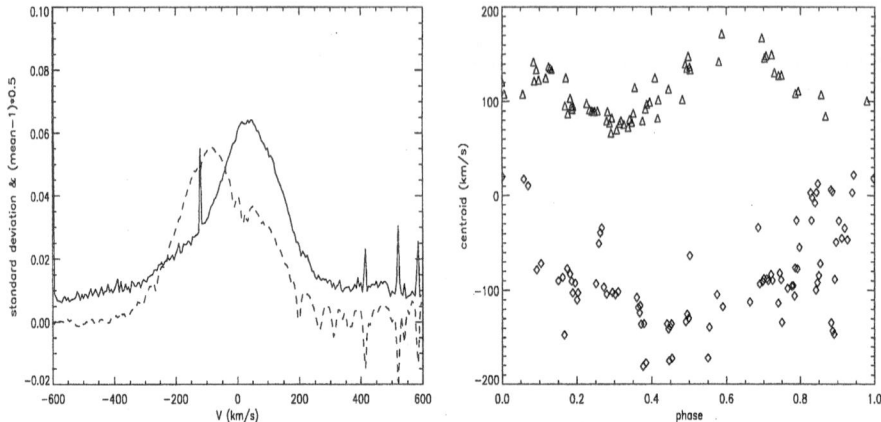

Fig. 2. *Left:* the standard deviation across the He I D3 line profile, for the whole series (full line); the mean profile is given for reference (dashed line). *Right:* centroid of the He I D3 blue component, rephased with a period of 43.6 hrs; the data are separated in two parts: from t=0 to t=65 hrs (triangles); from t=65 to t=210 hrs (diamonds); the data from t=0 to t=65 hrs are shifted by +200 km s^{-1} for clarity.

No signal was detected in the spectra of the V Stokes parameter for the average photospheric line, in any of the AB Aur spectra obtained with the instrument. The final 1-σ upper-limit for the strength of a radial field in a magnetic region covering 2.5% of the total stellar surface, and facing the observer, is of the order of 300 G.

4   The He I D3 Line

As in 1992, the variability of this line is amazing, and is shown in Fig. 2 (left), displaying the standard deviation across the line for the whole series.

The variability on the central part and red side of the line is in part due to a single dramatic event, around t=65 hrs, when a deep and broad absorption appears on the red side and in the central part of the line. This event is reminiscent of a similar one, observed during the MUSICOS 92 campaign, with the same characteristics. However, in addition to this event, we also note a strong variability present all along the series, as in 1992, with the following characteristics:

- The line has at least 2 separate components, one blue and one red.
- The blue component is always in emission. Its centroid varies in velocity, and its amplitude is also variable.
- The red component is most often in emission, but appears in absorption on several occasions, including during the dramatic event mentioned previously. Its centroid does not change significantly, but its intensity is highly variable.

We have separated the analysis of both components, by fitting each profile in the series by a sum of 2 gaussians, and measured their centroids.

The centroid of the red component is more or less constant all over the campaign, except near t=65 hrs, that is during the dramatic event mentioned earlier, while the blue component seems modulated.

The variations of the blue component centroid occur on a shorter time scale between t=0 and t=65 hrs than after t=65 hrs. A periodic modulation with a period near 45 hrs is suggested in the series after t=65 hrs, whereas a shorter timescale seems to prevail between t=0 and t=65 hrs. We used the "Information Entropy" method proposed by Cincotta et al. (1995) to search for periods in the He I data.

The results of this period search, applied to the centroid of the blue component before t=65 hrs reveal a clear minimum of the entropy near 43 hrs. The same method, applied to the remaining data (from t=65 hrs on), reveals a series of minima in the entropy, the most important one being near 43 hrs.

Figure 2 (right) presents the rephased data, both for the first and second part of the data. We note that the data of the second part of the campaign (after t=65 hrs) are indeed consistent with a simple periodicity with P=43.6 hrs, while those of the first part of the campaign (before t=65 hrs) are distributed on a double wave in the phase diagram calculated with the same period. Note finally that the moment when the double wave is changed into a simple wave coincides with the dramatic absorption event described earlier.

Finally, we measure strong variations in the total equivalent width of the He I D3 line, of the order of 70% during most of the campaign, and up to 200% during the absorption event at t=65 hrs. These variations are mostly due to those of the red part of the line. We find no clear periodicity in the variations of the equivalent width of the red part of the line, even studying separately the data strings from t=0 to t=65 hrs, and from t=65 hrs to the end. Besides, the variation of the red component is very much dominated by the strong event near t=65 hrs.

We have noticed that the equivalent width of the red half of the He I line (integrated redward from line center) is correlated with that of the red half of the photospheric lines, with a correlation coefficient of 0.76. On the other hand, the same analysis, performed on the blue side of the lines, up to -20 km s^{-1} for the photospheric lines and up to -100 km s^{-1} for the He I line, i.e. in a velocity range which is not affected by the reddest component, does not give any evidence of correlation between the photospheric lines and the He I line (correlation coefficient = 0.36).

5 The Hα Line

The Hα line of AB Aur was observed repeatedly in the past. It appears most often as a type II P Cygni profile, that is with an intense redshifted emission component and a blueshifted absorption component. Occasionally, this line exhibits a single component emission and no absorption, or a type III P Cygni profile, i.e. with an additional blueshifted emission component on the blue side of the absorption component.

We observe the three types of profiles during the campaign. The emission component varies much less than the absorption component. In the following, we will discuss only the absorption component of the line.

The variations in Hα seem to occur on a longer timescale than those of the photospheric and He I lines. The blue edge of the line tends to get bluer during the first 60 hours or so, then moves quickly to the red, before becoming bluer again until t=120 hrs. The blue edge then gets redder again and seems to remain constant until the end of the series.

In order to quantify these variations, we have measured the velocity in the blue wing at which the absorption component reaches $0.8F_c$, where F_c is the flux in the adjacent continuum, as well as the equivalent width between $v = -800$ and $v = -200$ km s^{-1}. These two quantities seem well correlated part of the time, from t=40 hrs to t=150 hrs. Outside this time interval, the velocity of the blue edge of the line remains more or less constant, while the equivalent width still varies. The variations of the equivalent width may be periodic. The information entropy method (Cincotta et al. 1995) indicates a period around 58 hrs.

6   Discussion

The behavior of AB Aur during the MUSICOS 96 campaign is far more complex than expected. It is out the scope of the present paper, presenting only preliminary results, to give a self-consistent picture of the atmospheric and envelope structures leading to such a complex variability. Only very preliminary remarks and interpretation are given below, and must be considered only as a starting point for further analysis and modelling.

The features that need to be accounted for in the modelling are summarized below:

1. the photospheric lines have a very asymmetric profile, including a blue absorption variable in position, modulated with a 35 hr period, and a red component in emission, at a more or less constant velocity, variable in amplitude, but with no apparent periodicity.
2. the He I D3 line exhibits at least two components, evolving differently.
3. the blue emission component of the He I D3 line occurs at a velocity which is modulated with a 43 hr period; the modulation appears as a double-wave for the first part of the campaign, then as a single-wave.
4. the red component, sometimes in emission and sometimes in absorption, does not vary significantly in velocity, but its intensity shows strong variations.
5. a dramatic event occurs about 65 hours after the start of the campaign, during which a strong wide absorption appears, centered at rest wavelength, both in the He I and the photospheric lines.

6. the variations of the equivalent width of the red component of the He I line are correlated with those of the red component of the photospheric lines; this behavior is not reproduced by the blue components.

7. the Hα line, presenting a P Cygni profile, is strongly variable, mostly in its blueshifted absorption component; on the other hand, the redshifted emission component is almost constant.

8. the variations of the blue component of Hα may be periodic, with a period of 58 hrs.

Strong perturbations of the photospheric layers are needed to explain the peculiar shape of the photospheric lines. In the following we discuss the absorption and emission components appearing in the photospheric profile after subtracting a rotational + turbulent profile (see Sect. 3). We note that our interpretation in terms of two components is certainly not unique. In particular, the photospheric line may be composed of a more complex set of blended components, originating from a complex set of inhomogeneous structures at the stellar surface, and still appearing in the end as two basic components in the resulting photospheric mean profile.

However, we give below possible interpretations of these two components, assuming that they are due to no more than two different features at the star's surface.

Since the blue absorption component and the red emission component behave differently, we first conclude that they originate from different parts of the stellar surface.

The velocity of the red component varies during the campaign (although less than the blue component), but we do not see any clear regularity in these variations. This velocity is positive, but remains smaller than the rotation velocity. We thus conclude that this component probably originates from the polar region, since it would be modulated by the star's rotation if it originated from lower latitudes, and that it is associated with a downflow occuring in this region. If the star is seen nearly edge-on, this implies important downward velocities. In the example given below with $i = 70°$, we would need velocities of the order of $90 \ \mathrm{km \ s^{-1}}$.

The velocity of the blue component behaves differently, with a periodic modulation between -40 and $-100 \ \mathrm{km \ s^{-1}}$, with a 35 hr period. We may reasonably assume that this corresponds to the rotation period of the star's photosphere. Note that this period is very close to that observed in the Ca II K line by Catala et al. (1986b), 32 hrs, also interpreted as the rotation period of the star. Some indication of a 35 hr period is also present in the photometric data of AB Aur of Gahm et al. (1993), although no firm conclusion on the periodicity of the observed variations was reached on the basis of these data. Thus the variations of the blue component may be due to a peculiar structure of limited area at the photospheric level or at the base of the wind.

If the rotation period of the star is indeed 35 hrs, and if its projected rotation velocity is $85 \ \mathrm{km \ s^{-1}}$, as we have assumed earlier, then we determine an inclination angle i of the order of $70°$, assuming a radius of $2.5 \ R_\odot$, adequate for

AB Aur.

Important velocity fields must be present in the line formation region to account for our observations. Structures with no velocity fields (such as temperature or abundance inhomogeneities for instance) would produce perturbations crossing the line profile from blue to red, and extending as far to the red as to the blue. This is not what we observe here. Besides, these structures must occur at high latitude, otherwise they would again produce perturbations extending significantly to the red for a significant amount of time. We have calculated that, if the inclination angle of the star's rotation axis is 70°, then velocity fields of the order of 150–200 km s^{-1} at a latitude of 80° would be necessary to explain the peculiar variability observed for the blue component. We need to qualify these conclusions, though, as the presence of the red component may hide a possible excursion toward the red of the blue component.

Ignoring this possible caveat for the moment, our observations of the photospheric lines can be tentatively explained by:

- (i) an expanding region with velocities reaching up to 150–200 km s^{-1} in layers deep enough to contribute significantly to the formation of photospheric lines; this region must be located at high latitude.
- (ii) a flow coming down to the pole, with velocities of the order of 90 km s^{-1}.

In the outflowing region, we may expect drastic changes in the density compared to the unperturbed stellar photosphere. Such density inhomogeneities will strongly affect the line formation, which may explain the large differences in the profiles of the photospheric lines of various species. As far as the inflow is concerned, we expect large temperature enhancements where the material hits the stellar pole, which again will affect strongly the formation of the lines.

The He I 5876 Å line is normally absent from the spectrum of an A0V star like AB Aur. Its mere presence, whether in absorption or in emission, indicates the existence of heated layers above the photosphere.

As for the photospheric lines, the He I line shows two components behaving quite differently. As before, we note the possibility that the observed two components may indeed reflect a more complex reality involving a larger number of blended components, but present below possible interpretations assuming that only two components are present and correspond to only two features in the wind of AB Aur.

Since the red component does not vary significantly in velocity ($v \approx 100$ km s^{-1}), we conclude that it may also correspond to material falling down onto the stellar pole. We note in this respect that the equivalent width of the red component of the He I line is well correlated with that of the red part of photospheric lines. We therefore conclude that both red components originate from the same phenomenon, probably an accretion to the pole affecting both the upper layers where it creates the red component of the He I line and the photospheric layers where it creates the red component of the photospheric lines. This accretion must be variable to account for our observations. In particular, a strong modification of its characteristics must have occured near $t = 65$ hrs, to give rise to the dramatic event that we described earlier. The exact nature of this modification is not known for the moment.

The blue component of the He I line seems modulated with a 43 hr period. It is also very tempting to interpret this periodicity as due to the stellar rotation. In the framework of this interpretation, the puzzle is to understand the period difference between the photospheric (35 hrs) and He I (43 hrs) line modulations. This result is reminiscent of the behavior of the Ca II K and Mg II h&k lines reported by Praderie et al. (1986) and Catala et al. (1986b), with periods of 32 hrs (Ca II K) and 45 hrs (Mg II h&k). An interpretation of these previous results was given by these authors in terms of a wind structure of fast and slow streams controlled by a surface magnetic field. The Ca II K line is modulated by this structure, which rotates with the rotation rate of the star. Further away in the wind, the structure is destroyed, and the lines originating from the outer regions of the wind, like the Mg II lines, are modulated by the rotation of the envelope at that distance.

This model was acceptable because the Mg II lines are formed at great distances from the stellar surface (up to 50 R_*, see Catala et al. 1984), where indeed the stream structures are likely to have merged due to shocks at the interface between fast and slow streams. With the present set of data, we show that the He I line too is modulated with a longer period than the photosphere. The similarity between this period and that of the Mg II lines is striking. However, unlike the Mg II lines, this line cannot be formed at such great distances from the star, since it originates from very excited levels. The new data presented here therefore contradict the model presented in the past.

We are led to conclude that:

(i) either the wind is indeed structured into fast and slow streams rotating with the star, but in this case the different periods found in the data correspond to the rotation at different latitudes on the star; we shall call this "the equatorial wind" model.

(ii) or the wind does not originate from the stellar surface, but e.g. from a circumstellar disk, and the different periods exhibited by the different lines correspond to the Keplerian rotation of the disk at different distances from the star; this model, that we may consider in spite of the various pieces of evidence against the presence of a thick accretion disk around AB Aur (Böhm & Catala, 1993, 1994) will be called "the disk" model in the following.

In the equatorial wind model, all wind lines are formed in a flow originating from equatorial regions of the star. This radial flow may be structured in fast and slow streams, which creates a modulation with the rotation period of the star's equatorial regions. This phenomenon would be at the origin of the modulation we see in the blue component of the He I D3 line. The equatorial regions would need to rotate more slowly than the high latitude regions, by as much as 20%. This is not surprising since the stellar wind is exerting a braking torque at the stellar surface, much more efficiently at the equator than at the pole. Since angular momentum is not easily redistributed in a non-convective structure like that of an A0V star, we expect a strong differential rotation at the star's surface, with the pole rotating faster than the equator. This model, inspired by that of Pogodin (1992), would also explain the long-term changes of the hydrogen Balmer line profiles reported by Beskrovnaya et al. (1991): if indeed the range of latitude affected by the wind varies, the line of sight may alternatively cross wind regions, creating a P Cygni profiles for these lines, and regions with little or no outflow, creating single emission profiles.

In the disk model, we assume that the wind originates from an accretion disk. We further assume that the disk wind has some azimuthal dependence, so that the periodic modulation of the lines is naturally due to the Keplerian rotation of the disk at the footpoint of the region of formation of the lines. Assuming a mass of 2.5 $M_\odot$ for AB Aur, we find that the blue component of the He I line, modulated with a period of 43.4 hrs, would be formed in the parts of the wind originating from the disk at 3.4 R_*. If we trust the 58 hrs period that we have tentatively identified in the Hα line, then the wind producing the absorption component of this P Cygni profile would originate from a distance of 4 R_*.

The preliminary analysis presented here is not sufficient to eliminate one or the other of these two different models.

References

Baranne A., Queloz D., Mayor M., et al. (1996): A&AS, 119, 373

Beskrovnaya N.G., Pogodin M.A., Tarasov A.E., Sherbakov A.G. (1991): Pis'ma Astron. Zh., 17, 825 (Sov. Astr. Lett., 17, 349)

Böhm T., Catala C. (1993): A&AS, 101, 629

Böhm T., Catala C. (1994): A&A, 290, 167

Böhm T., Catala C., Carter B., et al. (1996): A&AS, 120, 431

Catala C., Kunasz P.B., Praderie F. (1984): A&A, 134, 402

Catala C., Czarny J., Felenbok P., Praderie, F. (1986a): A&A 154, 103

Catala C., Felenbok P., Czarny J., et al. (1986b): ApJ, 308, 791

Catala C., Kunasz P.B. (1987): A&A, 174, 158

Catala C., Böhm T., Donati J.-F., Semel M. (1993): A&A, 278, 187

Catala C., Böhm T., Donati J.-F., et al. (1997): A&A, 319, 176

Cincotta P.M., Mendez M., Núñez J.A. (1995): ApJ, 449, 231

Donati J.F., Semel M., Carter B.D., et al. (1997): MNRAS, 291, 658

Donati J.F., Catala C., Wade G.A., et al. (1998): in preparation

Finkenzeller U., Mundt R. (1984): A&AS, 55, 109

Gahm G.F., Gullbring E., Fischerstrom C., et al. (1993): A&AS, 100, 371

Howarth I.D., Siebert K.W., Hussain G.A.J., Prinja R.K. (1997): MNRAS, 284, 265

Lignières F., Catala C., Mangeney A. (1996): A&A, 314, 465

Pogodin M.A. (1992): Pis'ma Astron. Zh. 18, 1066 (Sov. Astr. Lett., 18, 437)

Praderie F., Simon T., Catala C., Boesgaard A.M. (1986): ApJ, 303, 311

Flares and Circumstellar Material Around the Fast-Rotating Giant FK Comae

J.M. Oliveira[1,2], B.H. Foing[1,3], P. Sonnentrucker[1], P. Ehrenfreund[4], C. Schrijvers[5], and H. Henrichs[5]

[1] ESA Space Science Department, ESTEC
[2] Centro de Astrofísica da Universidade do Porto, Portugal
[3] Institut d'Astrophysique Spatiale, France
[4] Leiden Observatory, The Netherlands
[5] Astronomical Institute "Anton Pannekoek", The Netherlands

FK Comae is an ultra-fast rotating, solar-type giant star with $v \sin i = 162 \pm 3.5\,\mathrm{km\,s^{-1}}$, which is near break-up. It has a rotational period of 2.4 days. Spectra of Hα obtained in 1997 May with the ESA-MUSICOS spectrograph at the Isaac Newton Telescope (La Palma) are shown in Fig. 1. The variations in the shape of the multiple profiles from one rotational period to another seem to agree with the effect of active circumstellar structures that persists over a few rotations. The spectra from May 16 (dashed triple-dotted line) suggest the occurrence of a large flare; compared with spectra obtained at similar phases, an Hα equivalent width change from 5 to 11 Å is observed. No further trace of this emission was found on May 18 (dashed dotted line) in spectra taken at similar phases, which favours the interpretation of these variations in terms of a flare. The flare seems to have started already on May 15 (long dashed line), with a change in the equivalent width of Hα of 1 Å in 3.5 hours, which is consistent with the maximum observed after 1 day. The velocity fields measured for this large flare event range from 100 to 600 $\mathrm{km\,s^{-1}}$, which could indicate large distances or intrinsic flows reminiscent of coronal mass ejection.

Acknowledgements: JMO acknowledges the support of the *Fundação para a Ciência e Tecnologia* (Portugal) under grant BD9577/96.

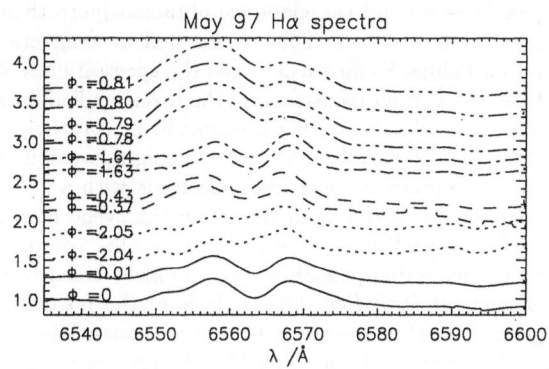

Fig. 1. Hα spectra obtained at INT from 1997 May 14–19. The change in Hα equivalent width on May 16 suggests the occurrence of a large flare.

First Results of the November 1996 MUSICOS Campaign on the O7.5III Star ξ Persei

H. Henrichs[1], J. de Jong[1], C. Catala[2], J.-F. Donati[2], J. Landstreet[3], B. Foing[4], J. Oliveira[4], P. Ehrenfreund[4], H. Stempels[5], C. Schrijvers[1], J. Telting[6], H. Cao[7], J. Hao[7], L. Huang[7], D. Yang[7], C. Mulliss[8], R. Dümmler[9], I. Ilyin[9], A. Hatzes[10], C. Johns-Krull[11], T. Böhm[12], A. Collier Cameron[13], N. Morrison[8], L. Kaper[12], T. Kennelly[14], J. Neff[15], T. Simon[16], E. ten Kulve[1], I. Tuominen[9], Y. Unruh[17], N. Walton[6]

[1] Astronomical Institute "Anton Pannekoek", Univ. of Amsterdam, Netherlands
[2] Observatoire des Midi-Pyrénées, Toulouse, France
[3] University of Western Ontario, London, Canada
[4] ESTEC, Noordwijk, Netherlands
[5] Uppsala Astronomical Observatory, Sweden
[6] Isaac Newton Group, La Palma, Spain
[7] Beijing Astronomical Observatory, China
[8] University of Toledo, Ohio, USA
[9] University of Oulu, Finland
[10] University of Texas at Austin, USA
[11] University of California, Berkeley, USA
[12] European Southern Observatory, Garching bei München, Germany
[13] St-Andrews University, Scotland, UK
[14] HAO, Boulder, Colorado, USA
[15] College of Charleston, South Carolina, USA
[16] Institute for Astronomy, University of Hawaii, USA
[17] University of Vienna, Austria

Abstract. We present the first results of the MUSICOS campaign on the O7.5III star ξ Persei, held in November 1996, which was aimed to study its wind variability, rotation, pulsation and magnetic field in order to study their mutual effects. During 10 days at 8 observatories around the globe we obtained more than 300 high-resolution optical spectra between 4100 and 8000 Å, as well as magnetic field measurements from Hawaii and La Palma. So far we analysed the spectral lines of Hα, He I λ5875 and O III λ5592. CLEANed Fourier transforms of the three studied lines yield a complicated multiperiod behaviour and indicate that the most likely rotation period is about 4 days. Combining these data with data from earlier campaigns, we find strong evidence in the photospheric lines for prograde non-radial pulsations with a period of 3.5 h. Since the pulsation period is much shorter than the dominant cyclic period in the stellar wind features (as found in the UV lines, recorded in an earlier campaign including the IUE satellite), we can conclude that pulsation is very unlikely the driving agent for the cyclic wind variations, at least for ξ Per. The analysis of the magnetic field measurements is still in progress. Whether magnetic fields are responsible for the observed wind modulation can therefore not be answered at the present stage, but remains still the most likely option.

1   Introduction

All O and many B stars show systematic variability in their stellar winds, mainly in the absorption part of the UV resonance lines of Si IV, C IV, N V and other wind lines like N IV $\lambda1718$. The most prominent features of variability are the *discrete absorption components* (DACs), which are migrating from low to high velocity towards the observer, with a recurrence timescale that can be interpreted as (an integer fraction of) the stellar rotation period.

The main unsolved issue is where the modulation comes from. A comparison of 4 timeseries of the Si IV profile of this star in subsequent years (see Kaper et al. 1998) shows that the variations are cyclical with a dominant period of 2 or 4 days, but the variability slightly differs from year to year. Such behaviour strongly suggests a corotating pattern in the wind, but the pattern itself apparently changes on a timescale of less than a year. Other stars show similar long-term behaviour. See examples in Kaper et al. (1996) and Kaper (1998).

The wind structures can be traced back to very low velocities: basically down to the $v\sin i$ value of the star (see for ξ Per Henrichs et al. 1994 or Kaper et al. 1996). This argues in favour of a model with corotating windstructures, originating at or near the stellar surface, similar to Corotating Interacting Regions in the Sun (CIRs, first proposed by Mullan 1984). In the radiative hydrodynamical computations by Cranmer & Owocki (1996) these spiral-like regions indeed emerged, giving rise to accelerating DACs in the spectral lines, very similar to what is observed. They did not need to specify the origin of the perturbation: either magnetic fields or non-radial pulsations (NRP) could, in principle, equally provide the required perturbation.

In order to find the origin of the observed wind variability we organized several campaigns on the O7.5III star ξ Persei, including magnetic field measurements. We chose this star because of its brightness and the very prominent DACs in the UV resonance lines with a suitable recurrence period of a few days. The timescales of pulsation, rotation and wind flow are all in the order of one day, which makes it particularly difficult to disentangle these effects, and a global network of observatories is needed to collect the necessary data. The strategy was to probe simultaneously the outer part of the stellar wind (UV resonance lines), the inner part of the wind (N IV 1718, Hα) and the stellar photosphere (optical O, Si and He lines). The deep photospheric lines are used to study pulsation behaviour by means of Doppler imaging techniques. The two most recent campaigns were in 1994, which included IUE and groundbased spectroscopy, and in 1996 (MUSICOS), with groundbased spectroscopy only.

A summary of the October 1994 campaign was presented by Henrichs et al. (1998). Here we describe the first results of the MUSICOS campaign of November 1996, which was after the IUE satellite was taken out of service. The major improvement with respect to the previous campaign was the use of echelle spectrographs covering 4100 to 8000 Å, with some 20 suitable spectral lines, among which a number of deep photospheric lines which were not covered in the 1994 campaign. (Only the Hα and He I 6678 lines were included. These two lines appeared seriously wind contaminated, and therefore not suitable for pulsation studies.)

With the MUSICOS spectropolarimeter at CFHT (Hawaii) and the SOFIN spectrograph at the NOT (La Palma) we also obtained (surface) magnetic field measurements. The instrumentation was about 5 times more sensitive than that used in October 1994 (see Henrichs et al. 1998). The analysis of the magnetic data is still in progress. Here we present the first results from 3 optical lines, chosen to span a range of conditions in and above the photosphere.

2   The Campaign

We observed ξ Per during 10 days from November 17 till 27, with continuous coverage over 6 days and less coverage over 4 days. High S/N (> 300) echelle spectra were obtained at seven observatories (see Table 1). In total we obtained 313 spectra, out of which 176 are presently reduced and analysed. We used the MIDAS echelle package with an improved background subtraction for the INT and Xinglong spectra. The spectra of CFHT and OHP were reduced with for those instruments dedicated software. We removed the telluric lines from Hα and He I. We extracted, normalized and rebinned the regions around Hα, He I λ5875 and O III λ5592 on a uniform grid of 15 km/s. This resolution is an optimized compromise between the various resolutions to yield a sufficiently high signal to noise ratio (300–500) for the kind of variations we are looking for. Most of the spectra taken simultaneously at different observatories match within the achieved accuracy.

Table 1. Overview MUSICOS November 1996 Campaign on ξ Per

Observatory	Location	Observers	Coverage	Spectra
CFHT	Hawaii, USA	Böhm, Catala, Donati Landstreet	4101–8138Å	36
INT	La Palma, Canary Islands	de Jong, Ehrenfreund, Foing, Oliveira, Stempels, Telting	4305–9490	50
McDonald	Texas, USA	Hatzes, Johns-Krull, Nefff	5430–6723	76
OHP	France	Henrichs, Schrijvers	3892–6817	42
Ritter Obs.	Toledo, USA	Mulliss	5440–6827	9
Xinglong	Beijing, China	Hao, Huang, Yang	5510–8389	53
Ondrejov	Czech Republic	Cao	6300–6700	6
NOT	La Palma	Dümmler, Ilyin	3580–10800	41

3   Results

The dynamic quotient spectra of Hα, He I 5875 and O III 5592 are shown in Fig. 1. The general parallel behaviour between the patterns in Hα, He I and O III is apparent, notably around $t = 8$ and $t = 9.2$ days. Note also the sharp, linear pattern around $t = 12$ in He I and O III (not in Hα), which appears only between -200 and 200 km/s ($= v \sin i$) and migrates very rapidly from blue to red.

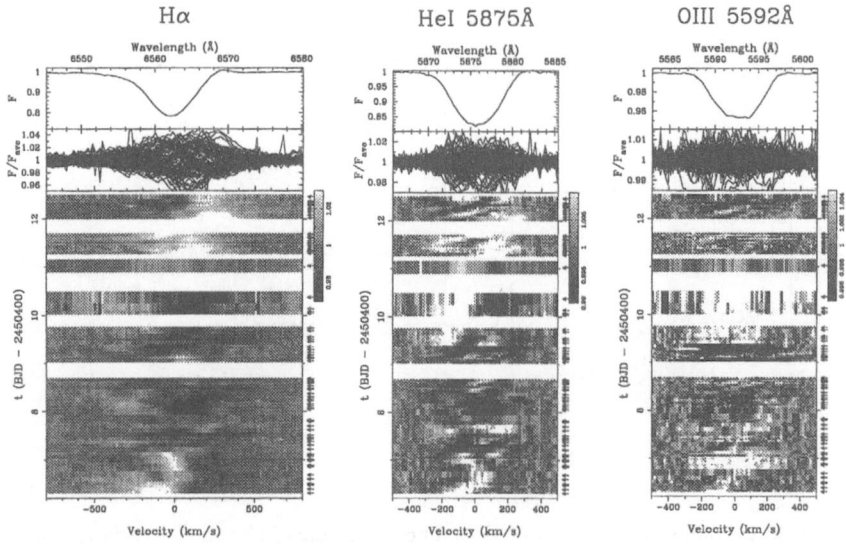

Fig. 1. The dynamic spectra of the lines Hα, He I 5875 and O III 5592. The horizontal axis is drawn in the stellar rest frame.

Period Analysis

We performed a CLEANed Fourier Transform period analysis on the 3 lines. Fig. 2 shows the powerdiagrams of Hα and He I. At this stage of the analysis we do not trust the period analysis for the O III line because of remaining systematic differences between simultaneous spectra from different observatories. In Hα the most prominent frequencies appear at 0.18(9) c/d (5.5 d), 0.44(6) c/d (2.3 d) and 1.42(6) c/d (0.70 d). The large uncertainty in the first period is due to the relatively short coverage. For He I we find similar periods: 0.20(6) c/d (4.4 d), 0.49(10) c/d (2.04 d), 1.02(11) c/d (0.98 d) and 1.47(12) c/d (0.68 d). The given uncertainties are calculated from the widths of the peaks in the powerdiagrams. This has to be compared with the period analysis of our previous campaign in October 1994, where we found periods of 2.08 d, 1.04 d and 0.67 d in the Hα, Si IV and N IV lines, respectively (see Henrichs et al. 1998).

Non-radial Pulsation

We have investigated the possibility that the fast moving absorption feature around $t = 12.2$ in the weak photospheric lines O III λ5592 and He I λ5875 is due to NRP. The steepness indicates a few-hour period. Such a period is, however, too short to determine with certainty from this campaign because of the sampling rate. After reanalysing previous observations of the weak photospheric He I λ4713 line taken in October 1989 (from Calar Alto and Kitt Peak) and joining these with the MUSICOS results, we found a very clear NRP period of 3.5h, $l = 3$ (or less likely $l = 4$) mode and 0.2% full amplitude (de Jong et al., in preparation).

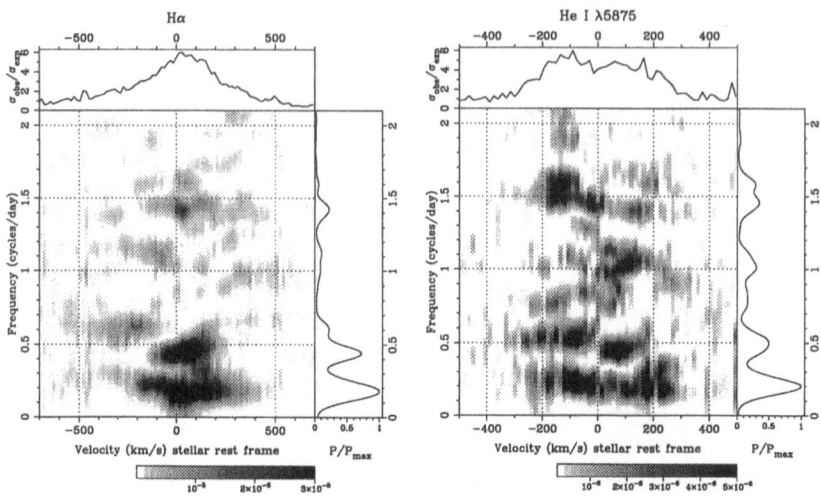

Fig. 2. Powerdiagrams of Hα and HeI. The upper panel shows the variance (which is the normalized sum over all frequencies). The right panel shows the power summed over the total velocity range.

Magnetic Fields

The bad weather conditions in Hawaii prevented the MUSICOS polarimeter attached to the CFHT from performing at its nominal conditions, which meant a loss corresponding to a factor of 10 in accuracy. The upper limits achieved were $\sigma(B_\parallel) = 100$ G, and $\sigma(B_\perp) = 300$ G, which was no improvement with respect to the $B_\parallel = 27 \pm 70$ G limit set by the October 1994 campaign (Henrichs et al. 1998). The analysis of the results from NOT spectra with the SOFIN instrument attached is still in progress.

4   Discussion and Conclusions

The dominant, well-known period of ~ 2 d of the UV wind lines is confirmed by the MUSICOS data in all studied (optical) lines, which therefore supports the hypothesis of a rotation modulation of the wind. It is interesting to note that the behaviour of the base of the wind looks similar, but not identical to what we observed in 1994 (see de Jong (1998) or Henrichs et al. (1998) for a dynamical spectrum). The inner wind structure has apparently slightly changed its configuration.

Although the long periods in Hα (5.5 d) and He I (4.4 d) are very uncertain, they represent a real timescale of variability, since the observed variations cannot be reproduced without such modulations. Within the errors these are compatible with a 4-day rotation period which is twice the period we found earlier in the UV and Hα (see Henrichs et al. 1998). This period would require a minimum radius of 16R$_\odot$ which falls within the range as calculated from atmospheric model fits for this star (see Puls et al. 1996). We therefore adopt the 4-day period as the rotation period.

Summarizing, the 3.5 h NRP period, as derived from earlier data, is too short to cause directly the 2- or 4-day wind modulation period. The combined effects of such pulsations and rotation on the wind need to be investigated.

The presence of small surface magnetic fields remains the most plausible hypothesis for modulating the stellar wind, but this is still to be demonstrated. A simple estimate yields that 50–100 G would be sufficient to perturb the wind.

Acknowledgements

We are very much indebted to the scheduling staffs of all observatories for their help to coordinate this campaign. Part of this research is supported by the Netherlands Organization for Scientific Research (NWO project nr. 781-71-053).

References

Cranmer, S.R. Owocki, S.P. (1996): ApJ, 462, 469

Henrichs, H.F., Kaper, L., Nichols, J.S. (1994): A&A, 285, 565

Henrichs, H.F., de Jong, J.A., Kaper, L., et al. (1998): *Proc. UV Astrophysics Beyond the IUE Final Archive*, November 1997, ESA-SP, in press

de Jong, J.A. (1998): this volume

Kaper, L., Henrichs, H.F., Nichols, J.S., et al. (1996): A&AS, 116, 257

Kaper, L. (1998): *Proc. UV Astrophysics Beyond the IUE Final Archive*, ESA-SP, in press

Kaper, L. Henrichs, H.F., Nichols, J.S., Telting, J.H. (1998): A&A, submitted

Mullan, D.J. (1984): ApJ, 283, 303

Puls, J., Kudritzki, R.P., Herrero, A., et al. (1996): A&A, 305, 171

Kaufer: I find your Hα dynamical spectra difficult to interpret since you have subtracted the time-averaged profile, which leads to pseudo-emissions and absorptions. Do you see any possibility to recover an undisturbed Hα profile so that we could judge whether absorption or emission features are running across the profile?

Henrichs: The average Hα profile is given in the top panel of Fig. 1. The superimposed absorption/emission variations are very small.

Prinja: (1) Is it possible that the 1.04d period in e.g. the UV lines is an alias of the 2.08d period, such that the observations really point to a "one event per rotation" scenario – as for example noted in ζ Pup and (possibly) HD93843 (O5 III; MEGA II)? (2) Are the v_{edge} changes in saturated C IV lines also phased on the 2.08d period?

Henrichs: (1) We indeed consider the 1.04 day period as a higher harmonic of the 2.04 d period. Adopting a 4-day rotation period we have here "two events per rotation". (2) No.

MUSICOS 1996 Campaign on the δ Scuti Star V480 Tau

Jinxin Hao[1,2], Huilai Cao[1], Ted Kennelly[3], and the MUSICOS team

[1] Beijing Astronomical Observatory, Beijing 100080, China
[2] Department of Astronomy, University of Toronto,
   60 ST. George Street, Toronto Ontario, Canada M5S 1A7
[3] Harvard Smithsonian CfA, 60 Garden Street, Cambridge, MA 02138, USA

1 Introduction

V480 Tau = 97 Tau (5.10v, A7IV-V) is a δ Scuti star in the Hyades cluster. Breger (1979) included it in his δ Scuti star list with a period of 0.042d and amplitude of 0.012mag. Unlike some well observed δ Scuti stars, it seems that this star was rarely studied since its discovery as a variable in 1967 (Breger 1979). In recent years, more and more δ Scuti stars have been observed spectroscopically enabling the identification of their pulsation modes. The δ Scuti variables represent potentially very good candidates for seismological investigations of the structure of main-sequence and slightly more evolved A- and F-type stars (e.g., Walker et al. 1987, Kennelly et al. 1992a, Kennelly et al. 1992b). Much attention has been paid to this type of variable stars in the MUSICOS project.

Why was V480 Tau selected as one of the targets for MUSICOS 1996 campaign? Originally, HR 8851 or τ Peg was proposed to be the δ Scuti target, as one of the four objects for the campaign (together with AB Aur, ξ Per and SU Aur). Later, another δ Scuti target was suggested (ϵ Cep) in order to coordinate the observations with the EVERIS project, even though it was a little bit late to observe this star in that season. Since the launch of the rocket was not on schedule, we finally selected V480 Tau, another target of EVERIS project. The idea was to supplement the MUSICOS ground-based spectroscopic observations with EVERIS space photometric observations. Unfortunately, the EVERIS project did not come true due to a failed rocket launch. But we still fixed on V480 Tau because this star was ever selected as one of the targets of a spectroscopic survey project for pulsation mode identification of δ Scuti stars (Kennelly 1996) and a few hour monitoring made in 1990 showed that prominent line-profile variations are present in this star (Fig.1).

2 Observations

The MUSICOS 1996 campaign was held from Nov. 15 to Dec. 2, and six telescopes (1.93m telescope Haute-Provence Observatory, INT La Palma, 2.1m telescope McDonald Observatory, CFHT Hawaii, Tillinghast 1.5m telescope F.L

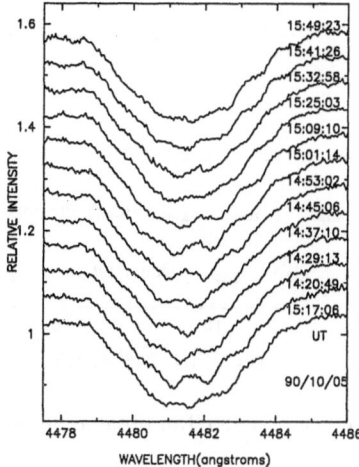

Fig. 1. The line profile variations detected on October 5, 1990, at the CFHT. The showed line is Mg II λ4481.

Whipple Observatory Arizona, and 2.16m telescope at Xinglong Station of Beijing Astronomical Observatory) were involved in this campaign. For the δ Scuti star V480 Tau, we are interested in its prominent metal lines, most of which can be found in the blue part of the optical spectrum. Since all instruments are echelle spectrographs, they cover a wide range of the blue spectrum. We aimed to obtain as many as possible line profiles which can be used for pulsation mode analysis. A total of 740 spectra were obtained from five telescopes (excluding CFHT) covering a time span of more than 7 days. All these spectra have a resolving power $R > 30.000$ and most of them have S/N ratio exceeding 100. The time coverage of the campaign on this star is shown in Fig. 2 and the window function for this dataset is displayed in Fig. 3. We notice that a very good time coverage was obtained for V480 Tau and this makes it possible to carry out reliable time-series analysis of radial velocity and "moving bumps" as well as a "Doppler Spatial" analysis of the line profiles enabling the identification of high-order non-radial pulsation modes (e.g., Gies & Kullavanijaya 1988, Kennelly et al. 1996, Telting & Schrijvers 1997, Hao 1997). Due to the limited size of the CCDs used in some of the spectrographs, gaps can be present between two adjacent orders of the echelle spectrum, resulting in the loss of some lines at some sites; therefore, the final time coverage depends on the line under consideration. Data reduction is currently under way, so that only part of the results are presented in this paper.

3 Data Reduction

The data obtained at F.L. Whipple observatory in Arizona and the INT in La Palma were just sent to me recently, the complete dataset of the campaign cannot

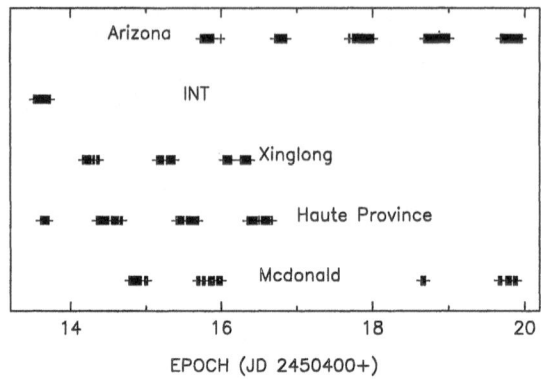

Fig. 2. General time coverage of V480 Tau obtained in MUSICOS 1996 campaign.

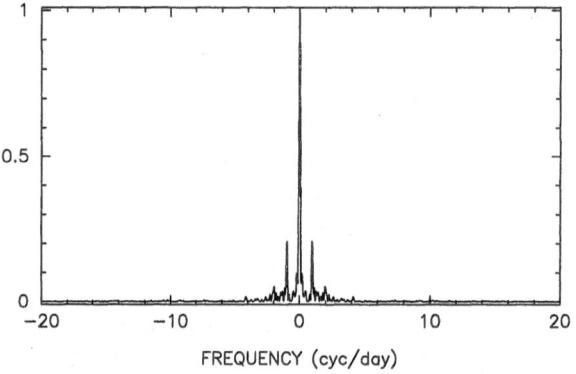

Fig. 3. Obtained window function for V480 Tau.

be presented here. The data obtained at Haute Provence Observatory (OHP), McDonald Observatory (MCD) and Xinglong Station of Beijing Astronomical Observatory (XLG) have been reduced. The used software is MIDAS. Here we present the profile variations of the Mg II λ4481 line. In order to have a good look of the moving bumps across the line profile, the data shown in Fig. 4 were smoothed and the high frequency noise has been removed (the data used as input for the Fourier analysis were not smoothed). It is clear that bumps are moving through the line profiles with time. The bump structure looks like the one of a high-order mode (about l=8). More accurate identification of the pulsation mode needs to wait for the moment that all data from this campaign are available.

4   Preliminary Frequency Analysis

The dynamical (residual) spectrum of the combined data from MCD and OHP is displayed in Fig. 5; its Fourier power spectrum is shown in Fig. 6. The residual

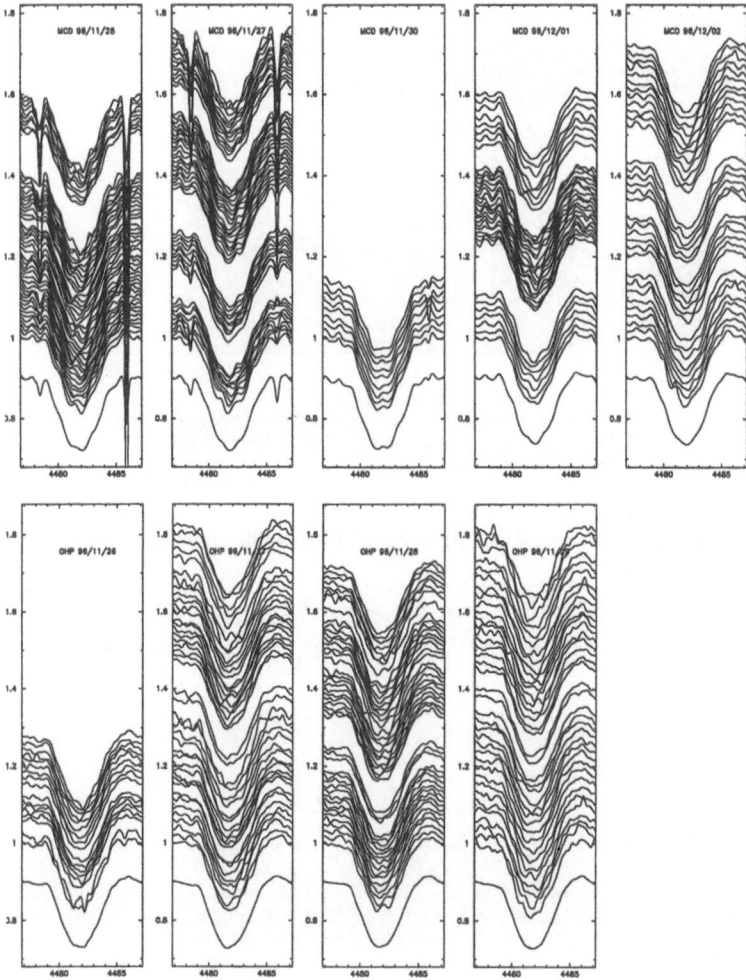

Fig. 4. Time series of the Mg II λ4481 line obtained at McDonald Observatory (top panel) and Haute Provence Observatory (bottom panel).

spectra were obtained by subtracting the mean profiles of each night at each site from the profiles of the same night at the same site in order to minimize the systematic differences from one site to another. We see that the highest peak in power appears around a frequency of 29.5 c/d. It appears that there are relatively low peaks at frequencies of 16 c/d and 36 c/d. The higher one is probably caused by the noise. But the lower one needs to be confirmed. Since the data reduction is not yet finished, we are not able to present a definite result of the frequency analysis; the result presented here can be seen as a first estimate.

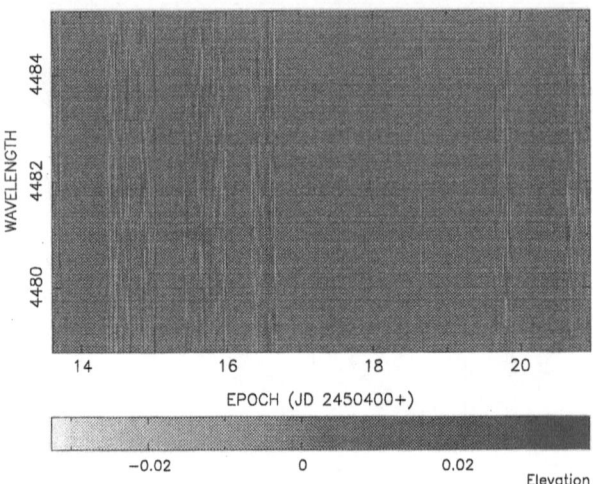

Fig. 5. Dynamical spectrum of the Mg II λ4481 line profile shown as a grey-scale image.

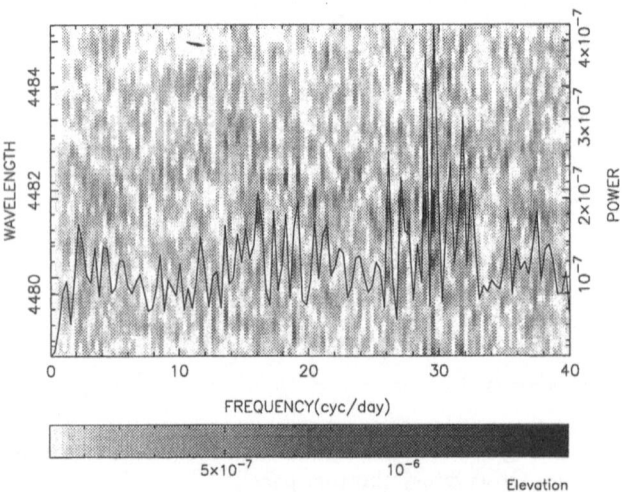

Fig. 6. Power spectrum of the time series shown in Fig. 5. The solid line overplotted is the mean power across the line profile.

5   Conclusions

Even though there is still a lot of work to do before the data reduction and analysis for this star is completed, we present some conclusions based on the (preliminary) results we have obtained:

1. Non-radial pulsations seem to be the cause of the observed line-profile variations in V480 Tau.

2. An $l \approx 8$ mode could possibly be identified with a prominent frequency of 29.5 c/d.

We expect to detect a low-order mode which could be responsible for the light variations with a period of 0.042 d. Concerning the future work on this star, we will try to organize a photometric campaign as soon as possible in order to obtain more fundamental parameters.

Acknowledgments

The exchange program between the Chinese Academy of Sciences and the Dutch Royal Academy Society that supported us for a one month visit to the Astronomical Institute of Amsterdam University and the attendance of this workshop is gratefully acknowledged.

References

Breger, M. (1979): PASP 91, 5
Gies, D.R., Kullavanijaya, A. (1988): ApJ 326, 813
Hao, J. (1997): ApJ (accepted)
Kennelly, E.J. (1996): private communication
Kennelly, E.J., Yang, S., Walker, G.A.H., Hubeny, I. (1992a): PASP 104, 15
Kennelly, E.J., Walker, G.A.H., Merryfield, W.J. (1992b): ApJ 400, L70
Kennelly, E.J., Walker, G.A.H., Catala, C., et al. (1996): A&A 313, 571
Telting, J., Schirjvers, C. (1997): A&A 317, 723
Walker, G.A.H., Yang, S., Fahlman, G.G. (1987): ApJ 20, L139

Telting: You have presented a method to derive the pulsational velocity field from a residual spectrum. However, this is only an intermediate step towards the identification of modes in terms of l and m values. Is there a unique transformation between the recovered velocity curve and the l and/or m values of the pulsation modes?

Hao: Theoretically, yes. Since the recovered velocity curve is an integration of a two-dimensional surface velocity, it is very difficult to distinguish from one to another when their l and m values are close. So, the problem of uniqueness still exists in practice.

Part 6

FUTURE PERSPECTIVE

Metallicity Dependence of Stellar Outflows and their Variability

Joachim Puls[1], Uwe Springmann[1], and Stanley P. Owocki[2]

[1] Universitäts-Sternwarte, Scheinerstraße 1, D–81679 München, Germany
[2] Bartol Research Institute of the University of Delaware, Newark, DE 19716, USA

Abstract. In this paper, we discuss the properties of radiation-driven winds as function of metallicity. We illuminate effects on mass-loss rate and terminal velocity as well as consequences for the wind-momentum luminosity relation. The effects are attributed to the influence of metallicity on the line-strength distribution function, and we show that not only the effective number of driving lines is modified, but also the slope of this function, corresponding to a change of the line-force multiplier parameter α.

In the second part, we concentrate on a special problem for thin winds, which are ubiquitous in a low-metallicity environment. We stress the importance of accounting for curvature terms of the velocity field. From theoretical arguments and first simulations, we show that thin winds should be subject to a stronger line-force instability, compared to similar objects with a solar-composition wind plasma, and that they should have a smaller mass-loss rate than derived by a standard approach using the Sobolev theory.

1 Introduction

In this paper, we will review certain aspects of the metallicity dependence of stellar outflows, as well as some recent results on the relation of metallicity and variability. Note, that we will consider only winds of massive stars, i.e., radiation-driven winds.

The ideal laboratories to study this dependency are the Magellanic Clouds, with typical stellar (iron-group) metallicities of order $0.3\ldots0.4\;\epsilon_\odot$ in the LMC and $0.1\ldots0.2\;\epsilon_\odot$ in the SMC.

Hot massive stars in the Clouds have been studied both from ground-based (to derive stellar parameters and mass-loss rates $\dot{M}$ from H_α) as well as from space-borne observatories (IUE, HST) to obtain wind parameters from UV P Cygni profiles. Without aiming for completeness, we mention here the work by Garmany & Conti (1984, IUE), Haser (1995), Walborn et al. (1995, HST) and Puls et al. (1996).

Theoretical predictions concerning the dependence of radiation-driven winds on different metallicities were given by Abbott (1982) and Kudritzki et al. (1987). As to be expected from the nature of radiative line-driving, it is predicted that $\dot{M}$, and also the terminal velocity v_∞, vary in concert with metallicity. For O-stars, a dependence of roughly $\dot{M} \sim \sqrt{(\epsilon/\epsilon_\odot)}$ has been found.

With respect to the metallicity dependence of wind *variability*, much less is known, both on the observational and on the theoretical side. The obvious reason for this lack of information is related to the requirements of variability studies (e.g., high resolution, high S/N and long observational runs), which present time allocation committees – and maybe also observers' interest – are not willing to spend for this kind of objective. (Note, however, the vast amount of data one has obtained for Galactic objects by IUE!)

To address the relation of variability and metallicity *observationally*, the same features as in the Galactic case have to be studied as a function of abundance, namely (cf. the various papers in this volume related to these topics):

- *Line profile variability* in the optical: to our knowledge, nothing is known for MC objects so far.
- *Discrete absorption components* in the UV: only a few cases have been detected on snapshots of MC O-stars, as is true for
- *Black troughs* in UV P Cygni profiles (for a discussion of both topics, see Walborn et al. 1995)
- X-rays as indicators for shocked wind material have not been detected yet in individual MC objects, due to the lack of spatial resolution.

Thus, this aspect seems to belong to the category of "future applications". However, at least some theoretical predictions can be made, which will be covered at the end of this paper, which is organized as follows. In part one, we concentrate on the reaction of radiative line driving on changes in metallicity, and review scaling relations. In part two, we discuss some phenomena predicted to arise in *thin* winds, which are ubiquitous in a low metallicity environment. This discussion will comprise both stationary and time-dependent effects.

2   Metallicity Dependence of Stellar Outflows

The dominating physical mechanism which initiates and accelerates the winds of (hot) massive stars is radiation line driving by thousands of (UV-) metal lines irradiated by a high flux, where the momentum of the accelerated metal ions is subsequently redistributed to the wind bulk plasma (hydrogen and helium) by Coulomb collisions.

Various aspects of this acceleration mechanism have been discussed, where we will refer here to important papers dealing with the *stationary aspect* of winds. The principal mechanism was first shown to work by Lucy & Solomon (1970), and Castor, Abbott & Klein (1975: CAK) provided an elegant way of solving the hydrodynamic equations. Abbott (1982), Friend & Abbott(1986) and Pauldrach, Puls & Kudritzki (1986: PPK) improved the theory quantitatively, and the present state of the art of constructing consistent wind models is discussed by Pauldrach et al. (1994) and Pauldrach et al. (1998). The momentum transfer by Coulomb collisions was investigated by Castor, Abbott & Klein (1976) and Springmann & Pauldrach (1992).

In the following, we will concentrate on the metallicity dependence of line driving, where some progress has been obtained by Abbott (1982) and Kudritzki et al. (1987), as mentioned in the introduction. In order to investigate this metallicity dependence, three fundamental points concerning the principal behaviour of line driving have to be understood.

2.1   An Expression for the Total Line Acceleration

1. The radiative line acceleration provided by scattering of photons in a *single* line (transition frequency ν_i) to wind material in a spherically expanding shell of size dr, mass $dm = 4\pi\rho r^2 dr$ and velocity $v(r) \ldots v(r) + dv$, is given by the average transferred momentum per unit time and dm, i.e.,

$$g_{\text{rad}}^i = \frac{<\Delta P>}{\Delta t dm} = \frac{1}{4\pi r^2 c^2} L_\nu \nu_i \frac{1}{\rho} \frac{dv}{dr} (1 - \exp(-\tau_S)) \tag{1}$$

(cf. also the contribution by S.P. Owocki, this volume). L_ν is the stellar luminosity in the frequency range $\nu \ldots \nu + d\nu$, and τ_S is the line optical depth (in Sobolev approximation)

$$\tau_S = \frac{\sigma_e v_{\text{th}}}{dv/dr} k_L; \quad k_L = \frac{\bar{\chi}_i \lambda_i}{\rho s_E v_{\text{th}}} \tag{2}$$

with k_L the line-strength of the transition, being roughly constant for the resonance lines and lines from meta-stable levels which predominantly drive the wind. $\bar{\chi}_i$ is the frequency integrated line opacity, $\sigma_e = s_E \rho$ is the Thomson-scattering opacity, and v_{th} the thermal velocity of a representative ion. The latter two quantities are used as reference values to obtain a line-strength definition which is *dimensionless*.

2. If we consider now the total line acceleration by summing up the individual contributions of each line, we find

$$g_{\text{rad}}^{\text{tot}} = \frac{s_E v_{\text{th}}}{4\pi r^2 c^2} \left\{ k_1 \int_{k_1}^{\infty} L_\nu \nu \, |dN| + \int_0^{k_1} k_L L_\nu \nu \, |dN| \right\}. \tag{3}$$

The first term inside the brackets accounts for the contribution by *optically thick* lines ($\tau_S \geq 1$) and depends, via k_1, only on the hydrodynamic structure. The second term gives the contribution by *optically thin* lines ($\tau_S < 1$) and depends, via line-strength k_L, on details of the level population at the considered depth point. k_1 is the "critical" line-strength where $\tau_S = 1$ is reached,

$$k_1 = k_L(\tau_S = 1) = \frac{dv/dr}{\rho s_E v_{\text{th}}}, \tag{4}$$

and $dN(\nu, k_L)$ the so-called line-strength distribution function, which gives the number of lines in the intervals $\nu \ldots \nu + d\nu$ and $k_L \ldots k_L + dk_L$.

3. As first realized empirically by CAK, and as later turned out to be true under quite general conditions in hot star winds, the line-strength distribution follows a power law

$$dN(\nu, k_L) = -N_o f_\nu(\nu) k_L^{\alpha-2} \, d\nu \, dk_L, \tag{5}$$

where $f_\nu(\nu)$ describes the frequential distribution (mostly dependent on T_{eff}) and N_o is proportional to the total number of lines. For most winds, the exponent α lies in the range $0.5 \ldots 0.7$.

Furthermore, it can be shown that this distribution is intimately related to the statistics of the *oscillator strength distribution*, since k_L depends linearly on the oscillator strength f_{ul} (with n_l the lower und n_u the upper level of the involved transition),

$$k_L \sim \frac{\bar{\chi}_i}{\rho} \sim \frac{n_l}{\rho} f_{ul}, \tag{6}$$

as well as on the excitation if transitions from metastable levels play an important role (for details, see Puls et al. 1998b). Note, that the dependence on line-strength is fairly equal in any considered frequency subinterval, so that the total statistics can be separated in a frequential and a line-strength part which are roughly independent from each other. Examples of line-strength distributions are given below.

If we finally use the distribution function (5) and evaluate the total line acceleration (3), we obtain the important result that both the optically thick and the optically thin contribution scale with $k_1^\alpha \sim (dv/dr/\rho)^\alpha$. Hence, the total line acceleration becomes a non-linear function of the hydrodynamic variables

$$g_{rad}^{tot} = \frac{s_E v_{th} N_o \int_0^\infty L_\nu \nu f_\nu(\nu)\, d\nu}{4\pi r^2 c^2} \left\{ \frac{1}{1-\alpha} k_1^\alpha + \frac{1}{\alpha} k_1^\alpha \right\} = \frac{L/cs_E}{4\pi r^2} k_{CAK} k_1^\alpha. \tag{7}$$

The exponent α (describing the steepness of the line-strength distribution) can be alternatively interpreted as the ratio of line force from optically thick lines to total force,

$$\alpha = \frac{g_{rad}^{thick}}{g_{rad}^{tot}}. \tag{8}$$

Note, that the first term in (7) is proportional to the so-called force multiplier parameter "k_{CAK}" introduced by CAK, and that k_1 is the inverse of their depth variable t.

2.2   Influence of Metallicity

Having introduced now the basics of line-driving from an ensemble of lines, it is easy to understand the effects of metallicity.

The Direct Effect. Due to its definition (2), the line-strength scales with metallicity (under the realistic assumption that the ionization balance is not heavily changed by modifying the abundances) as

$$k_L(\epsilon) \sim z \frac{n_l(\odot)}{\rho} \sim z k_L(\epsilon_\odot) \tag{9}$$

where z is the actual abundance ϵ relative to its solar value, $z = \epsilon/\epsilon_\odot$. Thus, the first effect of changing the metallicity is to modify the normalization of the line-strength distribution,

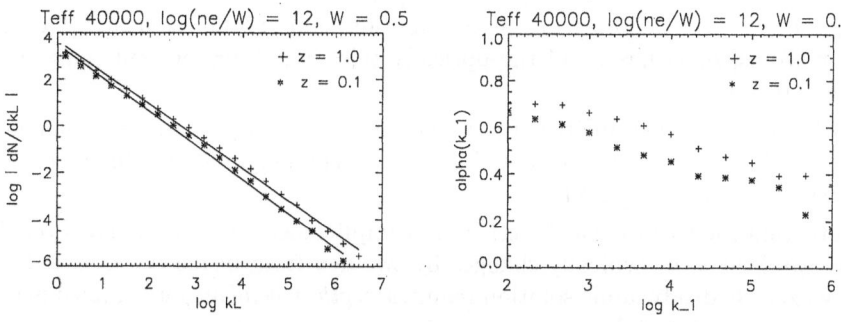

Fig. 1. Line-strength distribution function $(\log|\mathrm{d}N/\mathrm{d}k_{\mathrm{L}}|)$ for the wind of an O-type star with $T_{\mathrm{eff}} = 40000$ K, dilution factor $W = 0.5$ and ratio of electron-density to $W = 10^{12}$. Crosses: solar abundance; asterisks: 1/10 solar abundance. Overplotted is the linear regression to $\alpha - 2$, resulting in $\alpha = 0.63$ for $z = 1.0$ and $\alpha = 0.55$ for $z = 0.1$.

Fig. 2. α as function of k_1, calculated from Eq. 8, for the same model as on the left.

$$\mathrm{d}N(\nu, k_{\mathrm{L}})(\epsilon) = -N_{\mathrm{o}}(\epsilon)\, f_{\nu}(\nu)\, k_{\mathrm{L}}^{\alpha-2}\, \mathrm{d}\nu\, \mathrm{d}k_{\mathrm{L}}; \quad N_{\mathrm{o}}(\epsilon) = N_{\mathrm{o}}(\epsilon_{\odot}) z^{1-\alpha}, \qquad (10)$$

or, in other words, the total number of *contributing* lines is changed. Since the force-multiplier k_{CAK} is proportional to N_{o} (Eq. 7), this quantity also scales according to $k_{\mathrm{CAK}} \sim z^{1-\alpha}$. An example of this variation is given in Fig. 1, where we have plotted the frequency-integrated line-strength distribution function for a representative depth point in an O-type wind, both for solar and $z = 1/10$ abundance. Note the difference in the offset of the function (logarithmic scaling!).

The Indirect "α"-effect. Besides the obvious direct effect, we have to account for an additional complication: If we compare the α values derived by a linear regression to the log-log representation, then the lower metallicity wind yields a lower α! If one concentrates now on the deviation of actual and "power law" fitted line statistics (for both cases), one notes that the local α value is smaller than the mean for large line-strengths, whereas it is larger for low k_{L}-values. ($k_{\mathrm{L}} = 1$ corresponds to a line which has the same strength as Thomson-scattering.) The reason for this behaviour is two-fold. At the high k_{L}-end, the line-statistics becomes different since there are only few lines that are very strong, introducing a rather sharp cut-off corresponding to a steep (negative) slope. On the weak line side, we are simply running out of contributing lines due to the exponential factor controlling the excitation, so that the distribution becomes flatter there.

Under these conditions, we can calculate an *effective* value of α using Eq. 8 as a definition. This was done in Fig. 2, where we have plotted the ratio of optically thick to total line acceleration *as function of k_1*, i.e., by changing the border $\tau_{\mathrm{S}} = 1$. Again, two different metallicities were used, $z = 1$ and $z = 1/10$.

The result is obvious: *Thin and fast winds (with large k_1) as well as low metallicity winds tend to have lower effective α-values than high density or high metallicity winds!* This behaviour can be explained by:

a) the different weighing of line-distribution slope as function of k_1 (e.g., for thin winds, k_1 is large and the optically thick line force depends only on the low α regime).

b) For different metallicities, the maximum value of k_L varies in proportion to metallicity (cf. Fig. 1), which leads to a different weighing of α in the optically thick range $k_1 \ldots k_L(\max)$.

Note, that the variation as function of k_1 implies also a variation of α *throughout the wind*, since k_1 typically changes by two dex from inside to outside. Hence, any exact hydrodynamic solution requires *depth dependent* force-multipliers (cf. Kudritzki et al. 1998).

2.3   Consequences

Scaling relations. Including now the "finite disk" correction factor (Friend & Abbott 1986; PPK) and a factor accounting for ionization effects onto N_o, $\sim (n_e/W)^\delta$, introduced by Abbott (1982), with n_e the electron density and W the dilution factor), we can summarize the resulting scaling relations for $\dot{M}$ and v_∞ as function of metallicity, which arise if the *metal dependent* line force is used to calculate the hydrodynamic equations (cf. PPK and Kudritzki et al. 1989):

$$\dot{M} \sim (z^{1-\alpha})^{1/\alpha'} L^{1/\alpha'} (M(1-\Gamma))^{1-1/\alpha'} \tag{11}$$

$$v_\infty = \frac{\alpha}{1-\alpha} f(\alpha,\delta) v_{\mathrm{esc}} \tag{12}$$

$$\alpha' = \alpha - \delta. \tag{13}$$

M is the stellar mass, Γ the Eddington factor and v_{esc} the escape velocity. $f(\alpha,\delta)$ is a decreasing function of δ^{-1}, and has a value of roughly 2.2 if δ is small (cf. Kudritzki et al. 1989). By changing the metallicity and to first order, we thus find

$$\dot{M} \sim z^{\frac{1-\alpha}{\alpha'}}; \; v_\infty \sim \frac{\alpha}{1-\alpha}(z), \tag{14}$$

which, in case of O-star winds (small δ), yields the often quoted scaling relation for the mass-loss rate $\dot{M} \sim \sqrt{z}$. Note, however, that α and δ do heavily vary as function of spectral type and luminosity class (Kudritzki et al. 1998).

Observational proof. Since the terminal velocities scale with the metallicity dependent factor $\alpha/(1-\alpha)$, one might expect lower v_∞ in a low metallicity environment. This is just what has been found by comparing O-star terminal velocities in the Galaxy and the Clouds, cf. Fig 3! For a detailed discussion, we refer the reader to the papers by Garmany & Conti (1984), Kudritzki et al. (1987), Haser et al. (1993) and Walborn et al. (1995).

With respect to the meanwhile well-known wind-momentum luminosity relation (Kudritzki et al. 1995, Puls et al. 1996), our theoretical findings imply (leading terms only):

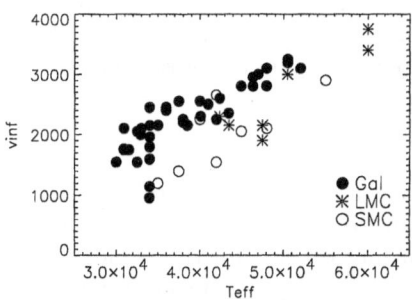 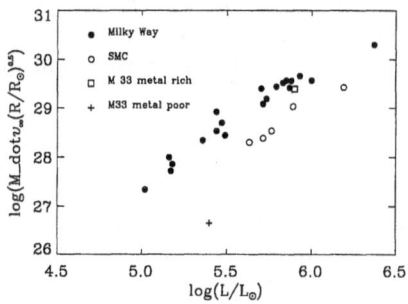

Fig. 3. Terminal velocities of O-type stars in the Galaxy and the Magellanic clouds. Data from Haser (1995) and Puls et al. (1996).

Fig. 4. Wind momentum (in cgs units) and luminosity of galactic and SMC supergiants and two A-supergiants in M33. Open square: M33 A-supergiant with galactic metallicity. Cross: Extremely metal poor A-supergiant in the outskirts of M33. (From McCarthy et al. 1995.)

$$\log(\dot{M} v_\infty R_*^{\frac{1}{2}}) \sim \frac{1-\alpha}{\alpha'} \log z + \frac{1}{\alpha'} \log L + \ldots \tag{15}$$

From the presently available data, it is clear that at least in the SMC a different offset is visible (due to the first term above, resulting from the "direct" effect; cf. Fig. 4, and also Puls et al. 1996; Kudritzki 1997). Whether there is actually a different slope (as a consequence of a reduced α') is not certain due to the small number statistics for SMC O-stars. To clarify the situation, more objects have to be analyzed. This work is well under way in our group.

3 Metallicity Dependence of Stellar Wind Variability

As stated in the introduction, the observational perceptions with regard to the topic of this section are meager, if present at all. From what we do know, nothing peculiar is happening, which should also be expected from a theoretical standpoint.

At first, we want to mention a number of important papers related to some theoretical aspects of instabilities in radiatively driven winds, where the so-called line-driven instability is the dominant one. Again, Lucy & Solomon (1970) were the first to recognize that the process of radiative line-driving itself may lead to a strong instability, and a number of papers have investigated both the consequences in the linear regime (e.g., Owocki & Rybicki 1984) as well as in the non-linear regime by numerical simulations (e.g., Owocki et al. 1988; Owocki 1992; see also Feldmeier 1995). An important aspect here is the reaction of the so-called *diffuse* radiation field onto perturbations, where a lot of progress has been made (Lucy 1984; Owocki & Rybicki 1985; Owocki & Puls 1996).

All of these investigations do not point to significant metallicity effects, besides the implication that the *average* mass-loss rate should be coupled with abundance in the same way as in the stationary picture discussed above.

3.1   The "Thin Wind" Problem

However, as it turned out lately, there may arise a certain problem for *thin* winds in general, i.e., either for winds of dwarfs, but even more for winds in a low metallicity environment which is an obvious domain for thin winds.

This problem was firstly encountered in an investigation concerning the effects of the diffuse radiation field on the line-driven instability by two of the authors (S.P.O. and J.P.), and has been reported in some detail elsewhere (Puls et al. 1998a). For reasons of brevity, we will mention here only the problem and the conclusion, before turning to some newer results.

The present doctrine of radiation-driven winds is the assertion that the Sobolev approximation can be applied with high precision when calculating radiation forces, as long as only the stationary aspect is considered. This statement was validated by PPK in a careful comparison between the structure arising from using the (modified) CAK approach and the use of radiation forces calculated "exactly" in the comoving frame. However, as it turned out, these comparisons have been performed only for quite dense winds.

If one alternatively compares important quantities of the line radiation field, e.g. the scattering integral, *in an optically thin continuum*, one will find significant differences between the Sobolev and the comoving solution *in the transonic region*. (Actually, only there!)

This discrepancy arises primarily from the sharp *curvature* of the velocity field around the sonic point, which marks the transition from exponential stratification of the hydrostatic atmosphere to the rapid acceleration of the supersonic wind outflow. This severely strains the standard Sobolev assumption of a constant velocity gradient, i.e. with no curvature. For an extensive discussion of the influence of curvature terms, we refer to Sellmaier et al. (1993).

Although the differences between Sobolev and comoving frame solutions seem to be dramatic in the transonic region, one has to account also for the stabilizing effect due to interactions with the continuum (conversion of line into continuum photons). As it turns out, actual deviations between Sobolev and exact approaches can survive only in cases of thin winds (i.e., in cases where the continuum is formed well below the sonic point), whereas for moderate and thick winds the continuum terms are sufficient to prevent the total line-radiation field from becoming dominated by such curvature effects.

One might now question the extent to which this discrepancy in thin winds will influence the line *acceleration*. To this end, we have calculated, for different wind strengths, the total line force arising from a typical *ensemble of lines*. By summing up the force components of individual lines, we have calculated the total acceleration both in the comoving frame and in the Sobolev approach.

For thick and intermediate-strength winds, we have found, consistent with the results published by PPK, only small or moderate differences. For thin winds, however, the comoving frame acceleration turned out to become much smaller than the Sobolev one, and for a number of models it was actually negative (i.e., inward directed) in the complete transonic region! (Cf. Fig. 10 in Puls et al. 1998a.)

From this result, one might expect that thin winds have a smaller mass-loss rate than predicted by the standard (CAK, PPK) Sobolev result. This notion bases on the fact that the line acceleration scales (non-linearly) with the inverse density, and, for smaller $\dot{M}$, the contribution of the weaker lines (which are less affected by the discussed process) increases, so that in this case the acceleration is sufficient again to overcome the effective gravity.

This speculation has to be checked by calculating consistent "thin-wind" models, where a first attempt shall be reported in the following. (Note that we are referring here to wind-densities which are still high enough to prevent the decoupling of the metal ions from the bulk flow (Springmann & Pauldrach 1992; Babel 1995, 1996).

3.2 Consistent Wind Models with a Realistic Treatment of the Diffuse Radiation Field

If one investigates the reaction of the line acceleration due to disturbances of the velocity field in a linear analysis and includes the effects of the line-radiation field to first order (cf. Owocki 1992; Owocki & Puls 1996), one finds the so-called bridging law (see also Owocki & Rybicki 1984):

$$\frac{\delta g_{\mathrm{rad}}}{\delta v} \sim \frac{\omega_{\mathrm{B}}}{\chi_{\mathrm{B}}^2 + k^2}\Big[k^2\Big(1 - \frac{2S_{\mathrm{L}}r^2}{I_{\mathrm{c}}R_*^2}\Big) + ik\chi_{\mathrm{B}}\Big], \tag{16}$$

where χ_{B}^{-1} is the bridging length joining the domain of high-frequency perturbations (large wave-number k) to low-frequency perturbations, ω_{B} the growth rate of this process, S_{L} the line source function of the considered transition and I_{c} the illuminating core intensity in case of an optically thin continuum.

In this expression, we have made use of the so-called "smooth source function" (SSF) approximation, where S_{L} is calculated in the Sobolev approximation assuming an unperturbed flow. The first term inside the bracket controls the growth of the line-instability, and the (imaginary) second part contains the phase information.

The term proportional to S_{L} represents the "line-drag" effect first discussed by Lucy (1984). Within the SSF method, this implies marginal stability at the wind base $r \to R_*$ – where $2(r/R_*)^2 S_{\mathrm{L}}/I_{\mathrm{c}} \approx 1$ in the Sobolev approximation – but still yields significant instability away from the stellar surface, where this term falls below unity.

More realistically, the source function is not necessarily smooth, but may contain gradients associated with the above-mentioned curvature terms in the mean velocity, or with the flow perturbations. Such gradient effects can be effectively taken into account using the "Escape-Integral-Source-Function" (EISF) method recently developed by Owocki & Puls (1996).

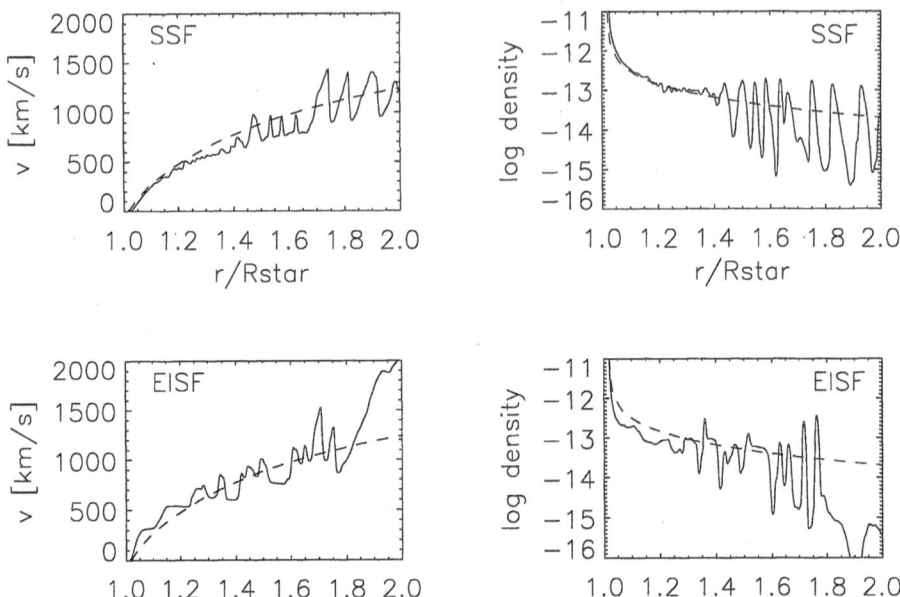

Fig. 5. Velocity and density structure of a stationary wind (dashed), compared with the snapshot of a time-dependent wind. Upper panel: line source function calculated by the SSF-method; lower panel: EISF-method (see text).

Fig. 5 shows the outcome of a (thin continuum) EISF-calculation and compares the resulting hydrodynamic stratification both with conventional models (dashed) as well as with models calculated with the previous SSF approach. For a detailed discussion, we refer to Owocki & Puls (1998), and mention here only the following points.

- The curvature terms in the transonic region discussed above modify the line radiation field in such a way that the source function decreases, yielding a value of $2(r/R_*)^2 (S_L/I_c \ll 1$. Thus, the line-drag can no longer work and the wind becomes unstable at its base!
- The phase relation between perturbed velocity and density is changed in the lower wind part, which is theoretically interesting, however has no important consequence for the gross wind behaviour (cf. Owocki & Puls 1998).
- Most important for our discussion, however, is the notion that by relaxing the Sobolev theory not only the instability, but also the line force itself is calculated in a realistic way, accounting consistently for curvature terms in the velocity field! As a result, the anticipated reduction of the line acceleration actually takes place, and we find a steeper velocity law at lower density. Compared to both the stationary and the SSF approach, we obtain a mass-

loss rate which is smaller, in agreement with our speculation above. In this case, we find a reduction by a factor of two.

Once again, we would like to stress that the latter effect is only present in cases of a missing stabilizing continuum, i.e., in thin winds.

4   Concluding Remarks

In this paper, we have discussed important effects of metallicity on radiation-driven winds. We have shown that, beside the obvious direct effect on the number of contributing lines, also the effective slope of the line-distribution function is changed, which has significant consequences for the acceleration: In thin winds (from dwarfs) or a low-metallicity environment, the force-multiplier parameter α is small, giving rise to lower terminal velocities and a steeper wind-momentum luminosity relation, compared to higher density winds of solar metallicity.

Additionally, we have discussed the "thin wind" problem, i.e., the importance of curvature terms in the transonic region, a problem which is inherent to winds of low metallicity. By means of the new EISF method, we have shown that the line-drag effect can no longer operate and the wind tends to become unstable even at its base. This behaviour may be related to the notoriously high X-ray luminosities of B main-sequence stars (e.g., Cassinelli et al. 1994, Cohen et al. 1997).

Finally, also the mass-loss rates of thin winds are reduced, compared to standard calculations. As was discussed by Puls et al. (1996, their Fig. 29), there seems to be some observational significance for this prediction. By comparison of observed and theoretically derived wind-momenta (standard approach!), it was actually found that the theoretical values are too high for thin winds driven by low-luminosity stars.

Of course, our latest results have to be checked carefully and further consequences have to be awaited. This work is presently under way.

References

Abbott D.C. (1982): ApJ, 259, 282
Babel J. (1995): A&A, 301, 823
Babel J. (1996): A&A, 309, 867
Cassinelli J.P., Cohen D.H., MacFarlane J.J., et al. (1994): ApJ, 421, 705
Castor J.I., Abbott D.C., Klein R.I. (1975): ApJ, 195, 157 (CAK)
Castor J.I., Abbott D.C., Klein R.I. (1976): in: *Physics des Mouvements dans les atmosphere stellaires*, Cayrel R., Steinberg M. (eds.), CNRS, Paris, p. 363
Cohen D.H., Cassinelli J.H., MacFarlane J.J. (1997): ApJ, 487, 867
Feldmeier A. (1995): A&A, 299, 523
Friend D.B., Abbott D.C. (1986): ApJ, 311, 701
Garmany C.D., Conti, P.S. (1984): ApJ, 284, 705
Haser, S.M. (1995): Thesis, Ludwig-Maximilians-Universität München
Haser S.M., Lennon D.J., Kudritzki R.-P. et al. (1994): Sp. Sci. Rev., 66, 179

Kudritzki, R.-P. (1997): in: *8th Canary Winter School, Stellar Astrophysics for the Local Group*, Aparicio A., Herrero, A. (eds.), MPA-Sonderdruck 1023, Garching, p. 1

Kudritzki R.-P., Lennon D.J., Puls J. (1995): in: *Proc. ESO Astrophysics Symposia, Science with the Very Large Telescope*, Walsh J.R., Danziger I.J. (eds.), Springer, Heidelberg, p. 246

Kudritzki R.-P., Pauldrach A., Puls J., (1987): A&A, 173, 293

Kudritzki R.-P., Pauldrach A., Puls J., Abbott D.C., (1989): A&A, 219, 205

Kudritzki R.-P., Springmann, U., Puls J., et al. (1998): PASPC, 131, 299

Lucy L.B. (1984): ApJ, 284, 351

Lucy L.B., Solomon, P. (1970): ApJ, 159, 879

McCarthy, J.K, Lennon, D.J., Venn, K.A., et al. (1995): ApJL, 455, L135

Owocki S.P. (1992): in: *The Atmospheres of Early-Type Stars*, Heber U., Jeffery S. (eds.) Springer-Verlag, Heidelberg, p. 393

Owocki S.P., Castor, J.I., Rybicki G.B. (1988): ApJ, 335, 914

Owocki S.P. & Puls J. (1996): ApJ, 462, 894

Owocki S.P. & Puls J. (1998): ApJ, submitted

Owocki S.P., Rybicki G.B. (1984): ApJ, 284, 337

Owocki S.P., Rybicki G.B. (1985): ApJ, 299, 265

Pauldrach A.W.A., Kudritzki R.-P., Puls J., et al. (1994): A&A, 283, 525

Pauldrach A.W.A., Lennon M., Hoffmann T.L. (1998): PASPC, 131, 258

Pauldrach A., Puls J. & Kudritzki R.-P. (1986): A&A, 164, 86 (PPK)

Puls J., Kudritzki R.-P., Herrero A., et al. (1996): A&A, 305, 171

Puls J., Kudritzki R.-P., Santolaya-Rey A.E. et al. (1998a): PASPC, 131, 245

Puls J., Springmann U., Lennon M. (1998b): A&A, in prep.

Sellmaier F., Puls J., Kudritzki R.-P., et al. (1993): A&A, 273, 533

Springmann U.W.E. & Pauldrach A.W.A. (1992): A&A, 262, 515

Walborn N.R., Lennon D.J., Haser S.M., et al. (1995): PASP, 107, 104

Ignace: From the modified CAK theory, the wind velocity distribution can be approximated by a $\beta = 0.8$–1.0 law. How does the β (i.e., time averaged) change in your thin winds? This naturally has implications for line profile synthesis.

Puls: Actually, this question cannot be answered so far, since the modified acceleration in the transonic region depends on a realistic description of the average diffuse radiation field. What I have shown was only a first simulation in the framework of the so-called "escape integral source function (EISF)" method. To answer your question more precisely, detailed calculations have to be awaited.

Non-LTE Departure Coefficients of Hydrogen

Jeroen van Gent

Astronomical Institute of Utrecht University, Princetonplein 5, 3584 CC Utrecht

Preliminary results of a study into the behaviour of the non-LTE departure coefficients of hydrogen in the outflowing atmospheres of hot stars are presented. Departure coefficients, b, define the ratio of the real population of an ionic level, n, to the LTE population, n_*: $b = n/n_*$.

Departure coefficients for hydrogen were calculated with the non-LTE code ISA-Wind (de Koter et al. 1993, A&A 277, 561) for a grid of model atmospheres. We find that the departure coefficients can parameterized in terms of the local temperature, local density, and effective temperature of the model. Computed values of b are illustrated in Fig. 1 (top row) as a function of the local atmosphere temperature, $T(r)$, along with the parameterized representation of the data obtained by fitting a fifth-order polynomial. These parameterizations were used as input to the Seispec code (Van Gent & Lamers, in prep.) to compute line profiles, which were subsequently compared with profiles generated from models computed with exact departure coefficients (Fig. 1, bottom row).

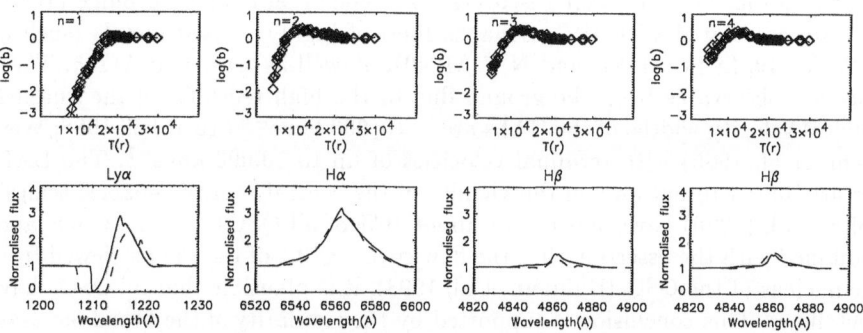

Fig. 1. Top: Hydrogen b-values and polynomial fits as a function of $T(r)$ for models with $T_{\mathrm{eff}} = 20\,\mathrm{kK}$, $R_* = 30\,\mathrm{R}_\odot$, $v_\infty = 1500\,\mathrm{km\,s^{-1}}$, and different mass loss rates. The influence of density produces a spread in the data, which is clearly visible for the $n = 4$ level. Bottom: Comparison of line profiles computed from the parameterizations (dashed) and exact values of b (solid). The Hβ profile on the far right incorporates a density-corrected version of the fit to b for $n = 4$.

Outflows in Quasars:
Radiative Acceleration and Variability

Nahum Arav

IGPP, LLNL, P.O. Box 808, L413, Livermore, CA 94550

Abstract. I give an overview of outflows associated with quasars and other Active Galactic Nuclei. Two topics are discussed in detail: 1) A promising observed signature of radiative acceleration "the ghost of Lyα." The best examples of the signature and model fits are shown and the radiative dynamics explained. This signature strongly suggests that the dominant acceleration mechanism for the outflows is radiative acceleration via resonance line scattering. 2) Variability in the outflows and its implications. Unlike stellar wind variability, the absorption features in AGN outflows have very stable velocity position and variability is only detected in the depth of the features. The mechanism responsible for this variability is probably changes in the ionizing flux of the AGN.

1 Introduction

Broad Absorption Lines (BALs) are the traditional indicators of quasars' outflows (see Fig. 1). They were discovered by Lynds (1967) and Burbidge (1970), and are associated with UV resonance lines of highly ionized metals (such as C IV λ1549, Si IV λ1397, and N V λ1240), as well as with Lyα λ1215. These lines are observable from the ground due to the high redshifts of the quasars. Typical velocity widths of the BALs are $\sim 10,000$ km s^{-1} (Turnshek 1988; Weymann et al. 1985) with terminal velocities of up to 50,000 km s^{-1}. The BALs are attributed to outflows in the vicinity of the central source ($\sim$parsec scales). Foltz et al. (1990) have shown that about 10% of all QSOs are BALQSOs and, combined with the assertion that the flow covers $\lesssim 0.2$ of the sky as viewed from the nucleus of the QSO (Hamann et al. 1993), it is plausible that all QSOs have BAL flows. This conclusion is supported by the similarity of the broad emission lines (BELs) in the spectra of BALQSOs and non-BALQSOs (Weymann et al. 1991). The BAL region is probably situated further away from the continuum source than the BEL region (Junkkarinen et al. 1983; Turnshek 1988).

A similar phenomenon is detected in the so-called "BAL-Seyferts" (Seyferts are the low luminosity cousins of quasars), where the velocity width of the absorption is typically less that 1000 km s^{-1}. NGC 4151 is a good example of such an object. An important recent development is the discovery of Mini BALs. Some of the narrower absorption systems seen in quasars were suspected for a long time to be associated with quasars' outflows and not with intervening cosmological systems (Anderson et al. 1987; Foltz et al. 1988). In the last two years it became clear that this is indeed the case. The evidence came mainly from high-resolution spectroscopy with the Keck telescope, which shows smooth absorption troughs,

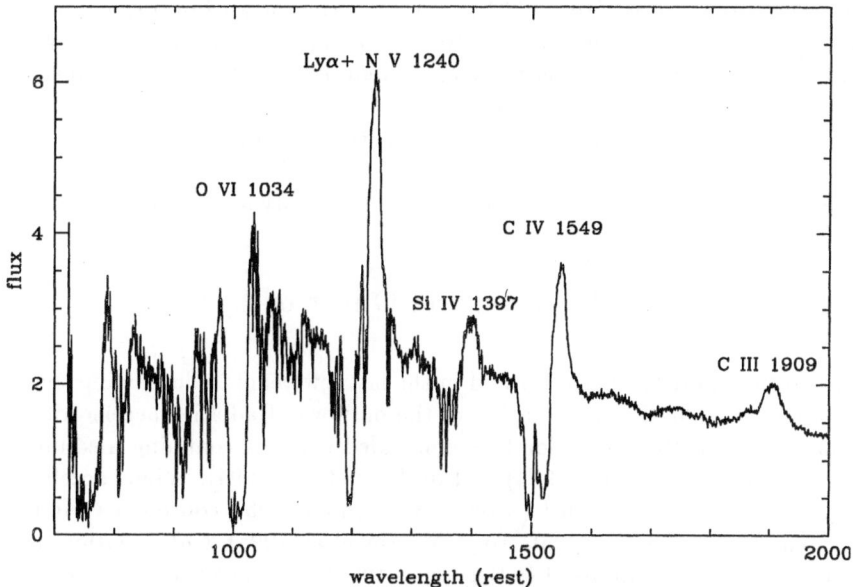

Fig. 1. Spectrum of BALQSO PG 0946+301, the wide absorption troughs are associated with some of the marked emission lines and arise from an outflow with $\Delta v \simeq 10,000$ km s^{-1}. The flux is in units of 10^{-15} erg s^{-1} Å^{-1}.

in sharp contrast to the ragged absorption profile associated with intervening absorption (Barlow & Sargent 1997; Barlow 1997; Hamann et al. 1997a), Evidence also comes from observed variability in these troughs (Hamann et al. 1997b). We refer to these systems as Mini BALs since they show evidence for a BAL-like flow, but their velocity width is much smaller.

Some of the outstanding questions in the field of AGN outflows are:

1. Where does the flow originate from and what is its geometry? Is it an equatorial disk wind (i.e., arise from the accretion disk and directed along it) or a jet-like polar-outflow?
2. Does radiative acceleration (i.e., using the central source photons to drive the flow) dominate the dynamics?
3. What are the ionization equilibrium and abundances (IEA) in the outflows?
4. What is the relationship between the different types of objects which show the outflows (BALs, BAL-Seyferts, and Mini BALs)?
5. Do all AGNs have such outflows which we observe only in a minority of the objects due to orientation effects? If so, why are radio-loud quasars so underrepresented in the BALQSOs sample compared to radio-quiet quasars?

For completion it is important to note that most absorption lines seen in quasars' spectra are associated with material which is not intrinsic to the AGN. These intervening absorption lines arise from matter situated at large distances from the central source ($10^3 - 10^9$ pc), thus giving us an important probe for the distribution and evolution of matter in the universe. Among these systems are: Lyα-forest lines, metal-line systems, Lyman-limit systems and damped-Lyα lines. I will not discuss intervening systems any further in this review.

In the remainder of this paper I will concentrate on two topics associated with AGN outflows: A promising signature for the dynamical dominance of radiative acceleration; and observations of absorption variability and its implications.

2   Radiative Acceleration: the Ghost of Lyα

Considerable effort has been devoted in the last few years to developing theoretical models of radiative acceleration of the outflows. Radiative acceleration via resonance line scattering of cloudlets embedded within a confining medium was investigated in Arav & Li (1994) and in Arav et al. (1994). Winds accelerated by radiation absorbed by dust (Scoville & Norman 1995), continuous winds accelerated by resonance line scattering (Murray et al. 1995), and radiation-driven magnetic disk winds (de Kool & Begelman 1995) have also been explored.

A possible signature for the acceleration mechanism, the so-called double-trough phenomenon, was observed by Weymann et al. (1991) and extensively studied by Korista et al. (1993). We developed a dynamical model (Arav & Begelman 1994; hereafter paper III) to explain the creation of such a signature. After acquiring the Korista et al. (1993) data, we identified what appears to be a clear signature of radiative acceleration — "the ghost of Lyα" (Arav et al. 1995, hereafter paper IV) — which is related to the double-trough phenomenon. The ghost appears as an absorption hump in the C IV λ1549 BAL and is centered around 5,900 km s^{-1} to the blue of the corresponding broad emission line (BEL). This velocity separation is the same as the velocity separation between the N v λ1240 and Lyα lines. Thus, when the N v BAL and the C IV BAL are plotted as a function of velocity (where zero velocity corresponds to the peaks of these BELs), the hump in the C IV BAL appears at the same velocity as the Lyα BEL and qualitatively resembles it.

The full details of the dynamical model and of each observed signature are given in Arav (1996, hereafter paper V). In Figure 1 we illustrate (and give a brief explanation for) the dynamical scenario that creates the ghost signature, and sketch its expected appearance based on our results from papers III and V. Figure 2 shows the data and model fits for the best cases of a ghost signature we have identified so far. In papers IV and V we: 1) presented model fits for "ghosts" in individual objects, 2) demonstrated the ability of our dynamical selection criteria to predict which object would show the signature, and 3) estimated that the probability for chance occurrence of the ghosts is $\lesssim 10^{-3}$. Each of these three lines of reasoning gives independent support for our dynamical scenario, and combined they make the strongest case yet for the nature of the acceleration mechanism of BALQSO flows: radiative acceleration via resonance-line scattering.

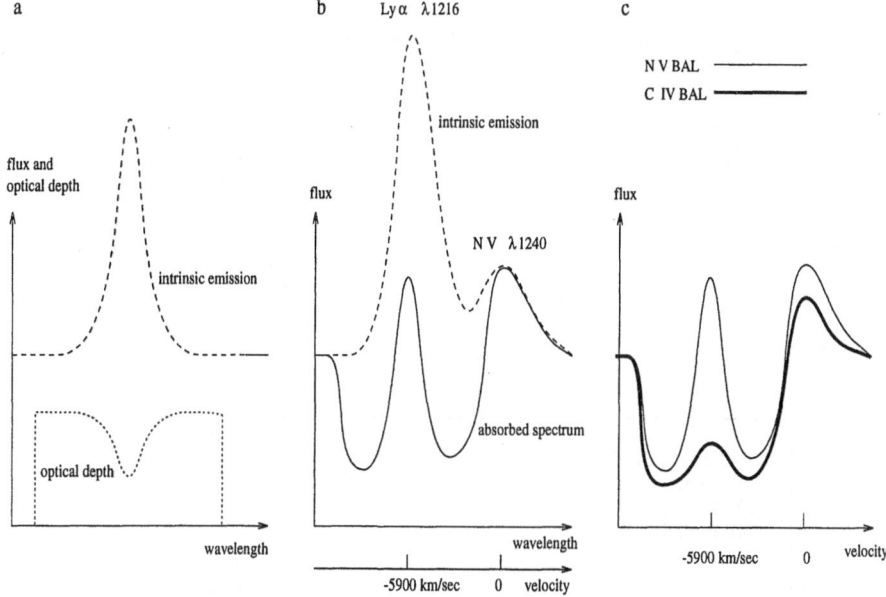

Fig. 2. Schematic illustration for the creation of the ghost of Lyα signature. For a constant νL_ν (where ν is frequency and L_ν is the luminosity per unit frequency), pure radiative acceleration creates an absorption trough with a constant optical depth. When the accelerating ions sample an increasing flux, their acceleration, relative to the constant νL_ν case, increases, and as a result the optical depth of the wind decreases. In panel *a* we show the optical depth changes that result from the exposure of the accelerating ions to a strong emission line. A very similar situation is found in quasars: panel *b* shows the spectral regime around a quasar's strongest emission line, Lyα. The N V ions are exposed to the strong Lyα flux and in their rest frame the peak of Lyα occurs at a velocity of $-5{,}900$ km s^{-1}. Since we do not know the intrinsic Lyα emission spectrum it is impossible to determine from the absorbed spectrum whether the optical depth has changed at the appropriate velocity interval. However, the N V ions share their acceleration with the rest of the flow through Coulomb collisions or the presence of magnetic fields, and thus gives the optical depth modulation to all the absorbing ions. Therefore, as we show in panel *c*, the C IV λ1549 BAL should show a lower optical depth around $-5{,}900$ km s^{-1} although there is no emission feature at the corresponding wavelength (1520 Å) in the quasar's rest frame. This hump in absorption around $-5{,}900$ km s^{-1} in the C IV BAL is what we termed "the ghost of Lyα."

3   Implications of the Ghost Signature

3.1   The Confinement Problem

If radiative acceleration is the dominant driving mechanism in all BAL outflows, there are interesting implications regarding the confinement of the BAL matter. First, let us briefly describe why confinement is necessary. Ionization equilibrium calculations show that the absorbing matter must have a certain ratio of ionizing-

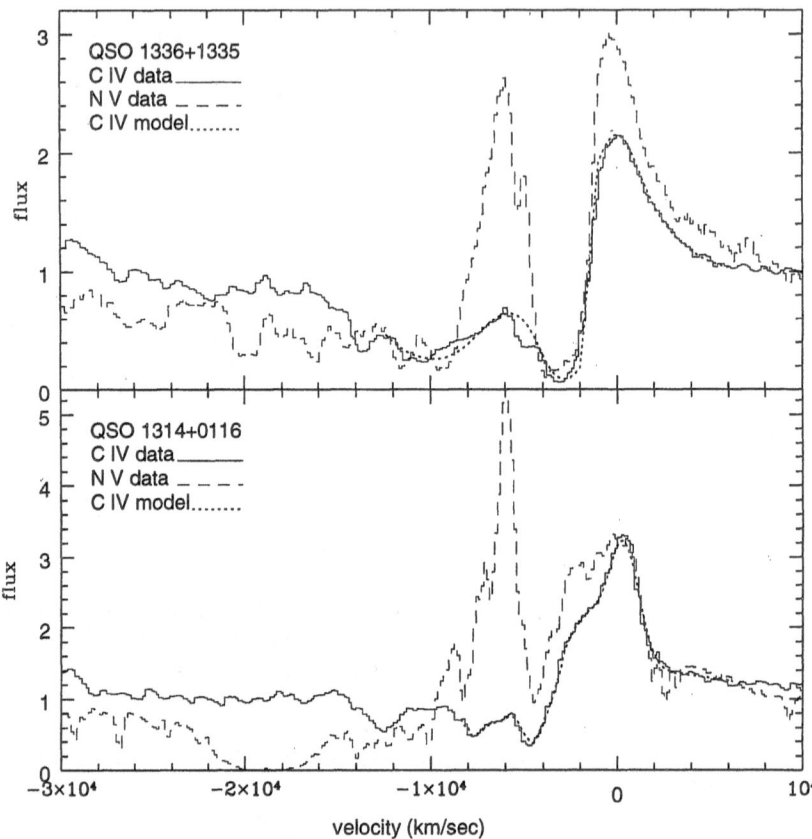

Fig. 3. The two best examples for the ghost of Lyα phenomenon. Data are plotted as flux vs. velocity relative to the quasar rest frame. The solid histogram shows the data for the C IV region where zero velocity corresponds to 1549 Å. The dashed histogram shows the N v–Lyα data where zero velocity corresponds to the rest wavelength of the N v line (1240 Å). The dotted lines show the model fits across the extent of the dynamical signature. Absorption structures at higher velocities are unrelated to the ghost signature and might be explained by other means (see Arav 1996, Fig. 3).

photon number density to hydrogen number density (the so-called ionization parameter U) in order to maintain the observed ion species. This ratio is 6–8 orders of magnitude smaller than it would be if the inferred column density was evenly spread in space along a length scale of 1 pc. Thus, we reach the conclusion that the absorbing material must be concentrated in small dense clumps and have a very small volume-filling factor, which necessitates the existence of a confining medium.

In paper I we have shown that confinement by a thermal plasma with $T \lesssim$

10^8 K (Krolik, McKee, & Tarter 1981) would require such a large mass flux in the confining medium that it would render radiative acceleration negligible. We can now reverse the argument: the compelling case for the dominance of radiative acceleration in the BAL outflows implies that the confining medium is nonthermal (e.g., magnetic field, cosmic rays, ram pressure). A somewhat more speculative assertion is that if the confinement of the BEL clouds is physically similar to that of the BAL clouds (which is the simplest assumption), then also the BEL clouds have a nonthermal confinement. We emphasize, however, that the creation of the ghost signature does not depend on the existence of a confining medium. If it will turn out that it is possible to have BAL outflows without a confining medium (see Murray et al. 1995), the radiative dynamics that produces the ghost (paper III and the formalism presented here) will still apply after some minor adjustments.

There is another interesting implication for the dominance of radiative acceleration in the BAL outflows. The contribution of the pressure gradient from the confining pressure to the total acceleration cannot be more than comparable to that of the resonance lines scattering. Otherwise, the ghost signature would be strongly reduced and might become undetected.

3.2   Consequences for Alternative Models of BALQSO Winds

Assuming that the ghost of Lyα is formed by the dynamics described in papers III and V, we can ask: what are the consequences for alternative models of BALQSO winds? The model which is affected most is the dust driven wind scenario of Scoville & Norman (1995). This model assumes that the BALs start as winds from asymptotic giant branch stars that orbit around the center of the AGN. These low velocity winds are exposed to the strong flux of the central source and thus are radiatively accelerated by dust absorption to the velocities observed in BALQSOs. The ghost signature suggests that the dominant source of acceleration is radiative acceleration via resonance-line scattering. Not more than about 30% of the acceleration can be contributed by dust absorption while accounting for the ghost. This implies that the dust-driven scenario should be modified in order to include the dominant effect of resonance-line scattering.

Disregarding the ghost signature the dust model has a few consequences that are difficult to reconcile with observations. In order to cover the continuum source along the entire velocity width of the BAL, the model invokes the presence of 10–20 stellar contrails along the line of sight. Each of these contrails has a dust optical depth of a few tenths. Therefore, the combined dust optical depth is most probably larger than three. However, evidence for strong dust extinction is not seen in most BALQSOs and the amount of dust extinction that arises from the model would make these objects very dim and therefore very hard to detect. Another prediction of the dust scenario is that most high-ionization lines should be very optically thick (of the order of 100, see their Fig. 8). This implies that all the high-ionization lines are highly saturated and that only different geometrical coverage changes the depth of absorption. However, there are many examples in which the depth of absorption is different among these lines. Finally, the model invokes turbulent broadening for the absorption associated with one contrail by a factor of 100 compared to the expected thermal width. This seems implausible since the contrail is supposed to be a very thin sheet and such strong turbulence along the thin axis of the sheet (as required by the model) would very probably destroy it.

The ghost phenomenon can fit nicely in the radiation-driven magnetic disk winds model (de Kool & Begelman 1995). This scenario explains both the origin and the confinement of the BAL material, while the main acceleration is still due to resonance-line scattering. Thus, it is completely consistent with the ghost scenario and might complement it.

The model of continuous winds accelerated by resonance line scattering (Murray et al. 1995) is also quite compatible with the ghost scenario, provided that its ionization equilibrium can be shown to be able to match the observations. This is again mainly due to the fact that the acceleration is due to resonance-line scattering. One difference is that the BEL region and the BAL region are cospatial in this model. This breaks the radial assumption for the photon flux in modeling the ghost. However, the effect would still be similar but the momentum extraction would have to be integrated on all incident angles. Since this effect occurs for all lines, it should not reduce the relative dynamical effect of the N v BAL appreciably.

4   Variability in AGN Outflows

Variability in AGN outflows is quite different than that observed in stellar winds. The most thorough study of BAL variability was conducted by Tom Barlow between 1988–1993. Here I summarize the main results and show the most dramatic variability observed, full details can be found in Barlow's Ph.D thesis (1993). One important result is the lack of change in the velocity position of the observed features. When variability is observed it is detected as changes in the depth of the absorption features and not in velocity shifts of identifiable features. To my knowledge there are no clear cases of velocity shifts in the available data spanning over two decades. HST observations of the C IV λ1549 absorption in NGC 4151 yielded the best upper limit for acceleration or deceleration of a specific absorption feature. Observing with the high-resolution spectrograph over a four years interval, Weymann et al. (1997) converted the lack of any velocity change in one specific absorption feature to an acceleration upper limit of 10^{-3} cm s^{-1}. In their words: about one millionth of the acceleration due to gravity on the surface of the Earth! The lack of velocity changes might be explained by assuming that although the flow accelerates, our line of sight always intersects the same velocity zone (see Arav 1996, Fig. 3). Such a scenario favors cylindrical symmetry associated with a disk wind.

Barlow found that from 23 objects 13 showed no or marginal BAL variability, 5 showed moderate variability and 4 showed strong variability. A weak correlation was found between BAL and continuum variability. This lends supports to the idea that the BAL variability is caused by ionization changes driven by

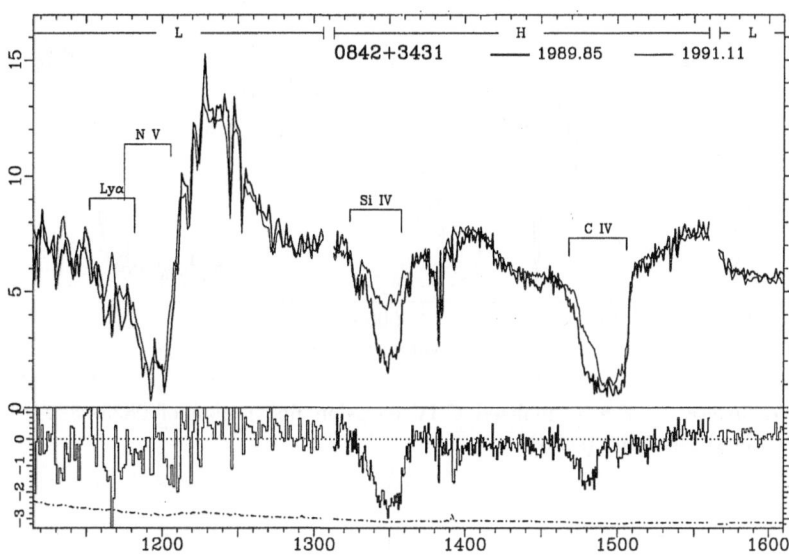

Fig. 4. BAL variability in Q0842+3431 (CSO 203). The spectrum is shown in the rest frame of the quasar and two different epochs are shown. The bottom panel shows the difference spectrum between the two epochs and the photon noise level. Note that the variability is entirely due to changes in the depth of absorption and not due to velocity motion of the features. This conclusion is also supported by a third epoch observation (1992.19) where troughs became deeper again.

ionizing flux changes. However, in some cases (including the strongest variability shown in Fig. 4) no correlation was observed between BAL and continuum variability.

Barlow also studied variability in narrow absorption systems close to the redshift of the quasar (the so called associated absorbers) and concluded that variability is detected in ~30% of these (Barlow 1997). Variability in these systems is the best evidence that they are indeed intrinsic absorbers (in this case Mini BALs) and not intervening ones. Figure 5 shows variability data with a time span of 10 years (3 years in the rest frame of the object) from a narrow absorption system (Aldcroft et al. 1997). Changes in the 6000 km s^{-1} trough and at the edges of the low velocity trough are evident.

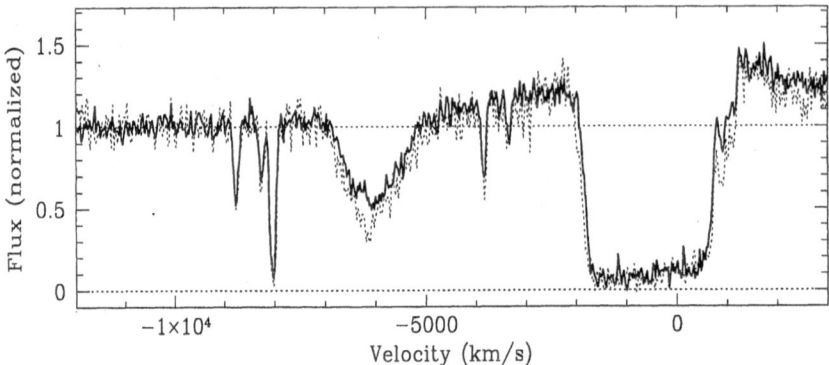

Fig. 5. Spectra of the normalized C IV emission line region of 0835+5804, in 1985 (dotted line) and 1995 (solid)

5 Conclusions

The relationship between stellar and quasar winds is still an interesting open question. On the one hand, like in hot stars, there is good evidence that the dynamics is dominated by radiative acceleration via line scattering. However, unlike the case of stellar winds the geometry of the flow and its boundary conditions are at present undetermined. There is also the issue of confinement (see § 3.1) that does not exist in stellar winds. Variability of quasar winds seems to be dominated by ionization changes and unlike in stellar winds absorption features are not changing their velocity on time scales up to several years. It might be that these differences are the result of macro physics: bolometric luminosity of 10^{46} erg s^{-1}, long timescales, flux arising from accretion, etc...; and that the microphysics is quite similar. However, this hypothesis is yet to be proven.

Acknowledgments

I thank Kirk Korista and Tom Barlow for giving me the data of their BALQSOs samples, Tom Aldcroft for supplying me with Figure 5, and Roger Blandford and Simon Morris for many useful comments and suggestions.

References

Aldcroft, T., Bechtold, L., Foltz, C. (1997): in "Mass Ejection from AGN", ASP Conf. Ser., Vol. 128, ed. N. Arav, I. Shlosman, and R.J. Weymann, p. 25
Anderson S.F., Weymann, R.J., Foltz, C.B., Chafee, F.H.
Arav, N., Li, Z.Y. (1994): ApJ, 427, 700
Arav, N., Li, Z.Y., Begelman, M.C. (1994): ApJ, 432, 62

Arav, N., Begelman, M.C. (1994): ApJ, 434, 479 (paper III)

Arav, N., Korista, T.K., Barlow, T.A., Begelman, M.C. (1995): Nature, 376, 576 (paper IV)

Arav, N. (1996): ApJ, 465, 717 (paper V)

Barlow, T.A. (1993): Ph.D thesis, University of California, San Diego

Barlow, T.A., Sargent, W.L.W. (1997): AJ, 113, 136

Barlow, T.A. (1997): ApJ, in preparation

Burbidge, E.M. (1970): ApJ, 160, L33

de Kool, M., Begelman, M.C. (1995): ApJ, 455, 488

Foltz, C.B., Chafee, F.H., Weymann, R.J., Anderson S.F. (1988): in STScI Symp. 2, QSO Absorption Lines: Probing the Universe, ed. S.C. Blades, D.A. Turnshek and C.A. Norman (Cambridge Univ. Press), p. 53

Foltz, C.B., Chafee, F.H., Hewett, P.C., et al. (1990): BAAS, 2, 806

Hamann, F., Korista, T.K., Morris, S.L. (1993): ApJ, 415, 541

Hamann, F., Barlow, T.A., Junkkarinen V.T., Burbidge, E.M. (1997a): ApJ, 478, 80

Hamann, F., Barlow, T.A., Junkkarinen V.T. (1997b): ApJ, 478, 87

Junkkarinen, V.T., Burbidge, E.M., Smith, H.E. (1983): ApJ, 265, 51

Korista, T.K., Voit, G.M., Morris, S.L., Weymann, R.J. (1993): ApJS, 88, 357

Lynds, C.R. (1967): ApJ, 147, 396

Murray, N., Chiang, J., Grossman, S., Voit, G.M. (1995): ApJ, 451, 498

Scoville, N., Norman, C. (1995): ApJ, 451, 510

Turnshek, D.A. (1988): in STScI Symp. 2, QSO Absorption Lines: Probing the Universe, ed. S.C. Blades, D.A. Turnshek, & C.A. Norman (Cambridge Univ. Press), p. 17

Weymann, R.J., Morris, S.L., Foltz, C.B., Hewett, P.C. (1991): ApJ, 373, 23

Weymann, R.J., Turnshek, D.A., Christiansen, W.A. (1985): in Astrophysics of Active Galaxies and Quasi-stellar Objects, ed. J. Miller (Oxford Univ. Press), p. 333

Weymann, R.J., Morris, S.L., Gray, M.E., Hutchings, J.B. (1997): ApJ, 483, 717

Author Index

ESO ASTROPHYSICS SYMPOSIA
European Southern Observatory

Series Editor: Philippe Crane